Fault Detection

Fault Detection

Edited by

Wei Zhang

Fault Detection,
Edited by Wei Zhang

CBS, Edition 2016

Republished by InTech

Janeza Trdine 9, 51000 Rijeka, Croatia

Additional hard copies can be obtained from orders@intechopen.com

Fault Detection, Edited by Wei Zhang
p. cm.
ISBN 978-953-307-037-7

Preface

Fault diagnosis technology is a synthetic technology, which relates to several subjects, such as modern control theory, reliability theory, mathematical statistics, fussy set theory, information handling, pattern recognition and artificial intelligence.

The United States is the first study to carry out fault diagnosis countries. Since 1961, the United States at the beginning of the implementation of the Apollo program, has witnessed a series by equipment failure led to the tragedy, therefore, in April 1967, at NASA's the idea of, by the Office of Naval Research (ONR), opened the American Society of Mechanical Failure Prevention Group (MFPG) the establishment of the General Assembly, began to systematically fault diagnosis sub-topic for research. In addition to MFPG, the American Society of Mechanical Engineers (ASME), Johns Mitchel company, Spire Corporation are all carried out relevant research.

In Europe, the United Kingdom machine health centers in the late'60s began to study the initial diagnosis. In addition, the Norwegian ship diagnostic techniques, sound detection system in Denmark, Sweden, SPM's bearing detection technology are all more advanced. Japan's fault diagnosis technology in the steel petrochemical sectors such as railways developed rapidly, and in the international market certain advantages.

In 1971, Massachusetts Institute of Technology Beard in his PhD thesis was first put forward the concept of fault detection filter, which is to use analytical redundancy instead of hardware redundancy approach and make the system self-organization through the system closed-loop stability, through the observer output to be systems.

Fault diagnosis method based on analytical redundancy is to be diagnosed by an object exists in the analytical redundancy and other priori knowledge analysis and processing, enabling detection of fault diagnosis, isolation, identification. In the same year, Mehra and Peschon published relevant papers in Automatica, which marked the beginning of fault diagnosis. In 1976, Willsky published the first articles on fault detection and diagnosis technology, an overview of the article.

In 1978, Himmelblau published the first book on the international level fault detection and diagnosis (FDD) technology in academic works. Since then a lot of academic institutions, government departments, universities and enterprises are involved in or the fault diagnosis technology research, and made a large number of results.

Fault diagnosis tasks, from low level to high, can be divided into the following aspects:

1. Failure Modeling: In accordance with a priori information and input-output relations, build a mathematical model of system failure, as a basis for fault diagnosis.

2. Failure Detection: From the measured or unmeasured variables estimation, to determine whether there was a fault diagnosis system. The main task of Fault Detection system is to determine whether there are failures. In general, any fault detection subsystem can not detect a variety of failures correctly 100 percent. Therefore, improving the correct fault detection rate and reduce the failure rate of omission (which occurs without the fault is detected) and false alarm rate (not failure but police) has been the interesting topic in the area of fault detection and diagnosis.

3. Fault Isolation: After the fault has been detected, the location of the fault source should be given. Fault isolation is also known as fault identification or fault location.

4. Fault identification: After a fault has been isolated, determine the time of fault occurred and time-varying characteristics of the fault.

5. Failure evaluation and decision-making: Determine the severity of fault and its impact on the diagnosis of the object and trends in the different conditions for different measures.

After several decades of development, the formation of fault diagnosis technology generally three types of methods, analytical model-based fault diagnosis method, signal processing-based fault diagnosis method and knowledge-based fault diagnosis method.

Dynamic system model-based fault diagnosis method developed the earliest and most in-depth. It needs to establish accurate mathematical model of the object. The advantage of this method is that it can fully use of the deep knowledge within the system, which will help the system fault diagnosis.

But in fact a complex engineering system is very difficult to obtain accurate mathematical model, and the system modeling errors and uncertainties disturbance and measurement noise is always inevitable.

When diagnosed analytical mathematical model of the object is difficult to be established, signal processing-based method is very effective. The method uses the signal model directly, such as the correlation function, higher-order statistics, spectrum, autoregressive moving average, wavelet techniques to extract the amplitude, phase, spectral characteristics of value, and analyses these characteristics in order to achieve fault detection. This approach avoided the difficulties of building an object model. Obviously, this method is not only suitable for linear system is also suitable for nonlinear systems. This method of mining the system information contained in the signal and the structure of the system is not concerned.

Knowledge-based fault diagnosis method and signal processing-based fault diagnosis method is similar. It does not require quantitative mathematical model. The difference is that it introduces a lot of information of the diagnosis object, in particular, can take advantage of expert diagnostic knowledge. It is a promising method of fault diagnosis, particularly in the field of nonlinear systems.

In this book, a number of much innovative fault diagnosis algorithms in recently years are introduced. These methods can detect failures of various types of system effectively, and with a relatively high significance.

Editor:
Wei Zhang

Contents

2

The use of the GLRT for revealing faults in atomic frequency standards

Emilia Nunzi
University of Perugia
Italy

1. Introduction

Atomic clocks are instruments widely employed in many synchronization systems. When such measurement instruments are inserted in complex systems as telecommunication networks, global satellite navigation systems, tests of fundamental physics of matter, an unexpected and anomalous behavior or a degradation of the performance of the clock may give rise to an error condition in the global system (Galysh et al., 1996); (Vioarsson et al., 2000). In some cases, as in the GPS system, the anomaly must be identified and a real-time alarm signal must be transmitted to user. Aerospace systems and navigation support systems used in the area of personal security, for both military and civil purposes, must satisfy strict requirements of the parameters relative to the integrity, reliability, availability and accuracy of the signal. The possible lacking of information concerning one of the above features, may imply a series of inefficiencies and system bugs for end-users.

Monitoring stability of atomic clock frequency data is important for guaranteeing the correct behavior of the electronic system where they are inserted. The principal application where the frequency stability monitoring is a challenging problem is in GNSS systems (like GPS or Galileo) where the overall system performance critically depends on performance of on–board clocks. When the clock behaves bad, thus the anomaly has to be detected fast in order to provide an adequate action for restoring the correct behavior of the clock. In the field of navigation system "integrity", most of the studies are related to satellite integrity (Bruce et al., 2000) and not specifically to that of the embedded clock, while recently a study on GPS clock integrity showed GPS clock strange behaviors (Weiss et al., 2006), asking for suitable new statistical tools for its characterization.

The scientific and industrial community has done a lot of efforts for the theoretical characterization of the behavior of atomic clocks. The purpose is to improve the accuracy and reliability of these instruments while reducing their size and cost. Since such objectives are often in contradiction, the scientific and industrial research is investigating innovative techniques to overcome the limits imposed by the technological development.

A common assumption when analyzing atomic clock data is that clock noise is stationary or that at least increments in the frequency values are stationary and thus data are examined by using stability analysis tools such as the Allan variance (IEEEstd, 1999); (D. W. Allan, 1987). The scientific literature has shown how this hypothesis cannot be always verified in reality, in particular in the application context of the satellite navigation or in experiments of fundamental physics of matter. If this hypothesis is not satisfied, the accuracy and reliability

of the classical fault detection techniques, and therefore of the whole system where the devices are inserted, are compromised (Vioarsson et al., 2000); (Bruce et al., 2000).
In this context, this book chapter collects and summarizes the last years proposals about the use of Generalized Likelihood Ratio test (GLRT) as a fault detection technique, complementary to the classical ones, for revealing faults from frequency data (E. Nunzi et al., 2007); (E. Nunzi et al., 2007); (E. Nunzi et al., 2008); (E. Nunzi & P. Carbone, 2008); (E. Nunzi & D. D'Ippolito, 2009); (E. Nunzi et al., 2009).
In particular, the GLRT (S. M. Kay, 1998), following the Neyman-Pearson (NP) approach is presented and its effectiveness is demonstrated when clock frequency data are subjected to jumps in the mean and/or in the dynamic range. This method is largely employed for revealing faults in industrial manufacturing processes or for supporting the decision making problems in many different applications fields.
Although the large number of scientific publications allows a simple interpretation of the GLRT outcome, the application of the GLRT to frequency data acquired from atomic clocks still need to be properly customized and metrologically characterized.
It follows that reasons for proposing the GLRT as an alternative method for revealing faults in atomic frequency can be summarized as it follows:

1. GLRT does not require the stationarity hypothesis on processed data;
2. GLRT gives a reliable outcomes also when the acquired data record is affected by missing data;
3. GLRT is easy to implement (since it based on the evaluation of the Maximum Likelihood Estimates (MLEs) of the data model parameters) (S. M. Kay, 1996);
4. GLRT functionalities can be extended for revealing also, in the meantime or separately, anomalous behavior other than mean and variance changes.

In order to give a comprehensive presentation of the problem and of the theory needed for applying the GLRT technique to frequency signals acquired from an atomic clock, the next section recalls the atomic clock frequency sample model adopted for the analysis of the fault occurrence and suitable for the application of the fault detection theory as indicated in (S. M. Kay, 1998). The evaluated GLRT detector is applied to both simulated and experimental data subjected to anomalous behavior in order to validate the presented statistical models and theory.

2. Mathematical model of frequency data

In this chapter, a single frequency sample, $y[\cdot]$, is modeled as a white Gaussian random variable with unknown mean and standard deviation, indicated with μ_0 and σ_0, respectively. When a N–length record of frequency data is collected, all samples are assumed to be independent and identically distributed (i.i.d.). Moreover, an anomalous behavior of the clock is defined as a change in the frequency size and/or dynamic range. As a consequence, the anomaly, when and if it occurs, can be modeled as a change in the mean and/or in the standard deviation of the statistical model (D. W. Allan, 1987); (E. Nunzi et al., 2007); (E. Nunzi et al., 2007); (IEEEstd, 1999).
The simple model assumed for atomic clock frequency behavior, allows the description of the parameters model change by means of two different statistical hypotheses indicated with $\mathcal{H}_0$ and $\mathcal{H}_1$. $\mathcal{H}_0$ denotes the assumption that the clock behavior respects the given model;

$\mathcal{H}_1$ represents the assumption that the mean and/or standard deviation have changed to μ_1 and/or σ_1. These hypotheses are formally described by the following equations:

$$\mathcal{H}_0 \quad : \quad y[n] \sim \mathcal{N}(\mu_0,\sigma_0), \quad n = 0,...,N-1; \tag{1}$$

$$\mathcal{H}_1 \quad : \quad y[n] \sim \begin{cases} \mathcal{N}(\mu_0,\sigma_0), & n = 0,...,n_0-1; \\ \mathcal{N}(\mu_1,\sigma_1), & n = n_0,...,N-1, \end{cases} \tag{2}$$

where n_0 is the unknown sample number at which the change happens when $\mathcal{H}_1$ is true, and N is the number of analyzed data.
Let us indicate with K_μ and σ_{0f} the additive frequency jump and the standard deviation factor, respectively. When $\mathcal{H}_1$ is true, i.e. it is true that processed data include some $(N-n_0)$ anomalous samples, thus the following identities are true:

$$\mu_1 = \mu_0 + K_\mu \tag{3}$$

$$\sigma_1 = \sigma_0 \cdot \sigma_{0f}. \tag{4}$$

It should be noticed that there are many unknown parameters: μ_0 and σ_0 (both under $\mathcal{H}_0$ and $\mathcal{H}_1$), μ_1, σ_1 and n_0 when (and if) a fault occurs. On the other hand, N is a parameter of the data acquisition process.
In order to improve text readability, vectors of the unknown parameters when $\mathcal{H}_0$ and $\mathcal{H}_1$ are true are introduced and they are, respectively, $\boldsymbol{\theta}_{H0}=[\mu_0,\sigma_0]$ and $\boldsymbol{\theta}_{H1}=[\mu_0,\sigma_0,\mu_1,\sigma_1,n_0]$.

3. Generalized likelihood ratio Test (GLRT): theory

When there are only two different models, the scope of the detection theory is the determination of the optimal criterion for identifying which of the models is the most likely to be underlying the given experimental data and when, in case, the clock model parameters have changed their values. The commonly employed decision–rule is based on the so called likelihood ratio test (LRT), which maximizes the detection probability (PD) for a given false alarm probability (PFA). It should be noticed that, in this context, PFA is the probability of deciding for $\mathcal{H}_1$ when $\mathcal{H}_0$ is true, and PD is the probability of deciding for $\mathcal{H}_1$ when $\mathcal{H}_1$ is true.
The LRT technique is based on the evaluation of the likelihood ratio, i.e. the ratio between the likelihood functions calculated when the hypothesis $\mathcal{H}_1$ is true and when the hypothesis $\mathcal{H}_0$ is true. On the basis of the Neyman-Pearson (NP) theorem, if this ratio is sufficiently large, the hypothesis $\mathcal{H}_0$ is rejected.
From a practical point of view, the NP approach is applied to the set of available data $\mathbf{y} = \{y[0],...,y[N-1]\}$ and it is based on the evaluation of the likelihood ratio of $\mathbf{y}$, $L_G(\mathbf{y};[\boldsymbol{\theta}_{H0},\boldsymbol{\theta}_{H1}])$, defined as the ratio between the likelihood function of $\mathbf{y}$ under $\mathcal{H}_1$, $p(\mathbf{y};\boldsymbol{\theta}_{H1},\mathcal{H}_1)$, and the likelihood ratio of $\mathbf{y}$ under $\mathcal{H}_0$, $p(\mathbf{y};\boldsymbol{\theta}_{H0},\mathcal{H}_0)$. If $L_G(\mathbf{y};[\boldsymbol{\theta}_{H0},\boldsymbol{\theta}_{H1}])$ is larger than a given value γ, i.e.:

$$L_G(\mathbf{y};[\boldsymbol{\theta}_{H0},\boldsymbol{\theta}_{H1}]) = \frac{p(\mathbf{y};\boldsymbol{\theta}_{H1},\mathcal{H}_1)}{p(\mathbf{y};\boldsymbol{\theta}_{H0},\mathcal{H}_0)} > \gamma, \tag{5}$$

thus, the NP approach decides for $\mathcal{H}_1$, otherwise $\mathcal{H}_0$ is assumed to be true. The value of γ is chosen on the basis of the target PFA value, α, that should be guaranteed a–priori and it can be evaluated by solving the following equation:

$$PFA = Pr\{L_G(\mathbf{y};[\boldsymbol{\theta}_{H0},\boldsymbol{\theta}_{H1}]) > \gamma;\mathcal{H}_0\} < \alpha. \tag{6}$$

It should be noticed that for evaluating the γ value that satisfies (6), the statistic of $L_G(\mathbf{y};[\boldsymbol{\theta}_{H0},\boldsymbol{\theta}_{H1}])$ is needed.
$L_G(\mathbf{y};[\boldsymbol{\theta}_{H0},\boldsymbol{\theta}_{H1}])$ is calculated by following the statistical model introduced in section 2, i.e. by considering that each measured value, $y[n]$, is the realization of an independent Gaussian random variable. It follows that the likelihood functions of the data vector $\mathbf{y}$ under $\mathcal{H}_0$ and $\mathcal{H}_1$ can be evaluated by multiplying the marginal likelihood function of each data sample, thus obtaining:

$$p(\mathbf{y};\boldsymbol{\theta}_{H0}) = \frac{1}{\left(2\pi\sigma_0^2\right)^{N/2}} \exp\left(-\frac{1}{2\sigma_0^2}\sum_{n=0}^{N-1}(y[n]-\mu_0)^2\right) \tag{7}$$

$$\begin{aligned} p(\mathbf{y};\boldsymbol{\theta}_{H1}) = & \frac{1}{\left(2\pi\sigma_0^2\right)^{n_0/2}\left(2\pi\sigma_1^2\right)^{(N-n_0)/2}} \\ & \exp\left(-\frac{1}{2\sigma_0^2}\sum_{n=0}^{n_0-1}(y[n]-\mu_0)^2 - \frac{1}{2\sigma_1^2}\sum_{n=n_0}^{N-1}(y[n]-\mu_1)^2\right) \end{aligned} \tag{8}$$

Since all values of the parameters vectors $\boldsymbol{\theta}_{H0}$ and $\boldsymbol{\theta}_{H1}$ are unknown and supposed to be deterministic, a GLRT technique, instead of a LRT, has been applied (S. M. Kay, 1998). Thus, parameter values in (5) have been replaced by their maximum likelihood estimates (MLEs), that will be indicated with a hat–sign on the symbol of the corresponding variable name. By indicating with $\widehat{\boldsymbol{\theta}}_{H0} = [\widehat{\mu}_{0\_H0},\widehat{\sigma}_{0\_H0}]$, the MLE of $\boldsymbol{\theta}_{H0}$ when $\mathcal{H}_0$ is true, with $\widehat{\boldsymbol{\theta}}_{H1} = [\widehat{\mu}_{0\_H1},\widehat{\sigma}_{0\_H1},\widehat{\mu}_{1\_H1},\widehat{\sigma}_{1\_H1},\widehat{n}_0]$ the MLE of $\boldsymbol{\theta}_{H1}$ when $\mathcal{H}_1$ is true, it follows that the GLRT test decides for $\mathcal{H}_1$ if $L_G\left(\mathbf{y};[\widehat{\boldsymbol{\theta}}_{H0},\widehat{\boldsymbol{\theta}}_{H1}]\right) > \gamma$, i.e.:

$$L_G\left(\mathbf{y};[\widehat{\boldsymbol{\theta}}_{H0},\widehat{\boldsymbol{\theta}}_{H1}]\right) = \frac{p\left(\mathbf{y};\widehat{\boldsymbol{\theta}}_{H0},\mathcal{H}_1\right)}{p\left(\mathbf{y};\widehat{\boldsymbol{\theta}}_{H1},\mathcal{H}_0\right)} > \gamma \tag{9}$$

where:

$$p\left(\mathbf{y};\widehat{\boldsymbol{\theta}}_{H0}\right) = \frac{1}{\left(2\pi\widehat{\sigma}_{0\_H0}^2\right)^{N/2}} \exp\left(-\frac{1}{\widehat{\sigma}_{0\_H0}^2}\sum_{n=0}^{N-1}(y[n]-\widehat{\mu}_{0\_H0})^2\right) \tag{10}$$

$$\begin{aligned} p\left(\mathbf{y};\widehat{\boldsymbol{\theta}}_{H1}\right) = & \frac{1}{\left(2\pi\widehat{\sigma}_{0\_H1}^2\right)^{\widehat{n}_0/2}\left(2\pi\widehat{\sigma}_{1\_H1}^2\right)^{(N-\widehat{n}_0)/2}} \cdot \\ & \cdot\exp\left(-\frac{1}{2\widehat{\sigma}_{0\_H1}^2}\sum_{n=0}^{\widehat{n}_0-1}(y[n]-\widehat{\mu}_{0\_H1})^2 - \frac{1}{2\widehat{\sigma}_{1\_H1}^2}\sum_{n=\widehat{n}_0}^{N-1}(y[n]-\widehat{\mu}_{1\_H1})^2\right). \end{aligned} \tag{11}$$

It can be shown that MLEs of the mean and of the standard deviation are, respectively, the sample mean and the sample standard deviation and that the MLE of n_0 is the sample number that maximizes the NP detector over the whole available data record (S. M. Kay, 1996). Thus, MLE estimates of the unknown parameters are:

$$\widehat{\mu}_{0\_H0} = \frac{1}{N}\sum_{n=0}^{N-1} y[n] \tag{12}$$

$$\widehat{\mu}_{0\_H1} = \frac{1}{\widehat{n}_0}\sum_{n=0}^{\widehat{n}_0-1} y[n] \tag{13}$$

$$\widehat{\mu}_{1\_H1} = \frac{1}{N-\widehat{n}_0}\sum_{n=\widehat{n}_0}^{N-1} y[n] \tag{14}$$

$$\widehat{\sigma}_{0\_H0} = \sqrt{\frac{1}{N}\sum_{n=0}^{N-1}\left(y[n]-\widehat{\mu}_{0\_H0}\right)^2} \tag{15}$$

$$\widehat{\sigma}_{0\_H1} = \sqrt{\frac{1}{\widehat{n}_0}\sum_{n=0}^{\widehat{n}_0-1}\left(y[n]-\widehat{\mu}_{0\_H1}\right)^2} \tag{16}$$

$$\widehat{\sigma}_{1\_H1} = \sqrt{\frac{1}{N-\widehat{n}_0}\sum_{n=\widehat{n}_0}^{N-1}\left(y[n]-\widehat{\mu}_{1\_H1}\right)^2} \tag{17}$$

$$\widehat{n}_0 = \operatorname{argmax}_{n_0=0,\ldots,N-1}\left\{L_G\left(\mathbf{y};\left[\widehat{\boldsymbol{\theta}}_{H0},\widehat{\boldsymbol{\theta}}_{H1}\right]\right)\right\}. \tag{18}$$

By analyzing and comparing firstly expressions (15) with (10), and thus equations (16)–(17) with (11), it follows that arguments of the exponential terms in (10) and (11) are both equal to $(-N/2)$. It follows that the condition to be tested by the GLRT can be simplified as:

$$\begin{aligned} L_G\left(\mathbf{y};\left[\widehat{\boldsymbol{\theta}}_{H0},\widehat{\boldsymbol{\theta}}_{H1}\right]\right) &= \frac{p\left(\mathbf{y};\widehat{\boldsymbol{\theta}}_{H0},\mathcal{H}_1\right)}{p\left(\mathbf{y};\widehat{\boldsymbol{\theta}}_{H1},\mathcal{H}_0\right)} \\ &= \frac{\left(\widehat{\sigma}^2_{0\_H1}\right)^{\widehat{n}_0/2}\left(\widehat{\sigma}^2_{1\_H1}\right)^{(N-\widehat{n}_0)/2}}{\left(\widehat{\sigma}^2_{0\_H0}\right)^{N/2}} > \gamma \end{aligned} \tag{19}$$

By taking into account that the logarithm is a monotonically increasing function, thus inequality (19) does not change if the logarithm is taken on both inequality sides. Thus the NP approach can be further simplified by solving the following equivalent inequality:

$$\begin{aligned} T\left(\mathbf{y};\left[\widehat{\boldsymbol{\theta}}_{H0},\widehat{\boldsymbol{\theta}}_{H1}\right]\right) &= \log\left(L_G\left(\mathbf{y};\left[\widehat{\boldsymbol{\theta}}_{H0},\widehat{\boldsymbol{\theta}}_{H1}\right]\right)\right) = \\ &= \frac{N}{2}\log\left(\frac{\widehat{\sigma}^2_{0\_H0}}{\widehat{\sigma}^2_{1\_H1}}\right) - \frac{\widehat{n}_0}{2}\log\left(\frac{\widehat{\sigma}^2_{0\_H1}}{\widehat{\sigma}^2_{1\_H1}}\right) > \gamma' = \log\gamma \end{aligned} \tag{20}$$

This expression shows that the detector is function of the three different MLE variance estimators. In order to characterize this NP detector, and to customize the parameters of the detection algorithm for revealing specifically faults of atomic clock frequency data, the statistical characterization of $T\left(\mathbf{y};[\widehat{\boldsymbol{\theta}}_{H0},\widehat{\boldsymbol{\theta}}_{H1}]\right)$ is needed. However, the three variance estimators are not mutually independent. In particular, $\widehat{\sigma}^2_{0\_H0}$ is a function of $\widehat{\sigma}^2_{0\_H1}$ and $\widehat{\sigma}^2_{1\_H1}$, both under $\mathcal{H}_0$ and $\mathcal{H}_1$. In fact, by requiring the energy of the sequence $\{y[0],y[1],...,y[N-1]\}$ to be equal to the sum of the energies of the two complementary subsequences $\{y[0],y[1],...,y[\widehat{n}_0-1]\}$ and $\{y[\widehat{n}_0],y[\widehat{n}_0+1],...,y[N-1]\}$, the following relationship is always true:

$$\widehat{\sigma}^2_{0H0} = \frac{\widehat{n}_0}{N}\widehat{\sigma}^2_{0H1} + \frac{N-\widehat{n}_0}{N}\widehat{\sigma}^2_{1H1} + \frac{\widehat{n}_0}{N}\frac{N-\widehat{n}_0}{N}\left(\widehat{\mu}_{0H1}-\widehat{\mu}_{1H1}\right)^2, \tag{21}$$

both under $\mathcal{H}_0$ and $\mathcal{H}_1$.

This mutual relationship between variance estimators under $\mathcal{H}_0$ and $\mathcal{H}_1$ makes difficult task the evaluation of the probability density function (and thus of the likelihood function) of (20). It follows that the evaluation of the γ threshold from (6) is a difficult task.

In order to derive a criterion for evaluating the γ value, a further theoretical analysis is performed.

By substituting (21) in (20), the detector expression is formally obtained as function of MLE estimators of parameters under $\mathcal{H}_1$:

$$\begin{aligned} T\left(\mathbf{y};[\widehat{\boldsymbol{\theta}}_{H0},\widehat{\boldsymbol{\theta}}_{H1}]\right) &= \frac{N}{2}\log\left(\frac{\widehat{n}_0}{N}\frac{\widehat{\sigma}^2_{0H1}}{\widehat{\sigma}^2_{1H1}} + \frac{\widehat{n}_0}{N}\frac{N-\widehat{n}_0}{N}\frac{\left(\widehat{\mu}_{0H1}-\widehat{\mu}_{1H1}\right)^2}{\widehat{\sigma}^2_{1H1}} + \frac{N-\widehat{n}_0}{N}\right) - \\ &- \frac{\widehat{n}_0}{2}\log\frac{\widehat{\sigma}^2_{0H1}}{\widehat{\sigma}^2_{1H1}}. \end{aligned} \tag{22}$$

Equation (22) shows that the GLRT detector $T(\mathbf{y})$ depends only on the MLE estimates under $\mathcal{H}_1$ by means of a function $g(\cdot)$, i.e. $T(\mathbf{y}) = g(\widehat{\theta}_{H1})$. By recalling the invariance property of the MLE (S. M. Kay, 1996), it follows that also $T(\mathbf{y})$ is an MLE estimate of a theoretical value, T_{teor} which can be evaluated by substituting the MLE estimates of parameters with the corresponding theoretical values. Thus, in order to evaluate expression of T_{teor}, MLE estimates in (22), $\widehat{\sigma}^2_{0H1}$, $\widehat{\sigma}^2_{1H1}$, $\widehat{\mu}_{0H1}$, $\widehat{\mu}_{1H1}$ and $\widehat{n}_0$, have been replaced by the corresponding theoretical value, i.e. by σ_0^2, σ_1^2, μ_0, μ_1 and n_0, respectively. Moreover, by exploiting relationships (3) and (4), it can be shown that the theoretical behavior of the GLRT detector is described by the following equation:

$$T_{teor} = \frac{N}{2}\log A + \left(\frac{N}{2}-\frac{n_0}{2}\right)\log\frac{1}{\sigma^2_{0f}} \tag{23}$$

where

$$A = \frac{K_\mu^2}{\sigma_0^2}\frac{(N-n_0)(n_0-1)}{(N-1)^2} + \left(\frac{n_0-1}{N-1} + \frac{N-n_0}{N-1}\sigma^2_{0f}\right). \tag{24}$$

This equation can be used in practical cases for evaluating a-priori the γ threshold in the comparison process, from the knowledge of: the length N of the data sequence, the initial standard deviation σ_0 value, the frequency jump, K_μ, the frequency variance change factor, σ^2_{0f}, and the number of anomalous samples, $(N-n_0)$, before the detection event.

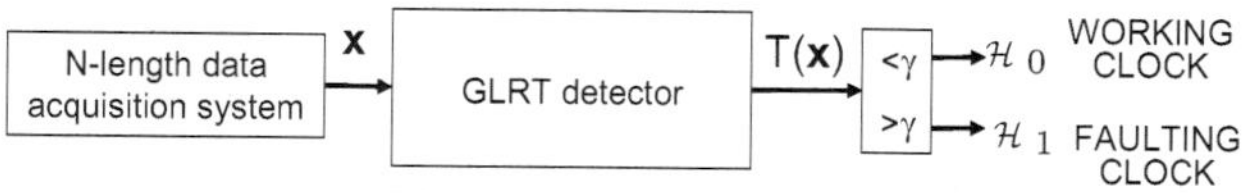

Fig. 1. Simplified block diagram of a GLRT scheme. $\boldsymbol{y}$ is the N–length available data sequence. $T(\boldsymbol{y})$ is the GLRT detector scalar output. γ is the threshold value used for assuming true $\mathcal{H}_0$ or $\mathcal{H}_1$.

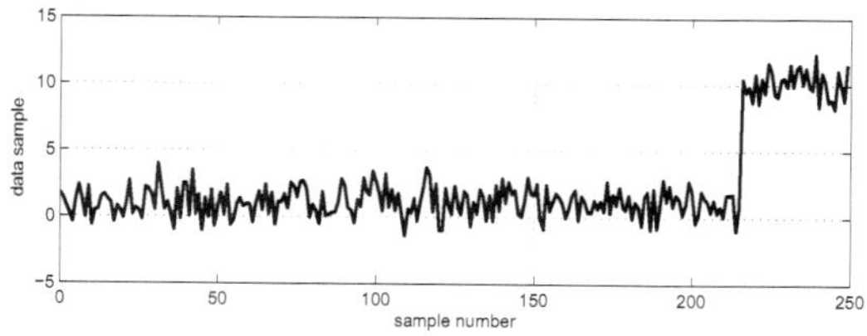

Fig. 2. Behavior of a normalized frequency data record, normally distributed, simulated by considering the following parameters: record length $M = 250$, $\mu_0 = 1$, $\sigma_0 = 1$, $K_\mu = 9$ (thus $\mu_1 = 10$), $\sigma_{0f} = 1$ (thus $\sigma_1 = \sigma_0$), $n_0 = 216$.

All these parameters values affect the GLRT performance and influence, in particular, the choice of the threshold value, γ, that is strictly dependent on the target PFA and that should be defined a–priori before the application of the GLRT. From a practical point of view, the use of the GRT is summarized in Fig.1: the available data sequence is processed by the GRT detector (20) and the corresponding scalar output is compared to a give γ' value in order to assume $\mathcal{H}_0$ or $\mathcal{H}_1$ true.

4. Simulation results

In this section simulation results are reported in order to validate theoretical results obtained in sec. 3. At first, theoretical formula (23) will be validated by means of Monte Carlo simulations and an application example is introduced in order to clarify its practical applicability. Moreover, the statistical characterization of the GLRT is presented by analyzing the behavior of PD versus PFA (i.e. the Receiver Operating Characteristics (ROCs)) . ROCs are evaluated by means of Monte Carlo simulations since the theoretical behavior of the detector is a difficult task, as already stated in sec.3.

4.1 Validation of (23)

In order to validate theoretical formula (23), Monte Carlo simulations on NREC=10000 records of simulated data, each of length M=250, have been performed. Each data record has been generated by considering a M–length sequence of Gaussian distributed data with initial mean and standard deviation values equal to $\mu_0 = 1$ and $\sigma_0 = 1$, respectively, and by setting a frequency jump $K_\mu = 9$ on $n_0 = 216$. For clarification purposes, the behavior of one data record versus the data sample index is shown in Fig.2.

For each data record, the GLRT detector (20) has been applied consecutively NREC times to N=200 data samples by following a First In First Out (FIFO) data organization. The behavior

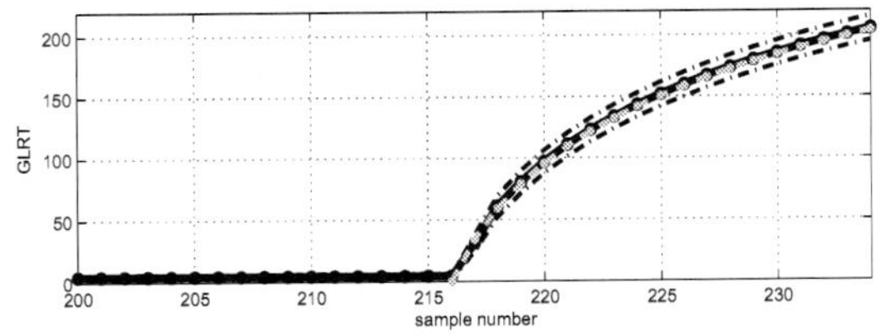

Fig. 3. Behavior of the averaged GLRT detector (solid black line) versus the index of the last acquired data sample, obtained by means of Monte–Carlo simulations. The average has been evaluated on $NREC = 10000$ GLRTs, each obtained by processing a simulated frequency data record with characterizing parameters equal to those of record shown in Fig.2 of data. The two dashed black lines represent the averaged GLRT plus and minus the corresponding sample standard deviation. Dashed gray line shows the behavior of T_{teor} given by (23) by using the same set of parameters values.

of the detector output, averaged over the number of record NREC, versus the index of the last acquired data sample, is shown in Fig.3 with a solid black line. Moreover, dashed black lines represent the averaged GLRT plus and minus the corresponding standard deviation evaluated on the same NREC data records. This figure shows that the GLRT detector is approximately equal to 0 if data are not affected by anomalies, i.e. until the last processed data sample is smaller than 216. When processed data include anomalous samples, thus the GLRT value increases.

For comparison purposes, the theoretical behavior of the detector described by eq.(23) versus the index of the last acquired data sample is also shown in Fig.3 with a dashed gray line for the same set of parameters values, i.e. $K_{\mu} = 9$, $\sigma_0 = 1$, $\sigma_{0f} = 1$, $n_0 = 216$. The theoretical behavior is close to the averaged GLRT output.

In order to give a quantitative characterization of the theoretical behavior, the relative displacement between the averaged GLRT evaluated by means of Monte Carlo simulations and the theoretical detector (i.e. between the solid–black and the dashed–gray lines in Fig.3) is shown in Fig.4(a) and the corresponding relative standard deviation is shown in Fig.4(b).

This figure shows that the error between the theoretical and simulated GLRT reduces as the number of samples with anomaly, and processed by the detector, increases. In particular, if $(N - n_0) > 2$ (i.e. sample index > 218), thus the error is smaller than 2% with a corresponding type A uncertainty, estimated by dividing the relative standard deviation by $\sqrt{NREC} = 100$, smaller than 0.1% (GUM, 1997). This result validates the expression of T_{teor}.

It follows that the theoretical expression (23) can be used for designing an accurate test procedure for revealing faults in data affected by frequency jump and/or by standard deviation variation. In particular, (23) gives the γ value to be used in the comparison process for a given set of: N, target readiness expressed in terms of $(N - n_0)$, target $K\mu$ and σ_{0f}.

4.2 An application example of (23)

In order to clarify how to use theoretical expression (23), let us consider an analysis problem on the simulated data sequence shown in Fig.2. In this case, the GLRT outcome applied to $N = 200$ data managed by using a FIFO strategy is shown in Fig.5. If the detector readiness is considered, in order to reveal a fault by using no more than $(N - n_0) = 4$ samples, by substituting in (23) values $N = 200$, $\sigma_0 = 1$, $n_0 = 196$, $\sigma_{0f} = 1$, $K_{\mu} = 9$, thus $T_{teor} = 95.37$. By

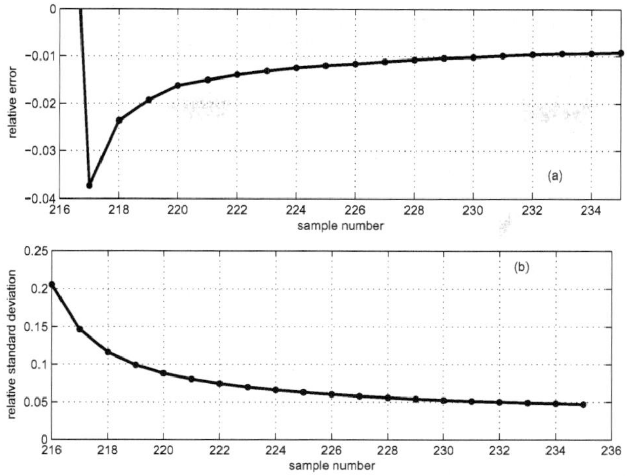

Fig. 4. Relative deviation (a) between the averaged GLRT obtained by means of Monte–Carlo simulations and T_{teor} shown in Fig.3 and corresponding relative sample standard deviation (b).

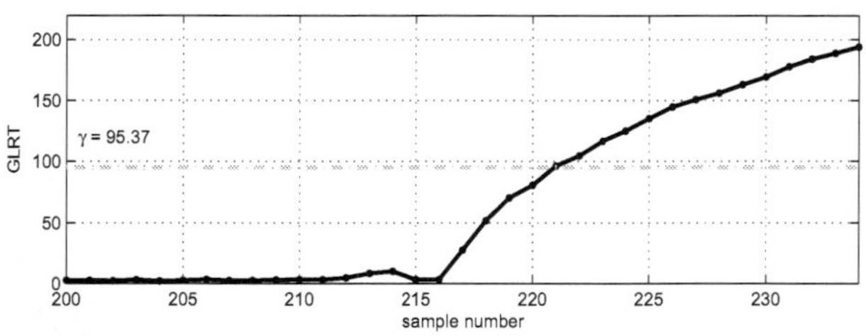

Fig. 5. Behavior of the GLRT detector applied to the data sequence shown in Fig.2 versus the index of the last acquired data sample.

setting $\gamma = T_{teor} = 95.37$, we can see in Fig.5 that the GLRT value is greater than γ for $n = 220$ and a warning signal should be emitted. This means that an additive frequency jump at least equal to 9 has occurred no more than 4 samples before the alarm signal emitted on the sample number 220.

Theoretical expression T_{teor} can be used also for analyzing the detector accuracy versus the number of employed anomalous samples $N - n_0$. In particular, Fig.6 shows the theoretical behavior of the GLRT with the $N - n_0$ when a data sequence of length $N = 200$ affected by a frequency jump equal to $K_m = 9$ is analyzed. This behavior clearly shows that the frequency jump is detectable also by using just one anomalous sample. The large value of the GLRT detector ($T_{teor} = 34$) when $N - n_0 = 1$ is mainly due to the large jump size considered in this application example and to the large $N = 200$ employed. It should be noticed that a large N value is due to the hypothesis that initially the clock is working properly and that a fault occurs after that the monitoring process starts. A large N value increases the accuracy of the MLE estimates of μ_0 and σ_0 since they are asymptotic Gaussian with N (S. M. Kay, 1996).

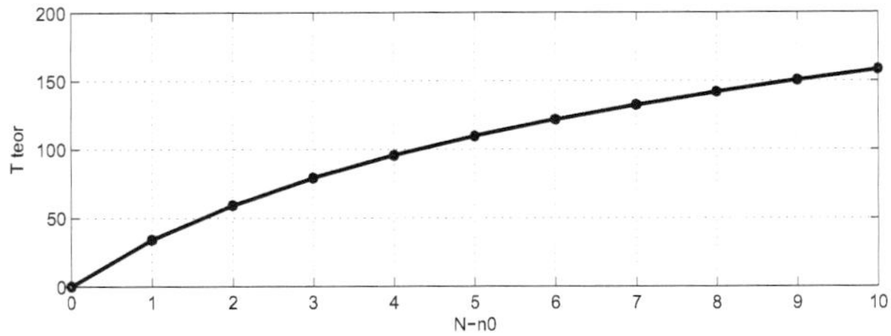

Fig. 6. Behavior of T_{teor} versus $N - n_0$ given by (23) obtained by setting characterizing parameters equal to those of record shown in Fig.2 of data.

4.3 Statistical characterization of the NP detector

In order to illustrate the properties of the NP detector when it is applied to data with parameters values typically employed in frequency standards and, in particular, to relate reasonable values for the threshold parameter γ to the corresponding PFA value, the test has been characterized by calculating the receiver operating characteristic (ROC), i.e. the behavior of PD versus PFA, for many γ and n_0 values, by means of Monte Carlo simulations (S. M. Kay, 1996); (S. M. Kay, 1998).

In particular, two sets of ROCs are presented here: the first one for analyzing the GLRT sensitivity (i.e.the detection capability by using no more than a given number of faulty samples), the other one for evaluating the GLRT readiness (i.e. the number of employed faulty samples for a given fault).

GLRT SENSITIVITY ANALYSIS

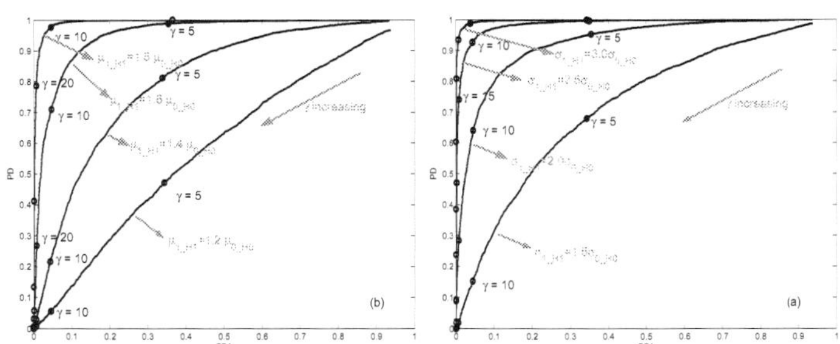

Fig. 7. ROCs obtained by considering 100 data samples and by setting $(N - n0) = 15$. (a) $\mu_1 = \mu_0$ and $\mu_1 = \{1.5\sigma_0, 2\sigma_0, 2.5\sigma_0, 3\sigma_0\}$. (b) $\mu_1 = \{1.2\mu_0, 1.4\mu_{0x}, 1.6\mu_0, 1.8\mu_0\}$ and $\sigma_1 = \sigma_0$.

Fig. 7(a) and (b) shows the ROCs when $\mu_1 = \mu_0$ and $\sigma_1 = \{1.5\sigma_0, 2\sigma_0, 2.5\sigma_0, 3\sigma_0\}$, and $\mu_1 = \{1.2\mu_0, 1.4\mu_0, 1.6\mu_0, 1.8\mu_0\}$ and $\sigma_1 = \sigma_0$, respectively, with $\mu_0 = 2.36 \cdot 10^{-11}$,

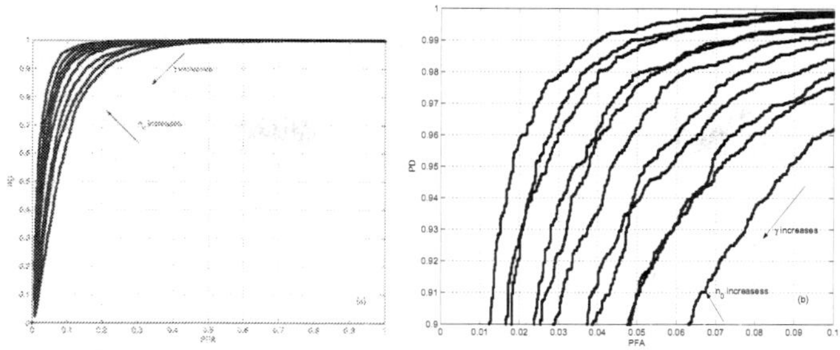

Fig. 8. Each line represents a ROC corresponding to a particular value of n_0 obtained by using a Monte Carlo approach based on 5000 pairs of data record each of length N=100 when: σ_1/σ_0=3, γ varies from 0 to 10. (a): n_0 ranges from 13 to 20. (b): n_0 ranges from 20 to 30.

$\sigma_0 = 1.046 \cdot 10^{-11}$, $N = 100$, and by setting $(N - n_0) = 15$. Circles indicate the γ value used for evaluating the corresponding PFA and PD. Fig. 7 can be employed to design a detection test that can track changes in the mean and in the standard deviation. As an example, one can see that a PD = 93% is obtained when the change in the mean is at least equal to a factor 1.8, and that the condition $T\left(\mathbf{y};[\widehat{\boldsymbol{\theta}}_{H0},\widehat{\boldsymbol{\theta}}_{H1}]\right) > \gamma = 10$ guarantees a PFA $< 8\%$. The same threshold value $\gamma = 10$ can be employed to detect a change in the standard deviation by a factor at least equal to 3.0, with PD $> 97\%$ and PFA $< 8\%$.

GLRT READINESS ANALYSIS

To this purpose, for each value of n_0 varying from 13 to 20, 5000 pairs of N–length data records, normally distributed, have been synthesized. For each record pair, the first sequence has been created by employing mean and standard deviation values respectively equal to $\overline{\mu}_0 = 2.3650 \cdot 10^{-11}$ and $\overline{\sigma}_0 = 1.0462 \cdot 10^{-11}$, that is by assuming $\mathcal{H}_0$ true. The second data set presents the first n_0 samples equal to those of the first record, while the last $N - n_0$ samples have been generated as normally distributed with mean $\overline{\mu}_0$ and $\sigma_1 = 3\sigma_0$. Thus, the second data record meets with $\mathcal{H}_1$. Both data records have been employed for calculating the corresponding value of $T\left(\mathbf{y};[\widehat{\boldsymbol{\theta}}_{H0},\widehat{\boldsymbol{\theta}}_{H1}]\right)$ when $\mathcal{H}_0$ is true, i.e. by applying (20) to the first record, and when $\mathcal{H}_1$ is true, i.e. by applying (20) to the second record.
Simulation results obtained by assuming N=100 are shown in Figs.8(a) and (b). In particular, each line in Fig.8(a) is the ROC corresponding to a particular n_0 value, ranging from 13 to 20, with γ varying from 0 to 15, as indicated by the figure label, when $\mu_0 = \mu_1 = \overline{\mu}_0$ and $\sigma_0/\sigma_1 = 3$ with $\sigma_0 = \overline{\sigma}_0$. Fig.8(b) shows the same ROC reported in Fig.8(a) but detailed for PFA<10%, PD>90% and n_0 ranging from 20 to 30. This figure shows that by processing a set of data of length $N = 100$, the fault detection occurs with PFA<5% and PD>95% only for n_0 larger than 25. Equivalently, under the same assumptions, it is possible to detect a fault within 25 samples from its occurrence. Moreover, for each n_0 value, this figure gives information on the range of the threshold values γ which can be used for revealing a fault.

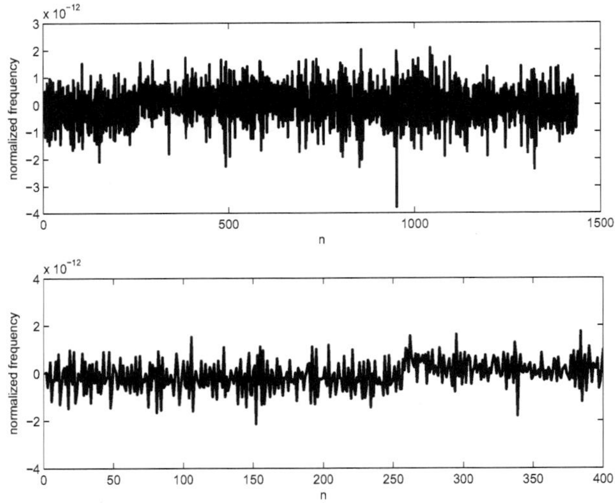

Fig. 9. (a): Behavior of normalized frequency data of atomic clock on satellite GPS 22 (downloaded by IGS). (b) Zoom of figure (a) on the first 400 experimental data samples.

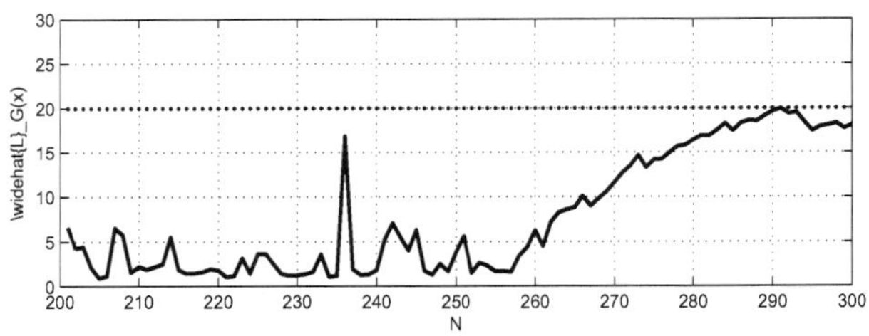

Fig. 10. Behavior of the GLRT detector applied to data shown in Fig. 9 (a) with $N = 200$.

5. Experimental results

Figure 9(a) shows the behavior of experimental normalized frequency data of GPS satellite number 22. Data have been obtained by the International GNSS Service (IGS) – Formerly the International GPS Service (J.M. Dow et al., 2005). The GLRT has been applied to these data by setting a processing window length $N = 200$. Thus, data samples have been processed by following a FIFO strategy and the processing result is shown in Fig.10. In particular, each time a new data sample is processed, the detector outcome is updated. The GLRT has a a spike on sample numbered as 236 while presents an increasing behavior after sample index 260. By requiring a small threshold value, i.e. $\gamma = 10$, warning signals are emitted on sample indexes 237 and 266. The first one is probably a false alarm, while the second is a true false alarm. If the γ value is increased (i.e. $\gamma = 18$), thus the false alarm probability is obviously reduced but the detection process requires a larger number of faulty samples in order to detect the fault

occurrence (Fig.10 shows that at least 22 faulty samples are needed since $\gamma = 18$ implies a fault detection on sample index 282).
A-posteriori analysis of data parameters confirm the validity of the GLRT technique. In fact, by zooming the behavior of the experimental frequency data on the first 400 samples (see Fig.9(b)) it is clear (and confirmed) that the first warning is a false alarm, while the second warning signal a fault detection signal and that the fault has occurred on sample numbered as 260 (as argued by a visual inspection of the GLRT behavior). In particular, the fault can be classified as a mean change. In fact, the arithmetic mean on the first 260 sample is equal to $\mu_0 = -2.21 \cdot 10^{-13}$ and on samples from 261 to 461 is equal to $\mu_0 = 1.96 \cdot 10^{-13}$. The corresponding standard deviation values are almost the same and equal to $\sigma_0 = \sigma_1 = 6 \cdot 10^{-13}$. Thus, we can asume $K_\mu = 4 \cdot 10^{-13}$ and $\sigma_{0f} = 1$.
By substituting these parameters values in the theoretical detector (23) and by requiring the fault detection within no more than 6 faulty samples, i.e. by using $N - n_0 = 6$, thus (23) gives $T_{teor} = 9.14$, which is a threshold value congruent with the analysis of the GLRT outcome shown in Fig.10.

6. Conclusion and future works

The GLRT detection algorithm for revealing faults from atomic clock frequency data has been analyzed. It has been shown that the GLRT algorithm efficiently detects anomalies when applied to atomic clock frequency data (E. Nunzi & P. Carbone, 2008)– (E. Nunzi et al., 2009). However, the mathematical expression of the detector depends on many parameters characterizing acquired data and previously analysis of the parameters dependencies have been based on ROCs evaluated by means of Monte–Carlo simulations. It follows that a proper customization of the GLRT technique to the analysis of data coming from atomic clocks is necessary and useful for designing the test procedure.
The information given by the GLRT method can be employed to detect the non stationary change in the clock behavior.
Future works are focused on the theoretical statistical characterization of the GLRT detector and on the designing of the test procedure for effectively reveals anomaly in real–time (i.e. while the data acquisition process is going on). Moreover, the MLE estimates used for evaluating the GLRT outcome could be employed for deriving information also on the identification of the occurred anomaly, although it is known MLE estimators are not optimal when a small number of faulty samples are employed (as required for this specific application).

Acknowledgment

This work has been supported by the Italian Ministry PRIN 2007 program (ARCADIA project: http://arcadia.diei.unipg.it).

7. References

D. W. Allan (1987) , "Time and frequency (time-domain) characterization, estimation and prediction of precision clocks and oscillators," *IEEE Trans. Ultrason. Ferroelectr. Freq. Control*,vol. UFFC-37, no. 6, pp. 647–654, Nov. 1987.

A. S. Bruce & A. J. Van Dierendonck & A. Jakab, J. Wiebe & B. Townsend (2000), "Detection of GPS Satellite Signal Failures in Satellite Based Augmentation Systems," Proc. of the *ION GPS Meeting*, 19-22 September 2000, Salt Lake City, Utah, USA (Institute of Navigation, Alexandria, Virginia), pp. 189-198.

J.M. Dow, R.E. Neilan, G. Gendt (2005), "The International GPS Service (IGS): Celebrating the 10th Anniversary and Looking to the Next Decade," *Adv. Space Res. 36* vol. 36, no. 3, pp. 320-326, 2005. doi:10.1016/j.asr.2005.05.125.

I. J. Galysh & D. M. Craig & W. G. Reid & J. A.Buisson (1996), "Performance Analysis Of The GPS Monitor Station Timing Subsystem Enhancement Program At The Naval Research Laboratory", Proc. *Precise Time and Time Interval*, 1996, Washigton, DC.

(1999) IEEE Standard Definitions of Physical Quantities for Fundamental Frequency and Time Metrology. Random Instabilities, IEEE Standard 1139, 1999.

S. M. Kay (1998), "Fundamentals of Statistical Signal Processing, Volume 2: Detection Theory," *Prentice Hall*, 1998.

S. M. Kay (1998), "Fundamentals of Statistical Signal Processing, Volume 1: Estimation Theory," *Prentice Hall*, 1998.

E. Nunzi & P. Carbone (2007), "The generalized likelihood ratio test for detecting anomalous behaviors of atomic clocks," (Invited paper) Proc. of *SPIE Noise and Fluctuations in Photonics, Quantum Optics, and Communications Conference*, 21-24 May 2007, Florence, Italy, vol. 6603, pages: 660310-1 - 660310-9.

E. Nunzi, L. Galleani, P. Tavella & P. Carbone (2007), Detection of Anomalies in the Behavior of Atomic Clocks, *Instr. and Meas., IEEE Transactions on*, Vol.XX, Apr. 2007, pp. 523–528.

E. Nunzi & P. Carbone (2008), Monitoring Signal Integrity of Atomic Clocks by Means of the GLRT, *Metrologia*, Vol. 45, Dec - 2008, S103-S107.

E. Nunzi and P. Carbone and P. Tavella (2008), "Fault detection in atomic clock frequency standards affected by mean and variance changes and by an additive periodic component: the GLRT approach," *Instr. and Meas. Tech. Conf.*, May 2008.

E. Nunzi & D. D'Ippolito (2009), "A Novel Theoretical Analysis of Fault Detection for Atomic Clock," *IEEE International Workshop on Advanced Method for Uncertainty Estimation in Measurement - AMUEM*, Bucharest, Romania, July 2009.

E. Nunzi & G. Barchi & U. Bartoccini (2009), "Methods and Tools for Frequency Jump Detection," *IEEE International Workshop on Advanced Method for Uncertainty Estimation in Measurement - AMUEM*, Bucharest, Romania, July 2009.

L. Vioarsson & S. Pullen & G. Green & P. Enge (2001), Satellite Autonomous Integrity Monitoring and its Role in Enhancing User Performance, Proc. of the *ION GPS Meeting*, 11-14 September 2001, Salt Lake City, Utah, USA (Institute of Navigation, Alexandria, Virginia), pp. 690-702.

M. Weiss, P. Shome, R. Beard (2006), "GPS signal integrity dependencies on atomic clocks," Proc. *38th Annual Precise Time and Time Interval (PTTI) Meeting*, Washington DC, Dec 2006, pp. 439-448.

(1997) Guide to the Expression of Uncertainty in Measurement. ISO 1997

Qualitative Fault Detection and Hazard Analysis Based on Signed Directed Graphs for Large-Scale Complex Systems

Fan Yang[1,2], Deyun Xiao[1] and Sirish L. Shah[2]
[1] *Tsinghua University*
[2] *University of Alberta*
[1] *China*
[2] *Canada*

1. Introduction

Nowadays in modern industries, the scale and complexity of process systems are increased continuously. These systems are subject to low productivity, system faults or even hazards because of various conditions such as mis-operation, equipment quality change, external disturbance, and control system failure. In these systems, many elements are interacted, so a local fault can be propagated and probably spread to a wide range. Thus it is of great importance to find the possible root causes and consequences according to the current symptom promptly. Compared with the classic fault detection for local systems, the fault detection for large-scale complex systems concerns more about the fault propagation in the overall systems. And this demand is much close to hazard analysis for the system risks, which is a kind of qualitative analysis in most cases prior to quantitative analysis.

The signed directed graph (SDG) model is a kind of qualitative graphical models to describe the process variables and their cause-effect relations in continuous systems, denoting the process variables as nodes while causal relations as directed arcs. The signs of nodes and arc correspond to variable deviations and causal directions individually. The SDG obtained by flowsheets, empirical knowledge and mathematical models is an expression of deep knowledge. Based on the graph search, fault propagation paths can be obtained and thus certainly be helpful for the analysis of root causes and sequences (Yang & Xiao, 2005a). And with development of the computer-aided technology, graph theory has been implemented successfully by some graph editors, some of which, like Graphviz (2009), can transform text description into graphs easily. Hence the SDG technology can be easily combined with the other design, analysis and management tools.

The SDG definition and its application in fault diagnosis were firstly presented by Iri et al. (1979). Ever since then, many scholars have contributed to this area, including modeling, inference, software development and applications. Many efforts have been particularly made to implement the methods and to overcome the disadvantages, such as spurious solutions. Here we recognize some representatives among them. Kramer & Palowitch (1987) used rules to describe SDG arcs, which shows that expert systems can be employed as a tool in this problem. Oyeleye & Kramer (1988) took into account the qualitative simulation for the SDG inference. Shiozaki et al. (1989) improved the SDG model by adding fault revealing time. Yu et al. (Chang & Yu, 1990; Yu & Lee, 1991) introduced fuzzy information for arc signs to describe the steady state gains. Maurya et al. (2003a, 2003b, 2006) described the modeling method based on differential equations (DEs) and algebraic equations (AEs), analyzed the initial and final responses based on SDGs, and studied the description and analysis of control loops. SDG method has been combined with other data-driven methods to improve the diagnosis accuracy (Vedam & Venkatasubramanian, 1999; Lee et al., 2006). At first, the inference is based on single fault assumption, but multiple fault cases attract more and more attention (Vedam & Venkatasubramanian, 1997; Zhang et al. 2005; Chen & Chang, 2007). Up to now, SDG method has been implemented in some software tools (Mylaraswamy & Venkatasubramanian, 1997; McCoy et al. 1999; Zhang et al., 2005) and applied in various industrial systems.

Aiming at SDG applications in the area of fault detection and hazard analysis, the problems of description and inference are most important. As the system extends, the time consumption of graph search is heavy, so the single-level SDG model should be transformed into hierarchical model to improve the search efficiency. The root cause can be searched in this model level by level according to the initial response of the system. In control systems and many other cases, cycles exist in the graph, resulting in the truncation or misleading to the search. Thus the theoretic fundamentals and dynamic features of SDGs should be studied. We have analyzed the fault propagation principles by operations of corresponding qualitative matrices and obtained some typical rules of control systems.

Moreover, fault detection is performed based on sensor readings, thus the sensor location strategy affects the performance of fault detection. Due to the economical or technical limitations, the number of sensors should be limited while meeting the demands of fault detection. This can be considered in the SDG framework. We analyze main criteria such as detectability, identifiability and reliability in the framework of SDGs and presented algorithms, in order to guarantee that the faults can be detected and identified, and to optimize the fault detection ability.

This chapter is organized as follows: first, the SDG description is reviewed and hierarchical model is indicated; then the fault propagation rules and inference approaches are summarized to lead to the successful application of fault detection and hazard analysis; some considerations about sensor location are introduced next; finally a generator set process in a power plant is modeled and analyzed to illustrate the proposed model and method.

2. Model Description of Signed Directed Graph

2.1 Basic Form of SDG Model

SDGs are established by representing the process variables as graph nodes and representing causal relations as directed arcs. An arc from node A to node B implies that the deviation of A may cause the deviation of B. For convenience, "+", "-" or "0" is assigned to the nodes in comparison with normal operating value thresholds to denote higher than, lower than or within the normal region respectively. Positive or negative influence between nodes is distinguished by the sign "+" (promotion) or "-" (suppression), assigned to the arc (Iri et al., 1979). The definition is as follows:

Definition 1: An *SDG model* γ is the composite (G,φ) of (1) a digraph G which is the quadruple $(N, A, \partial^+, \partial^-)$ of (a) a set of nodes $N=\{v_1, v_2, \cdots, v_n\}$, (b) a set of arcs $A=\{a_1, a_2, \cdots, a_m\}$, (c) a couple of incidence relations $\partial^+ : A \to N$ and $\partial^- : A \to N$ which make each arc correspond to its initial node $\partial^+ a_k$ and its terminal node $\partial^- a_k$, respectively; and (2) a function $\varphi: A \to \{+,-\}$, where $\varphi(a_k)$ $(a_k \in A)$ is called *the sign of arc* a_k .

Usually we use a_{ij} to denote the arc from v_i to v_j.

Definition 2: A *pattern* on the SDG model $\gamma=(G,\varphi)$ is a function ψ : $N \to \{+,0,-\}$. $\psi(v)$ $(v \in N)$ is called the *sign of node* v , i.e.

$$\psi(v)=0 \text{ for } |x_v - \bar{x}_v| < \varepsilon_v$$
$$\psi(v)=+ \text{ for } x_v - \bar{x}_v \geq \varepsilon_v$$
$$\psi(v)=- \text{ for } \bar{x}_v - x_v \geq \varepsilon_v$$

where x_v is the measurement of the variable v , $\bar{x}_v$ is the normal value, and ε_v is the threshold.

Definition 3: Given a pattern ψ on a SDG model $\gamma=(G,\varphi)$, an arc a is said to be *consistent* (with ψ) if $\psi(\partial^+ a)\varphi(a)\psi(\partial^- a)=+$. A path, which is consisted of arcs $a_1, a_2, \cdots, a_k$ linked successively, is said to be *consistent* (with ψ) if $\psi(\partial^+ a_1)\varphi(a_1)\cdots\varphi(a_k)\psi(\partial^- a_k)=+$.

2.2 Modeling Methods of SDGs

2.2.1 SDG modeling by mathematical equations

In general, SDGs can be obtained either from operational data and process knowledge, or mathematical models. If we have the differential algebraic equations (DAEs), then we can derive the structure and signs of the graph by specific methods (Maurya, 2003a).

A typical dynamic system can be expressed as a set of DEs

$$\frac{\mathrm{d}x_i}{\mathrm{d}t} = f_i(x_1, \cdots, x_n) \tag{1}$$

where $x_1, \cdots, x_n$ are state variables. By Taylor expansion near normal state, we get

$$\frac{\mathrm{d}x_i}{\mathrm{d}t} \approx f_i(x_1^0, \cdots, x_n^0) + \sum_{j=1}^{n} \left.\frac{\partial f_i}{\partial x_j}\right|_{x_1^0, \cdots, x_n^0} (x_j - x_j^0) \tag{2}$$

where $x_1^0, \cdots, x_n^0$ are normal states. Eq. (2) can be written as the following matrix form

$$\frac{\mathrm{d}}{\mathrm{d}t}\begin{bmatrix} x_1 \\ \vdots \\ x_n \end{bmatrix} \approx \begin{bmatrix} f_1\left(x_1^0,\cdots,x_n^0\right) \\ \vdots \\ f_n\left(x_1^0,\cdots,x_n^0\right) \end{bmatrix} + \begin{bmatrix} \frac{\partial f_1}{\partial x_1} & \cdots & \frac{\partial f_1}{\partial x_n} \\ \vdots & & \vdots \\ \frac{\partial f_n}{\partial x_1} & \cdots & \frac{\partial f_n}{\partial x_n} \end{bmatrix}_{x_1^0,\cdots,x_n^0} \begin{bmatrix} x_1 - x_1^0 \\ \vdots \\ x_n - x_n^0 \end{bmatrix} \tag{3}$$

The Jacobian matrix

$$\boldsymbol{J} = \begin{bmatrix} \frac{\partial f_1}{\partial x_1} & \cdots & \frac{\partial f_1}{\partial x_n} \\ \vdots & & \vdots \\ \frac{\partial f_n}{\partial x_1} & \cdots & \frac{\partial f_n}{\partial x_n} \end{bmatrix} \tag{4}$$

can be described by an SDG whose signs of arcs are defined as

$$\operatorname{sgn}\left(x_j \to x_i\right) = \operatorname{sgn}\left[\left. \frac{\partial f_i}{\partial x_j} \right|_{x_1^0,\cdots,x_n^0} \right] \tag{5}$$

if the nodes correspond to the state variables. Thus the SDG actually describes the direct influences or sensitivities between state variables.

In practical problems, the systems often have the following form as DEs:

$$a_n\left(\mathrm{d}^n x/\mathrm{d}t^n\right) + \cdots + a_2\left(\mathrm{d}^2 x/\mathrm{d}t^2\right) + a_1\left(\mathrm{d}x/\mathrm{d}t\right) + a_0 x = e \tag{6}$$

where x is the state and e is the disturbance. When $n = 1$, it is a first-order system:

$$\left(\mathrm{d}/\mathrm{d}t\right)x = -\left(a_0/a_1\right)x + \left(1/a_1\right)e \tag{7}$$

The step response is shown as Fig. 1(a). An arc is constructed from the node e to x with a sign sgn$[1/a_1]$ and a self-cycle on the node x with a sign -sgn$[a_0/a_1]$. For high-order systems, simplification can be made because the corresponding DE includes different order derivatives of the same variable, which can be avoided for the explicit physical meaning of the model. They can be approximated as first-order systems with delays:

$$\left(\mathrm{d}/\mathrm{d}t\right)x(t-\tau) = -\left(a_0'/a_1'\right)x(t) + \left(1/a_1'\right)e(t) \tag{8}$$

where τ is the equivalent pure delay. Its step response is shown as Fig. 1(b). The method of constructing SDGs is the same as the former one, and the delay can be embodied in dynamic SDGs (Yang & Xiao, 2006a).

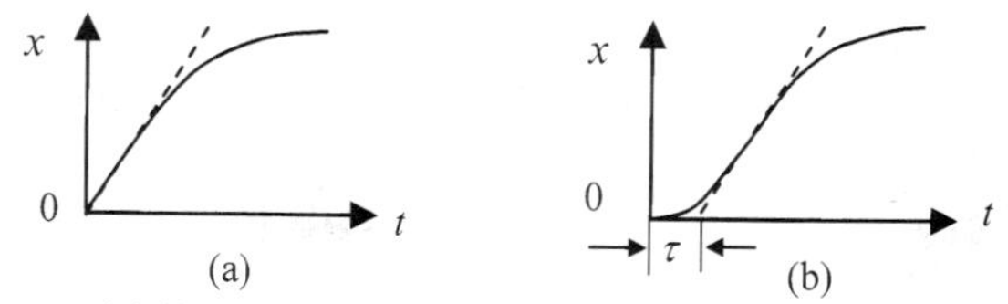

Fig. 1. Step response of different systems. (a) First-order system, (b) High-order system

Algebraic equations are usually included in the mathematical models as constraints which can also be transformed into SDGs (Maurya et al., 2003a) although they are noncausal in nature. Because there may be multiple perfect matchings between equations and variables, the corresponding SDGs may not be unique. Some treatment should be made to screen the unsteady or spurious SDGs (Oyeleye & Kramer, 1988; Maurya et al., 2003a).
For example, a tank system is shown as Fig. 2(a) where L is the level in the tank, R is resistance in the outlet pipe (can be manipulated by a valve), F_1 and F_2 are inlet and outlet flowrates respectively. The system is described as following DAEs:

$$F_2 = \frac{\alpha}{R}\sqrt{L} \tag{9}$$

$$A\frac{\mathrm{d}}{\mathrm{d}t}L = F_1 - F_2 \tag{10}$$

where A is the cross sectional area of the tank, and α is a constant. By the above method, the SDG is set up as Fig. 2(b).

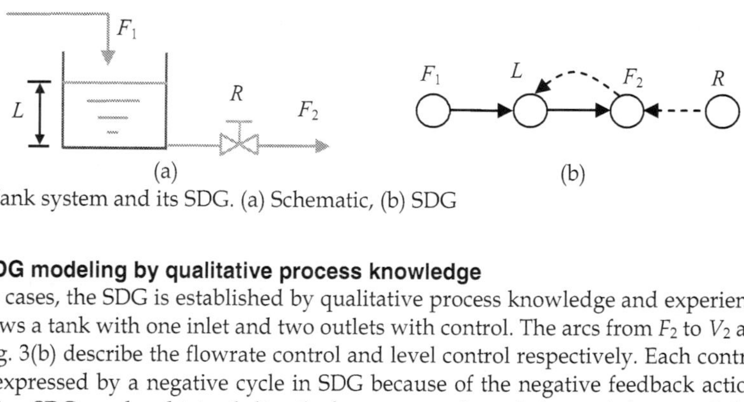

Fig. 2. Tank system and its SDG. (a) Schematic, (b) SDG

2.2.2 SDG modeling by qualitative process knowledge

In more cases, the SDG is established by qualitative process knowledge and experience. Fig. 3(a) shows a tank with one inlet and two outlets with control. The arcs from F_2 to V_2 and L to V_3 in Fig. 3(b) describe the flowrate control and level control respectively. Each control loop can be expressed by a negative cycle in SDG because of the negative feedback action. This qualitative SDG can be obtained directly from process knowledge and does need the exact mathematical equations. Sometimes the qualitative simulation and sensitivity experiments may also help. The SDGs obtained by this method often include indirect causalities besides direct ones, so the graph should be simplified and transformed so that all the arcs stand for direct causalities. Some rules are summarized by Yang & Xiao (2005b).

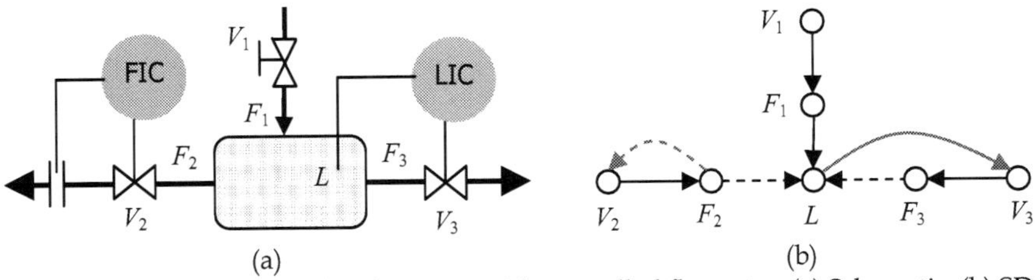

Fig. 3. Schematic and SDG of tank system with controlled flowrates. (a) Schematic, (b) SDG

Besides, P&ID diagrams and other flowsheets are very important topological process knowledge expression that can be standardized in XML (extensible markup language) format. It has been implemented in some commercial software products such as SmartPlant P&ID from Intergraph. The topology or connectivity obtained here includes both material flow and information flow, which are needed for SDG modeling. Although the granularity is entity-based, which is not enough for the variable-based SDG modeling, this kind of topological information is the fundamental of SDG and can be used as references as well (Thambirajah et al., 2009).

The SDG set up by the above methods can be validated by process data. For example, correlation is a necessary condition of causality, so the cross-correlation between every two measured variables can be used to validate the arcs in SDGs, and the directions can also be obtained by shifting the time series to find the maximal cross-correlation. Alternatively, probabilistic measure such as transfer entropy can be used to obtain the causality and directionality (Bauer et al., 2007).

In summary, the main steps of SDG modeling are: (1) Collect process knowledge, especially P&ID diagram and equations. (2) Set up the material flow diagraph by connectivity information between entities. (3) Choose the key variables and give them signs according to the process knowledge. (4) Add control arcs on the diagraph to constitute the SDG skeleton. (5) Add other variables and arcs to form the entire SDG. (6) Simplify and verify the SDG by graph theory. (7) Validate the SDG with process data and sensitivity experiments.

2.3 Hierarchical SDG Description of Large-Scale Complex Systems

Based on the decomposition-aggregation approach, a single-level SDG model can be transformed into a hierarchical model (Gentil & Montmain, 2004; Preisig, 2009). With this model, it is clear and easy to understand the system inherently. As such, the fault analysis method should also be modified from a centralized one to a distributed one.

The whole SDG model can be classified into 3 levels. If the scale of the whole system is too large, then more levels can be established, but 3-level model is enough for most cases. So we take it as a typical pyramid structure. The top level is called system level, where the system is divided into several sub-systems. Sometimes a large-scale system may include several independent sub-systems which can be dealt with separately. Also, in many cases, several components are operated in sequence or in parallel, with no recycle or other kind of interactions existed across the different components, then these components can also be regarded as sub-systems. Of course, if the SDG of the whole system is connected and cannot be separated, then it composes the only sub-system itself.

In the middle level, each control system is regarded as a super-node and the relations between control systems are expressed by arcs among controlled variables and a few important manipulated variables or other variables. The signs of arcs are determined according to the propagation rules to assure the consistency of the paths. The variables in some control loop and not appeared in other part of the system are usually invisible here. The SDG in this level is the backbone of the system which shows the main connectivity in the system flowsheet.

The bottom-level SDGs are the SDG units of all the control systems. The description is the most detailed qualitative expression because it shows the causalities between variables. Since most of the control systems are based on feedback actions, each SDG in this level usually contains at least a loop with various bias nodes attached on them.

2.4 Matrix Explanation of SDG Model

In this section, we look at the SDG model from another viewpoint. An SDG can be also described as an adjacency matrix $\boldsymbol{X}$ with the element 1/0 denoting the direct adjacency and direction between two variables. Actually it is the transpose of Jaccobian matrix in Eq. (4) with unsigned elements. By matrix computations, reachability matrix $\boldsymbol{R}$ can be obtained from $\boldsymbol{X}$, which shows the directed reachability from one variable to another, in which the element 1 means there are at least a path in the corresponding SDG (Jiang et al., 2008). It can be observed that the computation is just another form of graph traversal.

By simultaneous permutation of row and column (with variable order changed), $\boldsymbol{X}$ can be block triangulated as follows:

$$\boldsymbol{X}' = \boldsymbol{T}\boldsymbol{X}\boldsymbol{T}^T = \begin{bmatrix} \boldsymbol{A}_{11} & \boldsymbol{A}_{12} & \cdots & \boldsymbol{A}_{1m} \\ 0 & \boldsymbol{A}_{22} & \cdots & \boldsymbol{A}_{2m} \\ \vdots & \vdots & \ddots & \vdots \\ 0 & 0 & \cdots & \boldsymbol{A}_{mm} \end{bmatrix} \tag{11}$$

Each block in the diagonal denotes a sub-system with a partial order meaning that the sub-system with larger number can not reach the one with smaller number. It can also be explained by the reachability matrix which is definitely also block triangulated with the same order as:

$$\boldsymbol{R} = \left(\boldsymbol{X}' + \boldsymbol{X}'^2 + \cdots + \boldsymbol{X}'^n\right)^{\#} = \begin{bmatrix} \boldsymbol{B}_{11} & \boldsymbol{B}_{12} & \cdots & \boldsymbol{B}_{1m} \\ 0 & \boldsymbol{B}_{22} & \cdots & \boldsymbol{B}_{2m} \\ \vdots & \vdots & \ddots & \vdots \\ 0 & 0 & \cdots & \boldsymbol{B}_{mm} \end{bmatrix} \tag{12}$$

where the sign # is the Boolean equivalent (Mah, 1989). If the intersection block $\boldsymbol{B}_{ij}$ is a zero matrix, then the corresponding two sub-systems are independent (no arcs between them), otherwise they are in sequence. Thus we explain the decomposition between the top and middle level.

When we look at the relationship among control systems, we take a control loop as a super-node and add an arc from node i to node j, if the controller output of controller i can directly affect the controlled variable of controller j without going through controller output of any other nodes. This SDG as a part of the middle-level SDG is also named as control loop diagraph (Jiang, 2008).

For a feedback control system, there exists a loop in the corresponding SDG. Thus according to the controllability concept, all the variables within the loop are strongly connected, which can be found in the reachability matrix as a block with all the elements are ones.

Let us look at the tank example as Fig. 3 and get the adjacency matrix and reachability matrix by Eq. (12) as follows, both of which are block triangulated.

$$\boldsymbol{X}=\left[\begin{array}{cc|cc|ccc} 0 & 1 & 0 & 0 & 0 & 0 & 0 \\ 0 & 0 & 0 & 0 & 0 & 0 & 1 \\ \hline 0 & 0 & 0 & 1 & 0 & 0 & 0 \\ 0 & 0 & 1 & 0 & 0 & 0 & 1 \\ \hline 0 & 0 & 0 & 0 & 0 & 1 & 0 \\ 0 & 0 & 0 & 0 & 0 & 0 & 1 \\ 0 & 0 & 0 & 0 & 1 & 0 & 0 \end{array}\right], \quad \boldsymbol{R}=\left[\begin{array}{cc|cc|ccc} 0 & 1 & 0 & 0 & 1 & 1 & 1 \\ 0 & 0 & 0 & 0 & 1 & 1 & 1 \\ \hline 0 & 0 & 1 & 1 & 1 & 1 & 1 \\ 0 & 0 & 1 & 1 & 1 & 1 & 1 \\ \hline 0 & 0 & 0 & 0 & 1 & 1 & 1 \\ 0 & 0 & 0 & 0 & 1 & 1 & 1 \\ 0 & 0 & 0 & 0 & 1 & 1 & 1 \end{array}\right] \tag{13}$$

where the variable order is V_1, F_1, V_2, F_2, V_3, F_3, L. They are divided into 3 groups: inlet (V_1 and F_1), one outlet with flowrate control (V_2 and F_2) and another outlet with level control (V_3, F_3 and L). The elements of $\boldsymbol{R}_{22}$ and $\boldsymbol{R}_{33}$ are all ones because they are control loops, and the elements of $\boldsymbol{R}_{23}$ are ones showing the flowrate controller influences the level controller. Hence the control loop diagraph is consisted of two nodes corresponding to the two controllers and an arc corresponding to the influence between them. Moreover, if the variable order is changed to put V_1 and V_2 meaning the two controller outputs at the last, the corresponding block is just the adjacency matrix of the control loop diagraph. This is a useful property that links the concepts of SDG, control loop digraph and the matrices.

Matrix explanation helps us understand the SDG concept and its potential in applications. In fact, some results, such as propagation rules, are derived from matrix description.

3. Inference Approaches Based on SDGs

3.1 Fault Propagation Rules

Based on the SDG description, the fault propagation can be described qualitatively. There are two basic principles:

Proposition 1: The fault is propagated along the consistent paths.

Proposition 2: The node signs are determined by nodal balance, i.e. the sign on each node must be equal to the net influence on the node:

$$\psi(v_j)=\sum_i \varphi(a_{ij})\cdot\psi(v_i) \tag{14}$$

where the qualitative operation rules are as Table 1. Due to the loss of quantitative information, some signs can not be determined shown as '?' in the table, which causes the uncertainty in the solutions.

No.	sgn[x]	sgn[y]	sgn[x]+sgn[y]	sgn[x]·sgn[y]
1	0	sgn[y]	sgn[y]	0
2	±	sgn[x]	sgn[x]	+
3	±	-sgn[x]	?	-

Table 1. Qualitative operation rules

The logic on a node in SDGs is OR in nature because any input deviation can result in the node sign. In some cases, however, there are other types of logics, for example, the logic is AND, XOR or high/low-selective, or arcs or nodes are conditional, some necessary logic nodes should be added to the SDG (Yang & Xiao, 2007).

Proposition 1 can be easily understood. By testing the consistency one can find the fault propagation paths based on the measurements, which form a sub-graph of the original SDG, called cause-effect graph (Iri et al., 1979). On the other hand, one can predict the next step response based on the measured and assumed variables.

Proposition 2, however, may have some limitations because it is only suitable for the dynamic trends near the initial state. When a fault occurs, the response of variables can be divided into three stages - initial, intermediate, and final responses. In large-scale complex systems, the intermediate response is very complex, but in most cases, we concentrate only on the initial and final stages. For stable systems with fixed input, the final response is a steady state. Thus the input and exogenous disturbances are assumed as step functions to show abrupt changes.

Initial response is the first response just after the exogenous input changes. In dynamic systems expressed by DAEs, initial response is the nonzero response of system variables predicted by propagation through all the shortest paths in the corresponding SDG if we define the length of arcs in AE and DE portion by 0 and 1 respectively (Maurya, 2003a). Final response is the steady states of variables obtained after the dynamic period ends. It can be solved simply by setting the derivatives as zeros in DE portion of DAE. For the obtained AEs, the final response can be predicted by propagation through all the directed (acyclic) paths in the corresponding SDG. However, there may exist more than one perfect matching between equations and variables, thus there may exist more than one SDG corresponding to the AEs. If there is only one perfect matching, the above method is correct; otherwise, the result may be wrong because the results based on different perfect matchings are inconsistent. There is an exception, however, if an SDG corresponding to a perfect matching contains only negative cycles, then any perfect matching (for which the SDG contains only negative cycles) can be chosen and the final response can be decided using the above method (Maurya, 2003a).

3.2 Control Action Influences on Fault Propagation

3.2.1 SDG description and fault propagation analysis of single control loop

Control actions should be considered particularly because they are forced actions that are different from process property itself and they may cause the truncation or misleading of fault propagation. We discuss this problem using the general methods and obtain some special results (Maurya, 2003b, 2006).

In the bottom level, SDG models are established for all kinds of control systems among which the most common and basic one is single PID loop shown as Fig. 4. The deviation e of the set point r and the measurement x_m of the controlled variable x, is inputted into the controller whose output u goes to the actuator and thus effects the controlled plant through the manipulated variable q. Hence they compose a closed loop. Because the controlled variable may be affected by some disturbances or be coupled with other system variables, the exogenous plant and variable x_j are also added. Assume that controlled plant and the controller are both linear amplifiers, i.e. proportion elements, with the positive gain k and k_v respectively. The control law of PID controller is:

$$\begin{cases} u = u_P + u_I + u_D \\ u_P = k_c e \\ (d/dt)u_I = k_c e/\tau_I \\ u_D = k_c \tau_D \cdot (de/dt) \end{cases} \quad (15)$$

where, k_P is the positive proportion parameter, τ_I and τ_D are integral and differential time constant respectively.

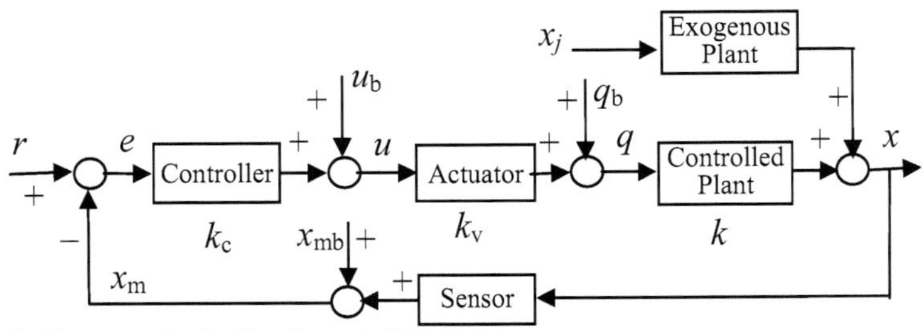

Fig. 4. Block diagram of a feedback control loop

According to the control law, the DAEs of the system are as follows:

$$x_m = x + x_{mb} \quad (16)$$

$$e = r - x_m \quad (17)$$

$$u_P = k_c e \quad (18)$$

$$(d/dt)u_I = k_c e/\tau_I \quad (19)$$

$$u_D = k_c \tau_D \cdot de/dt \quad (20)$$

$$u = u_P + u_I + u_D + u_b \quad (21)$$

$$q = k_v u + q_b \quad (22)$$

$$x = kq + a_j x_j \quad (23)$$

where subscript 'b' denotes bias. There are two perfect matchings between the equations and variables in AE portion, shown as Table 2, whose corresponding SDGs are shown as Fig. 5, in which the nodes with shadow are deviation nodes, arrows with solid and dotted lines denote signs "+" and "-" respectively. It is noted that the node de/dt is an individual node with special function, although it is the derivative of e. In applications, we generally assume that all changes on nodes are step functions, because the SDGs are only used to analyze the qualitative trends. Hence de/dt can be also replaced by e, but its effect is limited in initial response. Here the effect of de/dt on u_D is the same as the effect of e on u_P, but with shorter duration.

Equations	Matched variables in perfect matching No. 1	Matched variables in perfect matching No. 2
(16)	x_m	x
(17)	e	x_m
(18)	u_P	e
(20)	u_D	u_D
(21)	u	u_P
(22)	q	u
(23)	x	q

Table 2. Perfect matchings between the AEs and variables

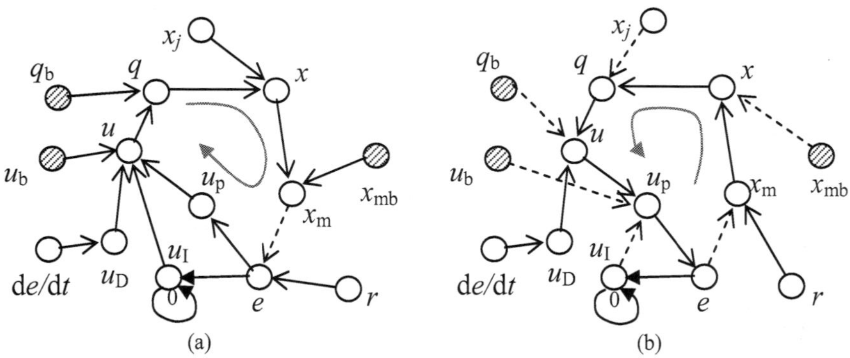

Fig. 5. Two SDGs of the PID control loop. (a) Case 1 (corrected), (b) Case 2 (spurious)

Eq. (23) describes the controlled plant, thus the arc direction should be from q to x according to the physical meaning, which shows the cause-effect relationship, so the case of Fig. 5(b) is removed. Moreover, if the plant shows some dynamic characteristic, for example, the left-hand of the equation is dx/dt, then the equation becomes a DE, hence there is only one perfect matching, and the case of Fig. 5(b) does not exist any more. Using Fig. 5(a), the initial response can be analyzed, for example, if the set point r increases, e, u_P, u, q, x and x_m will become "+" immediately, and u_I will become "+" gradually because the arc from e to u_I is a DE arc. This propagation path $r \to e \to u_P \to u \to q \to x \to x_m$ is consistent with the actual information transfer relations. Thus when we only consider the initial response of the system, the SDG of this control loop is obtained by transforming the blocks and links in block diagram into nodes and arcs while keeping the direction. However, in this example, no matter whether the case of Fig. 5(b) is reasonable, the analysis results of initial response by the two SDGs are the same because there are no positive cycles within them. We summarize the following rule:

Rule 1: The fault propagation path of the initial response in a control loop is the longest acyclic path starting from the fault origin in the path "set point → error → manipulated variable → controlled variable → measurement value → error", which is consistent with the information flow in the block diagram.
Final response is easier. The left-hand side of Eq. (19) is zero, so $e = 0$ in the steady state, which can be obtained from the concept. Hence u_P and u_D are both zeros. The above DAEs can be transformed into:

$$x_m = x + x_{mb} \tag{24}$$

$$x_m = r \tag{25}$$

$$u = u_I + u_b \tag{26}$$

$$q = k_v u + q_b \tag{27}$$

$$x = kq + a_j x_j \tag{28}$$

Now the perfect matching is exclusive and the corresponding SDG is shown as Fig. 6 that is the simplification of Fig. 5(b). There are two fault propagation paths: $r \to x_m \to x \to q$ and $x_j \to q \to u \to u_I$. If the set point r increases, then x_m, x, q, u and u_I will all increase in the steady state as long as the control action is effective. However, if only x_{mb} increases, then x_m will not be affected, but x will increase, that is the action of the control loop. We find that the Fig. 5(b) also makes sense for it reflects the information transfer relation in steady state. From the viewpoint of physical meaning, when control loop operates, the controlled variable is determined by the set point, and the controller looks like an amplifier with infinite gain, whose input equals to zero and whose output is determined by the demands. This logical transfer relation is opposite to the actual information relation.

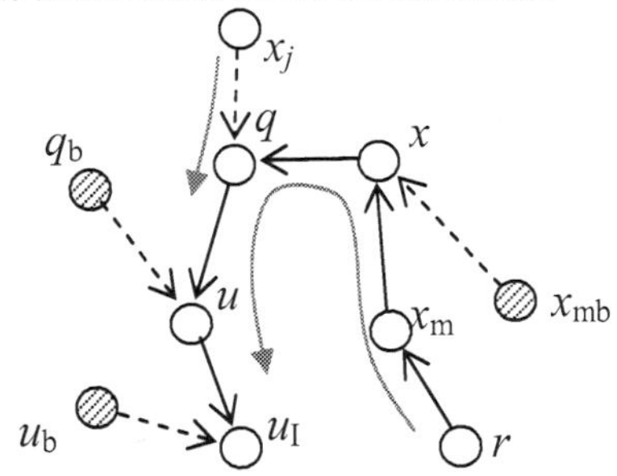

Fig. 6. Steady-state SDG of a PID control loop

Because the D action is only effective in the initial period, the fault propagation path of PI control is the same as the above one. Because of I action, some variables show compensatory response, for example, the response of x_m due to x_{mb} is limited in the initial stage. If there is only P action, then e is not zero in the steady state, thus u_I and related arcs in Fig. 5(a) are deleted, and both the initial response and steady-state response can be analyzed with this graph.
The rule of fault propagation analysis in steady state can be summarized as follows:

Rule 2: The fault propagation path of the steady-state response in a control loop is the path "set point → measurement value → controlled variable → manipulated variable" and "exogenous variable → manipulated variable".

When control loop operates, the above analysis shows the fault propagation principles due to the output deviation of sensor, controller, actuator and other exogenous variables. When control loop does not operate, there are two cases: (1) structural faults, e.g. the failure of sensor, controller or actuator causes the break of some arcs and the control loop becomes open, (2) excessive deviation causes the controller saturation, leading to the I action cannot eliminate the residual and let $e = 0$, which is similar with the P action case.

3.2.2 SDG description and fault propagation analysis of various control systems

Based on the above analysis of PID control loop, other control loops can be modeled as SDGs by the extension, combination, or transformation of the above SDG. Fault nodes are added according to the actual demands. Based on these models, fault propagation can also be shown explicitly.

Feedforward control is a supplement of feedback control. It is very familiar in actual cases, but it is easy to be treated according to the foregoing methods because it composes paths but not cycles, not leading to multiple perfect matchings.

Split-range control means the different control strategies are adopted in different value intervals. Here the sign of the arcs or even the graph structure may change with the variable values, which is realized by several controllers in parallel connection. This case is very hard for SDG to deal with. We have to do some judgments as making inference, and modify the structure or use conditional arcs to cover all the cases (Shiozaki et al., 1989).

Cascade control can be regarded as the extension of single loop case. It can be solved directly by AEs, or by the combination of two single loops. For example, the cascade control system in Fig. 7 has the steady-state SDG as shown in Fig. 8, where the controlled variable of the outer loop u_1 is the set point of the inner loop r_2.

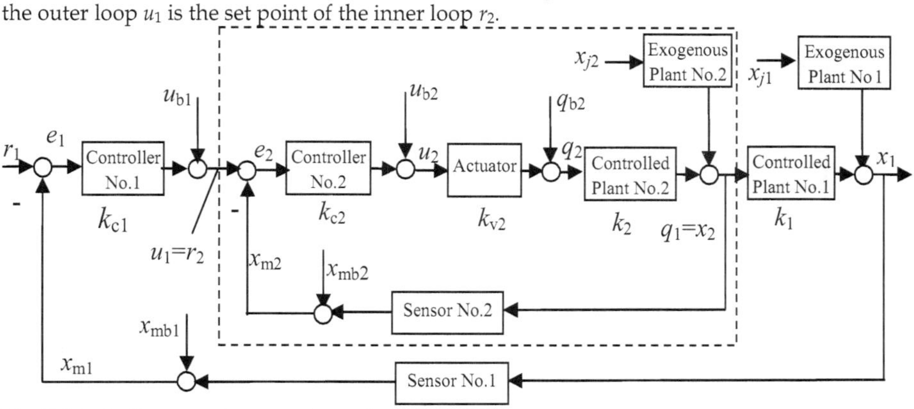

Fig. 7. Block diagram of a cascade control system

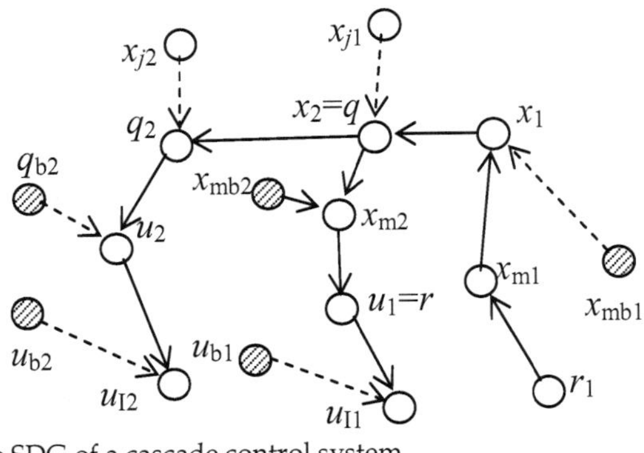

Fig. 8. Steady-state SDG of a cascade control system

Similar control methods are ratio control, averaging control, etc. Fig. 9 is a dual-element averaging control system whose objective is to balance two variables - level and flow, the block diagram of which is shown as Fig. 10. $P_x=P_L-P_F+P_S+c$, where P_x is the pressure signal of the adder output, P_L is the level measurement signal, P_F is the flow measurement signal, P_S is a tunable signal of the adder. In the simplest case, flow process and its measurement are both positive linear elements, and the level process is a negative linear element, so the steady-state SDG is shown as Fig. 11. Although there are several perfect matchings, SDG has only a negative cycle, thus we can analyze the fault propagation principle through the directed paths.

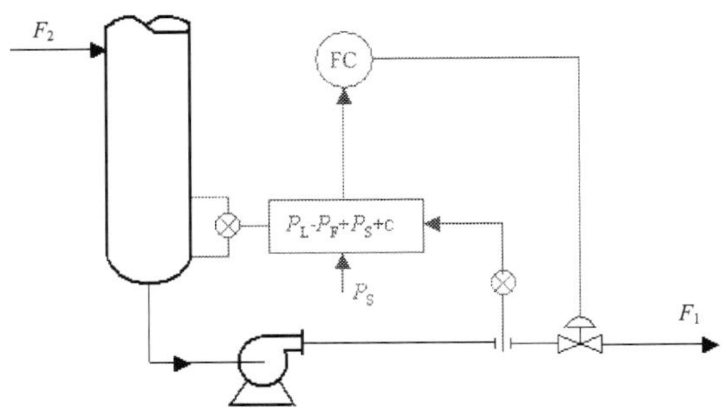

Fig. 9. A dual-element averaging control system

Thus we conclude:

Rule 3: The fault propagation path in a control system in steady state can be combined from the ones of single-loop by combining the same nodes and adding arcs by transforming AEs. In an industrial system, control systems play a special but important role. They compose information flow cycles in initial response but result in different flow in steady state response. Fig. 12 shows a system with a control loop. According to the above rules, the bias in x_1 propagates along the forward path (blue) in initial response while against the feedback path in steady state response.

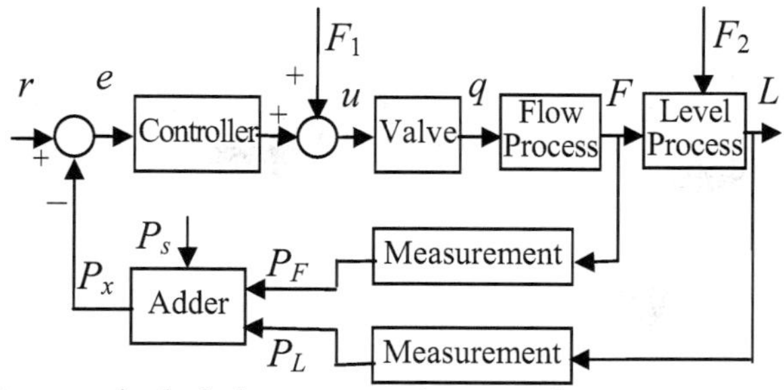

Fig. 10. Block diagram of a dual-element averaging control system

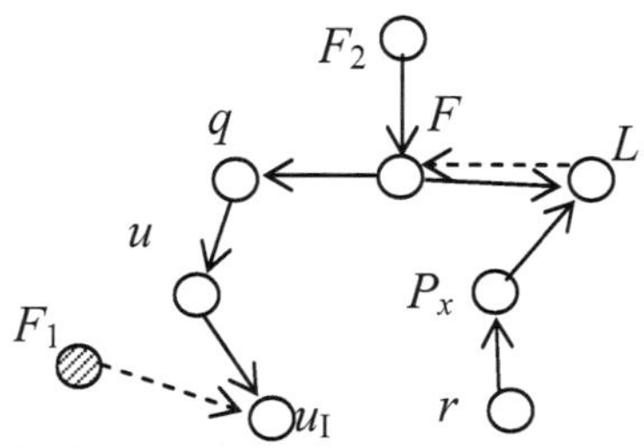

Fig. 11. Steady-state SDG of a dual-element averaging control system

The bias in x2 propagates along two paths until x3 and x4 in initial response, while PV and x4 restore to normal in the steady state because the steady state SDG changes the structure and directionality of the graph and thus PV becomes a compensatory variable.

Fig. 13. Control system's effect on fault propagation in a system. (a) Bias in x_1, (b) Bias in x_2

3.2.3 Example

In a boiler system, the three-element control of the boiler water level is widely used, in which the main controlled variable is water level. If we take steam flow and inlet flow into account, the control system is a feedforward-cascade system, as shown in Fig. 13. In the initial stage of the disturbance, the SDG is shown as Fig. 14(a), which can be derived by original DAEs. Certainly the initial fault influence follows this SDG. The control action, however, is enrolled and some deviations are restored to the normal region after a complex intermediate process. If the control action is successful, the fault may be blocked in the control loop and does not spread any more. But for some kinds of faults, the situation is different, because the control action makes the fault propagation path change. According to the method in the foregoing sections, we can construct the backbone (ignoring the bias nodes) of the steady-state SDG model as Fig. 14(b) which is quite different from Fig. 14(a). Similar with Fig. 9, other fault nodes can be added to Fig. 14(b) and thus we can find the steady-state fault propagation paths.

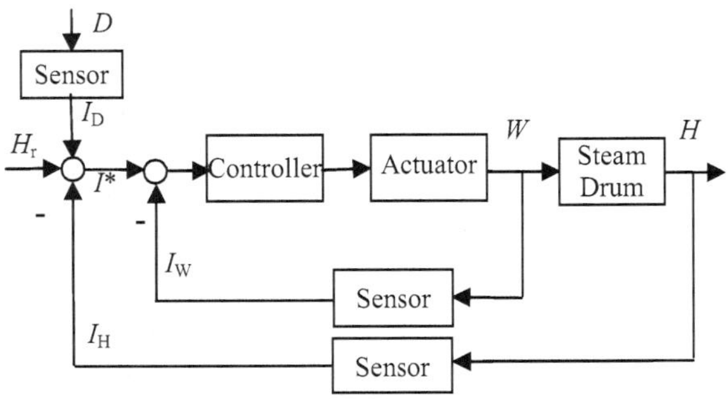

Fig. 13. Three-element control system

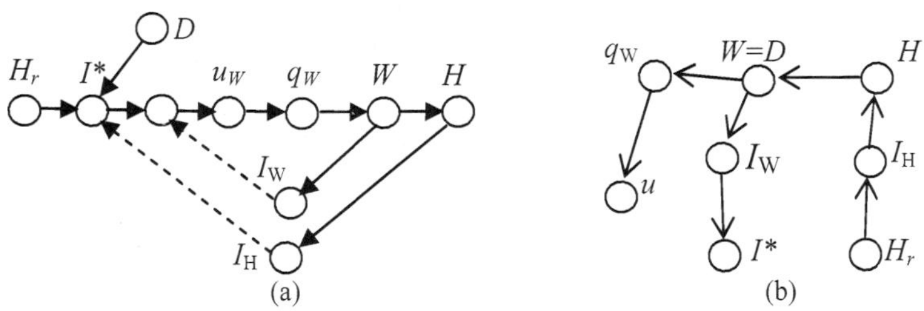

Fig. 14. SDGs of the three-element control system. (a) Initial response, (b) Steady state

3.3 Inference Approaches

In safety area, fault diagnosis and hazard assessment, especially hazard and operability analysis (HAZOP) are two different tasks. The former is to correctly find and identify the fault origin that is the cause leading to the symptom when fault occurs. It is based on measurements and is real-timed. While the latter, hazard assessment is to an off-line analysis whose purpose is to find the possible hazards due to all various causes. For this reason, we assume a series of departure nodes as fault origins, then analyze the possible consequences that are all the triggered departure nodes. Both fault diagnosis and hazard assessment need the interior mechanism of the system to express how the faults propagate. Thus the SDG model can be employed.

3.3.1 Graph traversal approach

The most common algorithm for searching the fault origin is depth-first traversal on the graph (Iri et al., 1979), which is a kind of efficient fault inference for both the single and multiple fault origin cases (Zhang et al., 2005). Its theoretical basis is nodal balance in Eq. (14). A depth-first traversal algorithm constructs a path by moving each time to an adjacent node until no further arcs can be found that have not yet been visited, the implementation of which is a recursive procedure.

For HAZOP purpose, forward traversal is applied from the assumed origin to predict all the variables based on consistency, which is deductive reasoning (Venkatasubramanian et al., 2000; Yang & Xiao, 2006b). For the fault detection purpose, backward traversal is applied within the causal-effect graph to find the maximal strongly connected component (Iri et al., 1979), which is abductive reasoning. Actually, the whole procedure includes two steps:

Step 1: Trace the possible fault origins back along the arcs.

Step 2: Make forward inference from these nodes to screen the candidates to choose which one is the real or most probable fault origin.

The time complexity of a traversal search is $O(n^2)$ in which n denotes the node number in the graph. When the system scale increases, the time for a traversal is too long to meet the demands of fault detection. Thus the model structure should be transformed from a single-level one to a hierarchical one. By this way, the search is first performed in the higher level to restrict the fault origin in a sub-system. Then the search is performed in the sub-graph of this sub-system.

For the hierarchical model, hierarchical inference from top to bottom is obtained naturally. The graph traversal is performed firstly in the higher level finding the possible super-node that includes the fault origin. Next perform the graph traversal in the lower level to restrict the possible location of the root cause. Assume the sub-system contains m control systems, and each control system contains k variables, then the time complexity of a traversal in a single-level model is $O(m^2k^2)$, and the time complexity in a 2-level model is $O(m^2+k^2) \ll O(m^2k^2)$. Thus the fault analysis in a hierarchical model has much higher efficiency.

Here the number of fault origin is assumed to be only one, that is, the reason that leads to the fault is only one (Iri et al., 1979). This is reasonable because multiple faults seldom appear at the same time (Shiozaki et al., 1985). For multiple fault origin cases, minimal cut sets diagnosis algorithm was presented (Vedam & Venkatasubramanian, 1997), where all possible combinations of overall bottom events should be input into the computer to explore and those which make the top events appear are the cut sets. This algorithm has the distinct disadvantage of low efficiency because of exponential explosion.

3.3.2 Other improved approaches

In order to utilize the system information more sufficiently, Han et al. (1994) used fuzzy set to improve the existing models and methods, but their method is not so convenient for on-line inference and is not applicable for dynamical systems. Some scholars introduced temporal evolution information such as transfer-delay (Takeda et al., 1995; Yang & Xiao, 2006a) and other kind of information into SDG for dynamic description. Probability is also proposed to model the system, which uses conditional probabilities of fault events to describe causes and effects among variables (Yang & Xiao, 2006c). Hence the inference is respect to the fault probability.

We can use Bayesian inference on the graph to calculate the probabilities; it is a direct method. Suppose that the node set of the probabilistic SDG is $V = E \cup F \cup H$, in which E is the subset of evidence nodes whose value or probabilities are known, F is the subset of query nodes whose probabilities are to be computed, and H is the subset of hidden nodes which is not cared about in the inference. The inference process of is to compute the conditional probability of x_F given the known x_E.

$$p(x_F \mid x_E) = \frac{p(x_E, x_F)}{p(x_E)} \tag{29}$$

where

$$p(x_E, x_F) = \sum_{x_H} p(x_E, x_F, x_H) \tag{30}$$

$$p(x_E) = \sum_{x_F} p(x_E, x_F) \tag{31}$$

To solve this problem, Bayesian formula and its chain rule should be used adequately, and also the junction tree algorithm can be used for multiple fault origin cases. This method could be used where there are distinct random phenomenon, both for fault detection and HAZOP (Yang & Xiao, 2006b), but the cycles in SDGs should be handled. The algorithm is the combination of depth-first search and junction tree algorithm, written as pseudo code:

```
BEGIN
 INTEGER i
 PROCEDURE DFS(v,u); COMMENT v is the father of u
 BEGIN
   NUMBER(v):=i:=i+1
   Calculate the probability of the father of v ; COMMENT junction tree algorithm
   FOR v has a father w with the probability more than the threshold DO
     BEGIN
      IF w is not yet numbered THEN
        BEGIN
        DFS(w,v);
     END;
   END;
   i:=0;
   DFS(s,0)                ; COMMENT s is an abnormal variable node
END;
```

On the other hand, rule-based inference (Kramer & Palowitch, 1987) is applicable when expert system is available. This method can be used to improve the inference accuracy with the appropriate rule description and operation. Rough set theory provides an idea of handling vague information and can be used to data reduction, thus it can be introduced to the fault isolation problem (a kind of decision problems) to optimize the decision rules. The decision algorithm is proposed by Yang & Xiao (2008a), in which the generation and reduction method of the rules are related to the structure of the SDG model. The main steps are listed as follows:

(1) List all the possible rules as *Table A* (as Table 3), with each row denoting a rule $\varphi \to \psi$, where φ denotes the values of the condition attributes are assumed and ψ denotes the decision to be obtained. For convenience, we can give each attribute value a notion.

Attributes **Q** / Objects **X**	Condition attributes **C**	Decision attributes **D**

Table 3. The framework of a decision table

(2) Try to delete each condition attribute in turn and test the consistency of the formula and obtain the reducts and the core. Delete all the elements except the cores and get *Table B*. There are several methods to test the consistency. For example,
  (a) Each condition class $E \in X \mid IND(\mathbf{C})$ has the same decision value.
  (b) For each object x, the condition class covering x is contained in the decision class covering x.
  (c) For every two decision rules $\varphi \to \psi$ and $\varphi' \to \psi'$, we have $\varphi = \varphi' \to \psi = \psi'$.

(3) Calculate the reducts of each rule by use of Table B, and get *Table C*.

(4) Delete redundant rules and thus get *Table D*.

(5) Educe the rules and the decision algorithm according to Table D.

The authors combine the algebraic and logical expression ways to achieve the purpose. Moreover, due to the convenience of expressing granularity, the decision algorithm is still applicable when the types of the faults of concern are changed or reformed.

4. Sensor Location Problem Based on SDGs

4.1 Performance Criteria of Fault Detection

4.1.1 The dynamic SDG and fault reachability

In actual systems, the fault propagation needs time, which effects the fault detection performance. So we take into account the fault propagation time for each branch and form the dynamic SDG (Yang & Xiao, 2008b). If the variable denoted by the node n_1 has a direct influence on the variable denoted by n_2, and after a time period for the fault propagation the fault revealed, then we define this time period τ (n_1, n_2) as the fault propagation time between n_1 and n_2, as shown in Fig. 15. Obviously, we have $\tau(n_1, n_2) \geq 0$. A path starting from n_1 and ending at n_m (denoted as l $(n_1 \mapsto n_m)$) holds the overall fault propagation time τ $(n_1 \mapsto n_m)$ which is the summation of time τ (n_i, n_j) of each branch in this path, as shown in Fig. 15. Note that this is a simplified treatment, which fits the case of pure propagation delay, but when the dynamic properties are complex, the overall time may slightly decrease due to the effects of intermediate transients (Yang & Xiao, 2006a).

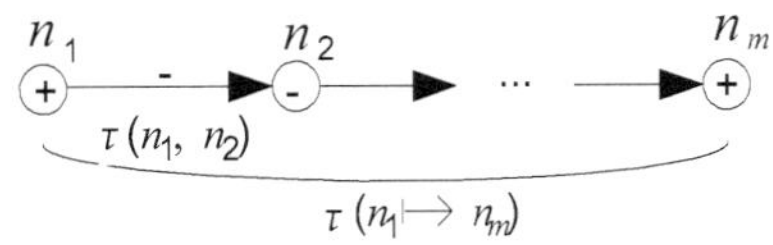

Fig. 15. Propagation time of a consistent branch and a consistent path

Because the nodes in SDGs are classified into two types - variables and fault origins, we denote them as n_is and f_js respectively. When a fault occurs, it is propagated along the consistent paths together with the time progress.

Definition 4: Starting from the fault node f, after the time t, the set of nodes affected by f is

$$R(f,t) = \{m : \exists l(f \mapsto m) \text{ and } \tau(f \mapsto m) \leq t\} \tag{32}$$

where t is the fault propagation time. If $n \in R(f, t)$, then we say, node n is *reachable* from fault f in time period t. Obviously, when time proceeds, the set of affected nodes expands, thus $R(f, t_1) \subseteq R(f, t_2)$, if $t_1 < t_2$.

The basic criteria of fault detection are detectability and identifiability to assure the faults be detected and identified from each other. The concepts here are the extension of the concepts in the framework of the SDG.

4.1.2 Fault detectability and detection time

A fault should be detected by at least one sensor in a short enough time period. Below is the definition.

Definition 5: If there exist at least one sensor located in the nodes of $R(f, t)$ (measuring the corresponding variables), then we say that the fault f is *detectable* in the time period t. The time needed to detect a fault by these sensors is called the detection time $T_D(f)$.

For each sensor, the time needed to detect a fault f can be calculated by shortest path algorithm. Among all these sensors, the shortest time is recorded as $T_D(f)$. The number of nodes with sensors in $R(f, t)$ is called the degree of detectability.

Based on the traditional SDG, only leaf nodes are needed to consider whether or not to locate sensors (Raghuraj et al., 1999). Then we have the following theorem.

Theorem 1: Based on the SDG, disregarding the cases that some variables cannot be measured, sensors need to be located only on the leaf nodes.

Corollary 1: In the framework of dynamic SDG with propagation time, the sensors need to be located on only leaf nodes of $R(f, t)$.

4.1.3 Fault identifiability and identification time

Different faults have different behaviors. Represented in the SDG, the reachable nodes are different. So we must put sensors on these different nodes to identify the different faults. Below is the definition.

Definition 6: If there exist at least one sensor on the nodes of $R(f_1, t)$ (measuring corresponding variables), and these sensor nodes are not within the nodes of $R(f_1, t)$, in other words, if there are sensors in the nodes of $I(f_1, f_2, t) = R(f_1, t) \cup R(f_2, t) - R(f_1, t) \cap R(f_2, t)$, then we say that the faults f_1 and f_2 are *identifiable* in the time period t. The time needed to identify two faults by these sensors is called the identification time $T_I(f_1, f_2)$.

Detectability and identifiability are two independent concepts. We can understand easily, when two faults are both detectable, they may not be identifiable. On the other hand, identifiability does not imply detectability generally, because we can place only one sensor to identify them too. But usually we assume that only when the faults are detectable, they can be considered for identifiability. Thus the identifiability condition is stronger. In Definition 3, $I(f_1, f_2, t)$, for two identifiable faults, must have more than one element. The number of nodes with sensors in $I(f_1, f_2, t)$ is called the degree of identifiability. Besides, we have

Proposition 1: $T_I(f_1, f_2) \geq \max\{T_D(f_1), T_D(f_2)\}$.

Proposition 2: The number of elements in $I(f_1, f_2, t)$ is not necessarily increasing monotonically with time t.

It should be noted that the signs of the nodes and branches can help identify different faults because some sensors are not only able to activate the alarm, but also indicate the direction of the departure from the normal values. For this case, we could change a node into two, one shows the higher reading, another shows the lower reading (Wilcox & Himmelblau, 1994). Then the above definition and the following rules can be applied.

4.1.4 Detectability and identifiability with multiple faults

Sometimes we also need to deal with the case of multiple simultaneous faults. It can be dealt with by node set transformation.

Here we take two faults as an example. If the faults f_i and f_j occur at the same time, their reachable node set is $R_i \cup R_j$, so we can take these C_n^2 node sets to be considered besides the sets of R_i, then the problem is transformed into the detectability and identifiability problems with a single fault.

Obviously, if each fault can be identified, but when several faults occur at the same time, they are not assured to be identified. How about the inverse proposition?

Theorem 2: If the case of n simultaneous faults can be identified, then the case of less than n simultaneous faults can be also identified.

4.1.5 Fault detection reliability

Detectability and identifiability are necessary conditions for fault detection. However the sensor readings are not always reliable, which affects the reliability of fault detection. Let F_is (i=1, 2, ..., n) and S_js (j=1,2,...,m) denote system faults and process variables measured by sensors individually. They can be shown as a bipartite graph with all the arcs directed from the fault set to the process variable set as shown in Fig. 16. Based on the detectablity criterion, there should be at least one arc departing from every fault node, and based on the identifiability criterion, the connected sensor nodes of different fault nodes should be different. The fault occurrence probabilities of the fault F_i is f_i, while the sensor missed alarm rate and false alarm rate of variable S_j is u_j and v_j. The influence relation from fault F_i to sensor S_j is denoted by reachability d_{ij} (0 or 1) where 1 means reachable and 0 means unreachable. Because of the causal relations between process variables, the reachability includes direct and indirect influences.

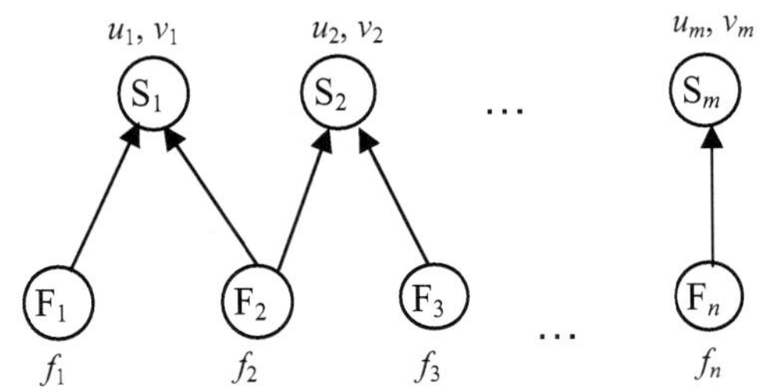

Fig. 16. Bipartite graph to show the relations between faults and sensors

As shown in Fig. 17, the confusion matrix reflects the true/false classification of alarms (Izadi eta al. 2009). The entries in the matrix are the number of true alarms (TA), false alarms (FA), missed alarms (MA) and true no-alarms (TN). These numbers can be obtained by experiments. The missed alarm rate of sensor S_j is u_j which can be calculated by MA/(TA+MA), and false alarm rate of v_j can be calculated by FA/(FA+TN). These rates are determined by the sensor quality and the threshold selection.

Hypothesized Class	True Class: Fault	True Class: No fault
Alarm	True Alarms (TA)	False Alarms (FA)
No alarm	Missed Alarms (MA)	True No-alarms (TN)

Fig. 17. Confusion matrix to show the terminology of missed alarms and false alarms

For each fault F_i, we should minimize its probability of not being detected. Because it is propagated to many other variables on which the sensors can also detect it, the undetectability of F_i occurs only when all the variables miss alarms. Besides, the redundant sensors on the same variables are also helpful for the improvement. We define the undetectability probability (Bhushan & Rengaswamy, 2002) of F_i as

$$U_i = f_i\left(\prod_{j=1}^{m}\left(u_j\right)^{d_{ij}x_j}\right) \tag{33}$$

where x_j is the integer number of sensors put on the variable j. If there is no sensor on variable j, x_j is zero. Obviously, when x_j with the corresponding nonzero d_{ij} increases, U_i decreases. So adding sensors will increase the reliability.

On the other hand, we think about the false alarm problem. For the variable S_j, adding a sensor with false alarm rate v_j (with respect to fault F_i) will be accompanied with the increase of the following false alarm probability

$$V_j = v_j\left(\prod_{i=1}^{n}(1-f_i)^{d_{ij}}\right) \tag{34}$$

which means the sensor reading gives the alarm even though no faults occur.
The calculations of missed alarms and false alarms are dual problems that adding sensors will reduce the undetectability whilst increasing the false alarm probability. Here the false alarm probability reflects the influence of a sensor's false alarm on the whole system.

4.2 Sensor Location Based on Fault Detectability and Identifiability

The purpose of the sensor location problem is to choose sensors and design the sensor location to meet the demands of fault detection. Neglecting the reliability problem, here we deal with the detectability and identifiability problems. In the framework of the static SDG, the problem can be solved directly. But in the framework of dynamic SDG, the arising times of various faults are various, and different faults may interact, so it is hard to analyze all the cases of fault propagation or even solve the sensor location problem in advance. One possible way is to embed the sensor location problem in the forward inference process as the following algorithm:

(1) Add fault node f_i to the evidence node set E and the reachable node set R_i. Set the inference system time T_{sys} to zero.
(2) Check if the evidence node set is empty. If it is empty, then go to the end, otherwise go on.
(3) From the evidence nodes, choose the nodes in the reachable node set for one forward step, and add them to the reachable node set R_E of the evidence nodes. Meanwhile, update their detection time $T_D(f_i)$ (detection time of the starting node of the branch plus the propagation time on the branch).
(4) From the nodes in R_E, choose one with the shortest detection time T_k and the nodes to be updated, N_{T_k}, at time T_k.
(5) $T_{sys} = T_{sys} + T_k$. Make forward inference from all the nodes in N_{T_k} for one step.
(6) Add N_{T_k} to the evidence node set E and reachable node set R_i.
(7) If an evidence node whose one-step reachable nodes are all updated, then delete this node from E.
(8) Placing sensors in the reachable node set R_i can assure the detectability of fault i, and placing sensors in $R_i \cup R_j - R_i \cap R_j$ can assure the identifiability of fault i and j.
(9) If a new fault has occurred, then add its corresponding node to the evidence node set E, and set T_{sys} as the current time. Go to step (2).

Note that the treatment in step (3) is not accurate because the detection time is just approximate. So we often increase the threshold of the degree of detectability and identifiability to assure performance is optimal.

4.3 Sensor Location Based on Fault Detection Reliability

The two criteria, detectability and identifiability should be met at first when deciding the sensor location. Besides above algorithm, Yang and Xiao (2008b) also proposed some useful rules to solve this problem in consideration of the propagation time, which is a stricter requirement than that mentioned above. The sensor location obtained has the minimum number of sensors required for fault detection. Since the increase of sensors will not destroy these criteria, the following optimization algorithm should be based on this location and try to find the crucial variables for putting additional sensors.

In the trade-off between false alarms and missed alarms, missed alarms are often considered to be more important because we do not want to lose a real fault. Thus the algorithm handles this criterion first. Meanwhile, we hope the false alarm rates to be as small as possible, so we integrate the treatment of false alarms into the whole algorithm.

If we consider all the faults, then we want to minimize the total undetectability probabilities for all the faults, each one of which is a probability that no sensors give the alarm for the corresponding fault. Thus we have the following optimization problem:

$$\min_{x_j}\left[\sum_{i=1}^{n} U_i\right] \tag{35}$$

This optimization problem cannot be solved at once (by branch and bound method or other methods) for the following reasons. First, this problem does not have a continuous solution space; instead it is an integer programming problem. Thus we should update the solution (x_j, $j=1,\ldots,m$) once at an integer. Secondly, the problem has constraints. For example, putting a sensor on a variable needs some cost, and the total cost should be limited within a range, so we have

$$\sum_{j=1}^{m} c_j x_j \le C_0 \tag{36}$$

where c_j is the cost to be paid when putting a sensor on variable j, and C_0 is the cost limit. Thirdly, the initial value of the problem is obtained according to the criteria of detectability and identifiability, and the x_js should not be negative, which can be regarded as another constraint. Sometimes we have more constraints such as the number limit of sensors. This algorithm is just used to reduce the undetectability by adding sensors at critical location. Thus the problem is solved by an iterative algorithm, and within each step we should only add 1 to one of the x_js and check the constraints. This is a heuristic algorithm.

Besides, false alarm problem can also be formalized as an integer optimization problem:

$$\min_{x_j}\left(\sum_{j=1}^{m} x_j V_j\right) \tag{37}$$

but this problem is accompanied with the undetectability optimization problem and is less important for most cases. Thus we do not take it as an individual problem but as a supplement to the above problem expressed by the following formulation

$$\min_{x_j}\left[\sum_{i=1}^{n} U_i + \alpha \sum_{j=1}^{m} x_j V_j\right] \tag{38}$$

where α is a constant coefficient.

When trying to reduce the undetectability by adding a sensor, one is concerned not with the total number of missed alarms but the number for each fault. Thus the summation in Eq. (35) can be replaced by a weighted summation, where the weights correspond to the importance. The weights are not impersonal or rational to obtain, so we can alternatively use the maximization to deal with the bottleneck which is the fault with maximal undetectability. Hence we have the following optimization problem as a combination of a minimaxization and a linear minimization

$$\min_{x_j}\left[\max_i(U_i)+\alpha\sum_{i=1}^{m}x_jV_j\right] \tag{39}$$

subject to

$$\sum_{j=1}^{m}c_jx_j \le C_0,\ \ x_j \in Z^+ \cup \{0\} \tag{40}$$

If less attention is paid to the false alarm rate, we can take its optimization as a constraint and just set a limit V_0 instead of optimization. Then we get the simplified algorithm:

(1) Initialization:
 (a) Get f_i, u_j and v_j by a priori knowledge and measurements.
 (b) Get d_{ij} from SDG or reachability matrix.
 (c) Get the minimal x_js according to the criteria of detectability and identifiability as the starting point.
 (d) Calculate V_j by Eq. (34).
 (e) Calculate V by summation of all the V_js with x_j is not 0.

(2) Calculate U_i and select the maximal one U_I.
(3) Let the set of js with d_{Ij} is 1 as $A_I = \max\{j \mid d_{Ij}=1\}$.
(4) Select the maximal u_J from A_I, i.e. $u_J = \max A_I$. If A_I is empty, stop. If there is more than one maximum element, select the one with smallest V_j.
(5) Put a sensor on variable J, $x_J \leftarrow x_J + 1$.
(6) Update the false alarm rate $V \leftarrow V + V_j$, and see if it is tolerable. If so, go on; if not, delete J from A_I and go to step (4).
(7) Check the constraints. If they are met, go on; if not, delete j from A_I and go to step (4).
(8) Go to step 2 and update the undetectability.

The algorithm is illustrated as a flow chart in Fig. 18.

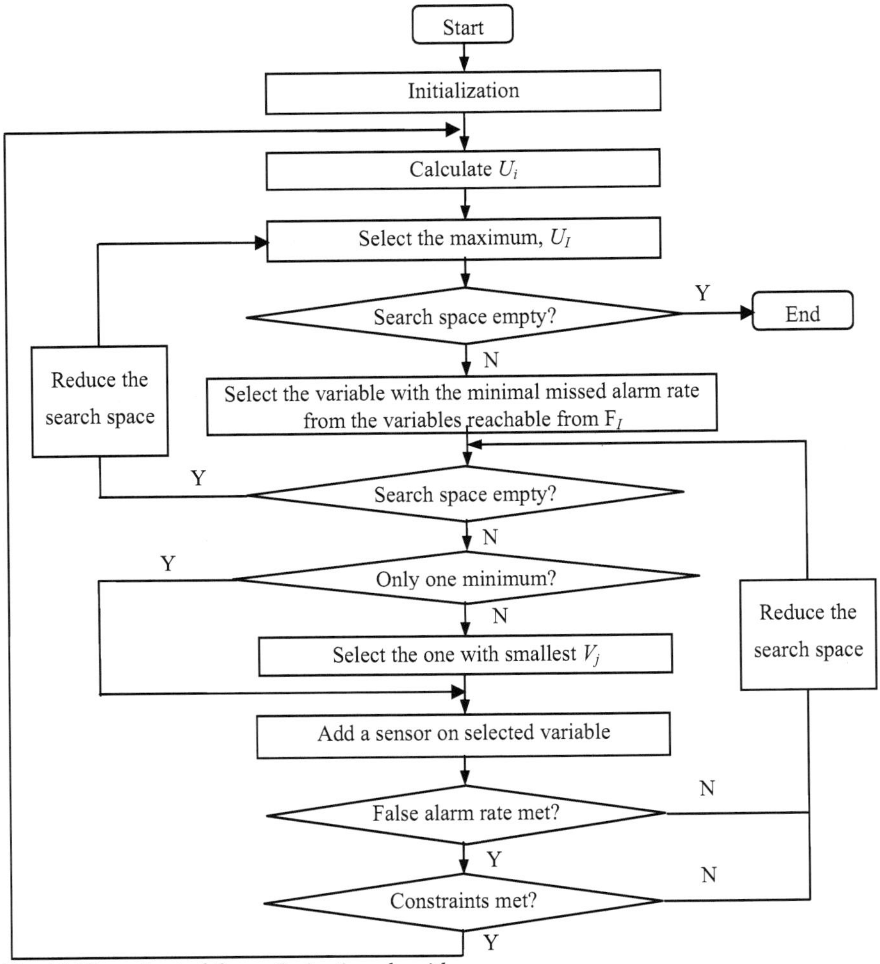

Fig. 18. Flow chart of the optimization algorithm.

5. Case Study

We take a 100 MW generator set process in a power plant as an example, which is composed of a typical natural-circulation steam boiler and a turbine. The system is operated and controlled by a DCS of MACS-II.

The process can be divided into several sub-systems such as water & steam system, coal & air system, and turbine system. The core flowsheet is shown in Fig. 19.

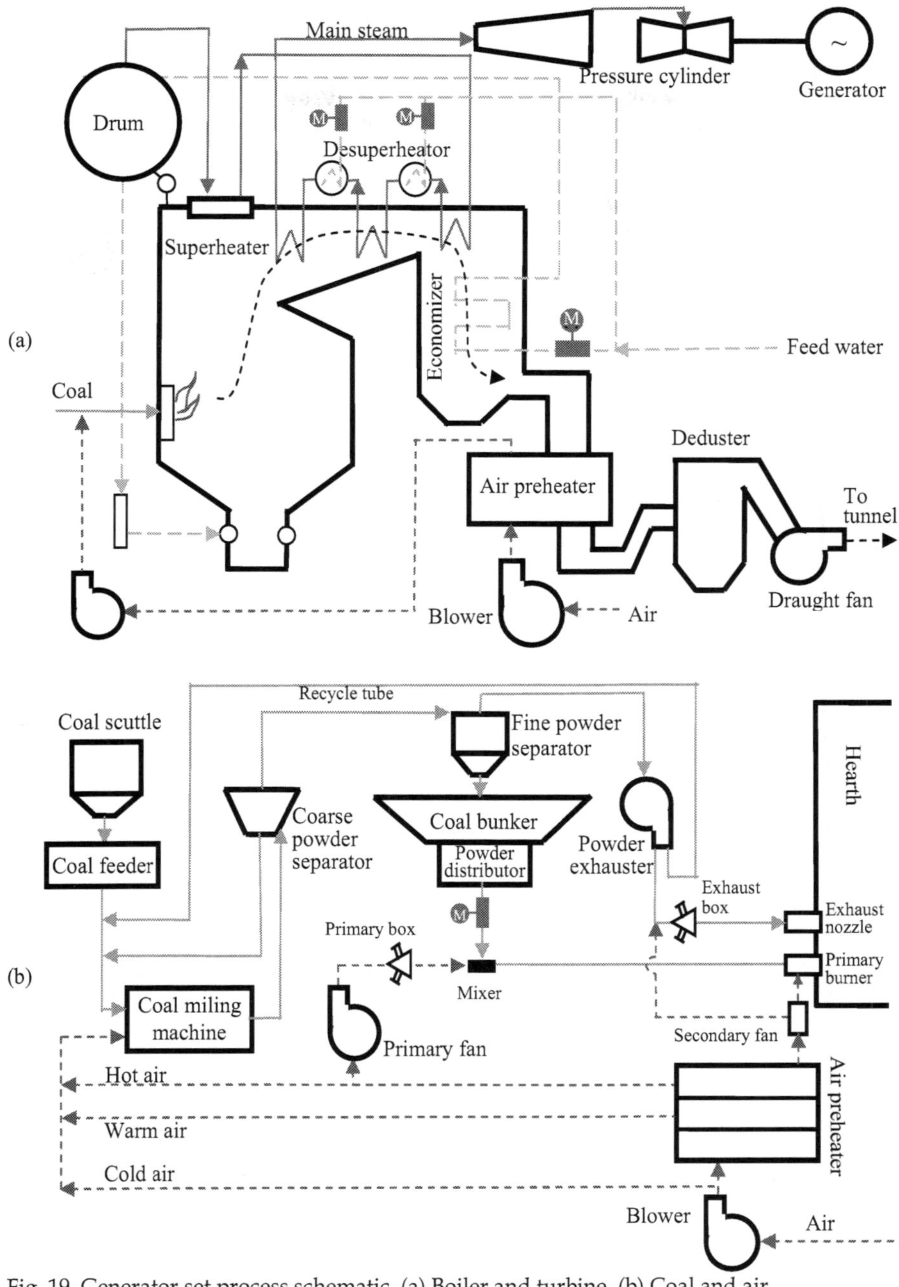

Fig. 19. Generator set process schematic. (a) Boiler and turbine, (b) Coal and air

5.1 SDG Model Description of the Generator Set Process

As listed in Table 4, the major variables are controlled in separated systems. Note that the control system of oxygen percentage in the smoke is not included here because it is so complex that it is usually operated manually. And the control systems of water level of condensator, deoxidizor and exchangers are also not included because they are operated in independent single control loops that can be separated from the whole graph. Thus, the top-level SDG model is composed of several super-nodes in which we are only concerned about the major ones corresponding to the 5 controlled variables.

Control system	Description	Controlled variable	Tag name	Operating value
S1	Water level control by manipulating inlet water flow	Water level of the top steam drum	L1	0 mm
S2	Overheated steam flow control by throttle valve to maintain the turbine rev	Turbine torque	M	
S3	Overheated steam temperature control by manipulating the cooling water valve	Overheated steam temperature	T1	535 °C
S4	Overheated steam pressure control by manipulating the transducer frequency of coal powder distributor	Overheated steam pressure	P2	8.83 MPa
S5	Hearth pressure control by manipulating the fan baffle	Hearth pressure	P0	-50 Pa

Table 4. Controlled variables in the generator set process

The other variables in the process are listed as Table 5. We model the middle-level SDG of the system shown as Fig. 20. In the bottom level, control systems in S2-S5 are single loops in nature, whose SDGs are shown as Fig. 21(a)-(d).

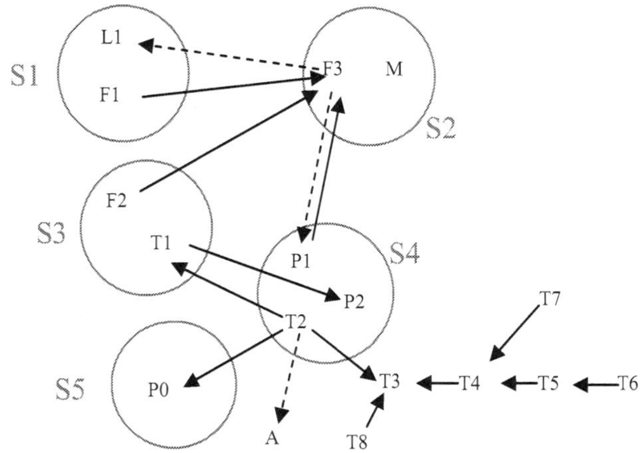

Fig. 20. Middle-level SDG model of the generator set process

Tag name	Variable	Operating value
F1	Inlet water flow	360 t/h
F2	Cooling water flow	10 t/h
F3	Overheated steam flow	370 t/h
F4	Exit smoke flow	
P0	Furnace negative pressure	
P1	Steam drum pressure	10 Mpa
P2	Superheated steam pressure	
P3	Inlet water pressure	12 Mpa
P4	Cooling water pressure	12 Mpa
P5	Primary total air pressure	1.4 kPa
P6	Blower exit air pressure	
P7	Coal powder exhauster exit air pressure	
P8	Coal powder exhauster inlet air pressure	
T2	Hearth temperature	
T3	Exit smoke temperature	150 °C
T4	Primary air temperature	
T5	Primary air exit temperature	
T6	Primary air inlet temperature	
T7	Coal milling machine exit air temperature	68 °C
T8	Inlet water temperature	215 °C
V1	Transducer frequency of coal powder distributor	
V2	Turbine rev	
N1	Turbine power	100 MW
N2	Turbine load	100 MW
A	Oxygen percentage in the smoke	5.8 %
C1	Cooling water valve	
C2	Primary fan baffle	
C3	Blower baffle	
C4	Coal powder exhauster baffle	
C5	Draught fan baffle	
C6	Inlet water valve	
C7	Main throttle valve	

Table 5. Other variables in the generator set process

In S1, the three-element control of the water level is used, in which the main controlled variable is L1. If we take steam flow F3 and inlet water flow F1 into account, the control system is a feedforward-cascade system, as shown in Fig. 22(a). In the initial stage of the disturbance, the SDG is shown as Fig. 22(b), which can be derived by DAEs. Certainly the initial fault influence follows this SDG.

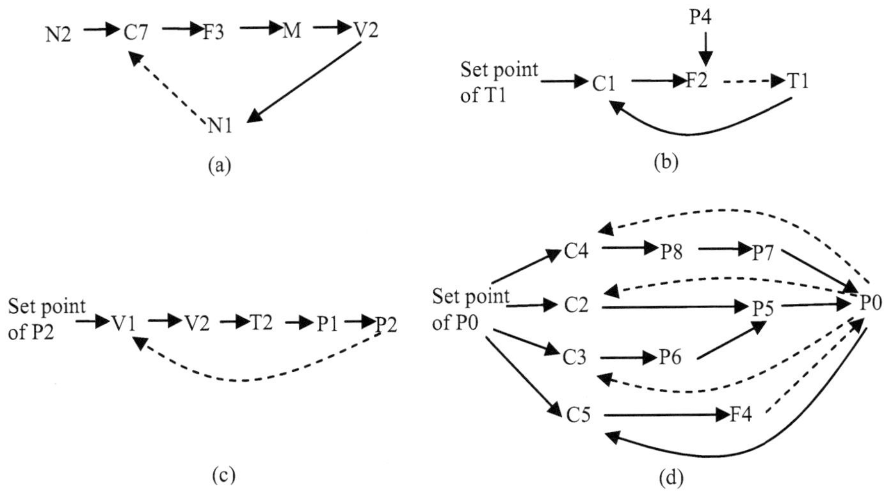

Fig. 21. Bottom-level SDGs of the generator set process. (a) S2, (b) S3, (c) S4, (d) S5

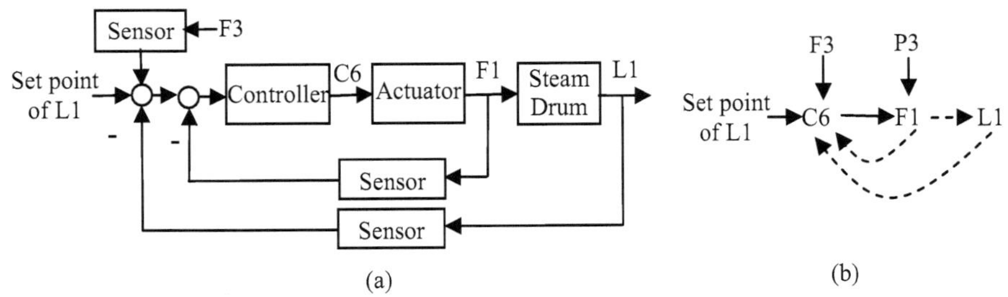

Fig. 22. Three-element water level control system. (a) Block diagram, (b) SDG

By combing the above sub-SDGs, the whole SDG of the generator process is shown as Fig. 23 where the red arcs stand for the control actions (Yang, 2008).

5.2 Fault Analysis of the Generator Set Process

When fault occurs, symptoms can be explained by SDG inference. Typical faults and their fault propagation paths are summarized as Table 6. Along the paths we can find the possible fault origins.

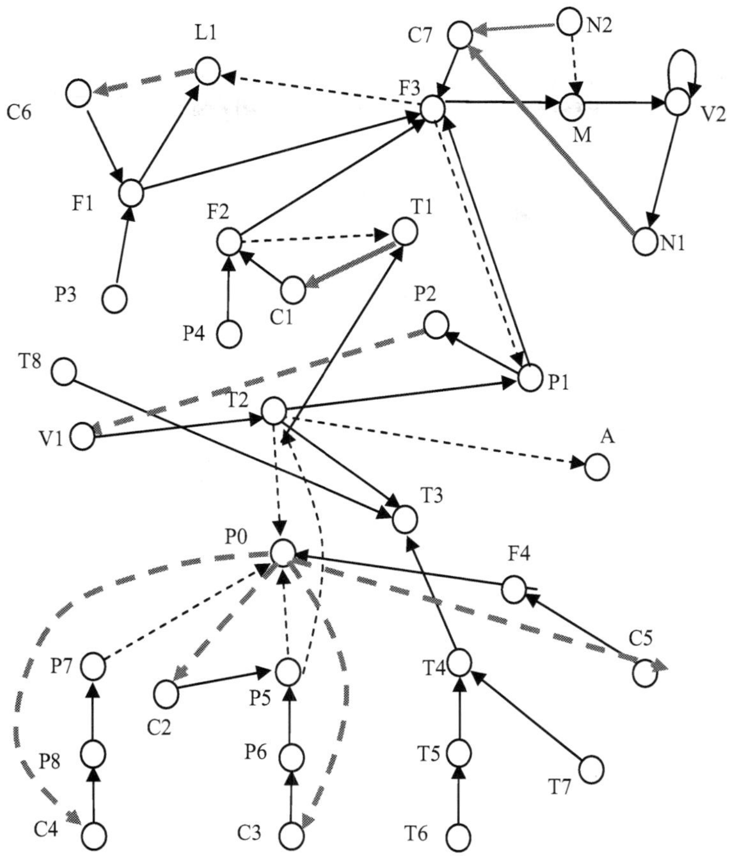

Fig. 23. Single-level SDG of the generator set process.

Fault	Fault propagation path in middle level	Fault propagation path in bottom level
Full of water in steam drum	S1	F1(+)→L1(+)
Lack of water in steam drum	S1	F1(-)→L1(-)→C6(+)
Too large of draught fan baffle	S5	C5(+)→F4(+)→P0(-)
Increase of load	S2	N2(+)→C7(+)→F3(+) M(-)→V2(-)→N1(-)→C7(+)
Change of coal quality	S4→S5 S4→T3 S4→S3→S2→S1	T2(+)→P1(+)→P2(+)

Table 6. Typical faults and their fault propagation paths

In the case of coal quality change, from the middle-level inference we find the fault origin is located in S4. Thus we just go on to make inference in the SDG of S4 and ignore other symptoms. If we make inference in a single-level SDG, then there are other paths (shown in Fig. 22) that are all redundant for fault origin search:

T2(+)→P0(+)

T2(+)→T3(+)

T2(+)→T1(+)→C1(+)→F2(+)→F3(+)→M(+)→V2(+)→N1(+)

T2(+)→T1(+)→C1(+)→F2(+)→F3(+)→L1(-)→C6(-)

However these paths are useful for hazard analysis. In Fig. 24, some control arcs are deleted compared with Fig. 23 because they are usually performed manually in the application. From the propagation path, we find the hearth temperature T2 is the key variable, so adding sensors on it can improve the fault detection reliability.

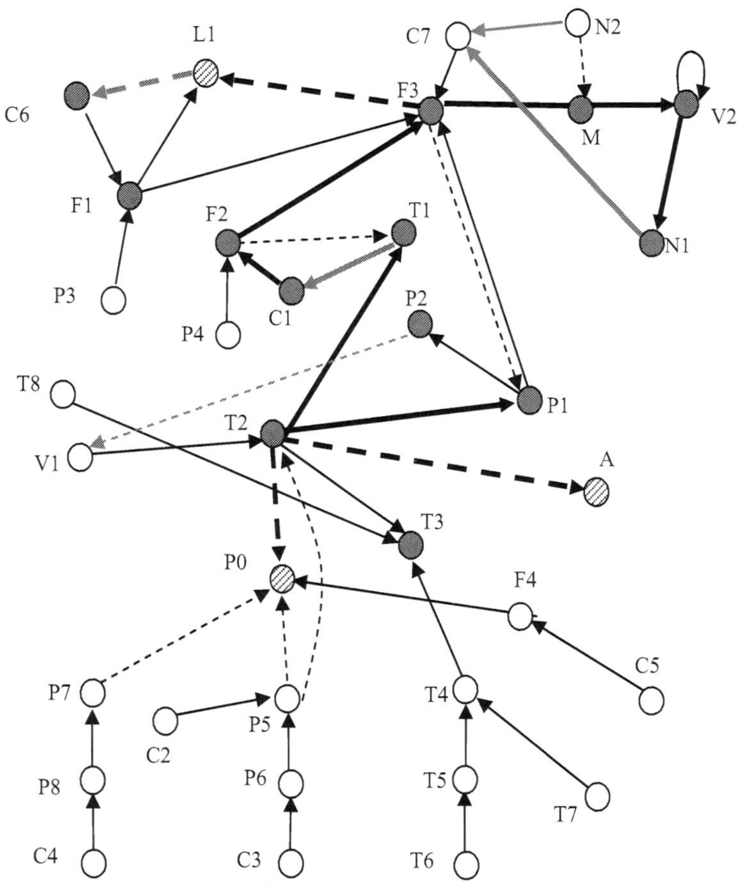

Fig. 24. Fault propagation when coal quality changes

6. Conclusion

In this chapter, after the introduction of the SDG concept and modeling methods, the inference approaches aiming at the fault detection and hazard analysis, especially the SDG description of control systems, have been analyzed from theory to practice.

The classical control methods on the basis of feedback idea are in common use, so the modeling and analysis of the systems under these control methods, have been discussed. When a control system is transformed into an SDG model, the direction of fault propagation in steady state may differ from the direction in initial response because of the control action. For a single-loop control system, the SDG is a directed path whose backbone is set point → measurement value → controlled variable → manipulated variable → controller output, which is also the fault propagation path and does not compose a loop. Based on this result, SDGs and fault propagation paths of various control systems can be obtained by the combination and connection of several single-loop control elements. Thus we do not have to list all the system equations when analyzing the actual problem, but only need to construct the local SDG for each separate control component and then combine them together, which is convenient for actual use. After analyzing the fault propagation paths in control systems, we can embed the resulted SDGs for initial response and final response into the SDG of the whole system and analyze the propagation paths considering the truncated or changed paths. This method enables the application of SDG method in large-scale complex systems.

It is to be noted that model description should meet the actual needs but does not need to be too accurate. For example, the SDG-based qualitative analysis of the three-element control of the boiler water level usually do not refer to the details, so usually we only construct a single loop to describe the major problem and ignore the minor ones. In large-scale complex systems, however, SDG models can be adopted to describe the interactions between different parts and reveal the propagation for the use of fault analysis; it is the advantage of SDG models.

In industrial systems, alarm monitoring design is a very important issue, among which the trade-off between missed alarms and false alarms should be treated appropriately. We should pay attention to two levels of design problems: In the local level, the threshold selection, data filtering and alarm triggering are the key problems to be solved. In the system level, topology expression and sensor location for alarm rationalization is essential. In this chapter, we have described and solved the sensor location problem aiming at the trade-off with the help of topology expressed by SDG. The optimization objective is expressed as the minimization of all the fault undetectabilities in the system. The false alarm rate is used as constraint as well as the cost limit.

Our future work may include: standard modeling method using XML-based process knowledge, modeling and validation of SDGs using process data, and combination of qualitative fault propagation and quantitative diagnosis.

7. Acknowledgment

The authors gratefully acknowledge the financial aid for this research by the National Natural Science Foundation of China (No. 60736026, 60904044), China Postdoctoral Science Foundation (No. 20080440386) and NSERC (Natural Sciences and Engineering Research Council of Canada) - Suncor - Matrikon - iCORE Industrial Research Chair Program.

8. References

Bauer, M.; Cox, J.W.; Caveness, M.H.; Downs, J. & Thornhill, N.F. (2007). Finding the direction of disturbance propagation in a chemical process using transfer entropy. *IEEE Transactions on Control Systems Technology*, Vol. 15, No. 1, January 2007, pp. 12-21, ISSN: 1063-6536

Bhushan, M. & Rengaswamy, R. (2002). Comprehensive design of a sensor network for chemical plants based on various diagnosability and reliability criteria. 1. Framework. *Industrial & Engineering Chemistry Research*, Vol. 41, No. 7, April 2002, pp. 1826-1839, ISSN: 0888-5885

Chang, C.C. & Yu, C.C. (1990). On-line fault diagnosis using the signed directed graph. *Industrial & Engineering Chemistry Research*, Vol. 29, No. 7, July 1990, pp. 1290-1299, ISSN: 0888-5885

Chen, J. Y. & Chang, C.T. (2007). Systematic enumeration of fuzzy diagnosis rules for identifying multiple faults in chemical processes. *Industrial & Engineering Chemistry Research*, Vol. 46, No. 11, May 2007, pp. 3635-3655, ISSN: 0888-5885

Gentil, S. & Montmain, J. (2004). Hierarchical representation of complex systems for supporting human decision making. *Advanced Engineering Informatics*, Vol. 18, No. 3, July 2004, pp. 143–159, ISSN: 1474-0346

Graphviz (2009). URL: http://www.graphviz.org

Han, C.C.; Shih, R.F. & Lee, L.S. (1994). Quantifying signed directed graphs with the fuzzy set for fault diagnosis resolution improvement. *Industrial & Engineering Chemistry Research*, Vol. 33, No. 8, August 1994, pp. 1943-1954, ISSN: 0888-5885

Iri, M.; Aoki, K.; O'shima, E. & Matsuyama, H. (1979). An algorithm for diagnosis of system failures in the chemical process. *Computers & Chemical Engineering*, Vol. 3, No. 1-4, 1979, pp. 489–493, ISSN: 0098-1354

Iri, M; Aoki, K.; O'Shima, E. & Matsuyama, H. (1980). A graphical approach to the problem of locating the origin of the system failure. *Journal of the Operations Research Society of Japan*, Vol. 23, No. 4, December 1980, pp. 295–311, ISSN: 0453-4514

Izadi, I.; Shah, S.; Kondaveeti, S. & Chen, T. (2009). A framework for optimal design of alarm systems, *Proceedings of 7th IFAC Symposium on Fault Detection, Supervision and Safety of Technical Processes*, Barcelona, Spain, July 2009

Jiang, H.; Patwardhan, R. & Shah, S. (2008). Root cause diagnosis of plant-wide oscillations using the adjacency matrix, *Proceedings of 17th IFAC World Congress*, pp. 13893-13900, ISBN: 978-3-902661-00-5, Seoul, Korea, July 2008, Elsevier, Amsterdam

Kramer, M.A. & Palowitch, B.L. (1987). A rule-based approach to fault diagnosis using the signed directed graph. *AIChE Journal*, Vol. 33, No. 7, July 1987, pp. 1607-1078, ISSN: 0001-1541

Lee, G.; Tosukhowong, T.; Lee, J.H. & Han, C. (2006). Fault diagnosis using the hybrid method of signed digraph and partial least squares with time delay: the pulp mill process. *Industrial & Engineering Chemistry Research*, 2006, Vol. 45, No. , pp. 9061-9074, ISSN: 0888-5885

Mah, R.S.H. (1989). *Chemical Process Structures and Information Flows*. Butterworth Publishes, ISBN: 0-409-90175-X, Boston

Maurya, M.R.; Rengaswamy, R. & Venkatasubramanian, V. (2003a). A systematic framework for the development and analysis of signed digraphs for chemical processes. 1. Algorithms and analysis. *Industrial & Engineering Chemistry Research*, Vol. 42, No. 20, October 2003, pp. 4789-4810, ISSN: 0888-5885

Maurya, M.R.; Rengaswamy, R. & Venkatasubramanian, V. (2003b). A systematic framework for the development and analysis of signed digraphs for chemical processes. 2. Control loops and flowsheet analysis. *Industrial & Engineering Chemistry Research*, Vol. 42, No. 20, October 2003, pp. 4811-4827, 2003, ISSN: 0888-5885

Maurya, M.R.; Rengaswamy, R. & Venkatasubramanian, V. (2006). A signed directed graph-based systematic framework for steady-state malfunction diagnosis inside control loops. *Chemical Engineering Science*, Vol. 61, No. 6, March 2006, pp. 1790-1810, ISSN: 0009-2509

McCoy, S.A.; Wakeman, S.J.; Larkin, F.D.; Chung, P.W.H. & Rushton, A.G. & Lees, F.P. (1999). HAZID, a computer aid for hazard identification. 1. The STOPHAZ package and the HAZID code: an overview, the issues and the structure. *Transactions of the Institution of Chemical Engineers. Part B*, Vol. 77, No. 6, November 1999, pp. 317-327, ISSN: 0957-5820

Mylaraswamy, D. & Venkatasubramanian, V. (1997). A hybrid framework for large scale process fault diagnosis. *Computer & Chemical Engineering*, Vol. 21, No. Suppl., May 1997, pp. S935-S940, ISSN: 0098-1354

Oyeleye, O. O. & Kramer, M. A. (1988). Qualitative simulation of chemical process systems: steady-state analysis. *AIChE Journal*, Vol. 34, No. 9, September 1988, pp. 1441-1454, ISSN: 0001-1541

Preisig, H. A. (2009). A graph-theory-based approach to the analysis of large-scale plants. *Computers & Chemical Engineering*, Vol. 33, No. 3, March 2009, pp. 598-904, ISSN: 0098-354

Raghuraj, R.; Bhushan, M. & Rengaswamy, R. (1999). Locating sensors in complex chemical plants based on fault diagnostic observability criteria. *AIChE Journal*, Vol. 45, No. 2, February 1999, pp. 310-322, ISSN: 0001-1541

Shiozaki, J.; Matsuyama, H.; O'Shima, E. & Iri, M. (1985). An improved algorithm for diagnosis of system failures in the chemical process. *Computers & Chemical Engineering*, Vol. 9, No. 3, pp. 285-293, ISSN: 0098-1354

Shiozaki, J.; Shibata, B.; Matsuyama, H. & O'shima, E. (1989). Fault diagnosis of chemical processes utilizing signed directed graphs-improvement by using temporal information. *IEEE Transactions on Industrial Electronics*, Vol. 36, No. 4, November 1989, pp. 469-474, ISSN: 0278-0046

Takeda, K.; Shibata, B.; Tsuge, Y. & Matsuyama, H. (1995). Improvement of fault diagnostic system utilizing signed directed graph–the method using transfer delay of failure. *Transactions of the Society of Instrument and Control Engineers*, Vol. 31, No. 1, January 1995, pp. 98–107, ISSN: 0453-4654

Thambirajah, J.; Benabbas, L.; Bauer, M. & Thornhill, N. F. (2009). Cause-and-effect analysis in chemical processes utilizing XML, plant connectivity and quantitative process history. *Computers & Chemical Engineering*, Vol. 33, No. 2, February 2009, pp. 503-512, ISSN: 0098-1354

Vedam, H. & Venkatasubramanian, V. (1997). Signed digraph based multiple fault diagnosis. *Computers & Chemical Engineering*, Vol. 21, No. Suppl., May 1997, pp. S655-S660, ISSN: 0098-1354

Vedam, H. & Venkatasubramanian, V. (1999). PCA-SDG based process monitoring and fault diagnosis. *Control Engineering Practice*, Vol. 7, No. 7, July 1999, pp. 903-917, ISSN: 0967-0661

Venkatasubramanian, V.; Rengaswamy, R. & Kavuri, S.N. (2003). A review of process fault detection and diagnosis. Part II: Qualitative models and search strategies. *Computers & Chemical Engineering*, Vol. 27, No. 3, March 2003, pp. 313-326, ISSN: 0098-1354

Venkatasubramanian, V.; Zhao, J. & Viswanathan, S. (2000). Intelligent system for HAZOP analysis of complex process plants. *Computers & Chemical Engineering*, Vol. 24, No. 9-10, October 2000, pp. 2291-2302, ISSN: 0098-1354

Wilcox, N.A. & Himmelblau, D.M. (1994). The possible cause and effect graphs (PCEG) model for fault diagnosis—I. Methodology. *Computers & Chemical Engineering*, Vol. 18, No. 2, pp.103-116, February 1994, ISSN: 0098-1354

Yang, F. (2008). *Research on Dynamic Description and Inference Approaches in SDG Model-Based Fault Analysis [PhD Dissertation]*, Tsinghua University, Beijing

Yang, F. & Xiao, D. (2005a). Review of SDG modeling and its application. *Control Theory and Applications*, Vol. 22, No. 9, October 2005, pp. 767-774, ISSN: 1000-8152

Yang, F. & Xiao, D. (2005b). Approach to modeling of qualitative SDG model in large-scale complex systems. *Control and Instruments in Chemical Industry*, Vol. 32, No. 5, October 2005, pp. 8-11, ISSN: 1000-3932

Yang, F. & Xiao, D. (2006a). Approach to fault diagnosis using SDG based on fault revealing time, *Proceedings of the 6th World Congress on Intelligent Control and Automation*, pp. 5744-5747, ISBN: 1-4244-0332-4, Dalian, China, June 2006, IEEE, New York

Yang, F. & Xiao, D. (2006b). Probabilistic signed directed graph and its application in hazard assessment, In: *Progress in Safety Science and Technology (Vol. VI)*, Huang, P.; Wang, Y.; Li, S.; Zheng, C. & Mao, Z. (Ed.), pp. 111-115, Science Press, ISBN: 7-03-018145-X, Beijing

Yang, F. & Xiao, D. (2006c). Model and fault inference with the framework of probabilistic SDG, *Proceedings of 9th International Conference on Control, Automation, Robotics and Vision*, pp. 1023-1028, ISBN: 1-4244-0341-3, Singapore, December 2006, IEEE, New York

Yang, F. & Xiao, D. (2007). Expressions of logic operators in signed directed graph and its applications in complex systems. *DCDIS Series B*, Vol. 14, No. S2, pp. 1260-1264, ISSN: 1492-8760

Yang, F. & Xiao, D. (2008a). SDG-based fault isolation for large-scale complex systems solved by rough set theory, *Proceedings of 17th IFAC World Congress*, pp. 7221-7226, ISBN: 978-3-902661-00-5, Seoul, Korea, July 2008

Yang, F. & Xiao, D. (2008b). Sensor location strategy in large-scale systems for fault detection applications. *Journal of Computers*, Vol. 3, No. 11, November 2008, pp. 51–57, ISSN: 1796-203X

Yu, C.C. & Lee, C. (1991). Fault diagnosis based on qualitative/quantitative process knowledge. *AIChE Journal*, Vol. 37, No. 4, April 1991, pp. 617-628, ISSN: 0001-1541

Zhang, Z.; Wu, C.; Zhang, B.; Xia, T. & Li, A. (2005). SDG multiple fault diagnosis by real-time inverse inference. *Reliability Engineering & System Safety*, Vol. 87, No. 2, May 2005, pp. 173-189, ISSN: 0951-8320

4

Customized Fault Management System for Low Voltage (LV) Distribution Automation System

M.M Ahmed, W.L Soo, M. A. M. Hanafiah and M. R. A. Ghani
University Technical Malaysia Melaka (UTeM)
Malaysia

1. Introduction

Supply disruption such as overloading will cause interruptions of electricity supply to customers. The technicians have to manually locate the fault point and this tedious work may last for extended periods of time. The other reasons are the lack of use of efficient tools for operational planning and advanced methodology for quick detection of fault, isolation of the faulty section and service restoration. Currently, fault detection, isolation and service restoration takes a long time causing the interruption of supply for a longer duration.

An active development phase of information technology has given significant impact to the distribution network fault management. There is a tendency towards fully automated switching systems with the introduction of Supervisory Control and Data Acquisition (SCADA) systems. The decision making feature of the fault management system is depend on SCADA system. SCADA can be used to handle the tasks which are currently handled by the people and can reduce frequency of periodic visit of technical personal substantially. SCADA is a process control system that enables a site operator to monitor and control processes that are distributed among various remote sites. The control functions are related to switching operations, such as switching a capacitor, or reconfiguring feeders. Once the fault location has been analyzed, the automatic function for fault isolation and supply restoration is executed. When the faulty line section is encountered, it is isolated, and the remaining sections are energized. This function directly impacts the customers as well as the system reliability.

This chapter presents the development of a fault management system for distribution automation system (DAS) for operating and controlling low voltage (LV) downstream system such as 415 V for three phase system and 240 V for single phase system. The fault management system is referred as the fault detection, fault location, fault isolation, electricity restoration and automatic operation and control. This done because system is equipped with automated equipment to detect earth fault and over-current faults and identify them accordingly. An embedded controller with Ethernet access is used as remote terminal unit (RTU) to act as converter for human machine interface (HMI) and to interact with the digital input and output modules. Supervisory Control and Data Acquisition (SCADA) is integrated with the RTU for automatic operating and controlling the distribution system. The laboratory results are compared with the simulation results to verify the results and make the achieved results credible.

2. Substation Model

Fig. 1 shows a typical compact substation which is also referred as Ring Main Unit (RMU).

RMU can be obtained by arranging a primary loop, which provides power from two feeders. Any section of the feeder can be isolated without interruption, and primary faults are reduced in duration to the time required to locate a fault and do the necessary switching to restore service. A 12KV, 630A, 20KVA RMU is supplying power supply to Low Voltage Feeder Panel. A three-phase, 1000KVA, 11/0.433 kV transformer is used to step down 11kV to 433V before supplying to Low Voltage Feeder Panel (LVFP).

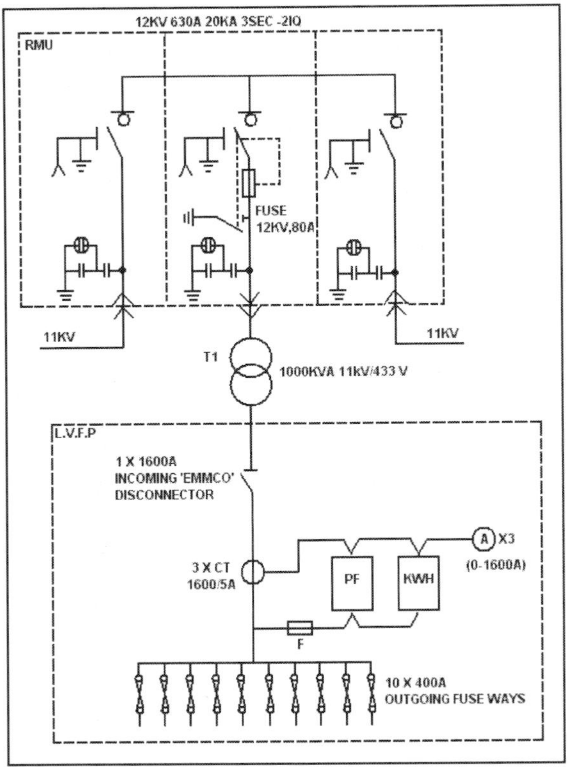

Fig. 1. Typical Substation

The outgoing loads are protected by fuses which have to be replaced if fault occurs. In this research project, fuses have been replaced by circuit breaker which can be manually or automatically controlled for switching operation and are not frequently replaced as shown in Fig. 1. The typical panel shown in Fig. 1 is using power factor meter, kilowatt hour meter and three ammeters to provide reading of power factor, kilowatt hour and three phase current values. Instead of using different types of meters to provide the reading, a single power analyzer is used in this research to provide the same reading and is able to send the data to the controller using modbus protocol. In this research, the service substation panel consists of four feeder points and each feeder is connected to customer service substation panel as shown in Fig.2.

In Fig. 2, the service substation block diagram consists of power line and control line. The power line shows the electricity power supply supplied to all the components in service substation panel. The controller and I/ O modules are supplied with dc power supply. Relay module received power supply from the digital output module. The control line shows the communication line between the controller and I/O modules.

Electromechanical relays are interfaces that control one electrical circuit by opening and closing its contacts in another circuit. An electromechanical relay is a relay with sets of contacts which are closed by magnetic force that includes magnetic coil and contacts. Electromechanical relays are limited to switching about 15A but are capable of switching several different circuits simultaneously.

Current transformers (CTs) are used in power installations for supplying the current to circuits of indicating instruments such as ammeter and wattmeter and protective relays. CTs are used so that ammeters and the current coils of other instruments and relays need not be connected directly to high-current lines. These instruments and relays are insulated from high currents. CTs also step down the current to a known ratio. The secondary winding is rated at 5A regardless of the current rating of the primary winding.

MK2200 which is the combination of overcurrent and earth leakage relay is used. The 630A, 50kVA, 415V MCCB is used as the main MCCB. The MCCBs are switched on or off by using customized solenoids.

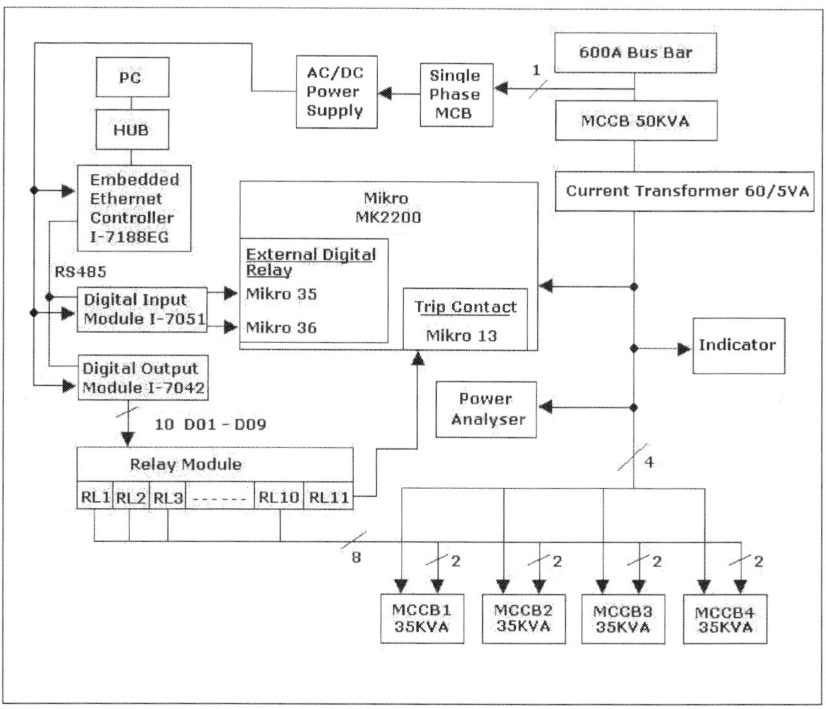

Fig. 2. Service Substation System Block Diagram

In Fig. 3, the customer service substation block diagram also consists of power line and control line. The power line starts from MCCB1 from the service substation panel. Digital input/output module is supplied with direct current power supply. The contactor module is connected to the ac power supply. Relay module received power supply from the digital input/output module. The control line shows that the digital input/output module is connected with the controller in the service substation panel.
Relays are connected to contactors to energize or de-energize loads. Contactors typically have multiple contacts, and those contacts are usually normally-open, so that power to the load is shut off when the coil is de-energized. The top three contacts switch the respective phases of the incoming three-phase AC power. The lowest contact is "auxiliary" contact. A normally-closed relays are connected to this auxiliary contact in series so that when the relays are opened, contactor coils will automatically de-energize, thereby shutting off power to the load.

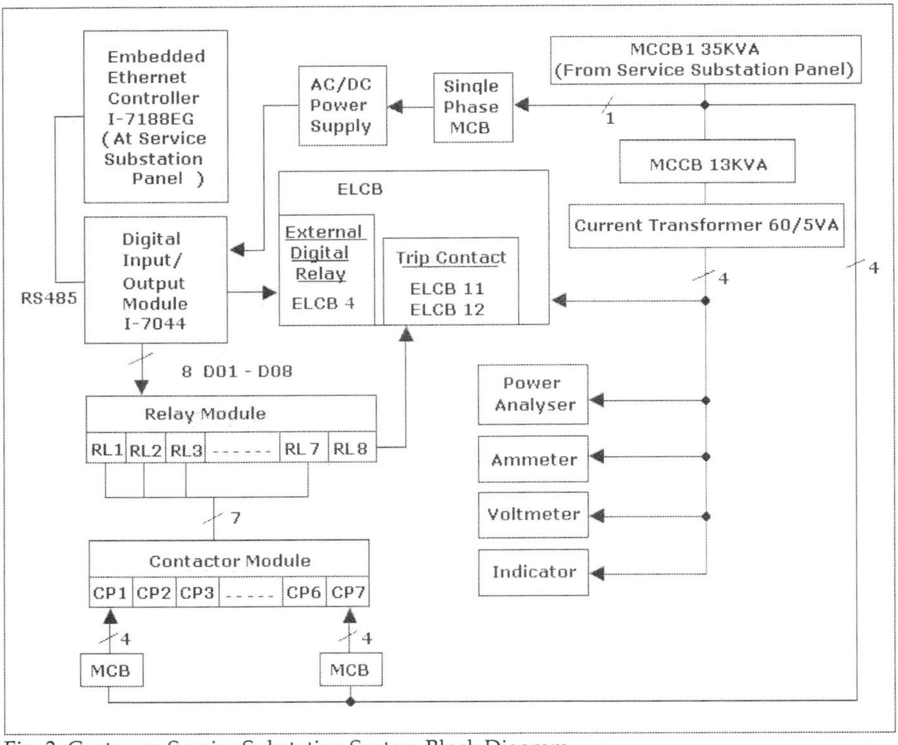

Fig. 3. Customer Service Substation System Block Diagram

3. Relays

In this research work, zero-phase sequence current transformer is connected to the earth leakage relay as show in Fig. 4. Earth Leakage Circuit Breaker (ELCB) works in the same way as the Residual Current Circuit Breaker (RCCB) and as such, must be accompanied by a circuit breaker (CB) or a fuse. Phase and neutral conductors are passed through a toroidal transformer, creating a magnetic field proportional to its current. In normal situations, the vector sum of the currents is zero even with unbalanced three-phase loads. The magnitude of the zero sequence current increases beyond its normal value when a ground fault is experienced. If the magnitude of the estimated zero sequence current is more than a threshold value in earth leakage relay, it is assumed that one or two phases are short circuited to ground. The short circuit between one phase and earth is probably the most common type of fault on low voltage electrical installations. This research project is focusing on earth fault.

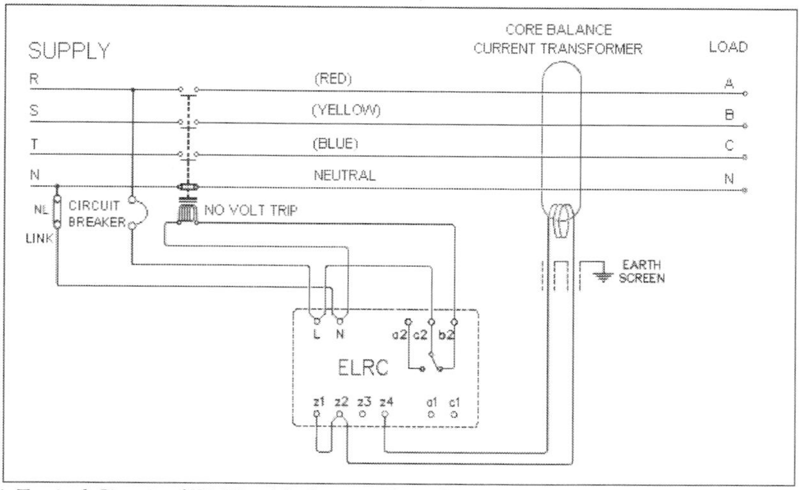

Fig. 4. Typical Circuit of ELCB with Current Transformer

A leakage current towards earth on one or more conductors at the downstream of the toroid causes an imbalance which is detected in the measurement winding and sent to an amplifier relay. The amplifier relay receives the signal from the ring current transformer and compares it with the preset threshold value. The relay output is turned on in the case where the detected value is higher than the preset threshold and lasts for a longer time than the preset tripping time value. The output remains in the on state until the relay is reset either manually or electrically. Generally, the relay output is fed to the shunt trip of a protective device such as CB which isolates the faulted circuit.

In service substation panel, ELCB is replaced by MK2200. Fig. 5 shows the pin assignments of MK2200. The MK2200 which combined overcurrent and Earth Fault (EF) relay is a digital microprocessor based relay. MK2200 relay provides three independent Phase Overcurrent (PO) elements and one non-directional Earth Fault (EF) element. These elements are connected to the current transformers of the feeders to be protected. PO protection and EF protection element consists of low-set stage and high-set stage for each PO elements. When the phase current exceeded the low-set stage or high-set stage, a signal is sent to the user through pre-assigned contact outputs and display panel.

MK2200 has six relay outputs of which 4 are user configurables. R1 contact is the dedicated trip contact and cannot be programmed. Either EF or overcurrent will activate this contact. The relay output remains in the on state until the relay is reset either manually or electrically. Generally, the relay output is fed to the shunt trip of a protective device such as CB which isolates the faulted circuit. Contact R2, R3, R4 and R5 are user configurable outputs. The sixth output contact internal relay failure (IRF) is also not user programmable. It is used to signal an internal failure of MK2200.
When the auxiliary power of MK2200 is switched on, the relay starts its operation. If the MK2200 is functioning normally, the IRF output is energized hence the normally close (NC) contact of the output will open and the normally open (NO) contact will close.
Fig. 6 shows the typical connection diagram of MK2200 with Current Transformer (CT). MK2200 works in the same way as the Residual Current Circuit Breaker (RCCB) and as such, must be accompanied by a Circuit Breaker (CB) or a fuse. Phase and neutral conductors are passed through a toroidal transformer, creating a magnetic field proportional to its current. In normal situations, the vector sum of the currents is zero even with unbalanced three-phase loads.

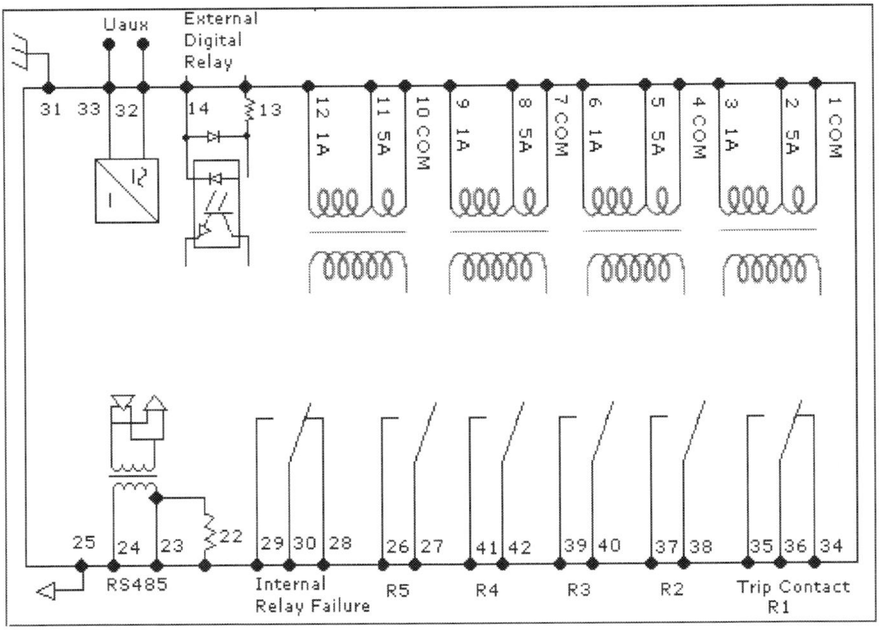

Fig. 5. Pin Assignment of MK2200

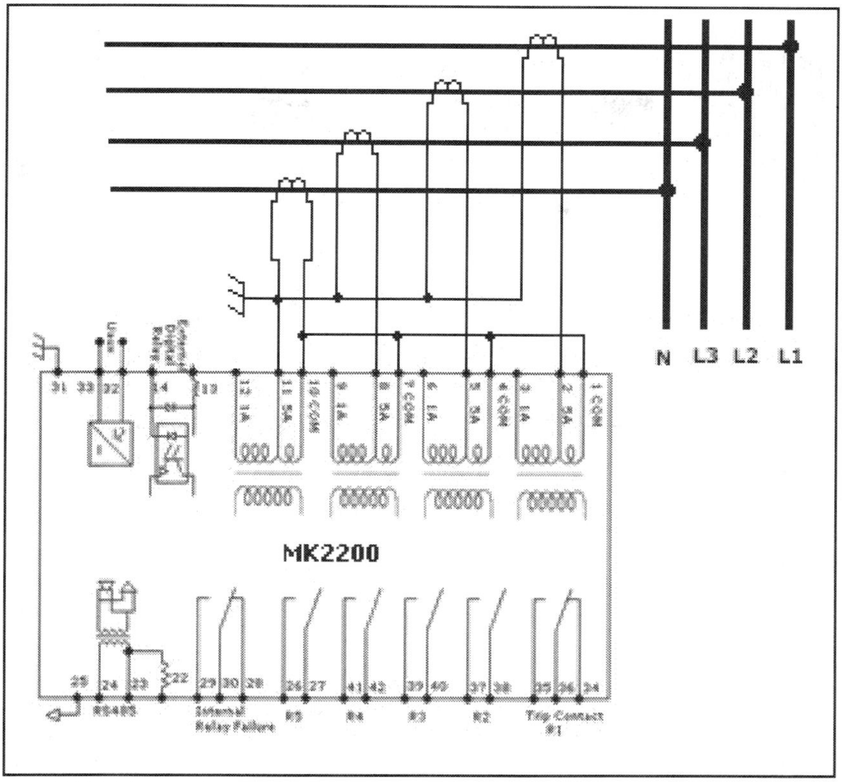

Fig. 6. MK2200 Typical Connection Diagrams

4. Fault Isolation Methods

The logic of controller is using this flowchart described in Fig. 7. By using a flowchart, the sequences of operations were determined. Based on the flowchart, there are five major actions which are the status of 'power input', mode state, the status of 'reset programme', execute Operation Logic Up-Counter (OLUC) and execute Operation Logic Down Counter (OLDC).

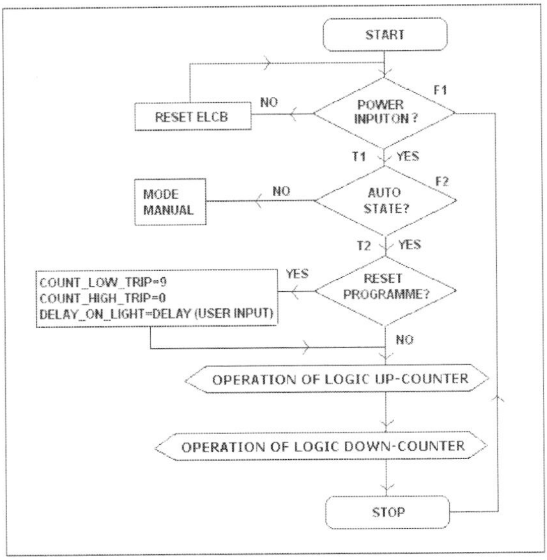

Fig. 7. General Flow Chart

First step is to check the power input whether it is turned on or turned off. If no faulty condition detected by ELCB, power input is turned on. When ELCB detects the fault condition, power input is turned off. In this case, the ELCB is resetted by using a delay timer and power input is turned on. There are two modes of operations. If automatic mode is selected, when fault occurs, the fault point is isolated automatically by activating the OLUC and OLDC. OLUC and OLDC are executed and only the fault point is isolated and the unaffected points are operated as normal condition. Once the fault point is operated as normal, the 'reset programme' button is pressed. This button resets back the counter to initial value and executes the OLUC and OLDC again.

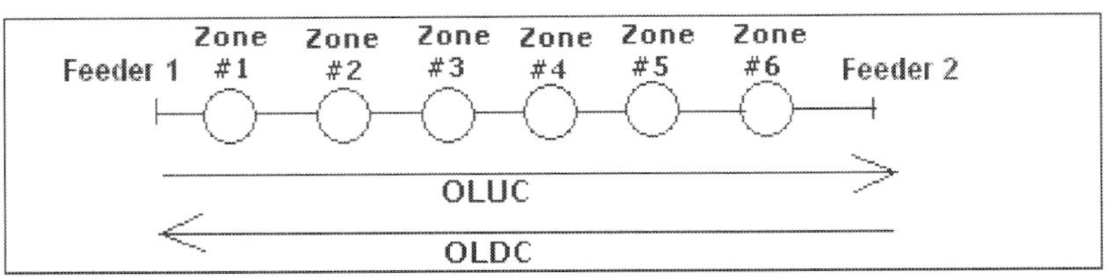

Fig. 8. OLUC and OLDC Descriptions

The general flow of these algorithms is described as follows. ALGORITHM: OLUC
Initial: TotalCountOLUC, TotalCountOLDC, NumOfFeeders

```
FOR ALL u e NumOfFeeders
    IF mode_auto & trip_counter_oluc > u

        FeederNo[u
        ]   =Switch
        On;       IF
        FeederNo[u
        ] = TRUE {
          TotalCountOLUC = TotalCountOLUC + 1;
        }
    }
}
```

ALGORITHM: OLDC
Initial: TotalCount, NumOfFeeders

```
FOR ALL u e NumOfFeeders
    IF mode_auto & OLUC_done & LastFeederOn & trip_counter_oldc < u
                                                                     {
        FeederNo[u]
        =Switch  On;
        IF
        FeederNo[u]
        = TRUE {
          TotalCountOLDC = TotalCountOLDC + 1;
        }
    }
}
```

The TotalCountOLUC and TotalCountOLDC are counters that will store the information of the fault points. This information was obtained during the second tripped of ELCB. As illustrated in Fig. 8, OLUC' checks the logic of up-counter which is from left to right while 'OLDC' checks the logic of down-counter which is from right to left. If the manual mode is switched on, when fault occurs, the checking is done manually by the operator. Fig. 9 shows the flowchart for OLUC and Fig. 10 shows the flowchart for OLDC. If the manual mode is switched on, when fault occurs, the checking is done manually by the operator. The developed GUI provides buttons to control the switching of the loads.

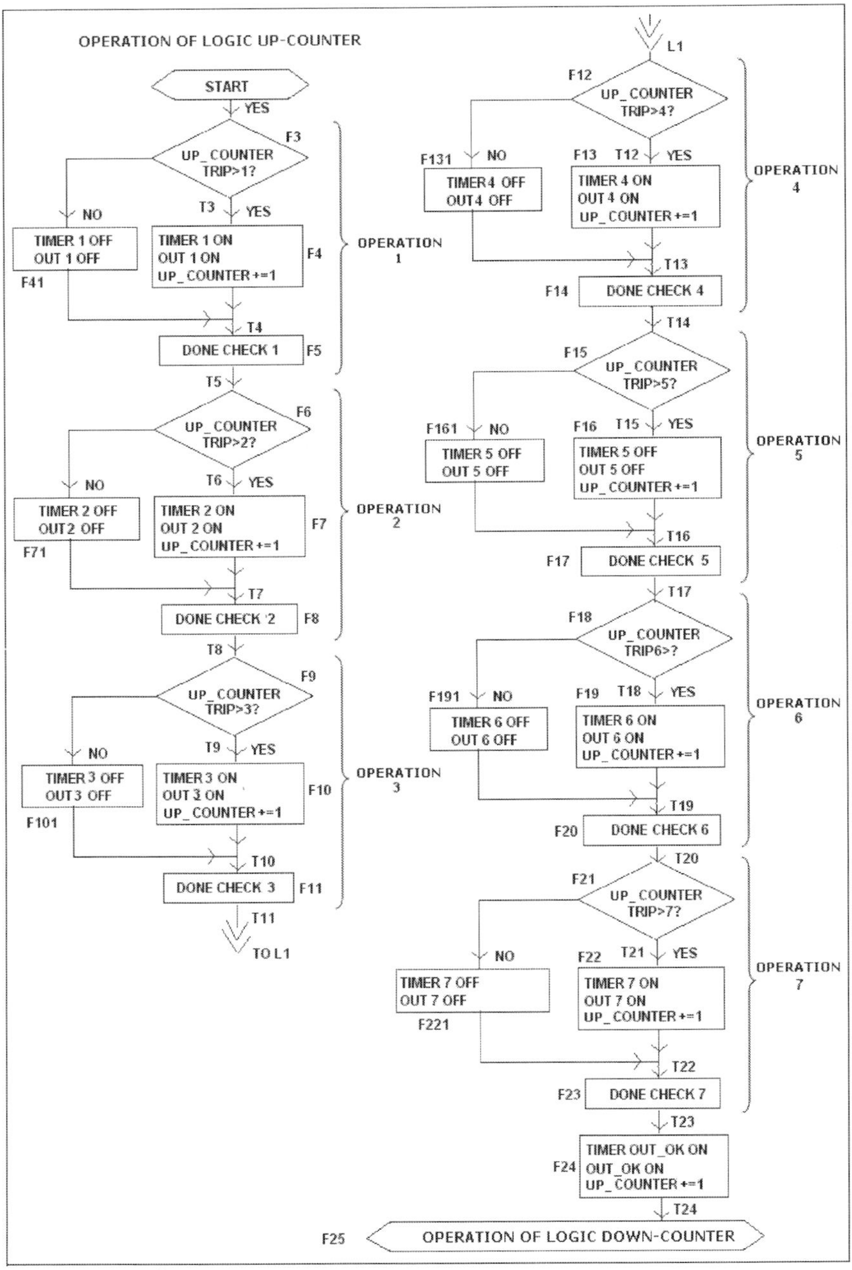

Fig.9: Flow Chart for OLUC

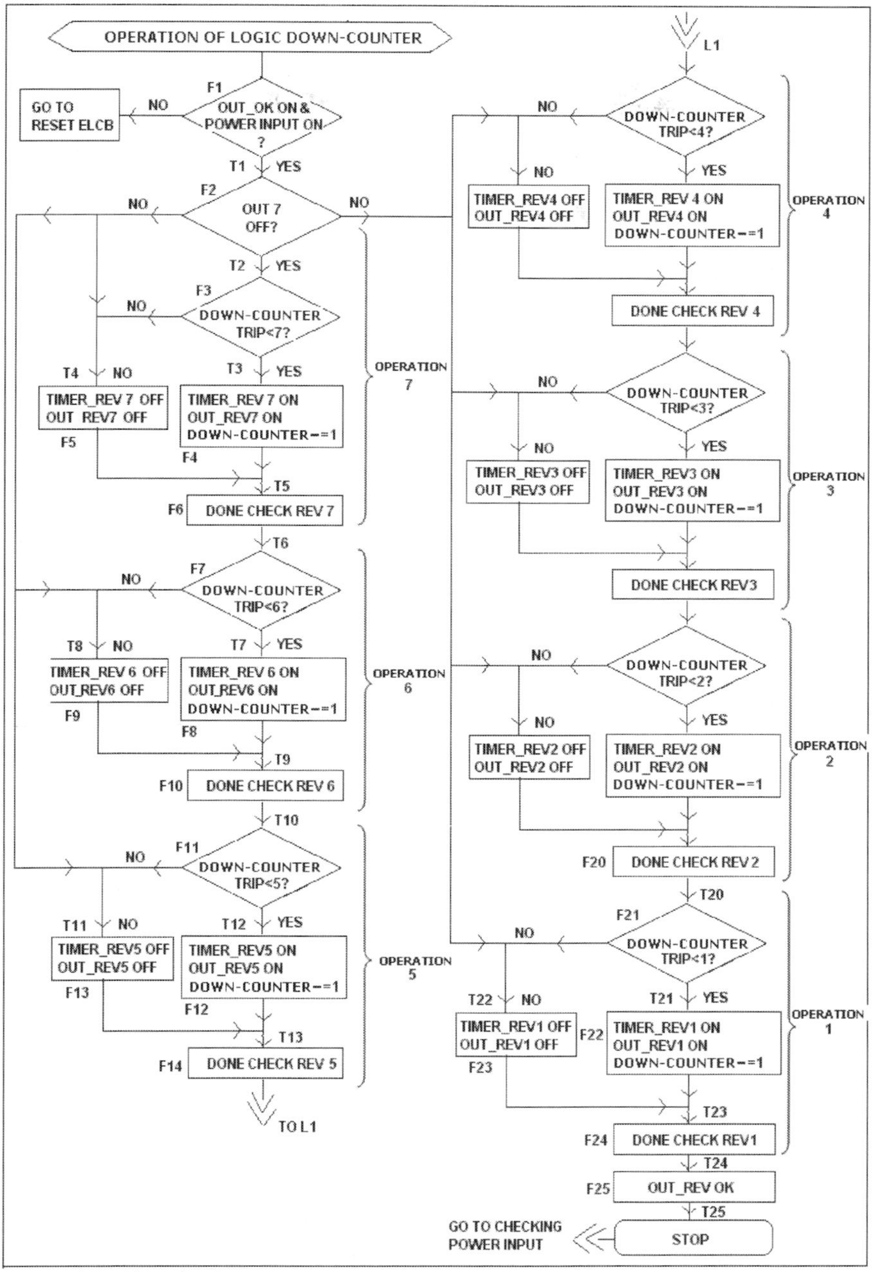

Fig. 10. Flow Chart for OLDC

5. Results and discussions

Both ELCB and MK2200 are fitted with test and reset buttons in order to conduct testing and commissioning of the relays. 32 bit microprocessor-based power quality analyzer is used to measure phase voltage and phase current of the incoming power supply to both panels. The power analyzer is connected using three phase, four-wire input connections for unbalanced loads. The setup procedures are done by using the keypad. It sends data to the controller using RS485 network. Table 1, Table 2, Table 3 and Table 4 are the results obtained by the following procedures described in the tables.

Customer Service Substation Panel															
No	Action at Customer Service Substation Screen	Result ■ On □ Off													
		HMI Symbols							Actual Loads						
		1	2	3	4	5	6	7	1	2	3	4	5	6	7
1	Click on "Reset Button"	■	■	■	■	■	■	■	■	■	■	■	■	■	■
2	Click on "Lamp1"	■	□	□	□	□	□	□	■	□	□	□	□	□	□
4	Click on "Lamp2"	■	■	□	□	□	□	□	■	■	□	□	□	□	□
5	Click on "Lamp3"	■	■	■	□	□	□	□	■	■	■	□	□	□	□
6	Click on "Lamp4"	■	■	■	■	□	□	□	■	■	■	■	□	□	□
7	Click on "Lamp5"	■	■	■	■	■	□	□	■	■	■	■	■	□	□
8	Click on "Lamp6"	■	■	■	■	■	■	□	■	■	■	■	■	■	□
9	Click on "Lamp7"	■	■	■	■	■	■	■	■	■	■	■	■	■	■

Table 1. Manual Mode Experimental Procedure-1

Service Substation Panel											
No	Action at Service Substation Screen	Result ■ On □ Off									
		HMI Symbols					Actual Loads				
		1	2	3	4	5	1	2	3	4	5
1	Click on "Reset Button"	■	■	■	■	■	■	■	■	■	■
2	Click on "Main MCCB"	■	□	□	□	□	□	■	□	□	□
4	Click on "Zone1"	■	■	□	□	□	□	□	□	□	□
5	Click on "Zone2"	■	■	■	□	□	□	□	□	□	□
6	Click on "Zone3"	■	■	■	■	□	□	□	□	□	□
7	Click on "Zone4"	■	■	■	■	■	□	□	□	□	□
8	Click off "Main MCCB"	□	□	□	□	□	□	□	□	□	□

Table 2. Manual Mode Experimental Procedure-2

By conducting the actual experiments, the appropriate delay timer for controlling the solenoid operation in service substation and also loads in customer panel are obtained. Fig. 11, Fig. 12, Fig. 13, Fig. 14 and Fig. 15 show graphs build using Excel to show the change of phase current for customer service substation panel and service substation panel during fault isolation operation done by the system. The duration time 1 is the time period that needed by the system to identify which load is the fault load. In this experiment, the Zone 4 and Zone 5 were chose to be the fault zones. Duration time 2 is the time period that is needed by the system to isolate the faulted load and restore electricity power supply to the rest of the healthy loads. Outage time is the duration of time that the customer experienced electricity power supply disruption. Reset time is the total time needed to restore electricity power supply to all the loads including the faulted load that already been repaired.

Service Substation Panel											
		Result ■ On □ Off									
	Action	HMI Symbols					Actual Loads				
No		1	2	3	4	5	1	2	3	4	5
1	Click on "Reset Button" at HMI screen	■	■	■	■	■	■	■	■	■	■
2	Click on "MK2200 Test Button" at panel when "Zone1" turned on	□	■	■	■	■	□	■	■	■	■
3	Click on "Reset Button" at HMI screen	■	■	■	■	■	■	■	■	■	■
4	Click on "MK2200 Test Button" at panel when "Zone2" turned on	■	□	■	■	■	■	□	■	■	■
5	Click on "Reset Button" at HMI screen	■	■	■	■	■	■	■	■	■	■
6	Click on "MK2200 Test Button" at panel when "Zone3" turned on	■	■	□	■	■	■	■	□	■	■
7	Click on "Reset Button" at HMI screen	■	■	■	■	■	■	■	■	■	■
8	Click on "MK2200 Test Button" at panel when "Zone4" turned on	■	■	■	□	■	■	■	■	□	■
9	Click on "Reset Button" at HMI screen	■	■	■	■	■	■	■	■	■	■

Table 3. Automatic Mode Experimental Procedure-1

Customer Service Substation Panel														
		Result ■ On □ Off												
		HMI Symbols							Actual Loads					
No	Action	1	2	3	4	5	6	7	1	2	3	4	5	6
1	Click on "Reset Button" at HMI screen	■	■	■	■	■	■	■	■	■	■	■	■	■
2	Click on "ELCB Test Button" at panel when "Lamp1" turned on	□	□	■	■	■	■	■	□	■	■	■	■	■
3	Click on "Reset Button" at HMI screen	■	■	■	■	■	■	■	■	■	■	■	■	■
4	Click on "ELCB Test Button" at panel when "Lamp2" turned on	■	□	□	■	■	■	■	■	□	■	■	■	■
5	Click on "Reset Button" at HMI screen	■	■	■	■	■	■	■	■	■	■	■	■	■
6	Click on "ELCB Test Button" at panel when "Lamp3" turned on	■	■	□	□	■	■	■	■	■	□	■	■	■
7	Click on "Reset Button" at HMI screen	■	■	■	■	■	■	■	■	■	■	■	■	■
8	Click on "ELCB Test Button" at panel when "Lamp4" turned on	■	■	■	□	□	■	■	■	■	■	□	■	■
9	Click on "Reset Button" at HMI screen	■	■	■	■	■	■	■	■	■	■	■	■	■

Table 4. Automatic Mode Experimental Procedure-2

10	Click on "ELCB Test Button" at panel when "Lamp5" turned on	■	■	■	■	□	□	■	■	■	■	■	□	■
11	Click on "Reset Button" at HMI screen	■	■	■	■	■	■	■	■	■	■	■	■	■
12	Click on "ELCB Test Button" at panel when "Lamp6" turned on	■	■	■	■	■	□	□	■	■	■	■	■	□
13	Click on "Reset Button" at HMI screen	■	■	■	■	■	■	■	■	■	■	■	■	■
14	Click on "ELCB Test Button" at panel when "Lamp1" turned on and when "Lamp3" turned on	□	□	□	■	■	■	■	□	□	□	■	■	■
15	Click on "Reset Button" at HMI screen	■	■	■	■	■	■	■	■	■	■	■	■	■
16	Click on "ELCB Test Button" at panel when "Lamp1" turned on and when "Lamp4" turned on	□	□	□	□	■	■	■	□	□	□	□	■	■
17	Click on "Reset Button" at HMI screen	■	■	■	■	■	■	■	■	■	■	■	■	■
18	Click on "ELCB Test Button" at panel when "Lamp1" turned on and when "Lamp5" turned on	□	□	□	□	□	■	■	□	□	□	□	□	■
19	Click on "Reset Button" at HMI screen	■	■	■	■	■	■	■	■	■	■	■	■	■

Table 5. Automatic Mode Experimental Procedure-2 Continue

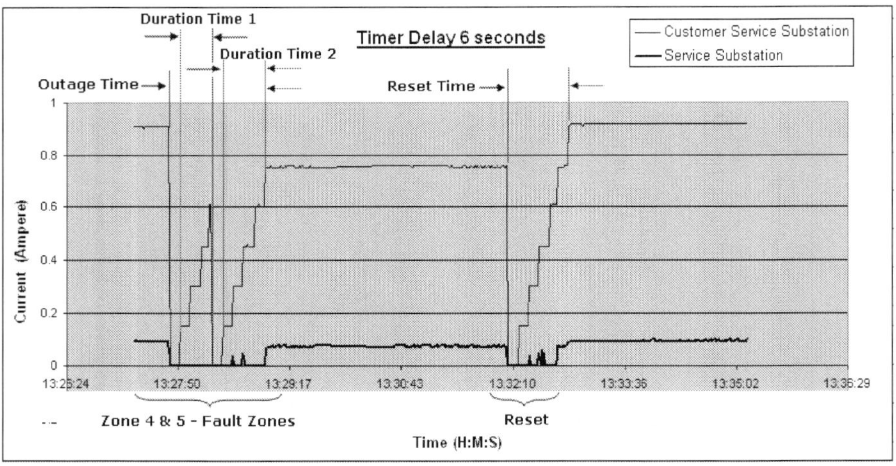

Fig. 11. Phase Current Graph for Delay Timer Six Seconds

Table 5 shows the total of restoration time for service substation panel. In the service substation panel, if the delay timer is set to fast, the solenoid will not be able to switch on or off the MCCB. The minimum delay timer needs to be set to five seconds for proper switching operation of the MCCBs by the solenoids. The results in Table 5 can only be applied using LKE LKS-600S MCCB and LKE LKS-100N MCCB with shunt trip operating time of five to fifteen minutes.

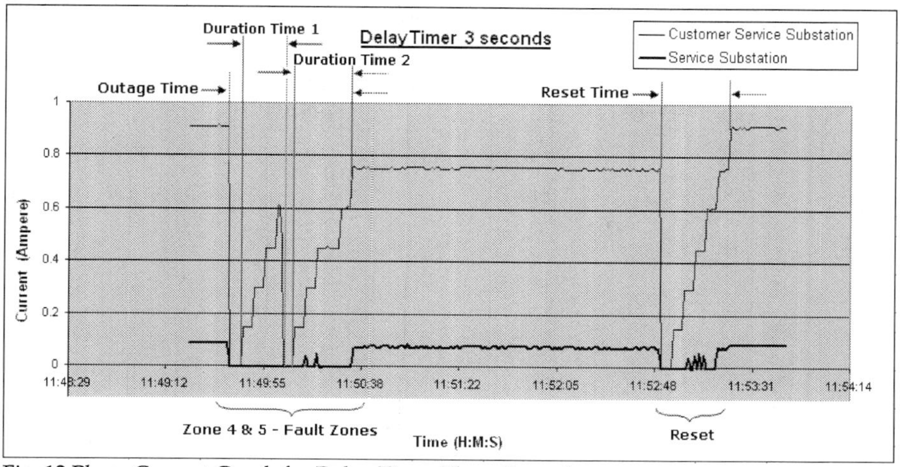

Fig. 12 Phase Current Graph for Delay Timer Three Seconds

Panel	Duration Time 1	Duration Time 2	Total of time
Service Substation	25 seconds	25 seconds	50 seconds

Table 6. Total Minimum Restoration Time for Service Substation Panel

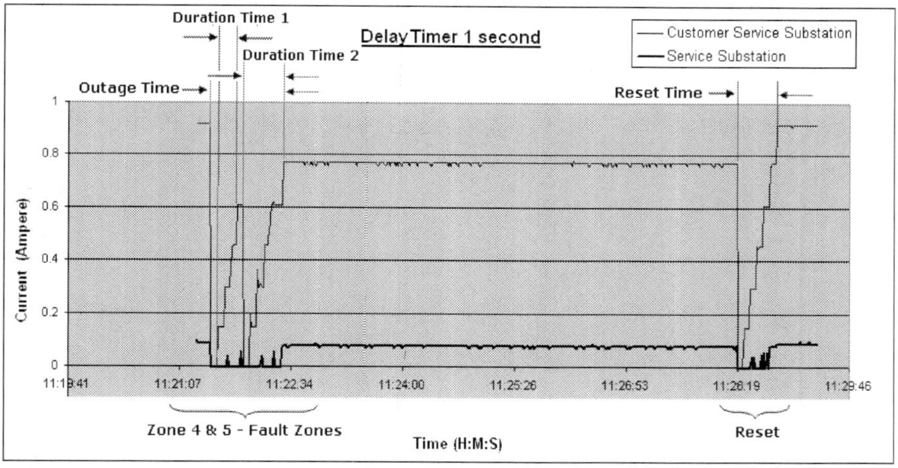

Fig. 13. Phase Current Graph for Delay Timer One Second

In Fig.15 shows the reset time when the delay timer in customer service substation is set to one hundred milliseconds. The delay timer was to short and fault was not able to be triggered manually during the experiment. In this case, duration time 1, duration time 2 and outage time were not displayed in Table 6. Table 6 shows the details of the duration time 1, duration time 2, outage time and reset time when the delay timer in customer service substation is set to six seconds, three seconds, one seconds, five hundred milliseconds and one hundred milliseconds.

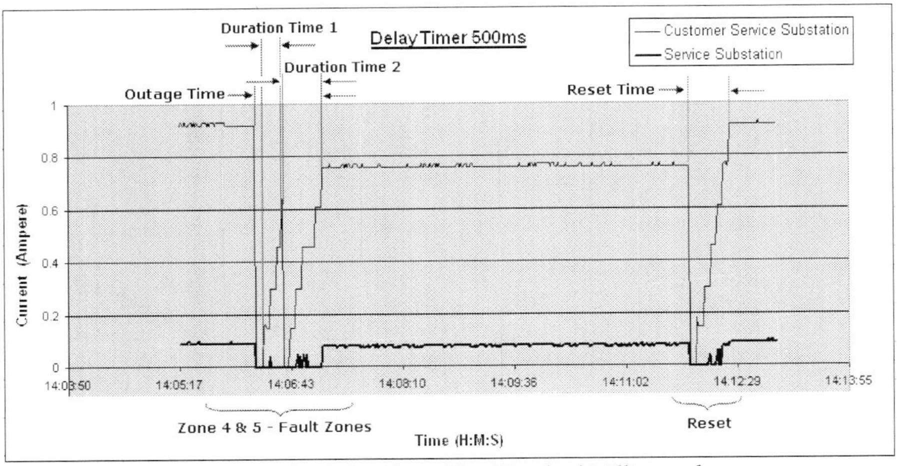

Fig. 14. Phase Current Graph for Delay Timer Five Hundred Milliseconds

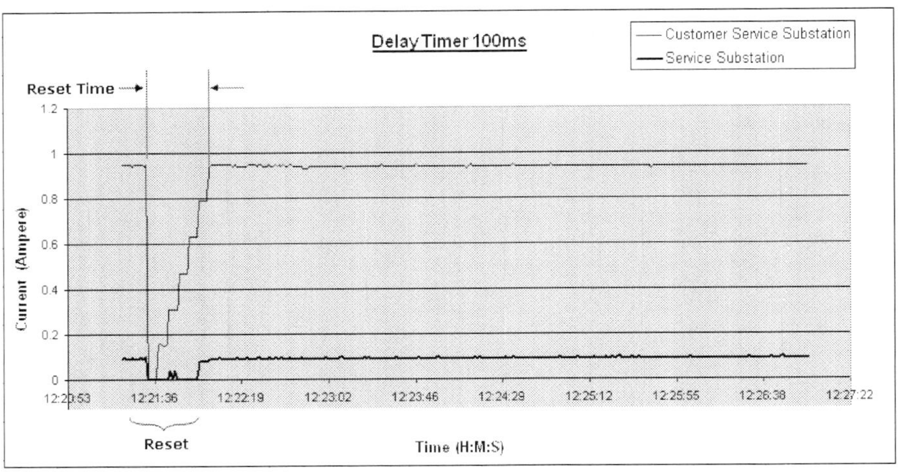

Fig. 15. Phase Current Graph for Delay Timer One Hundred Milliseconds

Table 7 shows the minimum total of restoration time for customer service substation panel and service substation panel. The ELCB minimum operation time is fifty milliseconds. In this case, the delay timer has to be set higher than fifty milliseconds. The minimum delay timer for customer service substation panel is one hundred milliseconds.

Delay Timer (Second)	Duration Time 1 (Second)	Duration Time 2 (Second)	Outage Time (Second)	Reset Time (Second)
6	24	30	54	42
3	12	15	33	21
1	4	5	15	7
0.5	2	2.5	10.5	3.5
0.1	-	-	-	0.7

Table 6. Restoration Time for Customer Service Substation Based on Delay Timer

Panel	Minimum Restoration Time (Second)
Customer Service Substation	0.7
Service Substation	50

Table 7. Minimum Total of Restoration Time

Experiment	Total of Experiment Steps	Total of Correct Operation	Total of Failure Operation	Percentage of Failure Operation (%)
Manual Mode Experimental Procedure-1	36	34	2	5.56
Manual Mode Experimental Procedure-2	32	36	0	0
Automatic Mode Experimental Procedure-1	27	27	0	0
Automatic Mode Experimental Procedure-2	57	56	1	1.75

Table 8. Percentage of Failure Operation

The system developed in this research proved to save the time needed to restore back the electricity supply after fault occurred. The restoration time definitely cannot be achieved by manually isolation done by technician. The failure percentages of the system for not able to operate correctly based on the experimental procedures are described in Table 8. The failure operations were during the manual mode of service substation to switch on Zone 2 and Zone 3. The reason of this failure is due to loose wiring at the digital output module. One failure was recorded during automatic mode of customer service substation when fault was simply triggered at Zone 2. The system was not able to switch on the other healthy loads during OLDC operation because loose wiring at the second feeder. After the causes of failures have been fixed, the experiments were continued and no operation failure was recorded. The percentages of failure operation in Table 8 are contributed by human factors.

6. Conclusion

In this research, a Customized SCADA is built to provide automatic fault isolation for low distribution system. The contribution of this research includes developing a complete fault isolation algorithm based on an open loop distribution system. Service Substation Panel, Customer Service Substation Panel and Customer Panel have been built to validate the proposed methodology.

In an open loop distribution system, two feeders are used to provide electricity power supply to the loads. Any section of the feeder can be isolated without interruption. The algorithm is written to check the fault point starting from one of the section feeders or OLUC algorithm and repeated with another section feeder or OLDC algorithm. At the beginning, this algorithm needs to clarify with which point is the fault point by supplying the power supply to each load after the fault is detected by the ELCB. When the fault point is being activated, the ELCB detects the fault and trip mechanism is operated. The algorithm will find the false point and reset the ELCB to restore the power supply to the loads. This time, only the un-faulted point will be restored. In Customer Service Substation panel, two contactors are used to activate one load. Although in Service Substation panel only uses one feeder, the same algorithms (OLUC and OLDC) are applied to control the switching operation of MCCBs by using customized solenoid. MCCBs have trip mechanism that is able to detect faults. MCCBs are different from the loads used in Customer Service Substation because they don't need to be accompanied by two switching devices to control their operations.

The HMI is capable to communicate with the I/O devices. An HMI for SCADA is developed in this research by using an embedded Ethernet controller as the converter to communicate with the I/ O devices. By integrating the ELCB and MK2200 into the SCADA system, the SCADA system is capable to respond to the faults by resetting both devices in order for the algorithm to check the fault point.

Based on the experimental results, the system correctly locates the fault point, isolates the fault point and reenergizes the un-faulted loads. However, during the fault isolation operation, the system has to detect the fault point by simply switching on the fault load. After the system acknowledges the fault point, the appropriate switching functions are executed. The developed system has a potential in reducing the outage time while comparing to the manual operation by the technicians and engineers.

From the analysis done, the outage times for both panels to locate fault and restore electricity power supply to healthy loads are 50.7 seconds. By assuming that it takes 1 hour for the technician to restore the electricity power supply, thus there is a 98.59% improvement in the outage time operation. This system will help the utility company to save money if the outage times are reduced. As described in Table 8, the failure percentage for the system in detecting the fault and isolating the fault point is none. This means that the system is reliable.

7. Future Research

The future research work should be aimed at developing of full scale Distribution Automation system, which can cover from primary substations to consumer level intelligent automation. Here are few recommendations which will lead new research projects in the future:

a) Automated solutions to the distribution system comprising automation-ready made building blocks such as a synthesis of state-of-the-art individual components (distribution switchgear, fault passage indication, RTU, local power supplies, and communications interfaces), integrated to form complete functional devices. These devices are assembled into a system to solve the complete control or automation needs of each distribution network.
b) Data collection of not only the status of CBs and relays but also status of each component in the panels. This will help to reduce black-out due to equipment malfunctions.
c) Develop a standard Distribution Automation software such as master distribution automation software, customer information system (CIS), trouble call management software (TCMS) and web based monitoring of distribution system.

8. References

Austerlitz, H. (2003). Data Acquisition Technique Using PCs. 2nd ed. California: Academic Press, 416p. ISBN: 0-12-068377-6

Clarke, G. (2004). Practical Modern SCADA Protocols: DNP3, 60870.5 and Related Systems, Massachusetts Newness, 1st ed., 743p. ISBN 07506 7995

Glover, J.D. & Sarma M.S. (2002). Power System Analysis and Design., California: Pacific Grove, 3rd ed., 527

p. ISBN 0-534-95367-0

Pabla, A.S. (2005). Electric Power Distribution, 2nd ed., New York: McGraw-Hill, 723 p. ISBN 0-07-144783-0

Lee, H.J. & Park, Y.M. (1996). A Restoration Aid Expert System for Distribution Substations. IEEE Transactions on Power Delivery, Vol. 11, pp.1765 -1769

Hsu, Y.Y & Huang, H.M. (1995). Distribution System Service Restoration using the Artificial Neural Network Approach and Pattern Recognition Method. IEEE Proceeding Generation Transmission

Distribution, Vol. 142, No.3, pp.251-256.

Hsiao, Y. T. & Chien, C. Y. (2000). Enhancement of Restoration Service in Distribution Systems using a Combination Fuzzy-GA Method. IEEE Transactions on Power Systems, vol. 15, no. 4, November

2000, pp. 1394-1400

Huang C.M (2003). Multiobjective Service Restoration of Distribution Systems Using Fuzzy Cause-Effect Networks. IEEE Transactions on Power Systems, Vol. 18, No. 2, pp.867 - 874

5

Faults Detection for Power Systems

Zeng Xiangjun,Wang Yuanyuan and Xu Yao
Changsha University of Science and Technology
China

1. Introduction

Power systems are the largest and most complex human made systems, where faults always occurred, in the world. Faults can cause personnel and equipment safety problems, and can result in substantial economic losses. In order to solve the problems, faults automatic detection, location and isolation must be employed. Most faults can cause large currents or voltages changing, and they are often detected by traditional protective relay. Whereas, some faults, such as high impedance faults, grounding faults of ineffectively earthed distribution systems, cause small currents and voltages changing and they are difficult to be detect using traditional protective relay. In this chapter, faults generated signals characteristics are investigated, special faults detection methods are developed, and their applications in power systems are presented.

2. Faults generated signals characteristics

2.1 Fundamental frequency signals characteristics

The three phase voltages and currents include fundamental frequency signals and many other different frequency harmonic signals. In this section the fundamental frequency signals is analyzed (Jingchao et al. 2003, Yang et al. 2003, Wang et al. 2006).

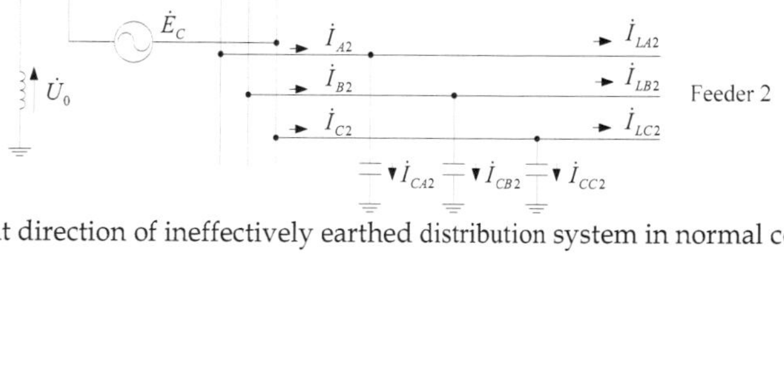

Fig. 1. Current direction of ineffectively earthed distribution system in normal condition

A typical ineffectively earthed distribution system is shown in Figure 1. The three phase fundamental frequency voltages named $\dot{E}_A$, $\dot{E}_B$ and $\dot{E}_C$ and fundamental frequency currents are symmetrical in normal condition. The neutral-to-ground voltage $\dot{U}_0$ is zero, and the phase-to-ground capacitances in Feeder 1 and Feeder 2 are C_1 and C_2 respectively. Taking Feeder 1 for example, the three phase fundamental frequency currents flowing from bus to feeder are $\dot{I}_{A1}$, $\dot{I}_{B1}$, $\dot{I}_{C1}$, the fundamental frequency capacitive currents are $\dot{I}_{CA1}$, $\dot{I}_{CB1}$, $\dot{I}_{CC1}$, and fundamental frequency load currents are $\dot{I}_{LA1}$, $\dot{I}_{LB1}$, $\dot{I}_{LC1}$.

The three phase fundamental frequency currents from bus to Feeder 1 can be calculated:

$$\begin{cases} \dot{I}_{A1} = \dot{I}_{CA1} + \dot{I}_{LA1} = j\omega C_1(\dot{E}_A + \dot{U}_0) + \dot{I}_{LA1} \\ \dot{I}_{B1} = \dot{I}_{CB1} + \dot{I}_{LB1} = j\omega C_1(\dot{E}_B + \dot{U}_0) + \dot{I}_{LB1} \\ \dot{I}_{C1} = \dot{I}_{CC1} + \dot{I}_{LC1} = j\omega C_1(\dot{E}_C + \dot{U}_0) + \dot{I}_{LC1} \end{cases} \tag{1}$$

The three phase fundamental frequency currents from bus to Feeder 2 can be calculated:

$$\begin{cases} \dot{I}_{A2} = \dot{I}_{CA2} + \dot{I}_{LA2} = j\omega C_2(\dot{E}_A + \dot{U}_0) + \dot{I}_{LA2} \\ \dot{I}_{B2} = \dot{I}_{CB2} + \dot{I}_{LB2} = j\omega C_2(\dot{E}_B + \dot{U}_0) + \dot{I}_{LB2} \\ \dot{I}_{C2} = \dot{I}_{CC2} + \dot{I}_{LC2} = j\omega C_2(\dot{E}_C + \dot{U}_0) + \dot{I}_{LC2} \end{cases} \tag{2}$$

If single-phase grounding fault happened in phase C of Feeder 2 (shown in Figure 2), the three phase fundamental frequency voltages of source are still symmetrical. But the neutral-to-ground voltage is not zero, it changes from $\dot{U}_0$ to $\dot{U}'_0$. The residual current is $\dot{I}_f$. Taking Feeder 1 for example, the three phase fundamental frequency currents are $\dot{I}'_{A1}$, $\dot{I}'_{B1}$ and $\dot{I}'_{C1}$ respectively, fundamental frequency capacitive currents are $\dot{I}'_{CA1}$, $\dot{I}'_{CB1}$ and $\dot{I}'_{CC1}$ respectively, and fundamental frequency load currents are $\dot{I}'_{LA1}$, $\dot{I}'_{LB1}$ and $\dot{I}'_{LC1}$ respectively.

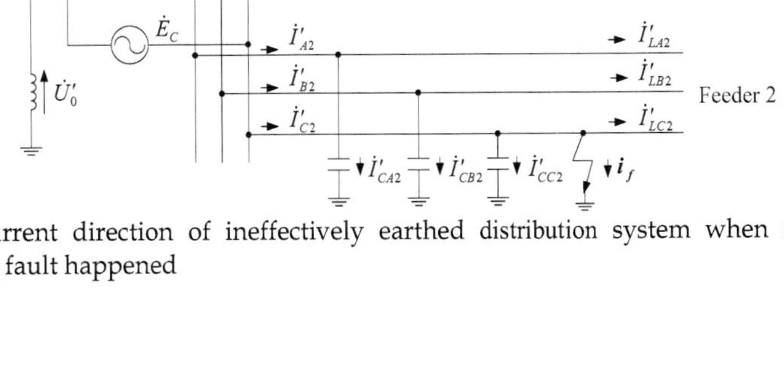

Fig. 2. Current direction of ineffectively earthed distribution system when single-phase grounding fault happened

The three phase fundamental frequency currents from bus to Feeder 1 can be calculated:

$$\begin{cases} \dot{I}'_{A1} = \dot{I}'_{CA1} + \dot{I}'_{LA1} = j\omega C_1(\dot{E}_A + \dot{U}'_0) + \dot{I}'_{LA1} \\ \dot{I}'_{B1} = \dot{I}'_{CB1} + \dot{I}'_{LB1} = j\omega C_1(\dot{E}_B + \dot{U}'_0) + \dot{I}'_{LB1} \\ \dot{I}'_{C1} = \dot{I}'_{CC1} + \dot{I}'_{LC1} = j\omega C_1(\dot{E}_C + \dot{U}'_0) + \dot{I}'_{LC1} \end{cases} \quad (3)$$

The three phase fundamental frequency currents from bus to Feeder 2 can be calculated:

$$\begin{cases} \dot{I}'_{A2} = \dot{I}'_{CA2} + \dot{I}'_{LA2} = j\omega C_2(\dot{E}_A + \dot{U}'_0) + \dot{I}'_{LA2} \\ \dot{I}'_{B2} = \dot{I}'_{CB2} + \dot{I}'_{LB2} = j\omega C_2(\dot{E}_B + \dot{U}'_0) + \dot{I}'_{LB2} \\ \dot{I}'_{C2} = \dot{I}'_{CC2} + \dot{I}'_{LC2} + \dot{I}_f = j\omega C_2(\dot{E}_C + \dot{U}'_0) + \dot{I}'_{LC2} + \dot{I}_f \end{cases} \quad (4)$$

Although the three phase fundamental frequency currents contain fundamental frequency load currents, the fundamental frequency load currents change little before and after fault happening. It can be assumed:

$$\begin{cases} \dot{I}_{LA1} = \dot{I}'_{LA1}\,,\ \dot{I}_{LA2} = \dot{I}'_{LA2} \\ \dot{I}_{LB1} = \dot{I}'_{LB1}\,,\ \dot{I}_{LB2} = \dot{I}'_{LB2} \\ \dot{I}_{LC1} = \dot{I}'_{LC1}\,,\ \dot{I}_{LC2} = \dot{I}'_{LC2} \end{cases} \quad (5)$$

Comparing (1), (2), (3), (4) and taking account of (5), the changing of the fundamental frequency currents can be calculated.
For Feeder 1:

$$\begin{cases} \Delta\dot{I}_{A1} = \dot{I}'_{A1} - \dot{I}_{A1} = j\omega C_1(\dot{U}'_0 - \dot{U}_0) \\ \Delta\dot{I}_{B1} = \dot{I}'_{B1} - \dot{I}_{B1} = j\omega C_1(\dot{U}'_0 - \dot{U}_0) \\ \Delta\dot{I}_{C1} = \dot{I}'_{C1} - \dot{I}_{C1} = j\omega C_1(\dot{U}'_0 - \dot{U}_0) \end{cases} \quad (6)$$

For Feeder 2:

$$\begin{cases} \Delta\dot{I}_{A2} = \dot{I}'_{A2} - \dot{I}_{A2} = j\omega C_2(\dot{U}'_0 - \dot{U}_0) \\ \Delta\dot{I}_{B2} = \dot{I}'_{B2} - \dot{I}_{B2} = j\omega C_2(\dot{U}'_0 - \dot{U}_0) \\ \Delta\dot{I}_{C2} = \dot{I}'_{C2} - \dot{I}_{C2} = j\omega C_2(\dot{U}'_0 - \dot{U}_0) + \dot{I}_f \end{cases} \quad (7)$$

The differences of the changing of the fundamental frequency currents of Feeder 1 between the phases can be calculated as:

$$\begin{cases} \Delta\dot{I}_{AB1} = \Delta\dot{I}_{A1} - \Delta\dot{I}_{B1} = 0 \\ \Delta\dot{I}_{BC1} = \Delta\dot{I}_{B1} - \Delta\dot{I}_{C1} = 0 \\ \Delta\dot{I}_{CA1} = \Delta\dot{I}_{C1} - \Delta\dot{I}_{A1} = 0 \end{cases} \quad (8)$$

For Feeder 2:

$$\begin{cases} \Delta \dot{I}_{AB2} = \Delta \dot{I}_{A2} - \Delta \dot{I}_{B2} = 0 \\ \Delta \dot{I}_{BC2} = \Delta \dot{I}_{B2} - \Delta \dot{I}_{C2} = -\dot{I}_f \\ \Delta \dot{I}_{CA2} = \Delta \dot{I}_{C2} - \Delta \dot{I}_{A2} = \dot{I}_f \end{cases} \tag{9}$$

From the above analysis, the following results are obtained:

a. In the sound feeder, the differences of the changing of the fundamental frequency currents between the three phases are zero under the ideal conditions.
b. In the faulted feeder, the differences of the changing of the fundamental frequency currents between the healthy and faulted phases are very large, equaling the residual current. However, it is zero between healthy phase currents.

In order to reduce the measuring error and the unbalance effect, the changing of the fundamental frequency currents ($\Delta \dot{I}$) is applied, which can be calculated by the following methods(Zeng et al. 2001):

a. $i(n) = i(n) - i(n-T)$
b. $i(n) = i(n) + i(n-T/2)$
c. $i(n) = i(n) - 2i(n-T) + i(n-2T)$
d. $i(n) = i(n) + i(n-T/2) - i(n-T) - i(n-3T/2)$

Where, n is the sample-time in digital relay, T is the system period (20ms when system frequency is 50Hz). Among the above methods, method d is usually employed.

2.2 Transient signals characteristics

2.2.1 High frequency characteristics

High frequency transient has high propagation velocity, and a large extent of attenuation, which lead to distortion on the process of transmission along the line. The transmission characteristics of different frequency components will result in the distortion of traveling waves, and the phenomenon is called dispersion, which decreases the mutation of initial traveling wave, influences the effective identification of initial traveling wave.

By analyzing the modulus of the frequency characteristics of the parameters, earth simulator component has the most serious attenuation and phase shift in the process of transmission, in that the earth simulator component is seriously influenced by zero-order inductance and resistance. The zero-order inductance and resistance is closely related with frequency because of skin effect. With obviously decreased of frequency, the zero-order inductance is decreased, while the zero-order resistance is increased, which cause attenuation coefficient and wave velocity have great changes. The higher frequency is included in earth simulator component, the more serious attenuation generated during transmission. Line model has greatly influenced by zero-order inductance and resistance, but these parameters have smaller impact of frequency. So attenuation coefficient and wave velocity of Line model are far less impact to earth simulator component. Therefore, dispersion of traveling wave is mainly caused by earth simulator component.

Seen from the fault type, single-phase grounding will generate large quantity of earth simulator component, which is greatly influenced by dispersion; two-phase earth short circuit has weak earth simulator component, which has smaller dispersion; two or three-phase short circuit has minimal impact, which has the smallest dispersion.

In practical application, the frequency band range of traveling wave is valued between 10 kHz to 1MHz. In this range, line model of traveling wave velocity has little influence by frequency changes, which has limited impacts for fault location. While high frequency component is serious attenuated, which to some extend decreased the amplitude of initial traveling wave.

Frequency characteristics of each modulus parameters, frequency variation curve of modulus velocity and frequency variation curve of modulus wave impedance are shown in Figure 3, 4 and 5 (Guo. 2007).

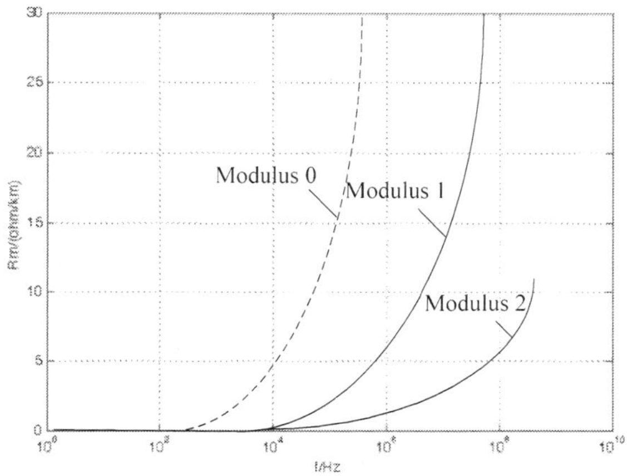

(a) Resistance frequency dependent character of each modulus parameters

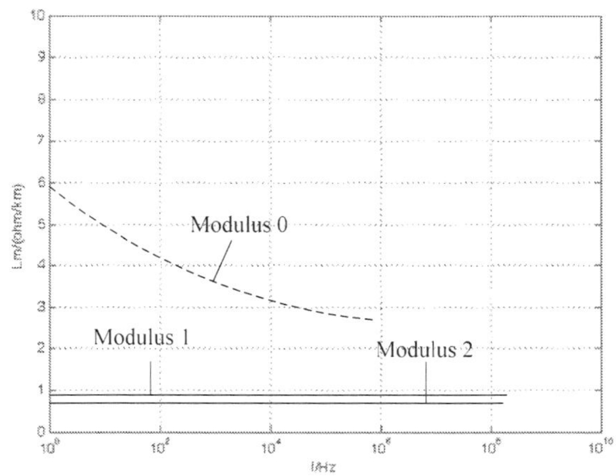

(b) Inductance frequency dependent character of each modulus parameters

Fig. 3. Frequency characteristics of each modulus parameters

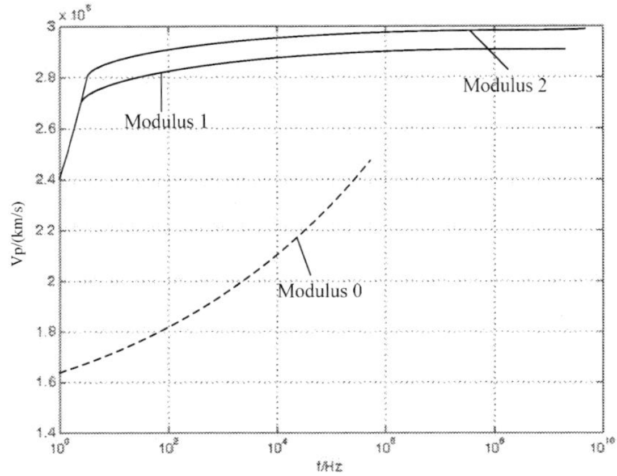

Fig. 4. Frequency variation curve of modulus velocity

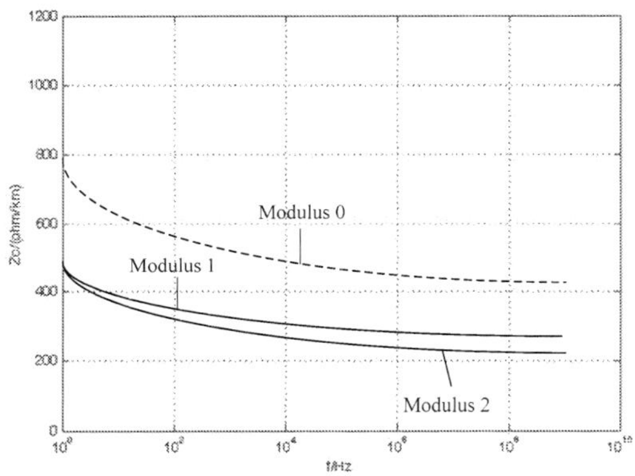

Fig. 5. Frequency variation curve of modulus wave impedance

2.2.2 Transient signals analysis with fourier transform

A signal $x(t)$ can be represented in the form(Wang et al. 2002):

$$x(t) = \int_{-\infty}^{\infty} F(\omega) e^{j\omega t} d\omega \tag{10}$$

Where $F(\omega)$, the Fourier transform of $x(t)$, is defined as:

$$\hat{x}(\omega) = \frac{1}{2\pi} \int_{-\infty}^{\infty} x(t) e^{-j\omega t} dt \tag{11}$$

The Fourier transform $\hat{x}(\omega)$ depend on the global properties of $x(t)$.

Short time Fourier transform (STFT) is the Fourier transform of signal $x(t)$ multiplied by a window function translating in time by b:

$$STFT(\omega,b)=\int_{-\infty}^{\infty}x(t)\overline{w}(t-b)e^{-j\omega t}dt \quad (12)$$

Where $\overline{w}$ denotes the conjugate of window function w. It can be rewritten as:

$$STFT_x(\omega,b)=e^{-jwb}\int_{-\infty}^{\infty}x(t)\overline{w}(-(b-t))e^{j\omega(b-t)}dt \quad (13)$$

Therefore, short time Fourier transform of a signal is that the signal is firstly through a band pass filter at analyzing frequency ω and then is modulated to zero frequency.

2.2.3 Transient signals analysis with wavelet transform

Proper signal analysis becomes a key issue. Analyzing singly in the time or frequency domain is not sufficient to capture the faults that spread in a wide band of frequencies. Faults of these types require analysis which is localized in both the time and frequency domains. The wavelet transform is an excellent signal analysis and decomposition tool for such signals using constant bandwidth analysis.

A Function $\Psi(t)\in L^2(R)$ is called a basic wavelet or mother wavelet, if the Fourier Transform satisfies equation (14) (Zhang et al. 2004):

$$\int_{-\infty}^{\infty}\frac{\left|\psi^*(\omega)\right|^2}{|\omega|}d\omega<\infty \quad (14)$$

Where $\psi^*(\omega)$ is the Fourier Transform of function $\psi(\omega)$. With its dilation and translation, we can get equation (15):

$$\psi_{a,b}(t)=\frac{1}{\sqrt{|a|}}\psi\left(\frac{t-b}{a}\right),b\in R,a>0 \quad (15)$$

This is called continuous wavelet relaying on factors a and b. Where a is called dilation factor and b is called translation factor. The continuous wavelet translation of a function $f(t)\in L^2(R)$ will be equation (16):

$$W_f(a,b)=<f,\psi_{a,b}>=|a|^{\frac{1}{2}}\int_R f(t)\psi\left(\frac{t-b}{a}\right)dadb \quad (16)$$

The function can be restructured by equation (17):

$$f(t)=\frac{1}{C_\psi}\int_{-\infty}^{\infty}\int_{-\infty}^{\infty}\frac{1}{a^2}W_f(a,b)\psi\left(\frac{t-b}{a}\right)dadb \quad (17)$$

Continuous wavelet transform (CWT) has perfect localization both in time domain and in frequency domain. With $|a|$ decrease, $\psi_{a,b}(t)$ focuses on the part of high frequency. It means the narrow in time window but high resolution. This performance makes CWT an advanced tool on fault detection.

When used in practice cases, especially in implementation in computer, continuous wavelet has been discrete. In fact, continuous wavelet and continuous wavelet transform of a signal can both be discrete. Supposing $a=a_0^j$ and $b=ka_0^jb_0$ $j\in Z$, equation (18):

$$\psi_{j,k}(t) = a_0^{-j/2}\psi(a_0^{-j}t - kb_0) \tag{18}$$

The decomposition and its reconstruction of the function $f(t)$ in discrete wavelet transform are shown in equation (19) and (20):

$$C_{j,k} = \langle f, \psi_{j,k} \rangle = \int_{-\infty}^{\infty} f(t)\psi_{j,k}(t)dt \tag{19}$$

$$f(t) = C\sum_{-\infty}^{\infty}\sum_{-\infty}^{\infty} C_{j,k}\psi_{j,k}(t) \tag{20}$$

Different wavelet transform may give different analysis results to the system. To most vibration signal, both CWT and discrete wavelet transform (DWT) can be used to detect faults.

There is a significant difference between wavelet transform and Fourier transform. The Fourier based functions are indexed by a single frequency parameter ω whereas the wavelet ones are indexed by two parameters, scale a and time translation b. Therefore, Fourier transforms describe the global properties of $x(t)$ while wavelet transforms describe the local properties of $x(t)$ in the neighborhood of each time translation b.

Short Time Fourier transform, which is the function of two variables: time translation b and analyzing frequency ω, shares with wavelet transform the same property of localization. It is customary to think of short time Fourier transform as the frequency analysis of a signal in a short time, particularly in the case of rectangular window, but the short time Fourier transform of signal $x(t)$ depends not only on the signal but also on the choice of the window.

2.2.4 Transient signals analysis with Hilbert transform

When we have decomposed the signal into a series of IMFS, Hilbert transform can be carried out on each intrinsic mode function (IMF) to get a series of instantaneous frequency $f_i(t)$ (Xiao'an et al. 2008).

For each IMF $c_i(t)$, its corresponding Hilbert transform is defined as:

$$Y_i(t) = \frac{1}{\pi}\int_{-\infty}^{+\infty}\frac{c_i(\tau)}{t-\tau}d\tau \tag{21}$$

Thus an analytic signal $Z_i(t)$ can be constructed as:

$$Z_i(t) = c_i(t) + jY_i(t) = a(t)e^{i\theta(t)} \tag{22}$$

Where $a(t) = [c_i(t)^2 + Y_i(t)^2]^{\frac{1}{2}}$; $\theta(t) = \arctan(Y_i(t)/c_i(t))$.

Instantaneous frequency of $c_i(t)$ is defined as:

$$f(t) = \frac{1}{2\pi}\frac{d\theta(t)}{dt} \tag{23}$$

2.2.5 Transient signals analysis with Hilbert-Huang transform

Hilbert-Huang Transform (HHT) is used for analyzing non-stationary and nonlinear signal. Compared with Fourier transform and wavelet transform, it is not necessary to select parameters in HHT, and the results of HHT absolutely lie only on the character of the signal itself. The transform has only one result.

HHT is based on two processes: empirical mode decomposition (EMD) and Hilbert transform. The central part of the HHT is EMD which is a sifting process to decompose a signal into a number of IMFs. Each IMF must satisfy the following two conditions: *a)* In the whole data set, the number of extrema and the number of zero-crossings must either equal or differ at most by one; *b)* At any point, the mean value of the envelope defined by local maxima and the envelope defined by the local minima is zero.

Any signal $s(t)$ can be decomposed as follows (Xiao'an et al. 2008):

a. When $s_{11}(t) = s(t)$, subscript i of signal $s_{ij}(t)$ expresses the decomposing order, and subscript j expresses the sifting times in the i order. Find all the local extrema, including maxima and minima, then connect all the maxima and minima of signal $s(t)$ using smooth cubic splines to get its upper envelope and lower envelope.

b. Subtracting mean $m_{ij}(t)$ of these two envelopes from the signal $s_{ij}(t)$ to get their difference: $h_{ij}(t) = s_{ij}(t) - m_{ij}(t)$.

c. If the sifting result $h_{ij}(t)$ meets the two criteria of an IMF, carry out step *d*; otherwise, treat $s_{i(j+1)}(t) = h_{ij}(t)$, and repeat steps *a* and *b*.

d. The i th IMF of signal $s(t)$ is: $c_i(t) = h_{ij}(t)$. Calculate the residual signal: $r_i(t) = s_{i1}(t) - c_i(t)$.

e. If the residue becomes monotonic, the final order $n = i$; Otherwise, treat $s_{(i+1)1}(t) = r_i(t)$, repeat steps *a, b, c* and *d*.

The original signal $s(t)$ can thus be expressed as follows:

$$s(t) = \sum_{i=1}^{n} c_i(t) + r_n(t) \tag{24}$$

Where $c_i(t)$ is i th IMF component, and $r_n(t)$ is the residue.

In summary, the empirical mode decomposition is similar with wavelet decomposition. The tree of empirical mode decomposition is shown in Figure 6. For the first IMF $c_1(t)$ contains the highest frequency component of the signal, it can be applied to fault detection (Li. 2005).

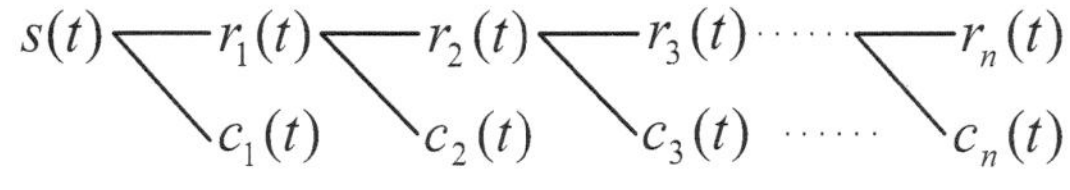

Fig. 6. Tree of empirical mode decomposition

2.3 Sensors and their characteristics analysis

Predicting or locating the faults as soon as possible is very important to maintain power grid reliable operation. Moreover, faults often accompanied by ultrasonic, electromagnetic field distribution and temperature changing, etc. Based on the change of these physical parameters, faults can locate quickly and accurately, so that the repair work can be carried out in the minimum time.

2.3.1 Sensors for faults generated ultrasonic

The basic idea of ultrasonic sensors is quite simple: They transmit acoustic waves and receive them after interaction of ultrasonic wave and investigated process. On it's arrival at the receiver the ultrasound signal carries the information about the parameters to be measured (and unfortunately many other parameters too - which demands compensation). In addition, software algorithms based on models for the ultrasonic propagation and the interaction between the ultrasonic wave and physical or chemical variables of interest are employed for analysis the measuring. Furthermore, ultrasonic measurements are only meaningful when state parameters like temperature, pressure etc. are measured simultaneously with the ultrasonic parameters at high accuracy for compensation.

Ultrasound covers a frequency range from 20 kHz to about 1 GHz. For technical applications the range 20 kHz to 10 MHz is the most important one (Hauptmann et al. 2001). The attractive features of ultrasonic sensors can be summarized as follow:

a. Non-invasive measurement.
b. In-line measurement.
c. Rapid response, usually a fraction of a second.
d. Low power consumption.
e. Excellent long term stability.
f. High resolution and accuracy.

At present, ultrasonic technique has been applied to the detection and location of discharges in liquid-immersed dielectric structures such as gaps, coil stacks, and reactors. However, in some cases it probably can be extended to solid structures, if higher frequencies are used, and also to gas-filled structures by using lower frequencies. It is applicable for almost any type of voltage testing. It does not interfere with normal current and voltage measurements, and indicates the presence and location of sparks and corona. In addition, it is relatively insensitive to external disturbances and is not influenced by the test piece capacitance. Therefore, it can serve as a valuable adjunct to other methods of discharge detection and measurement (Cooper et al. 1984).

The most challenging issues facing ultrasonic sensors are:

a. The exact knowledge of the acoustic properties of the substances is necessary for most ultrasonic measurements.
b. Substances under investigation must be acoustically transparent for transmission and some reflection techniques.
c. Ultrasonic measurements are highly disturbed when gas bubbles in liquids are present.
d. Ultrasonic signals tend to be complicated and need relatively complex signal processing.
e. Only integral information along the entire sound path is delivered.
f. Increase of the attenuation of sound with frequency.

It is known that corona and spark breakdowns produce both audible and ultrasonic pressure waves in the medium surrounding the discharge. The pressure waves have an intensity which is determined by the rate of energy release and the nature of the medium in which the discharge occurs. These pressure waves propagate radically outward in all directions, and generally contain a wide range of frequency components of up to several hundred kilocycles in liquid. In gases, the higher frequencies rapidly are attenuated, leaving only vibrations in the audible sound range. However, in liquid and in some solids, the attenuation of the higher frequencies is not as severe, and the electric discharges will produce pressure variations which have rich ultrasonic components, as well as audible sound.

The ultrasonic partial discharges measurement technique consists of two types, namely the contact type and the non-contact type. The contact-type measurement method uses a direct contact between the sensor and the equipment measured; for common transformer, the ultrasonic sensor was mounted on the wall of transformer to detect the ultrasonic signal and corona noise. But for some special transformer, such as epoxy-resin transformer, its coil surface relies on epoxy resin to insulate it from the air; it will be more difficult to use the contact-type measurement method in the field before the said phenomenon is resolved. The non-contact type measurement, although with a less degree of sensitivity than that of the contact type, experiences no such shortcoming mentioned above.

The sound wave's per-unit area energy and acoustic pressure will decrease as the propagation distance increases. Due to the rather weak partial discharges signals, their propagation distance is not far in the air and is easily interfered by the environment; hence ultrasonic measurement devices must be placed as closely to the measured object as possible. We will position the ultrasonic device, at various angles, 1m away from the object to be measured. During the measurement, we adopt partial discharges ultrasonic measuring devices with measurement frequency ranging from 20 to 100 kHz. To achieve the goal of identification of type and magnitude of fault, we design a partial discharges ultrasonic signal retrieving and analyzing system to proceed signal analysis; the whole analysis flowchart is shown in Figure 7 (Chen et al. 2005).

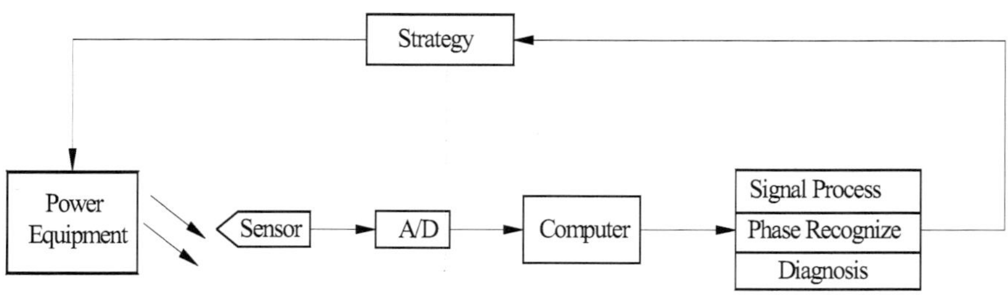

Fig. 7. Ultrasonic measurement flow

Step a. Measuring ultrasonic signals and, after analog–digital conversion, filing them into computer and proceeding noise identification.

Step b. Obtaining the field power-source signal as the reference signal; comparing ultrasonic signals with the reference signal to observe the discharges' sound wave characteristics.

Step c. Using a 60 Hz signal's angle as the reference, draw partial discharges signals into 3-D diagrams with the angle, the period, and the magnitude as the coordinates to observe partial discharges angles' characteristics.

Step d. Using a 60 Hz signal's angle as the reference; drawing partial discharges signals into polar-coordinate diagrams with the magnitude and the angle as the coordinates to observe partial discharges angles' characteristics.

Step e. Draw zone-identification patterns and them with basic discharge patterns to determine the equipment's abnormal discharge type.

2.3.2 Sensors for faults generated electromagnetic field distribution

Electromagnetic Pulse (EMP) Measurements

The sensor factor is defined here as the time-domain relation between magnetic or electric field ($H(t)$ or $E(t)$) and the sensor output voltage $V(t)$ (Middelkoop 1991):

$$F_h(t) = H(t)/V(t) \quad \text{[A/m/V], for a B-sensor} \tag{25}$$

$$F_e(t) = E(t)/\int_{\tau_i} V(t)dt \quad \text{[V/m/V], for a } \overset{\circ}{D}\text{-sensor} \tag{26}$$

Where τ_i is an integration time constant, $F_h(t)$ and $F_e(t)$ represent the magnetic and electric sensor factors, respectively, $B = \mu \cdot H$ and $\overset{\circ}{D} = \delta E / \delta t$.

The sensor factor characterizes the sensitivity of the sensor to the field. It is the function of an equivalent surface or equivalent length of the sensor. For calibration purposes it is recommended that sensors with unambiguously calculable equivalent surfaces and lengths be used. The dipole antenna to measure the electric field when combined with an integrator and the short-circuited loop antenna to measure the magnetic field are suitable sensors.

Figure 8 (Middelkoop R. 1991) shows a relatively simple closed circular loop (CCL) magnetic field sensor for which the sensor factor is calculated. The current that is induced in the closed loop by a magnetic field is measured using a current transformer clamped onto the loop. It can be shown that the current in the loop equals

$$I(t) = \mu_0 A H(t)/L \tag{27}$$

Where $A = \pi(a-b)^2$ is the effective area, μ_0 is the free-space permeability ($\mu_0 = 4\pi \cdot 10^{-7} V \cdot s / A \cdot m$) and L is the inductance of the loop. If the loop width $2b$ is small compared to its radius a , $b^2 << a^2$ and small compared to the wavelength, then

$$L = A\mu_0 \{\ln(8a/b) - 2\} \tag{28}$$

The relation between the induced current $I(t)$ and the magnetic field $H(t)$ is therefore only dependent on the loop dimensions. In practice, however, a time-dependent response $R(t)$ is introduced by the current clamp. The insertion impedance of the clamp causes a low-frequency cutoff, commonly designated as "sensor droop." This droop influences in particular the late time response of the field sensor (in practice > 500ns). The effect can be corrected if the cutoff frequency is known. The sensor factor becomes

$$F_h(t) = \frac{a\{\ln(8a/b) - 2\}}{\pi R(t)(a-b)^2} \quad \text{[A/m/V]} \tag{29}$$

A hollow spherical dipole (HSD) sensor that acts as an electric field sensor is shown in Figure 9. (Middelkoop R. 1991). In this case one half of the dipole is replaced by a conducting

plane, so the sensor can be used asymmetrically, e.g., on the bottom plate of a TEM cell.
The current $I(t)$ is equal to the change in surface charge per unit of time of the sphere as a result of a time varying electric field perpendicular to the base plate. It can be shown that

$$I(t) = 3\pi r^2 \varepsilon_0 \frac{dE(t)}{dt} \tag{30}$$

Where r is the sphere radius and ε_0 the free-space permittivity ($\varepsilon_0 = 8.85 \cdot 10^{-12} A \cdot s / V \cdot m$).
When the sensor is connected to a resistor R, the output voltage of the sensor is

$$V(t) = I(t) \cdot R = 3\pi r^2 R \varepsilon_0 \frac{dE(t)}{dt} \tag{31}$$

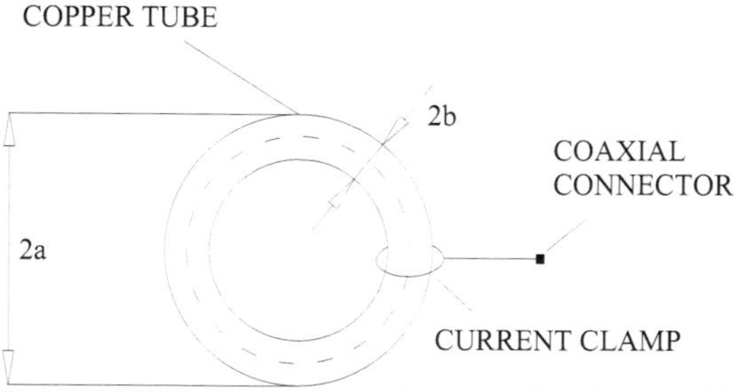

Fig. 8. Closed circular loop (CCL) sensor. 2a = diameter; 2b = loop width.

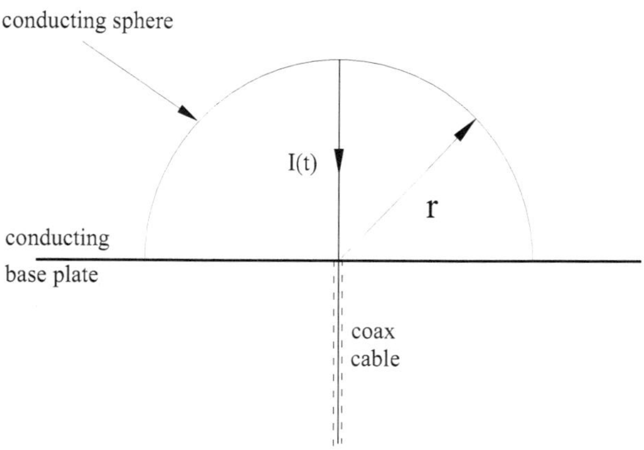

Fig. 9. Hollow spherical dipole (HSD) sensor

The output voltage is related to the time-derivative of the electric field. The sensor factor is found to be

$$F_e(t) = \frac{E(t)}{\int_{\tau_i} V(t)dt} = \frac{E(t)}{3\pi r^2 R\varepsilon_0 \frac{1}{\tau_i} \int_0 \frac{dE(t')}{dt'} dt'} = \frac{\tau_i}{3\pi r^2 R\varepsilon_0} \quad [\text{V/m/V}] \tag{32}$$

The sensor factor depends on some constants, the sphere radius r, resistor value R and the integration time τ_i. Integration is necessary to relate the output voltage $V(t)$ to the electric field $E(t)$. This integration can be done by using a passive or active integrating circuit in the output signal line of the sensor. Also, a numerical integration is possible.

Electromagnetic (EM) Field Measurement

The task of EM fields measuring we can find in international EMC standard EN 61000-4-3,which describes measurement of electric devices immunity against RF fields (Dixon and Dutcher 1990). The electromagnetic environment is determined by the strength of the electromagnetic field, is not easily measured without sophisticated instrumentation nor is it easily calculated by classical equations and formulae because of the effect of surrounding structures or the proximity of other equipment that will distort and/or reflect electromagnetic waves. Tested equipment is subjected to field strength of 3 V/m or 10 V/m with homogeneity -0dB/+6dB from 80 MHz to 1000 MHz. This frequency range is 80% amplitude modulated with a 1 kHz sine wave. The signal generator provides the modulated frequency at a step rate of 1% of fundamental to the RF amplifier. The RF amplifier provides the necessary power to the antenna to establish the field levels as monitored by the field probe. The anechoic chamber, where the test is performed, is calibrated according to the criteria as per EN 61000-4-3 for 16 points given by geometrical arrangement in Figure 10 (2006, Bittera et al. 2006).

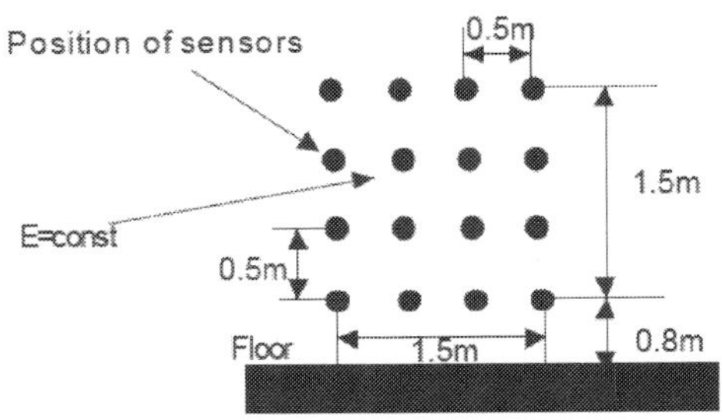

Fig. 10. Geometrical arrangement of measuring points

Then main task is setting the proper value of field strength at the place of tested device; or measuring of EM field. Because field strength is monitored by probe without tested device at test place result of measurement is influenced only by the presence of the probe, this process is not very easy and quality of evaluation depends just on the field probe.

a. EM field probe

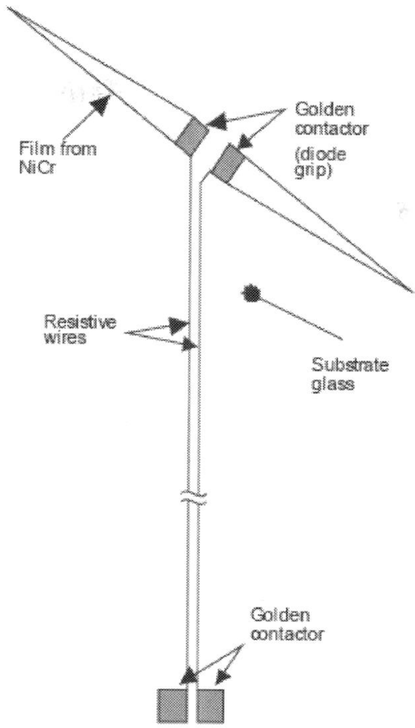

Fig. 11. Measuring resistive dipole(Bittera et al. 2006)

The task of field probe is to transform the electrical field into directly measurable quantity as e.g. DC voltage is. The heart of probe is sensor, which is often realized as a dipole with travelling wave. This sensor works in an ultra-wide frequency range, it has a sufficient sensitivity and its producing is not very difficult. Voltage induced on a dipole is directly proportional to strength of incident EM field. This voltage is rectified and transferred via resistive wires to electronic interpretation unit (EIU) that is located in a sufficient distance from sensor. Scheme of sensor can be seen in Figure 11, to the golden contact we connect EIU, it converts signal into digital form and transfers it to the controlling computer. In fact, probe consists of three independent sensors due to isotropic properties of the whole probe. In term of measurement the field around sensors is interested for us. This field can be affected by metal parts of probe as enclosure of EIU or sensors are.

b. Model of the probe

It is necessary to know how can be the EM field influenced by presence of the probe during the measurement. To solve this problem, we have to design such a model, which can represent real behavior of the probe and is also designed regarding to the used method. In the case EM field is excited by point source and waves propagate to surroundings and the main task is to calculate the strength of EM field, or current distribution on attendant structures in analyzed area eventually. Such problem leads integral equation that can be described formally

$$\iiint_V \nabla X dxdydz = Y \tag{33}$$

Where Y is source and X unknown function V analyzed area and ∇ is the Hamiltonian operator.
It means that using unknown function X can be solved, which in the case represents EM field distribution in V and Y is the feed of EM field source. Solution of the function leads to integral equation of first kind, which has not any analytical solution. Hence, we can transform it to integral equation of second kind or solve this equation numerically. However transformation to the integral equation of second kind is very difficult, complex and just approximate, so numerical solution, which is quite accurate, is preferred. One of the most popular methods is method of moments that is based on transformation of integral equation to system of linear equations - matrix equation and it can be solved easily using computer. This principle is used by much commercial software that solves EM problems, e.g. FEKO, NEC, etc.
Method of moment's principle is based on dividing analyzed structures to the small parts, called segments. If one has more segments it means that one needs more long time calculation, but has more accurate results. So it is necessary to strike a balance between number of segments and calculation accuracy. To get the model of probe all the metal parts (case of EIU and sensors) were replaced with segments with properties (dimensions, electrical properties) as real probe has (see Figure 12). We consider that probe is situated in infinitely large space without any caterers and it is incident by plane wave with linear polarization. The distance between sensors and case of EIU can be changed, in case of probe the distance is h = 30cm.

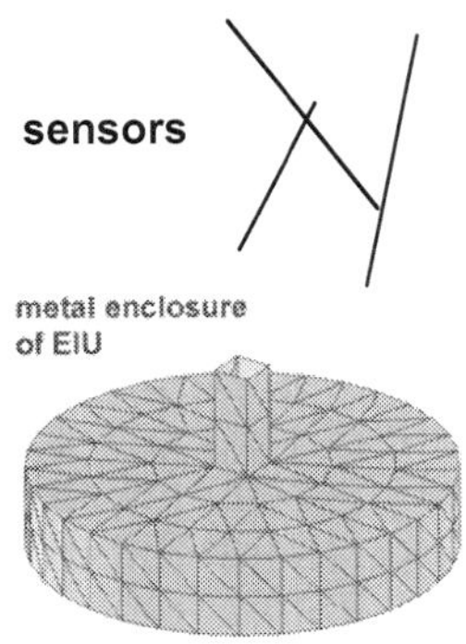

Fig. 12. Model of the probe in simulation(Bittera et al. 2006)

Spatial Electromagnetic Field Sensor

This is made up into a structure in which an electric-current sensor that detects the change in the spatial magnetic field due to the electric current passing through the overhead transmission/distribution line and that is comprised of a core and coil and a voltage sensor that detects by a plate electrode the line voltage capacitance-divided by the spatial charge between an electrically charged portion of the overhead transmission/distribution line and a plate electrode insulated from the earth, both, are accommodated in a hermetically sealed sensor container, such as Figure 13.

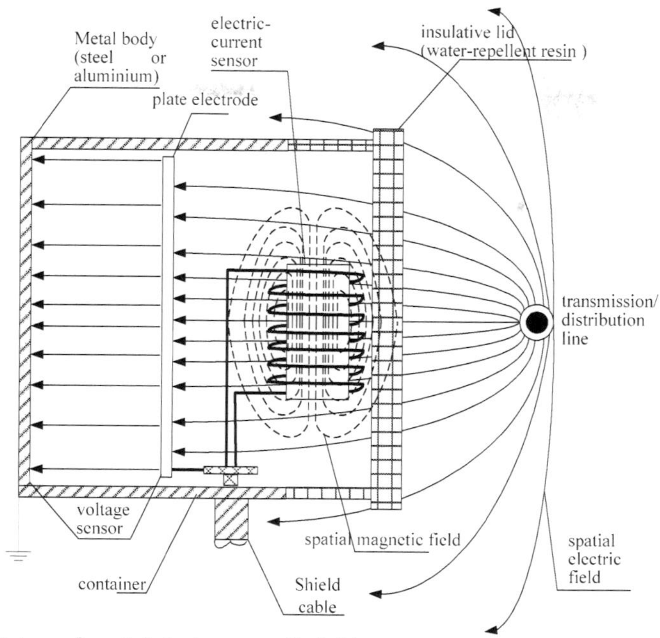

Fig. 13. Container of spatial electromagnetic field sensor

A noncontact sensor comprising a container including a container main body having an opening portion at one side and an isolative lid member that hermetically seals said opening portion of said container main body, and a voltage sensor disposed in said container in the way in which a lid member side is opposed to an electrically charged portion of an overhead wire line, being insulated from the earth, detecting a voltage that is induced in a plate electrode due to a spatial charge between said electrically charged portion and the plate electrode within said container, characterized in that said lid member has a structure in which a water-repellent layer constructed using a water-repellent resin is provided on the whole or a part of its at least exposed surface (Nakamura 2008).
The above-described "container main body" is a container in which a plate electrode and the like are accommodated and shielded so that they are not affected by a surplus magnetic field and a surplus electric field. The container main body like this can be made of metal such as stainless steel and aluminum. Also, the container may be a resin-made container on which an electrically coating material and the like are coated so as to add an effect of shielding an electric field, etc.

2.3.3 Sensors for faults generated temperature changing

Distributed fiber-optic temperature sensors (DTS) were first proposed in the early 1980s and have been commercially available since 1987. They are instruments which use an optical fiber, both as a temperature sensor and as a means of bringing the information back from the sensor to the terminal equipment.

Because distributed temperature sensors use optical fibers as the sensing and communications element, they can be used in electrically noisy environments without any problems of interference. The fact that a single fiber is able to replace many thousands of thermocouples simplifies wiring considerably and thus allows the technology to be used in applications where space, weight or wiring costs preclude traditional point sensors. An additional benefit of the technology is that the sensors can be made entirely from dielectric materials and in that sense are intrinsically safe(Hartog 1995).

The operation of distributed temperature sensors is based on the optical time domain reflectometry (OTDR) technique, in which a short pulse of light is launched into the fiber and the return signal is analyzed. The time from the launching of the pulse can be mapped directly into distance along the fiber---in a similar way to the measurement of target range in radars. The signal consists of light scattered by the glass in which the interrogating pulse is traveling. Some of the wavelength components (in particular the so-called Raman anti-Stokes band) in the return signal are temperature-sensitive and these are used to determine the temperature along the fiber. In order to separate the temperature variations from other factors which also influence the Raman backscatter signal, several referencing methods are used, including comparison with other wavelengths which are less sensitive to temperature and repeating the measurement, but launching into the opposite end of the fiber (the latter then being installed in a loop configuration). In this way, an accuracy of +0.3°C can be achieved.

The most advanced application for distributed temperature sensing in the power supply industry is in the monitoring of high voltage power cables. The capacity of a power transmission cable is limited by the increase in core conductor temperature caused by resistive heating as the current is increased. The maximum operating temperature is limited by degradation mechanisms in the insulation materials used. Traditionally, cables have been rated from detailed models which take into account the design of the cable, but also the ability of the soil surrounding the cable to dissipate the heat. Thus a cable running through a dry soil is less able to dissipate its self-generated heat than one laid in wet conditions.

The soil conditions are not necessarily known at all points along the cable at all times. Moreover, other factors affect the temperature along the cable, such as the proximity of other cables or services (e.g. district heating). The presence of these services may not always be known at the time when the cable is planned. Of course, it is the temperature of the hottest point along the cable which limits its capacity and the location of the hottest point is not always known. Energy cables are therefore usually operated below their maximum capacity, even at times of peak demand.

A cable with an optical fiber distributed sensor, either integrated within the cable, or laid in physical contact with the sheath can significantly reduce the uncertainty in determining the core temperature along the cable. This allows the cable to be operated with confidence at loads much closer to the maximum design value. The practical benefits are that knowledge of the temperature profile helps the utilities to meet peaks in demand with a smaller number of cables. Ultimately, the approach may allow cable sizes to be selected less conservatively, with cost savings well in excess of the cost of the monitoring equipment.

The temperature profile of cables can be measured by temperature sensor. The temperature distribution is extremely non-uniform; the peaks in temperature in this case are believed to be caused by the cable crossing other underground services. Once their location is known, only a few hot spots need be monitored by the operator, although it is possible for the relative magnitude of the hot spots to vary according to seasonal variations in the load of the cable and of neighboring services, and the distributed nature of the monitoring thus remains of value throughout the life of the cable.

3. Fault detection methods

3.1 Fault detection methods with fundamental frequency signals

3.1.1 Faults detection with harmonic current comparison

A typical ineffectively earthed distribution system is shown in Figure 14. And its zero sequence equivalent circuit with single-phase grounding fault is shown as Figure 15. The capacitance to earth of the Feeder 1 is C_1, and that of the Feeder n is C_n. The Feeder j is the faulted feeder. Peterson-coil inductance is L_n and zero sequence voltage of the faulted point is $\dot{U}_{0f}$ (Zeng et al. 2007).

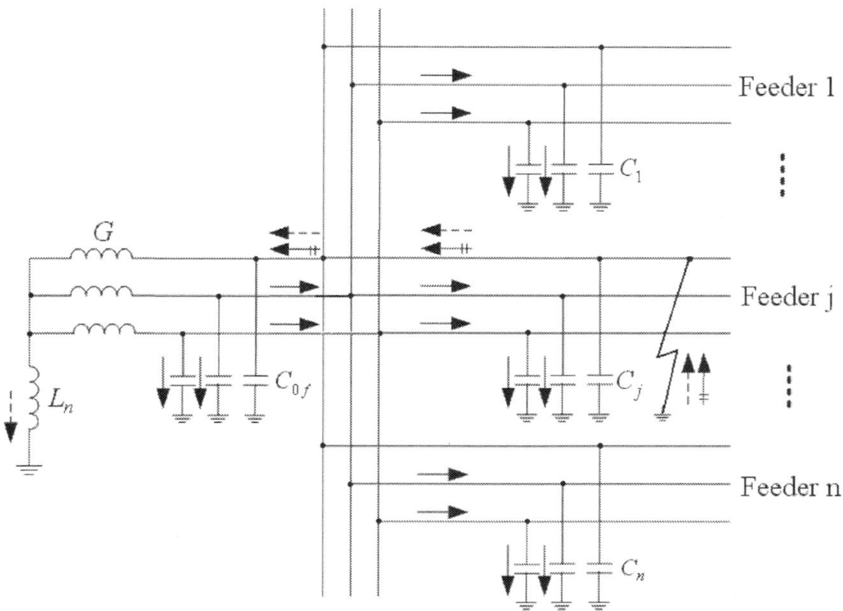

Fig. 14. Ineffectively earthed distribution system with single-phase grounding fault

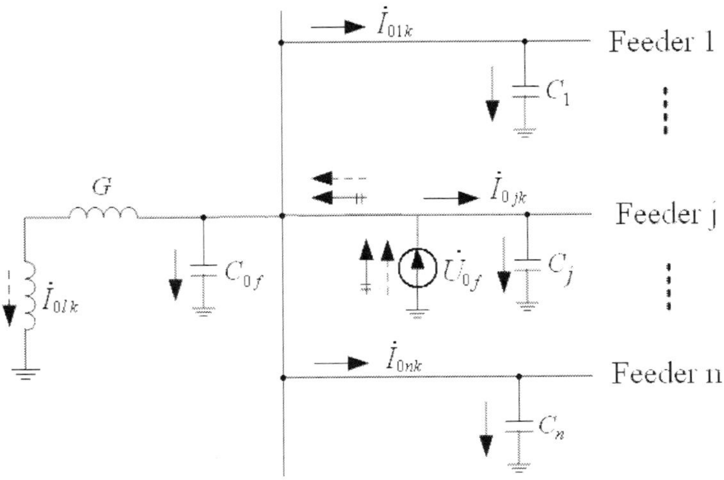

Fig. 15. Zero sequence equivalent circuit of ineffectively earthed distribution system with single-phase grounding fault

Grounding faults generate transient signals. They include many different frequency harmonic components. The components are produced in the faulted point, flow to the source and normal feeders. From Figure 15,

$$\dot{I}_{0lk} + \dot{I}_{ojk} + \sum_{i=1,i\neq j}^{n} \dot{I}_{oik} = 0 \tag{34}$$

Namely

$$\dot{I}_{ojk} = -(\dot{I}_{0lk} + \sum_{i=1,i\neq j}^{n} \dot{I}_{oik}) \tag{35}$$

Where, $\dot{I}_{olk}$ is the kth harmonic current of the zero sequence component in the Peterson-coil, $\dot{I}_{oik}$ is the kth harmonic current of the zero sequence component in the Feeder i, $\dot{I}_{ojk}$ is the kth harmonic current of the zero sequence component in the faulted feeder (i =1,2,…,n; j=1,2,…,n).

The inductive reactance will increase and the capacitive reactance will decrease along with the increasing of the harmonic order. Considering the zero sequence current, when the power system capacitive current flowing in the normal feeder is much larger than the inductive current flowing in the Petersen-coil, as $\sum_{i,i\neq j}^{n} \dot{I}_{0iM} >> \dot{I}_{0lM}$, the Mth harmonic inductive current compensated by the Peterson-coil can thus be ignored. Namely the current of the Mth harmonic component in the faulted feeder is almost the sum of the current in sound feeders, and the Mth harmonic component direction in the faulted feeder is just opposite to that in the sound feeders. The faulted feeder can thus be detected by utilizing

the Mth harmonic component. The feeder with M^{th} harmonic current component larger than other feeders and direction opposite to other feeders is the faulted feeder.
The M is often selected as 5, and the fifth harmonic component is often used to detect the grounding feeder in China. In practice, power transformer also generators 5th harmonic. It sometimes disturbs the 5th harmonic generated by grounding faults and has influenced on the fault detection. Moreover, the 5th harmonic current is relation to fault arc and resistance. Different type faults have different 5th harmonic, and have different fault detection characteristics. So the 5th harmonic current based grounding fault detection has some limits.

3.1.2 Faults detection with signal injection

A signal current is injected to the grounding fault phase of faulted feeder through bus voltage transformer. The signal frequency is between n and $n+1$ times the power fundamental frequency 50Hz:

$$n \cdot 50 < f_0 < (n+1) \cdot 50 \tag{36}$$

The signal frequency is set as 225Hz shown as Figure 16 (Zeng et al. 2007).

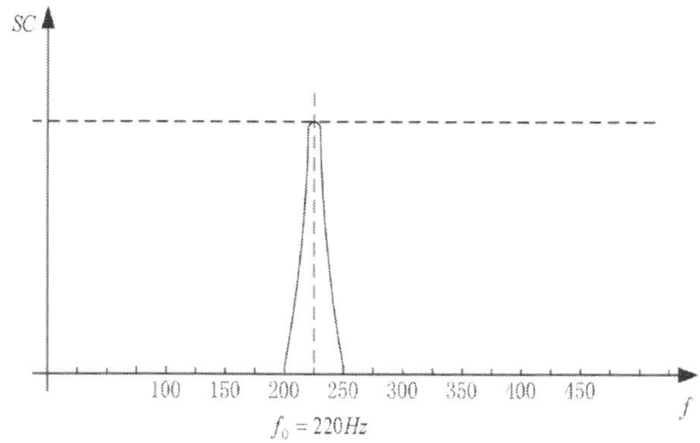

Fig. 16. Frequency character for current

A special designed detector is installed in every feeder to detect the injected signal. In normal conditions, the signal magnitude in every feeder is in direct proportion to its capacitance to earth. In grounding fault conditions, the signal is flowing to the fault point from the signal source, and the signal magnitude in every normal feeder is almost zero. The faulted feeder can thus be detected by measuring injected signal magnitude.

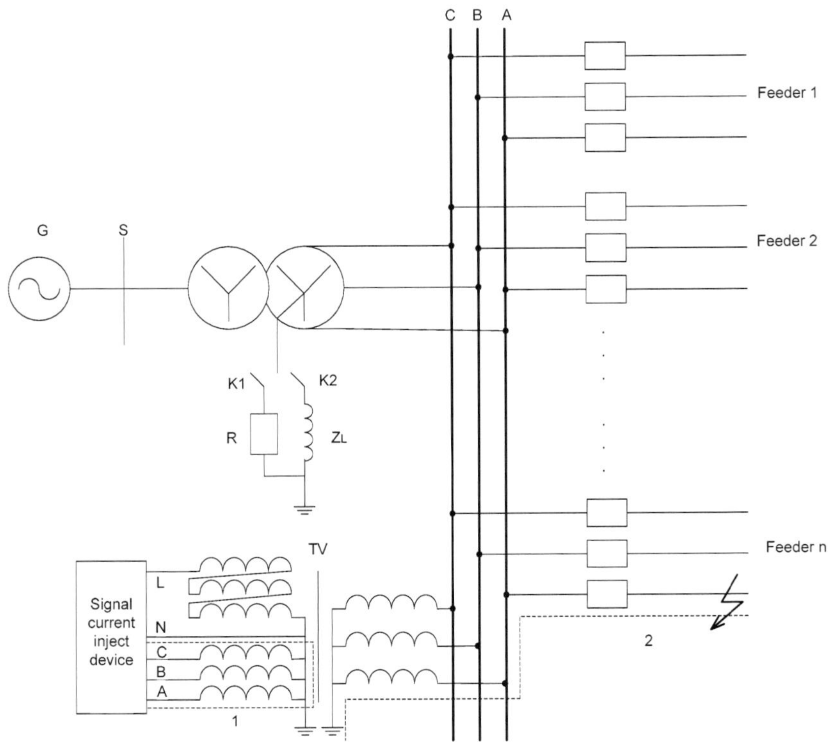

Fig. 17. Signal injection circuit

The fault detection scheme can be described as below. When the distribution system is in normal condition, no tested signal is injected to the system, and the signal detector tested signal magnitude is zero. When grounding fault occurs, the fault phase voltage A will decrease to zero, for example phase A fault, whereas the voltage of the normal phase B and phase C will rise up to the line voltage, namely $\sqrt{3}$ times the phase voltage. The signal injector device then inject signal from the phase A secondary of transformer to the power system. And the signal current flows to the fault point (shown as Figure 17 and 18); the signal in every normal feeder is almost zero. The faulted branch can also be detected by tracking the injection signal.

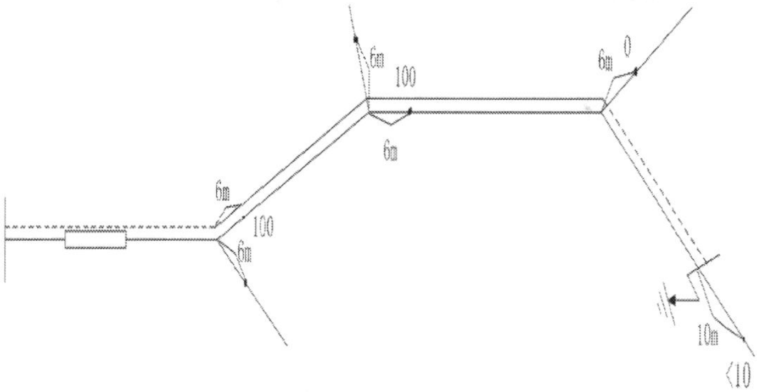

Fig. 18. Injection signal flowing from source to fault point

Feeder No.	Grounding Fault Times	Correct Fault Detection Times	Time Saving for Fault Isolation (h)
1	4	4	2
2	3	3	1.5
3	5	5	2.5
4	6	6	3
5	7	7	2
6	9	9	2
7	1	1	2
8	3	3	1.3
10	2	2	2
11	7	7	2
12	5	5	2
13	1	1	1.25

Table 1. Fault detection results for Sanjiang Substation

The signal injection based fault detection method has been applied in more than 2000 distribution systems. An operation statistic results in the Sanjiang Substation (Zengchen Power Company) is shown as Table 1. All faulted feeder can be detected correctly, and faults isolation time can be saved. The signal injection based detection method has some advantages:

a. The injected signal is small, and it only flows into the grounding fault point. It has no influence for power system normal operation.
b. The special signal current detector is installed near each feeder to detect the injected signal, it is easy to install.

c. When grounding faults occurs, the faulted point can also be determined if the operator holds the special signal current detector moveing along the faulted feeder to find the point which the injected signal current disappears.

Whereas, arcing faults is difficult to detect by the method because of the harmonics interference caused by earth faults. And high impedance grounding fault is also difficult to detect, because the signal is too small to detect in the faulted feeder.

3.1.3 Faults detection with phase current difference

In three-phase-symmetry systems, grounding faults caused changing currents have been analyzed in section 2.1. The difference between three phase changing currents in sound feeder is nearly zero. The difference between the healthy and fault phase changing currents in fault feeder is very large, equaling to the residual current. And the difference between healthy phase changing currents in the fault feeder is also relatively small, almost equaling to zero. If the difference is larger than a threshold, grounding fault can be detected in the tested feeder.

The technique of sampled value difference protection has been widely applied in compute based protections due to its inherent simplicity and good performance, which uses instantaneous sampled values to calculate the difference of currents, and checks whether samples satisfy the operating criterion or not. The number of satisfied samples in a system frequency cycle is larger than a threshold, fault can be detected (Wang et al. 2007, Wang et al. 2006).

The difference protection is implemented with different phase sampling values comparing. The operation criterion is:

$$i_d(k) \geq K_1 i_r(k) + i_{d0} \tag{37}$$

Where K_1 is restraint coefficient, k is sampling number, i_{d0} is pickup current, $i_r(k)$ is instantaneous current, and $i_d(k)$ is instantaneous difference current which can be calculated by:

$$i_d(k) = \max(|\Delta i_A(k) - \Delta i_B(k)|, |\Delta i_B(k) - \Delta i_C(k)|, |\Delta i_C(k) - \Delta i_A(k)|) \tag{38}$$

Under normal conditions, the difference between healthy phase changing currents would be equal to zero. Considering the influence of measuring error, it would be a small value. In order to limit the influence, i_{d0} can be obtained by:

$$i_{d0} = |\Delta i_B(k) - \Delta i_C(k)| \tag{39}$$

In order to improve the protection precision, the sampling rate of the protection is set as 32 samples per cycle. Taking the half cycle for example, set R=16. During a period after the fault, instantaneous difference current ($i_d(k)$) is calculated by (38). When $i_d(k)$ is satisfied operation criterion (37), grounding fault can be detected in the protected feeder. Namely, if there are S samples satisfying (37) during the continuous R sample's judgments in half system frequency cycle after fault inception, the protection would operate. As long as the number of the (37) satisfied sampling number is less than S, the protection would not mal-operate.

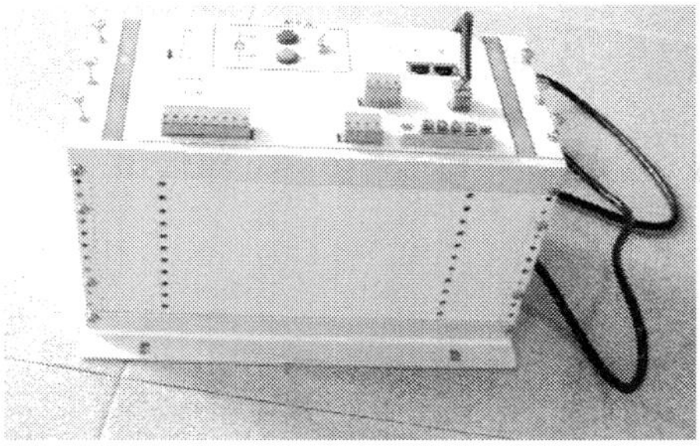

Fig. 19. Prototype of FTU

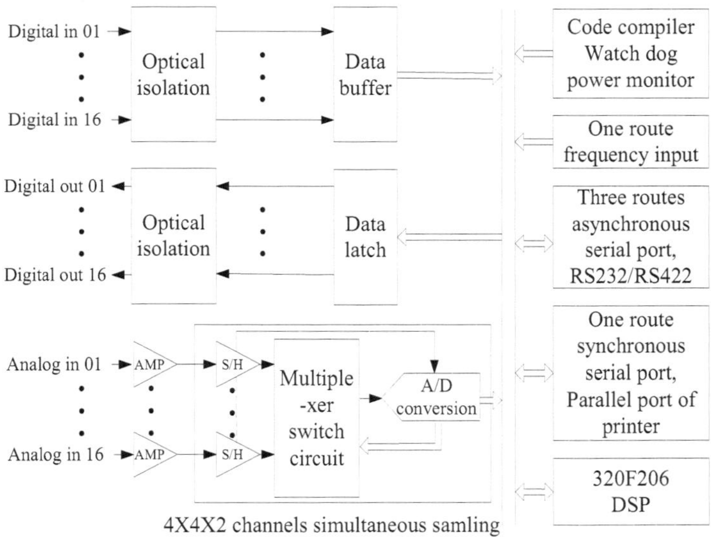

Fig. 20. Principle of FTU hardware frame

The protection scheme only utilizes the measured voltages and currents in the protected unit, and does not need the parameters of the distribution system or other unit. It can be implemented on feeder terminal units (FTU) with three-phase voltages and three-phase currents sampling in the distribution automation systems. The FTU prototype (shown as in Figure 19) has been developed. It uses with DSP TMS320F206, and its hardware frame is showed as Figure 20 (Zeng et al. 2008).

The phase current difference based detection method has been tested in laboratory. High impedance faults and arcing faults can be detected with high precision. Not enough phase current difference based fault detectors have applied in power system.

3.2 Fault detection methods with transient signals

When a fault occurs on a transmission line, the voltage at the point of fault suddenly reduces to a low value. This sudden change causes a high frequency transient that propagates away from the fault point in both directions along the line at a velocity almost equal to the speed of light. The high frequency transient is called traveling waves which can be used for fault point detection. When the initial traveling wave reaches an impedance discontinuity, such as a substation, a part of the wave is reflected back along the line towards the fault. The remainder is transmitted through the substation into the network. These traveling waves continue to be divided into reflected and transmitted waves and their amplitude attenuates with time until finally a new equilibrium is reached.

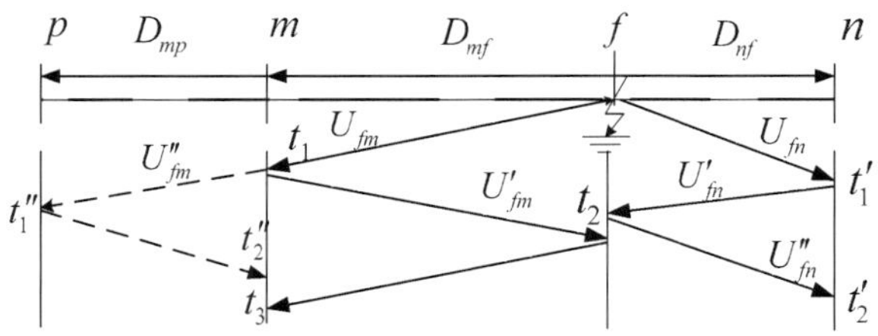

Fig. 21. Transport network of traveling wave in transmission line

Supposing that the fault occurs at *f*, the transport network is shown in Figure 21 (Zeng. 2000). The initial traveling wave propagates away from the fault in both directions along the line, and produces refraction and reflection at point of *m*, *n* and *p*. U_{fm}, U_{fn} are the initial traveling wave from fault point *f*; U'_{fm}, U'_{fn} are reflection traveling wave at point *m*, *n*; U''_{fm} is refraction traveling wave at point *m* (the refraction traveling waves from point p and n are not signed in Figure 21).

Taking point m as an example, reflection coefficient α_m and refraction coefficient β_m can be expressed as follows:

$$\alpha_m = \frac{Z_1 - Z_m}{Z_1 + Z_m} \tag{40}$$

$$\beta_m = \frac{2Z_1}{Z_1 + Z_m} \tag{41}$$

Where, Z_1 is equivalent wave impedance of non-fault line which is directed connected with bus *m*, Z_m is equivalent wave impedance of fault line, generally valued for 300Ω.

3.3 Fault detection methods based on information fusion or AI technology

Different fault detection methods usually operate using different fault information. They have different applications with different degrees of precision and robustness. In practical operation, a single ground fault protection scheme has poor precision in ground fault detection. If all kinds of fault information can be integrated and fused, the effect of the disturbing signal can be reduced, inherent limitation of the single protection scheme can be eliminated and the precision and robustness of fault detection can be improved .

Information fusion or AI technology is an intelligent information processing method, and it is the process of combining data to refine state estimates and predications. The purpose of information fusion is to produce information from different sources in order to support the decision-making process (Bossé et al. 2006).

3.3.1 Fault detection methods based on Neural Networks

Neural networks (NN), which are parallel distributed information processing units with different connection structures and processing mechanism, are particularly suitable to link the different variables of a power system where the relationship between the independent and the dependent variables are not easily quantifiable. Neural networks are robust to input and system noises, have learning capabilities, and can perform in real time (Lippmann 1987). Indeed, neural networks can be viewed as nonlinear adaptive system identification units, which rely on pattern recognition for the identification procedure (Widrow and Winter 1988).

The NN structures used in the fault detection are of the multilayer perceptron type, trained using the backpropagation algorithm. The basic philosophy of neural networks learning procedure is presented in Figure 22, where the weights are updated based on the error generated through the comparison of neural output and the target output. The typical single layer net is shown in Figure 23. The input vectors are selected variables derived from the transducer signals.

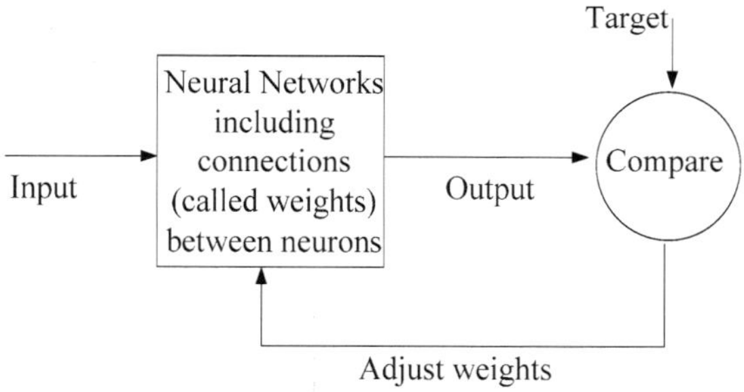

Fig. 22. Neural Networks philosophy(Awais 2003)

Training a network by backpropagation involves three stages:

a. The feedforward of the input training pattern
b. The back propagation of the associated error
c. The adjustment of the weights

During the learning process, the NN weights are adapted in order to create the desired output vectors. For learning process, the symptom-fault map is required. There is also the possibility of a hybrid learning process, simulated data for healthy and faulty machines. In this way, the NN can learn the health condition (Awais 2003).

The time domain signals both in healthy and faulty machines are passed through the low pass filter. Then the time domain signal is changed to frequency domain by fast Fourier transform (FFT). The magnitude from FFT is used as input vector for training the NN which defines the target as healthy or faulty. After training NN, we have the weight and bias for using to detect the faulty machine.

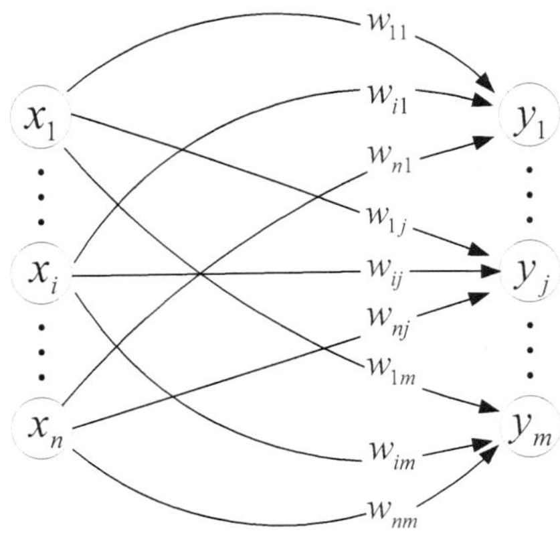

Fig. 23. A single-layer neural net

The dynamics of the system with out faults is represented by (Sreedhar et al. 1995):

$$\dot{x} = Ax + Bu + \Psi(x,u) \tag{42}$$

The dynamics of the system due to the occurrence of the fault is represented by

$$\dot{x} = Ax + Bu + \Psi(x,u) + \Phi(x,u) \tag{43}$$

Considering a sigmoid neural network whose input-output characteristics are described by:

$$y = \hat{\psi}(q,\dot{q},\tau;\hat{\theta}) \tag{44}$$

Where $(q,\dot{q},\tau) \in \Re^n \times \Re^n \times \Re^n$ is the input to the network, $y \in \Re^n$ is the output of the network and $\hat{\theta} \in \Re^p$ represents the adjustable weights of the network. Let the weights $\hat{\theta}(0) \in \hat{\theta}_0$ of the selected neural network be initialized such that

$$\hat{\psi}(q,\dot{q},\tau;\hat{\theta}_0) = 0, \quad \forall q,\dot{q},\tau \tag{45}$$

corresponding to the no-failure situation. Note that this can be achieved simply by initializing the output weights of the network to zero(Vemuri et al. 1998).

Artificial neural networks have been widely used partly because of their multi-input parallel processing capabilities, which are most suitable for real time applications. A large number of input variables can be simultaneously fed to a multi-input neural network. Despite the increase in the number of input nodes, the computation time of the network remains the same because neural nets perform parallel processing. Thus, increasing the number of input nodes does not affect the neural network processing speed. Besides, increasing the number of inputs nodes increases the robustness of the neural networks with respect to measurement noise. Moreover, once designed, the internal structure of artificial neural networks can be easily changed, if modifications or additions need to be made. The network can be updated, without much difficulty, by merely retraining it a few times more after the network structure is modified (Chow and Yee 1991).

3.3.2 Fault detection methods based on Fuzzy Algorithm

The output result of the single ground fault protective relay is either fault or no fault. If 1 is applied to represent fault, and 0 is applied to represent no fault, the output value of the ground fault protection can be chosen from the data collection {0, 1}. The principle of fuzzy ground fault protection is described as following. The subjection degree of element to data collection is extended to choose any value from data section [0, 1], comparing to choosing from data collection {0, 1} formerly. Output uncertainty of different ground fault protective relays can be expressed by the subjection function. And the faulted feeder can be identified by clarifying the fuzzy results.

The subjection degree function μ_i of an arbitrary ground fault protective relay i can be selected according to practical instance, such as normal function, trigonometric function, trapezoid function, S function and break-line function etc. As shown in Figure 24, the break-line subjection degree function is selected, where S is the tripping threshold of the general ground fault protective relay, S_L is the possible lower limit of tripping threshold in the case of fault, S_H is the possible upper limit of tripping threshold in the case of non-fault. Because the result of a single fault protection scheme isn't reliable in the section [S_L, S_H], this section is defined as fuzzy section. The possibility of fault in this section can be described by fuzzy subjection degree.

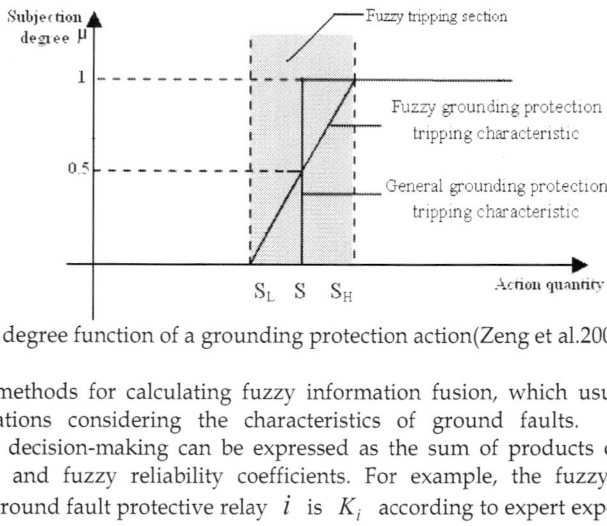

Fig. 24. Subjection degree function of a grounding protection action(Zeng et al.2004)

There are many methods for calculating fuzzy information fusion, which usually adopt additional calculations considering the characteristics of ground faults. The gross reliability of fault decision-making can be expressed as the sum of products of different subjection degree and fuzzy reliability coefficients. For example, the fuzzy reliability coefficient of the ground fault protective relay i is K_i according to expert experience. It's value domain is from zero to one. By addition calculation, the information fusion output of total ground fault protective relays is

$$Y = \sum_{i=1}^{n} K_i \mu_i \tag{46}$$

The clarifying process of fuzzy reliability degree is converting fuzzy reliablity degree into practical output judgement (fault or no fault) according to fuzzy principle. For example, when $Y \geq 0.5$, there is fault happening in this distribution line; when $Y < 0.5$, there is no fault happening in this distribution line.

The scheme for information fusion is shown as Figure 25.

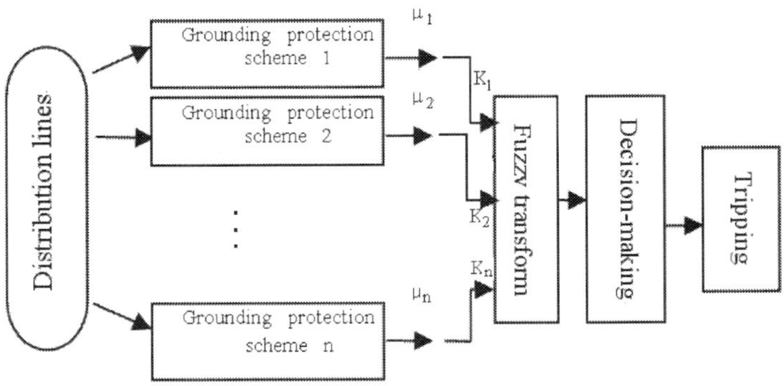

Fig. 25. Schemes of fuzzy grounding fault detecting(Zeng et al.2004)

3.3.3 Fault detection methods based on Genetic Algorithm

Genetic algorithms (GA) are probabilistic search techniques inspired by the "survival of the fittest" principle of the neo-Darwinian theory of natural evolution, and by the mechanisms of natural genetics. This algorithm looks for the fittest individual from a set of candidate solutions called population. Basic operations are selection, reproduction, crossover and mutation. Parent selection gives a higher probability of reproduction to the fittest individuals. During crossover some reproduced individuals cross and exchange their genetic characteristics. Mutations may occur in a small percentage and cause a random variation in the genetic material, thus contributing to introduce variety in the population. Fitness function of each individual is changed by crossover and mutation operators. The selection operator decides whether an individual survives in next generation or not. Roulette wheel and tournament selection are two common selection methods. The steps of genetic algorithm are depicted in Figure 26 (Amaral et al. 2007, Aydin et al. 2008).

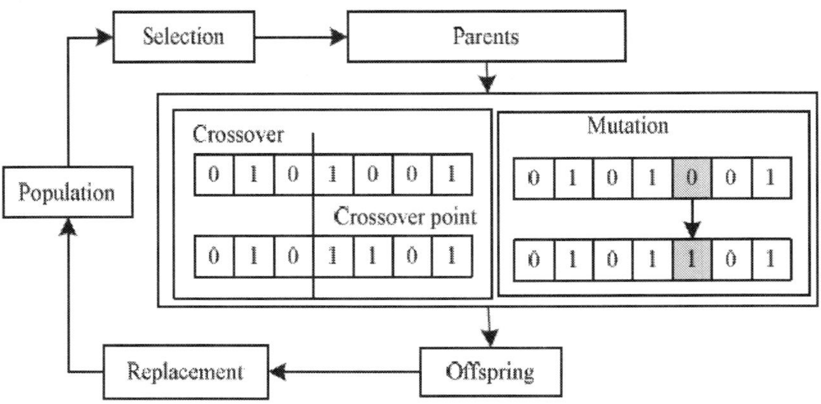

Fig. 26. Steps of Genetic algorithm

In every generation, a new set of artificial creatures P(t) is randomly generated using bits and pieces of the fittest members of the previous generation P(t-1). The process is repeated until the last generation P(t) does not satisfy the application requirements. A simplified sketch of the algorithm is shown in Figure 27 (Betta et al. 1996, Betta et al. 1998).

```
Procedure GA
  begin
  t:=0,
  initialize P(t);
  evaluate structures in P(t);
  while termination condition not satified do
      t:=t+1;
      select P(t) from P(t-1);
      recombine structures in P(t) using mutation and
      crossover;
      evaluate structures in P(t)
  end
  end.
```

Fig. 27. Sketch of the Simple Genetic Algorithm

For the robust ℓ_1 optimization problem the chromosome is constructed by formulating matrices A_e, W, P, H and N into a single vector Θ such that

$$\Theta = [V_{ec}(A_e)^T V_{ec}(W)^T V_{ec}(P)^T V_{ec}(H)^T V_{ec}(N)^T] \tag{47}$$

The search region is then defined by establishing upper and lower limits $\underline{\Theta}$ and $\overline{\Theta}$ such that

$$\underline{\theta}_{ij} \le \theta_{ij} \le \overline{\theta}_{ij} \tag{48}$$

If the stability criterion is not satisfied, a penalty is added to the cost such that

$$if \begin{cases} \max[\lambda_i(A_a)] < 1, & J = J \\ otherwise, & J = J + penalty' \end{cases} \tag{49}$$

where $\lambda_i, i \in (1, 2, \ldots, m + 2n)$ are the eigenvalues of the augmented system Aa (Curry et al. 2001, Curry and Collins Jr 2005).

4. Application of fault detection methods

4.1 Fault detection application in power system online monitoring

Many mining power systems utilize ineffectively grounded sources to restrict the residual current of single-phase earth fault in order to reduce outage and shock hazard. In practice, with the system expansion, topology changing and insulation aging, the potential residual current and zero sequence voltage for earth fault vary dynamically and some arcing earth faults can easily cause over-voltage and induce multiple faults. In order to improve the system safety, on-site condition monitoring, safety evaluation for earth fault are proposed in this section.

System states can be classified into normal secure state, alert state, incipient fault state and fault state. In alert state and incipient fault state, some preventive actions need to be carried out. But in fault state, the faulty feeder section needs to be isolated.

Normal secure state

No fault exists in industry power system. Even if earth fault occurs, no high overvoltage and no large residual current will be caused, and arcing fault will be self extinguished.

Alert state

No fault exists in industry power system. But the neutral point-earth-impedance does not match to the system capacitance to earth. If earth fault occurs, the residual current or zero sequence voltage will be very high, so that some arcing fault will easily induce multiple faults. In order to improve it, neutral-point-earth-impedance adjustment (e.g. Petersen-coil tuning) is required.

Incipient fault state

Some very high impedance earth fault exists because of insulation aging, and so on. Although it does not cause overvoltage and affect system normal operation directly, it can create personnel and equipment safety problems, and has an adverse environmental impact. So some preventive actions need to be carried out.

Fault state

Earth fault occurs. It causes overvoltage and affects system normal operation directly. The faulty feeder section is required to be isolated as soon as possible.

Based on the above indexes, insulation intensity (dissipation factor), potential over-voltage and potential fault residual current, and the safety for earth fault in ineffectively earthed systems can be evaluated. When the safety indexes are over their threshold, an alarm will be given to the operator to change the system operation methods or carryout preventive maintenance. As shown in Figure 28, many parameters are measured online in normal mining power systems. When zero sequence voltage (U_0) is larger than its high threshold (U_{0set1}) which is often set as $30\% U_{ph}$, earth fault occurs. The fault feeder section can be detected by over-current protection with the injection current.

If a feeder's signal current (e.g. feeder *k*) is bigger than its threshold (I_{skset}),

$$I_{sk} > I_{skset} \tag{50}$$

this feeder is detected as the earth fault feeder.

The fault can be isolated automatically with distribution automation equipment. It works as one of remedial actions.

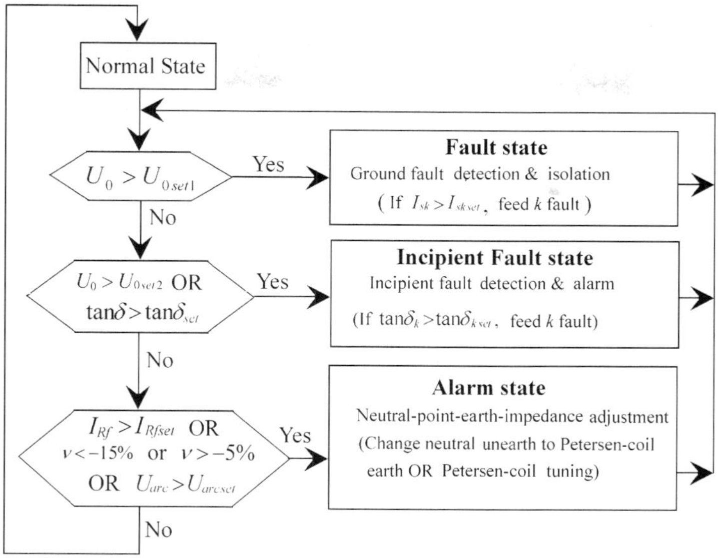

Fig. 28. On-site safety evaluation and security enhancement scheme (Zeng et al.2003)

If no fault existing, incipient fault will be tested. When zero sequence voltage (U_0) is larger than its low threshold (U_{0set2}) which is often set as 10% U_{ph}, or the total dissipation factor is larger than its threshold ($\tan\delta_{set}$) which is often set as 8%, incipient fault is detected. The feeder, whose dissipation factor is bigger than its threshold, is the high impedance earth fault feeder. An alarm is sent to the operator to maintain this incipient fault feeder.

If no fault and incipient fault existing, potential hazards will be evaluated. Supposed that some kind of earth fault occurs, it is assessed whether or not the potential residual current and overvoltage are out of their limits. In addition, resonance deviation in resonance earthed system is measured on-line. If some potential hazards for earth fault exist or resonance deviation is out of its limit, an alarm will be given to the operator to adjust neutral-point-earth-impedance. Some preventive action is usually carried out, such as: changing neutral unearthed or high resistance earthed to Petersen-coil earthed, Petersen-coil tuning, etc (Srivani and Vittal 2008).

4.2 Fault detection application in Power system condition-based maintenance

Condition-based maintenance (CBM) aims to detect latent failures on a basis of processing huge amount of information and then take actions to remove them (Liu et al. 2006). The main idea of CBM is to monitor the health of critical machine components and system almost continuously during operation and maintenance actions based on the assessed condition. If done correctly, CBM has the benefits such as reducing catastrophic failures, minimizing maintenance and logistical cost, maximizing system security and availability and improving reliability.

A CBM system usually has four major functional modules, namely feature extraction, diagnostics, prognostics and decision support. Figure 29 illustrates the relationship between these modules (Zhang et al. 2007).

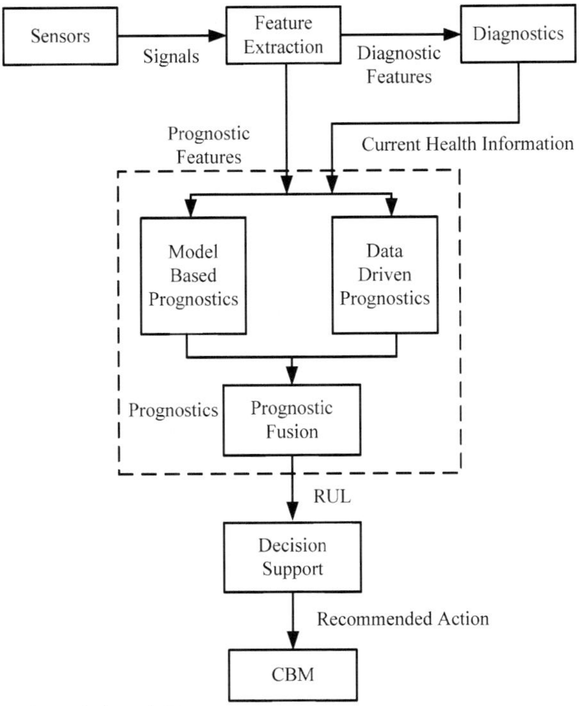

Fig. 29. Functional modules of CBM system

Fault prognostics is the process to project the current health state of equipment into the future taking into account estimates of future usage profiles, thus estimates the remaining useful life (RUL) of machine. The existing prognostic methods can generally be classified as two major categories, namely model based and data driven approaches. Compared to model based approaches, data driven ones do not need the prior knowledge (physical model or domain expert experiences) about system fault condition, thus makes it an effective approach in practical applications. The data-driven approaches are based upon statistical and learning techniques from the theory of pattern recognition.

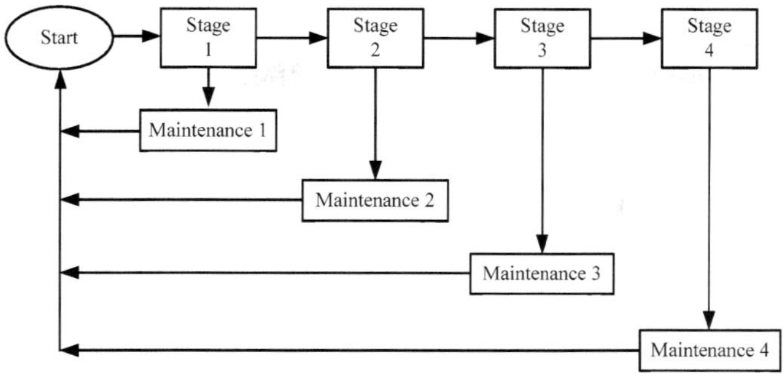

Fig. 30. condition based diagnosis and maintenance methodology

A procedure for diagnosis and condition based maintenance for power transformers is presented in (Setayeshmehr et al. 2004). The condition based diagnosis process uses some of the current diagnosis methods (e.g. Oil test, DGA, PD, etc.) to determine the condition of the transformer. The condition based diagnosis and maintenance methodology is shown in Figure 30 (Setayeshmehr et al. 2004).
Reference (Zhang, Li, Yu and Gao 2007) presents a fault prognostic algorithm based on a generic wavelet neural networks (WNN) architecture whose training process is based on genetic algorithm. The paper used MIMO WNN to conduct modeling task. The structure of wavelet neural networks is shown in Figure 31.

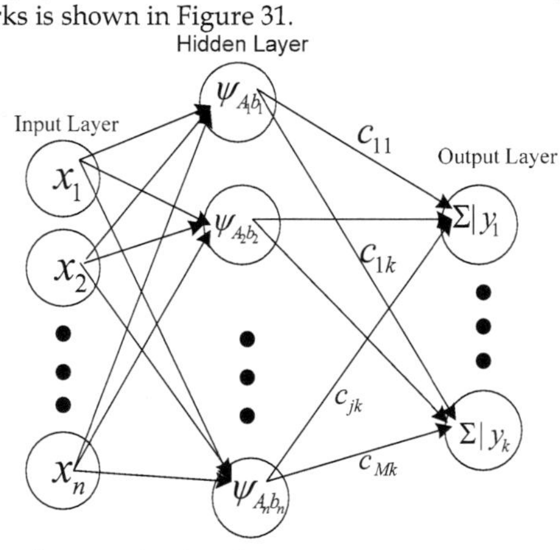

Fig. 31. Structure of wavelet neural networks

A MIMO WNN can be written as:

$$Y = [\Psi_{A_1 b_1}(X)\Psi_{A_2 b_2}(X)\cdots\Psi_{A_M b_M}(X)]C \tag{51}$$

The proposed WNN is employed to learn the input-output relationship using the GA. The training process of WNN includes following 8 procedures:

a. Representation

b. Fitness definition
c. Population initialization
d. Fitness evaluation
e. Selection
f. Crossover
g. Mutation
h. Stopping criteria

4.3 Fault detection application in transformer protection

Power transformers are important devices in an electrical energy system for supplying electricity. In order to obtain a high reliability level from an operation of the power transformer, a precise protection scheme is required. Generally, the transformers can be protected by overcurrent relays, pressure relays and differential relays depending on purposes. For differential protection, the differential current, which is generated by a comparison between the primary current and the secondary current detected via current transformers, is required. The differential protection is aimed at detecting internal faults in transformer windings. In a normal operation or in a fault condition due to the external short circuits, the differential current is relatively small, and the differential relay should not function. However, there are some factors that can cause a needless operation of the differential protection. To avoid the malfunction, the discrimination between internal faults, magnetizing inrush current and external short circuit current is required (Ngaopitakkul et al. 2005).

4.3.1 The method using short time Fourier transforms

A novel approach using the short-time Fourier transform (time frequency analysis tools) for fault detection during impulse testing of power transformers is described in (Al-Ammar et al. 2008). The neutral and/or capacitive transferred currents which are recorded during an impulse test can be directly analyzed with this approach. The primary objective of time-frequency analysis is to be able to define a function that will describe the energy density of a signal simultaneously in time and frequency, and is commonly used in applications to speech, sonar and acoustic signals. Among the few tools, STFT happens to be very common and popular, because the concept behind it is simple yet powerful. The basic idea of STFT is to slice up the signal into suitable overlapping time segments (using windowing methods) and then to Fourier analyse each slice to ascertain the frequencies contained in it. The accumulation of such spectra indicates how the spectrum is varying in time and is called the spectrogram. It is assumed that frequency information is associated with the time index in the middle of each slice of windowed data. STFT of a continuous-time signal $x(t)$ is defined as:

$$STFT(t,\omega)=\int_{-\infty}^{\infty}x(\tau)w(\tau-t)e^{-j\omega\tau}d\tau \qquad (52)$$

where $w(t)$ is the window function and $x(t)$ is the signal to be transformed. $STFT(t,\omega)$ is a complex function representing the phase and magnitude of the signal over time and frequency. The window $w(t)$ is slid along the time domain to establish $STFT(t,\omega)$. In order to estimate the "local" frequency contents at time , the signal $x(t)$ has to be isolated in the vicinity of time t, and then perform a FFT analysis.

4.3.2 The method using wavelet transforms

Signal processing using wavelet theory has emerged as a powerful tool over the past ten years and has led to significant developments in data analysis, data compression, image and speech processing, multi-resolution analysis etc. The WT like the Fourier transform decomposes a given signal into its frequency components, but differs in providing a non-uniform division of the frequency domain. In addition, unlike the Fourier transform which gives a global representation of the signal, WT provides a local representation in both time and frequency. These results from the fact that the analyzing basis functions in the case of the Fourier transform (namely sines and cosines) extend over infinite time, whereas they are compactly supported functions in the case of WT, thus giving them the localization property. This property greatly facilitates analysis of non-stationary signals, transient detection etc. A mathematical definition of WT follows (Satish 1998):

Let $x(t)$ denote a continuous-time finite energy signal, then WT of $x(t)$ is defined as:

$$WT(a,b) = \int_{-\infty}^{\infty} x(t) g_{(a,b)}(t) dt \tag{53}$$

where

$$g_{(a,b)}(t) = |a|^{(-1/2)} g((t-b)/a) \tag{54}$$

is called the base function or mother wavelet. a, b (real, $a \neq 0$) are the dilation and translation parameters, respectively. A restriction on the choice of $g(t)$ is that it must have a zero average value and be of short duration, which, mathematically, is called the admissibility condition on $g(t)$. Daubechies' wavelet, Morlet wavelet and Harr wavelet are some examples of popularly used functions for $g(t)$.

4.4 Fault detection application in generator protection of Multi-Generator-System

So far the most common generator fault style is stator winding single-phase ground fault, which brings about inter-phase failures and inter-turn faults easily. If the stator ground fault cannot be detected and isolated duly, this fault brings about overheating in the fault point and burns down the stator stick and iron core. So the stator ground fault has very bad influences on the generator normal operating, and we need highly accurate and reliable protection of stator single-phase ground fault to ensure the generator operates safely and reliably. Small or medium power generation systems are Multi-Generator-System (MGS) with more than one generator directly connecting to a bus and operating in parallel commonly. If stator ground fault happened in one of the generators operating in parallel, the generator which have happened single-phase ground fault should be detected and isolated by generator protective relay.

4.4.1 The method based on the grounding leakage current

The scheme utilizing the grounding leakage current of the fundamental component and the third harmonic fault components can detect HIGF. And the faulted generator can be detected and located by comparing the difference of leakage current of the fundamental component and third harmonic fault components between generator neutral and terminal side of each phase.

The features of grounding leakage current can be summarized as follow:

a. In the faulted generator, the differences of the leakage current variations between the sound and faulted phases are almost equal to the residual current. On the other hand, the differences of the leakage current variations between sound phases are is almost equal to zero.
b. In the healthy generator, the differences of the leakage current variations between any two phases are also zero under ideal condition.

Hence, protection against generator stator ground fault can be carried out through comparing the differences of leaking current variations. A difference larger than a threshold would indicate the presence of a ground fault.

The operation criterion is:

$$I_d \geq KI_r + I_{d0} \tag{55}$$

Where, K is the restraint coefficient, I_{d0} is the pickup current, I_r represents restraining current, I_d represents the maximum of the difference of the grounding leakage current between phases and it can be calculated:

$$I_d = \max(|\Delta I_{AB}|, |\Delta I_{BC}|, |\Delta I_{CA}|) \tag{56}$$

Under normal conditions, the differences of the grounding leakage current between healthy phases would equal to zero. Considering the influences of the measuring error, the differences of the grounding leakage current between healthy phases would be a small value. The setting value of I_{d0} can be set to the minimum of the differential current between phases which can stands aside, namely $|\Delta I_{BC}|$. If single-phase earthed fault happen in phase A, I_{d0} can be obtained by:

$$I_{d0} = |\Delta i_B(k) - \Delta i_C(k)| = |\Delta i_{BC}| \tag{57}$$

The restraining current I_r can be set to the minimum of the grounding leakage current of three phases, namely the single-phase ground leakage capacitive current.

$$I_r = \min(|\Delta I_A|, |\Delta I_B|, |\Delta I_C|) = 3\omega C_1 \Delta U_3' \tag{58}$$

The generator ground fault current equals the vector sum of the each generator grounding leakage current, the grounding current flowing through the Peterson-coil or resistance of the generator neutral side, and the capacitive current of the external facilities such as external transformers. The restraint coefficient K can be determined based on the neutral grounding modes, compensation factor, and the number of generators operating in parallel. The influence of the Peterson-coil to the third harmonic current is very small. If there are 2 generators operating in parallel, the restraint coefficient K can be set to 1. If there are 3 generators operating in parallel, the restraint coefficient K can be set to 1.5. If there are 4 generators operating in parallel, the restraint coefficient K can be set to 1.8.

Only the grounding leakage current of the fundamental component and third harmonic component on the neutral and terminal sides of each phase of the generator stator need be measured by the protection scheme. All kinds of single-phase ground fault can be detected with high reliability by the scheme. The scheme has high sensitivity and robustness (Xia et al. 2007).

4.4.2 The method based on the fault resistance measurement

The faulted generator can be detected with the differences of the leakage current. If the difference is larger than a pre-set threshold value, the generator is considered to have a ground fault. Unfortunately, the threshold is difficult to set. In order to improve its performance, a fault-resistance based protection scheme is proposed as follows.
The fault resistance can be calculated from fault voltage and fault current:

$$R_f = \frac{\Delta \dot{U}_0'}{\dot{I}_f} = \frac{\Delta \dot{U}_0'}{\Delta \dot{I}_{A1} - \Delta \dot{I}_{B1}} \tag{59}$$

To implement the fault-resistant based protection scheme, three phase calculated resistances are defined as:

$$R_{dA} = \left| \frac{\Delta \dot{U}_0}{\Delta \dot{I}_A - (\Delta \dot{I}_B + \Delta \dot{I}_C)/2} \right| \tag{60}$$

$$R_{dB} = \left| \frac{\Delta \dot{U}_0}{\Delta \dot{I}_B - (\Delta \dot{I}_A + \Delta \dot{I}_C)/2} \right| \tag{61}$$

$$R_{dC} = \left| \frac{\Delta \dot{U}_0}{\Delta \dot{I}_C - (\Delta \dot{I}_A + \Delta \dot{I}_B)/2} \right| \tag{62}$$

The generator calculated resistance is defined as:

$$R_d = \min(R_{dA}, R_{dB}, R_{dC}) \tag{63}$$

With these definitions, three phase calculated resistances of the faulted generator are:

$$R_{dA1} = \left| \frac{\Delta \dot{U}_0'}{\Delta \dot{I}_{A1} - (\Delta \dot{I}_{B1} + \Delta \dot{I}_{C1})/2} \right| = \left| \frac{\Delta \dot{U}_0'}{-\dot{I}_f} \right| = R_f \tag{64}$$

$$R_{dB1} = \left| \frac{\Delta \dot{U}_0'}{\Delta \dot{I}_{B1} - (\Delta \dot{I}_{A1} + \Delta \dot{I}_{C1})/2} \right| = \left| \frac{2\Delta \dot{U}_0'}{\dot{I}_f} \right| = 2R_f \tag{65}$$

$$R_{dC1} = \left| \frac{\Delta \dot{U}_0'}{\Delta \dot{I}_{C1} - (\Delta \dot{I}_{A1} + \Delta \dot{I}_{B1})/2} \right| = \left| \frac{2\Delta \dot{U}_0'}{\dot{I}_f} \right| = 2R_f \tag{66}$$

So, the calculated resistance in the faulted generator is:

$$R_{d1} = \min(R_{dA1}, R_{dB1}, R_{dC1}) = R_f \tag{67}$$

The calculated resistance in the normal generator is:

$$R_{d2} = \min(R_{dA2}, R_{dB2}, R_{dC2}) \to \infty \tag{68}$$

If the calculated resistance is less than its threshold R_{set}, ground fault is deemed to have occurred in the generator stator winding.
To improve protection reliability, two protection criteria are applied. One criterion is low

resistance protection, presented as above. If the calculated resistance is less than its threshold R_{set}, ground fault is deemed to have occurred in the generator stator winding. The threshold R_{set} is usually set from 1kΩ to 8kΩ. The larger the generator capacity, the higher is threshold setting. The other protection criterion is through resistance comparing: the generator in the MGS which has the lowest calculated resistance is the faulted unit.

The protection scheme is shown in Figure 32. It has the following steps:

a. The generator stator currents are monitored in the neutral and on the terminal sides in real time. The neutral point voltage is also monitored. The parameters are sampled by A/D converters.
b. If the zero sequence voltage is larger than its threshold, ground fault is detected, and the above proposed ground fault protection is initiated.
c. The faulted generator is the unit with its calculated resistance values less than its threshold R_{set}, or with the lowest calculated resistance among the units in the MGS.
d. The signal identifying the unit with the ground fault is sent to operator and the faulted generator is tripped.

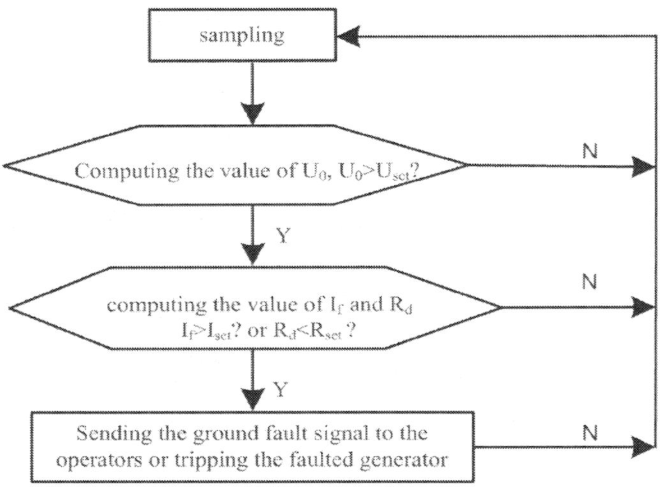

Fig. 32. Resistance protection scheme

4.5 Fault detection application in line protection

Grounding faults generate voltage and current traveling wave. They spread from the fault point to power station, and then to other sound feeders. The transmission procedure can be described as Figure 33. The traveling wave will fade down during spreading abroad.

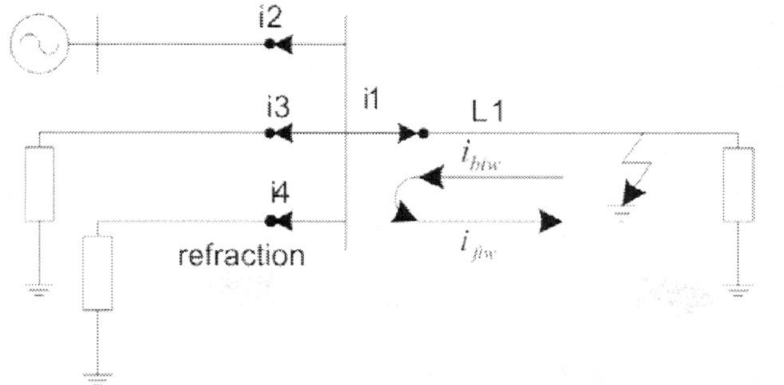

Fig. 33. Fault line selection based on traveling wave current

The reflecting occurs at the point (bus for the example) where impedance does not match. The traveling wave flows in the fault line is the strongest, which is the superposition of the initial inverted traveling wave (i_{btw}) and forward traveling wave (i_{ftw}) whose direction change after reflecting. Whereas the traveling wave in the sound lines is the refraction component of the traveling wave at the fault point and fault lines. So the traveling wave in the sound lines is very small, and its polarity is just reverse to the traveling wave in the fault line.

The traveling wave caused by fault can be detected in each feeder, and the traveling wave in the fault feeder is the largest one. The polarity of traveling wave in the fault feeder is just opposite to the polarity of traveling wave in sound feeders. The fault feeder can thus be detected by comparing the magnitude and polarity of traveling wave caused by distribution system faults in all feeders. The zero sequence components are often used.

Fig. 34. Traveling wave detection device

In order to detect the faults generated traveling wave, two types of traveling wave sensors are developed. They are current traveling wave sensor and voltage traveling wave sensor. The current sensor is a coil with magnetic ring iron core (as shown in Figure 34). A transient voltage suppressor (TVS) and a resistance divider are connected to the output of the coil. The output voltage is controlled to be under 50V. The sensor suppresses power-frequency signals and amplifies high-frequency signals above 10 kHz. It is installed at the grounded

line of capacitive equipment (such as CVT, transformer bushing, wall bushing) to capture the current traveling-waves flowing from the equipment to earth, and the grounded line is passing through the core. This installation does not influence the normal operation of the power system.
The traveling wave based detection method has been tested in laboratory. High impedance faults and arcing faults can be detected with high precision. Not enough traveling wave based fault detectors have applied in power system.

4.6 Fault detection application in ineffectively earthed distribution systems

Reliability and safety are always the two important aspects in the design and operation of industry power systems. Unscheduled outage can create personnel and equipment safety problems, have an adverse environmental impact, and can result in substantial economic losses. Ineffectively earthed systems can limit earth fault current and operate indefinitely with an earth fault on one phase, eliminating the need for an immediate shutdown. Thus many industrial power systems have been operated with floating neutral or high resistance earthed neutral.
The earth fault current in ineffectively earthed systems is usually not more than ten amperes. It is difficult to detect ground faults and isolate the faulted feeder. In order to improve them, some feeder terminal unit (FTU) based ground fault protectors are developed as follow:

4.6.1 Zero sequence overcurrent protection

Figure 35 shows zero sequence current based earth fault detectors (EFD) installed in a small distribution system.

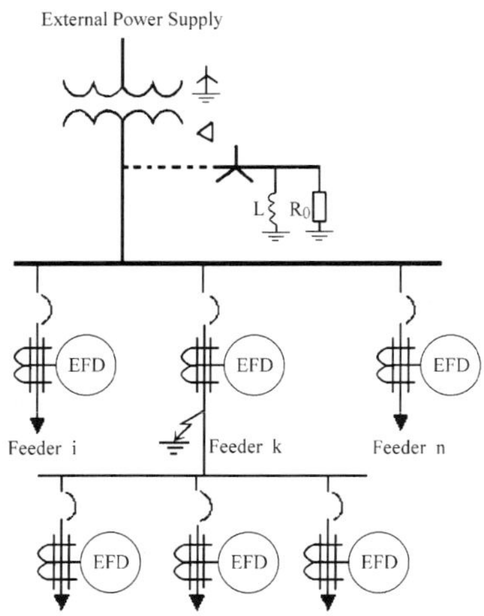

Fig. 35. Zero sequence current based earth fault detectors (EFD) installed in a small distribution system.

If the zero sequence current (I_0) in a feeder is larger than its threshold, this feeder is identified as the ground fault feeder. The threshold of feeder m (I_{0mset}) is usually set as the magnitude of the capacitive current (I_{cm}) in feeder m caused by other feeder direct earth faults, and it is sure that I_r is larger than the least value (I_{0m}) that can be measured accurately by the digital measurement unit (Zeng et al. 2004).

$$I_{0mset} = K_{0m} Max(I_{cm}, I_{0m}) \tag{69}$$

Where, K_{0m} is the reliability coefficient.

The zero sequence current in the faulted feeder k detected by EFD is the sum of charging currents of all other sound feeders (feeder i and n) and the neutral point current. Whereas, in the occurrence of other feeder earth fault the zero sequence current in the sound feeder k is its own capacitive charging currents and its extended feeders (feeders a, b and c). In some cases, the charging capacitive currents in the feeder k and its extended feeders (feeders a, b and c) are possibly bigger than that in the other feeders. The zero sequence current in feeder k caused by its own earth fault is less than that caused by other feeder earth fault. The zero sequence overcurrent protection is thus difficult to be set.

4.6.2 Negative sequence current protection

In Figure 36 (Zeng et al. 2001), an ineffectively earthed system with n feeders has an earth fault on feeder k. The system is a symmetrical radial network with only one source. When earth fault occurs, the distribution capacitance between the three phases of the system and ground is considered. From Figure 36 (b), the positive sequence current ($\dot{I}_1$), the negative sequence current ($\dot{I}_2$) and the zero sequence current ($\dot{I}_0$) through the fault point are equal to one third of fault current ($\dot{I}_f$).

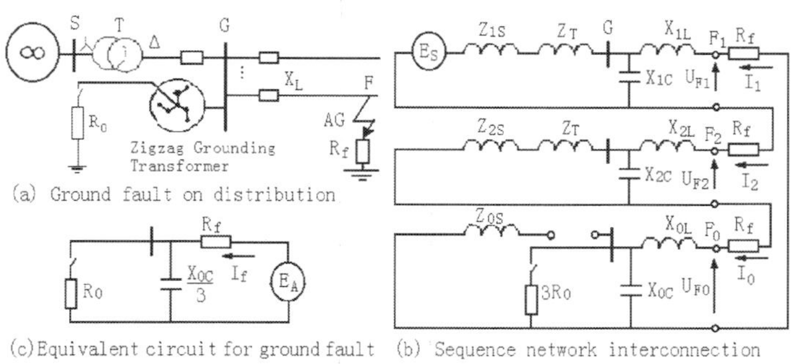

Fig. 36. Sequence network interconnection for "A" phase-to-ground fault

$$\dot{I}_1 = \dot{I}_2 = \dot{I}_0 = \frac{1}{3}\dot{I}_f = \frac{\dot{E}_A}{3R_f + Z_0 + Z_1 + Z_2} \tag{70}$$

Where: E_A is the faulted phase voltage (supposed fault is in phase A), Z_0 is the zero sequence impedance of this network, Z_1 is the positive sequence impedance, R_f is the fault resistance, and Z_2 is the negative sequence impedance.

Z_1, Z_2 can be neglected because they are much lower than Z_0 (as Figure 36 (c)), so Eq.70 can be rewritten as:

$$\dot{I}_1 = \dot{I}_2 = \dot{I}_0 = \frac{1}{3}\dot{I}_f = \frac{\dot{E}_A}{3R_f + Z_0} \tag{71}$$

The negative sequence current, generated in the fault point, flows through the whole network, reaches the sources and the loads, and then returns. As the negative sequence impedance of the source Z_{2S} is much lower than that of the loads, most negative sequence current in the fault point flows to the source. Thus the negative sequence current in the earth fault feeder ($\dot{I}_{2k}$) or in the source ($\dot{I}_{2s}$) is approximately equal to one third of the fault current, and it is much larger than that in sound feeder i ($\dot{I}_{2i}$).

$$\left|\dot{I}_{2k}\right| \approx \left|\dot{I}_{2s}\right| \approx \left|\dot{I}_2\right| = \frac{1}{3}\left|\dot{I}_f\right|, \quad \left|\dot{I}_{2k}\right| >> \left|\dot{I}_{2i}\right| \approx 0 \tag{72}$$

Considering the zero sequence current in the faulted feeder:

$$\dot{I}_{0k} = \frac{1}{3}\dot{I}_f - \dot{I}_{C\,k} \tag{73}$$

Where, $I_{C\,k}$ is the capacitive current in the faulted feeder k.

Comparing Eq.72 and Eq.73, the negative sequence current is usually slightly bigger than the zero sequence current in the faulted feeder.

$$\left|\dot{I}_{2k}\right| \geq \left|\dot{I}_{0k}\right| \tag{74}$$

In sound feeder i, the zero sequence current I_{0i} is approximately equal to its capacitive current I_{ci}. It is usually larger than the negative sequence current, which is approximately zero (Eq.72). Negative sequence current is therefore more sensitive in detecting earth fault feeder than zero sequence current.

Moreover, energy oscillations usually exist in zero sequence circuit between restrikes during arc-grounding fault, but not in negative sequence circuit because the negative sequence impedance of source is so small that energy is quickly faded away after arc extinguishing. Protection schemes based on negative sequence current hence have a better performance in detecting arcing fault than that based on zero sequence current.

The principles for earth fault detection based on negative sequence current are as follow:

a. Negative Sequence Current Based Overcurrent Protection

If the magnitude of negative sequence current component in a feeder is larger than its threshold, this feeder is identified as the earth fault feeder.

$$\left|\Delta\dot{I}_{2m}\right| > I_{2mset} \tag{75}$$

The threshold of feeder m (I_{2mset}) is usually set as the least value that can be measured accurately by digital measurement unit, and it is sure that I_{2mset} is larger than the magnitude of negative sequence current component in feeder m caused by other feeder directly earth fault.

b. Difference between Negative Sequence Current and Zero Sequence Current Based Protection

From Eq.74, when the component of negative sequence current is slightly bigger than that of zero sequence current in a feeder, it is identified as the faulted feeder.

$$\left|\Delta \dot{I}_{2m}\right| > K_0\left|\Delta \dot{I}_{0m}\right| \tag{76}$$

In order to improve fault detection sensitivity, a reliability coefficient K_0 is used, which is usually set as 0.5.

c. Negative Sequence Current Based Differential Protection

The component of negative sequence current in the faulty feeder is approximately opposite to that in the source, while that in a sound feeder is approximately equal to zero, and is much lower than that in the source. The faulted feeder can therefore be detected, if the difference between the component of negative sequence current in a feeder and that in the source is smaller than its threshold.

$$\left|\Delta \dot{I}_{2m} + \Delta \dot{I}_{2S}\right| < I_{2msset} \tag{77}$$

The threshold of feeder m (I_{2msset}) is usually set as a value that is larger than the sum of the magnitudes of negative sequence current component in all sound feeders caused by a directly earth fault of feeder m.

d. Phase Difference between Negative Sequence Current and Fault Phase Voltage Based Protection

If the phase difference between the negative sequence current in a feeder and the fault phase voltage phasor is near to zero (in same direction), this feeder can be identified as the faulted feeder.

$$\theta_{set1} < \arg(\Delta \dot{I}_{2m}) - \arg(\dot{U}_A) < \theta_{set2} \tag{78}$$

In order to improve fault detection sensitivity, θ_{set1}, θ_{set2} is usually set as -60^0 and 60^0 respectively.

e. Fault Point Energy Dissipation Based Protection

If the energy function (in Eq.79) of a feeder is larger than its threshold, this feeder can also be identified as the faulted feeder.

$$W_m(t) = 3\int_{t1}^{t2} u_A \Delta i_{2m} dt > W_{Set} \tag{79}$$

Where, $t1$ is the time of earth fault initiation, ($t2$-$t1$) is about one cycle (20mS in 50Hz). W_{Set} is usually set as a small positive value.

Among the above fault detection schemes, scheme a, b, c, and e only demand to measure the local current or voltage. These schemes are convenient to be implemented in Remote Terminal Unit (RTU), Feeder Terminal Unit (FTU) and Intelligent Electronic Device (IED) in Distribution Automation (DA) systems.

5. Reference

(2006) Electromagnetic compatibility (EMC)-Part 4-3: Testing and measurement techniques-Radiated, radio-frequency, electromagnetic field immunity test, Third.

Al-Ammar E, Karady GG, Sim HJ (2008) Novel technique to improve the fault detection sensitivity in transformer impulse test. IEEE Transactions on Power Delivery 23:717-725

Amaral JLM, Amaral JFM, Morin D, Tanscheit R (Year) An immune fault detection system with automatic detector generation by genetic algorithms. In: Proceedings of The 7th International Conference on Intelligent Systems Design and Applications, ISDA 2007

Awais MS, S.; Ahmed, N.; Shahid, S. (2003) A comparative study of feed forward neural networks and radial basis neural networks for modeling Tokamak fusion process. Multi Topic Conference, 2003 INMIC 2003 7th International:6

Aydin I, Karakose M, Akin E (Year) Artificial immune inspired fault detection algorithm based on fuzzy clustering and genetic algorithm methods. In: CIMSA 2008 - IEEE Conference on Computational Intelligence for Measurement Systems and Applications Proceedings

Betta G, Liguori C, Pietrosanto A (Year) Use of genetic algorithms for advanced instrument fault detection and isolation schemes. In: Conference Record - IEEE Instrumentation and Measurement Technology Conference

Betta G, Liguori C, Pietrosanto A (1998) An advanced neural-network-based instrument fault detection and isolation scheme. IEEE Transactions on Instrumentation and Measurement 47:507-512

Bittera M, Harťanský R (2006) EM field measurement aspects for electromagnetic compatibility purposes. MEASUREMENT SCIENCE REVIEW 6:14-17

Bossé E, Guitouni A, Valin P (Year) An essay to characterise information fusion systems. In: 2006 9th International Conference on Information Fusion, FUSION

Chen LJ, Tsao TP, Lin YH (2005) New diagnosis approach to epoxy resin transformer partial discharge using acoustic technology. IEEE Transactions on Power Delivery 20:2501-2508

Chow M-y, Yee SO (1991) Methodology for on-line incipient fault detection in single-phase squirrel-cage induction motors using artificial neural networks. IEEE Transactions on Energy Conversion 6:536-545

Cooper JH, South WH, Shimshock JF (Year) FIELD TEST EXPERIENCE AND DIAGNOSTIC TECHNIQUES ON 1200-KV GAS INSULATED SUBSTATION COMPONENTS. In: International Conference on Large High Voltage Electric Systems, Paris, Fr

Curry T, Collins EG, Jr., Selekwa M (Year) Robust fault detection using robust estimation and fuzzy logic. In: Proceedings of the American Control Conference

Curry TD, Collins Jr EG (Year) Robust l1 design of a multivariable PI controller using a real-coded genetic algorithm. In: Proceedings of the American Control Conference

Dixon DS, Dutcher C (Year) Feasibility of conducting unit level EMI tests in a system level environment. In: IEEE International Symposium on Electromagnetic Compatibility, Washington, DC, USA

Guo F (2007) Deep Research on both-terminal travelling wave fault location. Shandong University Master's Thesis, Ji Nan, Shandong University, China:34-37

Hartog A (1995) Distributed fibre-optic temperature sensors: technology and applications in the power industry. Power Engineering Journal 9:114-120

Hauptmann P, Hoppe N, Puettmer A (Year) Ultrasonic sensors for process industry. In: Proceedings of the IEEE Ultrasonics Symposium, Atlanta, GA

Jingchao Y, Xianggen Y, Deshu C, Ze Z (Year) Study on the Operating Characteristic of Sampled Value Differential Protection. In: Canadian Conference on Electrical and Computer Engineering

Lippmann RP (1987) INTRODUCTION TO COMPUTING WITH NEURAL NETS. IEEE ASSP magazine 4:4-22

Li T, Zhao Y, Li N, Fen G, H. G (2005) A new method for power quality detection based on HHT. Proceedings of CSEE, China 25:52-56

Liu Z, Zhou JZ, Zou M (Year) Condition based maintenance system of hydroelectric generating unit. In: Proceedings of the IEEE International Conference on Industrial Technology

Middelkoop R (1991) Time-domain calibration of field sensors for electromagnetic pulse (EMP) measurements. IEEE Transactions on Instrumentation and Measurement 40:455-459

Nakamura Y (2008) Noncontact sensor. Patent No:US 7,330,033 B2

Ngaopitakkul A, Kunakorn A, Ngamroo I (Year) Discrimination between external short circuits and internal faults in transformer windings using discrete wavelet transforms. In: Conference Record - IAS Annual Meeting (IEEE Industry Applications Society)

Satish L (1998) Short-time Fourier and wavelet transforms for fault detection in power transformers during impulse tests. IEE Proceedings: Science, Measurement and Technology 145:77-84

Setayeshmehr A, Akbari A, Borsi H, Gockenbach E (Year) A procedure for diagnosis and condition based maintenance for power transformers. In: Conference Record of IEEE International Symposium on Electrical Insulation

Sreedhar R, Fernandez B, Masada GY (Year) Neural network based adaptive fault detection scheme. In: Proceedings of the American Control Conference

Srivani SG, Vittal KP (Year) On line fault detection and an adaptive algorithm to fast distance relaying. In: 2008 Joint International Conference on Power System Technology POWERCON and IEEE Power India Conference, POWERCON 2008

Vemuri AT, Polycarpou MM, Diakourtis SA (1998) Neural network based fault detection in robotic manipulators. IEEE Transactions on Robotics and Automation 14:342-348

Wang X, Qian Q, Chen W (Year) Analyzing fault-induced transients with wavelets. In: Proceedings of the IEEE Power Engineering Society Transmission and Distribution Conference, Chicago, IL

Wang Y, Zeng X, Ma H, Xia Y, Xu Y (Year) Grounding fault protection with sampling value difference in ineffectively earthed power systems. In: 2006 International Conference on Power System Technology, POWERCON2006

Wang Y, Zeng X, Su S (Year) Grounding fault protection with phase current difference for ineffectively earthed power systems. In: Conference Record - IAS Annual Meeting (IEEE Industry Applications Society)

Widrow B, Winter R (1988) NEURAL NETS FOR ADAPTIVE FILTERING AND ADAPTIVE PATTERN RECOGNITION. Computer 21:25-39

Xia Y, Zeng X, Wang Y, Ma H (Year) Stator winding grounding fault protection with third harmonic current difference for multi-generator systems. In: Conference Record - IAS Annual Meeting (IEEE Industry Applications Society)

Xiao'an Q, Xiangjun Z, Xiaoli Z, Zewen L (Year) Traveling wave based distribution lines fault location using Hilbert-Huang Transform. In: Conference Record - IAS Annual Meeting (IEEE Industry Applications Society), Edmonton, AB

Yang JC, Yin XG, Chen DS, Zhang Z, Wang ZH (Year) The Study of Sampled Value Differential Protection. In: Proceedings of the IEEE Power Engineering Society Transmission and Distribution Conference

Zeng X, Guo Z, Su S (Year) Ground fault protection for ineffectively earthed power systems. In: 2004 International Conference on Power System Technology, POWERCON 2004

Zeng X, Li KK, Chan WL, Su S, Wang Y (2008) Ground-fault feeder detection with fault-current and fault-resistance measurement in mine power systems. IEEE Transactions on Industry Applications 44:424-429

Zeng X, Li KK, Chan WL, Yin X (Year) Novel techniques for earth fault feeder detection based on negative sequence current in industry power systems. In: Conference Record - IAS Annual Meeting (IEEE Industry Applications Society), Chicago, IL

Zeng X, Xia Y, Ma H, Wang Y (Year) Grounding faulted feeder detection methods applied in Chinese ineffectively earthed distribution systems. In: Conference Record - IAS Annual Meeting (IEEE Industry Applications Society), New Orleans, LA

Zeng X (2000) Reasearch on advanced principles of power lines fault detection & fault location and their implementing with information fusion. Doctoral Dissertation, Wuhan, Huazhong University of science & technology, China 19-21.

Zeng X (2003)On-Site Safety Evaluation for Earth Fault in Mining Power Systems.IEEE TRANSACTIONS ON INDUSTRY APPLICATIONS, 39(6): 1563-1569

Zhang L, Li X, Yu J, Gao Z (Year) A genetic training algorithm of wavelet neural networks for fault prognostics in condition based maintenance. In: 2007 8th International Conference on Electronic Measurement and Instruments, ICEMI

Zhang Y, Lee SR, Zhao H, Wang B, Bingqing L (Year) Research on wavelet analysis in fault signal processing. In: International Conference on Signal Processing Proceedings, ICSP, Beijing

6

Fault Detection and Isolation Scheme Based on Parity Space Method for Discrete Time-Delay System

Hongyu Wang, Zuohua Tian, Songjiao Shi and Zhenxin Weng
Shanghai Jiaotong University
China

1. Introduction

In recent years, fault detection and isolation (FDI) problem in dynamitic system has been paid more and more attention. A great number of methods for FDI have been proposed (Chow & Willsky, 1984; Frank & Ding, 1997; Chen & Patton, 1999; Patton et al., 2000; Venkatasubramanian et al., 2003). All of the FDI schemes are concerned with a core stage: the generation of the residual signals. The difference between the measurement of the system and its estimation is called residual, whose values are zero or near to zero when no fault occurs while differ distinctly from zero otherwise. Appropriate decisions such as the occurrence, magnification, type, location, etc. of the faults are called fault isolation, which are achieved by residual evaluation.

In the field of analytical model-based FDI techniques, the analytical redundancy relations of the system are used to create residual signal. The approaches can be roughly classified into observer-based approaches and parameter estimation approaches. Parity space approaches have been proved to be structurally equivalent to the observer-based though the design procedures differ (Gertler, 1991). However, the parity space methodology using the temporal redundancy has its advantages, especially in the discrete system. This method was firstly generalized by the (Chow & Willsky, 1984).

Time delays are inherent in many real physical processes (i.e. mechanical and chemical processes, long transmission lines in pneumatic systems, power and water distribution networks, air pollution systems etc.) Over the past two decades, analysis and synthesis of dynamic time-delay systems have attracted a great deal of interests (Dugard & Verriest, 1997; Yang & Saif, 1998). However, there are relative fewer research results on FDI of time-delay systems (KRATZ et al., 1998; Zhong et al., 2004).

This paper proposes a method to deal with the FDI problem for the linear discrete-time systems with delays. The results in (KRATZ et al., 1998) are extended. Both fault detection and fault isolation method are proposed. The occurrence of the fault can be detected timely and the position of the fault can be located exactly. A numerical example is given to illustrate the design method at the end.

2. Mathematical Preliminaries

A time delay operator ∇ is defined according to (KRATZ et al., 1998). $\nabla f(k) = f(k-1)$ for any discrete-time function f. It is easy to understand that $\nabla\nabla f(k) = \nabla^2 f(k) = f(k-2)$, $\nabla^W f(k) = f(k-w)$.

Consider a linear discrete time-delay system described by

$$\begin{cases} \boldsymbol{x}(k+1) = \sum_{i=0}^{v} \boldsymbol{A}_i \boldsymbol{x}(k-i) + \boldsymbol{B}[\boldsymbol{u}(k) + \boldsymbol{f}_a(k)] + \boldsymbol{E}_d \boldsymbol{d}(k) \\ \boldsymbol{y}(k) = \boldsymbol{C}\boldsymbol{x}(k) + \boldsymbol{F}_d \boldsymbol{d}(k) \end{cases} \tag{1}$$

where $\boldsymbol{x}(k) \in R^n$ is the state vector, $\boldsymbol{u}(k) \in R^p$ is the control vector, $\boldsymbol{y}(k) \in R^m$ is the output vector, $\boldsymbol{f}_a(k) = [f_{a1}(k), \cdots, f_{ap}(k)]^{\mathrm{T}}$ stands for the actuator faults. $f_{ai}(k), i = 1, \cdots, p$ is corresponding to the ith actuator fault. $\boldsymbol{d}(k) \in R^l$ is disturbance vector, $\|\boldsymbol{d}(k)\| < \bar{d}$. $\boldsymbol{A}_i (i = 0, \cdots, v)$, $\boldsymbol{B}$, $\boldsymbol{C}$, $\boldsymbol{E}_d$ and $\boldsymbol{F}_d$ are constant matrices with appropriate dimensions. Integers $v \geq 0$ denotes the number of time delays in the state.

Using the operator ∇ , the system (1) can be rewritten as

$$\begin{cases} \boldsymbol{x}(k+1) = \boldsymbol{A}(\nabla)\boldsymbol{x}(k) + \boldsymbol{B}[\boldsymbol{u}(k) + \boldsymbol{f}_a(k)] + \boldsymbol{E}_d \boldsymbol{d}(k) \\ \boldsymbol{y}(k) = \boldsymbol{C}\boldsymbol{x}(k) + \boldsymbol{F}_d \boldsymbol{d}(k) \end{cases} \tag{2}$$

where,

$$\boldsymbol{A}(\nabla) = \boldsymbol{A}_0 + \boldsymbol{A}_1\nabla + \cdots + \boldsymbol{A}_v\nabla^v \tag{3}$$

3. Parity Space Residual Generation for Fault Detection and Isolation

The task of FDI is to design a residual signal which is zero or near to zero in a fault free case and non-zero when a fault occurs in the monitored system. Time delay implies that the state of the system for the next time step is not only determined by the current state but also concerned with the state of the former intervals. The recursion of equation (2) from time instant $k-L$ to time instant k yields

$$\boldsymbol{y}_L = \boldsymbol{H}_{o,L}\boldsymbol{x}(k-L) + \boldsymbol{H}_{u,L}\boldsymbol{u}_L + \boldsymbol{H}_{d,L}\boldsymbol{d}_L + \boldsymbol{H}_{fa,L}\boldsymbol{f}_{aL} \tag{4}$$

where

$$\boldsymbol{y}_L(k)=\begin{bmatrix} \boldsymbol{y}(k-L) \\ \boldsymbol{y}(k-L+1) \\ \vdots \\ \boldsymbol{y}(k) \end{bmatrix} = \underbrace{\begin{bmatrix} \underbrace{\begin{bmatrix} \Delta^L & & \\ & \ddots & \\ & & \Delta^L \end{bmatrix}}_{m} & & & \\ & \underbrace{\begin{bmatrix} \Delta^{L-1} & & \\ & \ddots & \\ & & \Delta^{L-1} \end{bmatrix}}_{m} & & \\ & & \ddots & \\ & & & \underbrace{\begin{bmatrix} 1 & & \\ & \ddots & \\ & & 1 \end{bmatrix}}_{m} \end{bmatrix}}_{\triangleq \Delta_{Lm}} \begin{bmatrix} \boldsymbol{y}(k) \\ \boldsymbol{y}(k) \\ \vdots \\ \boldsymbol{y}(k) \end{bmatrix} \in R^{(L+1)m},$$

$$\boldsymbol{u}_L(k)=\begin{bmatrix} \boldsymbol{u}(k-L) \\ \boldsymbol{u}(k-L+1) \\ \vdots \\ \boldsymbol{u}(k) \end{bmatrix} = \Delta_{Lp}\begin{bmatrix} \boldsymbol{u}(k) \\ \boldsymbol{u}(k) \\ \vdots \\ \boldsymbol{u}(k) \end{bmatrix} = \in R^{(L+1)p},\ \boldsymbol{f}_{aL}(k)=\begin{bmatrix} \boldsymbol{f}_a(k-L) \\ \boldsymbol{f}_a(k-L+1) \\ \vdots \\ \boldsymbol{f}_a(k) \end{bmatrix} = \Delta_{Lp}\begin{bmatrix} \boldsymbol{f}_a(k) \\ \boldsymbol{f}_a(k) \\ \vdots \\ \boldsymbol{f}_a(k) \end{bmatrix} \in R^{(L+1)p},$$

$$\boldsymbol{d}_L(k)=\begin{bmatrix} \boldsymbol{d}(k-L) \\ \boldsymbol{d}(k-L+1) \\ \vdots \\ \boldsymbol{d}(k) \end{bmatrix} = \Delta_{Ll}\begin{bmatrix} \boldsymbol{d}(k) \\ \boldsymbol{d}(k) \\ \vdots \\ \boldsymbol{d}(k) \end{bmatrix} \in R^{(L+1)l},\ \boldsymbol{H}_{o,L}=\begin{bmatrix} \boldsymbol{C} \\ \boldsymbol{CA}(\nabla) \\ \vdots \\ \boldsymbol{CA}^L(\nabla) \end{bmatrix} \in R^{(L+1)m\times n},$$

$$\boldsymbol{H}_{d,L}=\begin{bmatrix} \boldsymbol{F}_d & & & & \\ \boldsymbol{CE}_d & \boldsymbol{F}_d & & 0 & \\ \boldsymbol{CA}(\nabla)\boldsymbol{E}_d & \boldsymbol{CE}_d & \boldsymbol{F}_d & & \\ \vdots & \vdots & \vdots & \ddots & \\ \boldsymbol{CA}^{L-1}(\nabla)\boldsymbol{B} & \boldsymbol{CA}^{L-2}(\nabla)\boldsymbol{B} & \vdots \ \cdots & \boldsymbol{CE}_d & \boldsymbol{F}_d \end{bmatrix} \in R^{(L+1)m\times(L+1)l},\ \boldsymbol{H}_{fa,L}=\boldsymbol{H}_{u,L}$$

Define the following parity space:

$$\boldsymbol{P}_L=\left\{\boldsymbol{v}_L \middle| \boldsymbol{v}_L\boldsymbol{H}_{o,L}=0\right\} \tag{5}$$

where $\boldsymbol{v}_L \in R^{1\times(L+1)m}$ is row vector. Vectors belong to parity space are called parity vectors. Residual signals can be created by the folowing equation:

$$r_L(k)=\boldsymbol{v}_L\left[\boldsymbol{y}_L(k)-\boldsymbol{H}_{u,L}\boldsymbol{u}_L(k)\right]\in \boldsymbol{R} \tag{6}$$

Substituting equation (6) to equation (4) yields:

$$r_L(k)=\boldsymbol{v}_L\boldsymbol{H}_{d,L}\boldsymbol{d}_L(k)+\boldsymbol{v}_L\boldsymbol{H}_{fa,L}\boldsymbol{f}_{aL}(k) \tag{7}$$

It should be noted that the parity vectors $\boldsymbol{v}_L$ satisfying equation (5) are not unique, and the corresponding residual signals $r_L(k)$ are not unique. The freedom of the $\boldsymbol{v}_L$ can be used to creat specific residual signals, so as to fulfill specific design purpose.
The parity vectors $\boldsymbol{v}_L$ can be described as $\boldsymbol{v}_L = [v_1, v_2, \cdots, v_{(L+1)m}]$. Substituting it to equation (7), the terms corresponding to disturbance $\boldsymbol{v}_L \boldsymbol{H}_{d,L} \boldsymbol{d}_L(k)$ and faults $\boldsymbol{v}_L \boldsymbol{H}_{fa,L} \boldsymbol{f}_{aL}(k)$ can be respectively expanded as follows:

$$\boldsymbol{v}_L \boldsymbol{H}_{d,L} \boldsymbol{d}_L(k) = \psi_1 d_1(k) + \psi_2 d_2(k) + \cdots + \psi_l d_l(k) \tag{8}$$

$$\boldsymbol{v}_L \boldsymbol{H}_{fa,L} \boldsymbol{f}_{aL}(k) = \omega_1 f_{a1}(k) + \omega_2 f_{a2}(k) + \cdots + \omega_p f_{ap}(k) \tag{9}$$

Where $\psi_1, \psi_2, \cdots, \psi_p$, $\omega_1, \omega_2, \cdots, \omega_p$ are polynomials corresponding to $v_1, v_2, \cdots, v_{(L+1)m}$ and ∇ .
The successful detection of a fault is followed by the fault isolation procedure which will distinguish (isolate) a particular fault from others. While a single residual signal is sufficient to detect faults, a set of residuals (or a vector of residual) is usually required for fault isolation. According to (Chen & Patton, 1999), a commonly used scheme in designing the residual set is to make each residual sensitive to all but one fault, i.e.

$$\begin{cases} r_1 = R(f_{a2}, \cdots, f_{a3}) \\ \vdots \\ r_i = R(f_{a1}, \cdots, f_{ai-1}, f_{ai+1}, \cdots, f_{ap}) \\ \vdots \\ r_p = R(f_{a1}, \cdots, f_{ap-1}) \end{cases} \tag{10}$$

where $R(\cdot)$ denotes some functional relation, which works as the residual generator. This is defined as a generalized structured residual set. The isolation can be performed by the following logic:

$$\left.\begin{aligned} &\|r_i\| \le th^i \\ &\|r_j\| > th^j \ \text{for}\ j = 1, \cdots, i-1, i+1, \cdots p \end{aligned}\right\} \Rightarrow f_{ai} \ne 0 \tag{11}$$

where th^i means the fault isolation threshold to the corresponding fault.
To achieve the so called generalized structured residual set, let $\omega_1, \omega_2, \cdots, \omega_p$ satisfy the folowing equatinos:

$$\begin{cases} \omega_1 \ne 0 \\ \vdots \\ \omega_{i-1} \ne 0 \quad ,i = 1, 2, \cdots, p \\ \omega_i = 0 \\ \omega_{i+1} \ne 0 \end{cases} \tag{12}$$

Solving the equations (12) respectively can achieve a set of parity vecters $\boldsymbol{v}_{Lfai}$ $i = 1, 2, \cdots, p$,witch lead to a set of residual signals $r_{Lfai}(k)$, $i = 1, 2, \cdots, p$ by equation (6). When the ith actuator fault occurs($f_{ai}(k) \ne 0$), the corresponding residual signal $r_{Lfai}(k)$ is not affected, while the other residual signals $r_{Lfa1}(k)$, $r_{Lfa1}(k)$;$\cdots$, $r_{Lfai-1}(k), r_{Lfai+1}(k), r_{Lfap}(k)$ are affected. The isolation can be fulfilled by equaton (11).

4. Numerical Example

To illustrate the design process of the proposed mothed and verify its effectiveness, the following numerical example is demonstrated. Consider a time delay system of the form (1),

$$\begin{cases} \begin{bmatrix} x_1(k+1) \\ x_2(k+1) \\ x_3(k+1) \end{bmatrix} = \begin{bmatrix} 0.15 & 0.324 & 0 \\ 0.102 & 0.258 & 0 \\ 0 & 0 & 0.412 \end{bmatrix} \begin{bmatrix} x_1(k) \\ x_2(k) \\ x_3(k) \end{bmatrix} + \begin{bmatrix} 0 & 0.265 & 0 \\ -0.406 & 0 & 0.308 \\ 0 & -0.252 & 0.124 \end{bmatrix} \begin{bmatrix} x_1(k-1) \\ x_2(k-1) \\ x_3(k-1) \end{bmatrix} \\ \quad + \begin{bmatrix} 1 & 0 & 0 \\ 0 & 1 & 0 \\ 0 & 0 & 1 \end{bmatrix} \left(\begin{bmatrix} u_1(k) \\ u_2(k) \\ u_3(k) \end{bmatrix} + \begin{bmatrix} f_{a1}(k) \\ f_{a2}(k) \\ f_{a3}(k) \end{bmatrix} \right) + \begin{bmatrix} 1 \\ 1 \\ 1 \end{bmatrix} d(k) \\ \begin{bmatrix} y_1(k) \\ y_2(k) \\ y_3(k) \end{bmatrix} = \begin{bmatrix} 1 & 0 & 0 \\ 0 & 1 & 0 \\ 0 & 0 & 1 \end{bmatrix} \begin{bmatrix} x_1(k) \\ x_2(k) \\ x_3(k) \end{bmatrix} + \begin{bmatrix} 1 \\ 1 \\ 1 \end{bmatrix} d(k) \end{cases}$$

Where $\boldsymbol{d}(k) = 0.01 \times randN(0,1)$, $randN(0,1)$ stands for zero mean, uint Gauss noise. Using the operator ∇, the system (1) can be rewritten into the form (2), the system matrix is:

$$\boldsymbol{A}(\nabla) = \begin{bmatrix} 0.15 & 0.324 + 0.265\Delta & 0 \\ 0.102 - 0.406\Delta & 0.258 & 0.308\Delta \\ 0 & -0.252\Delta & 0.412 + 0.124\Delta \end{bmatrix}$$

In order to fulfill the FDI, a parity space with $L = 2$ is established, the matrix $\mathbf{H}_{o,L}$, $\boldsymbol{H}_{u,L}$, etc. can be computed using the software MAPLE. Solving the equation (5) can get the parity vecter:

$$v_L = \begin{bmatrix} t_1 \\ \left(\begin{array}{l} 1.767\nabla t_1 + 2.160 t_1 - 0.03768 t_2 - 0.7173\nabla^2 t_2 - 0.6968\nabla t_2 + 0.2520\nabla t_3 \\ -0.01221 t_4 - 0.2357\nabla t_4 - 0.1901\nabla^3 t_4 - 0.4170\nabla^2 t_4 - 0.1074\nabla^2 t_5 - 0.009721 t_5 \\ -0.1798\nabla t_5 + 0.1133\nabla t_6 + 0.2068\nabla^2 t_6 + 0.1807512\nabla^3 t_6 \end{array} \right) \\ \left(\begin{array}{l} -0.308\nabla t_2 - 0.412 t_3 - 0.124\nabla t_3 - 0.09979\nabla t_4 - 0.08162\nabla^2 t_4 - 0.2064\nabla t_5 - \\ 0.03819\nabla^2 t_5 + 0.06224\nabla^2 t_6 - 0.1023\nabla t_6 - 0.1697 t_6 \end{array} \right) \\ \left(\begin{array}{l} -6.667 t_1 - 0.6800 t_2 + 2.7067\nabla t_2 - 0.3703 t_4 + 0.6968\nabla t_4 + 0.7173\nabla^2 t_4 + 1.104\nabla t_5 \\ -0.2774 t_5 - 0.6821\nabla^2 t_6 + 0.1714\nabla t_6 \end{array} \right) \\ t_2 \\ t_3 \\ t_4 \\ t_5 \\ t_6 \end{bmatrix}^T \tag{13}$$

Where $t_1, t_2, t_3, t_5, t_5, t_6 \in R$ and $t_1, t_2, t_3, t_5, t_5, t_6$ are not all zeros. v_L is a row 9 dimensions vector with 6 fredom.

In order to achieve fault detection and isolation, a generalized residual set consists in three residual signals should be created as show in Fig.1.

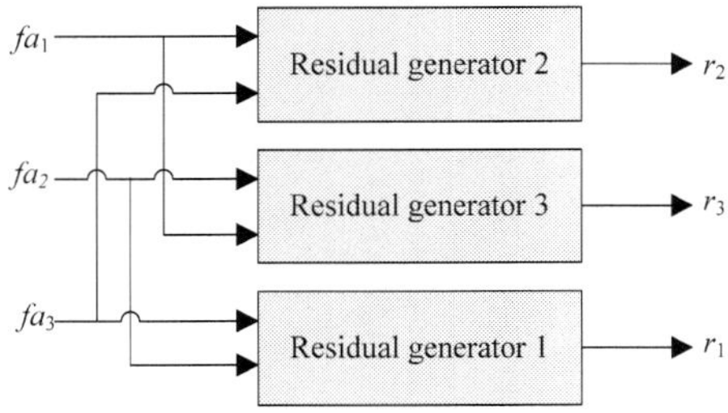

Fig. 1 Generalized residual set

In Fig.1, f_{ai} stands for the ith actuator fault. Residual singal r_1 is sensitive to f_{a2} and f_{a3} while insensitive to f_{a1}. The situation of r_2 and r_3 are the same with r_1.
Substituting v_L to equation (7), and expanding the terms corresponding to the actuator fault $\boldsymbol{v}_L\boldsymbol{H}_{fa,L}\boldsymbol{f}_{aL}(k)$ into equation (9) yields, $\boldsymbol{v}_L\boldsymbol{H}_{fa,L}\boldsymbol{f}_{aL}(k)=\omega_1 f_{a1}(k)+\omega_2 f_{a2}(k)+\omega_3 f_{a3}(k)$. Where $\omega_1,\omega_2,\omega_3$ are polynomials corresponding to t_1,t_2,t_3,t_5,t_5,t_6 and ∇. Using equation (12), let

$$\begin{cases}\omega_1=0\\ \omega_2\neq 0\\ \omega_3\neq 0\end{cases}$$

t_1,t_2,t_3,t_5,t_5,t_6 can be achieved by solving the above equation, and substituting them into (13) can get the parity vector corresponding to the first actuator fault f_{a1}:

$$v_{L\,fa1}=\begin{bmatrix}0\\ 0.2520\Delta+0.07762\Delta^2\\ -0.4120-0.2509\Delta-0.03819\Delta^2\\ -0.1020+0.4060\Delta\\ -0.2580\\ 1\\ 0\\ 1\\ 0\end{bmatrix}^T$$

Substituting $v_{L\,fa1}$ to equation (6) can get generalized residual signal $r_{Lfa1}(k)$ corresponding to f_{a1}, denoted by r_1.

$$r_1=(-0.1020\Delta+0.40600\Delta^2)y_1(k)+(1-0.2580\Delta+0.2520\Delta^3+0.07762\Delta^4)y_2(k)+$$
$$(\Delta-0.4120\Delta^2-0.2509\Delta^3-0.03819\Delta^4)y_3(k)+(-\Delta)u_2(k)+(-\Delta^2-0.3080\Delta^3)u_3(k)$$

By the same process, let

$$\begin{cases} \omega_1 \neq 0 \\ \omega_2 = 0, \\ \omega_3 \neq 0 \end{cases} \begin{cases} \omega_1 \neq 0 \\ \omega_2 \neq 0 \\ \omega_3 = 0 \end{cases}$$

The residual signals correspond to f_{a2} and f_{a3} can be designed, denoted by r_2 and r_3.

Now, the disign of the generalized residual set for actuator fault detection and isolation is accomplished. The result of the simulation is show in Fig. 2 to Fig. 5.

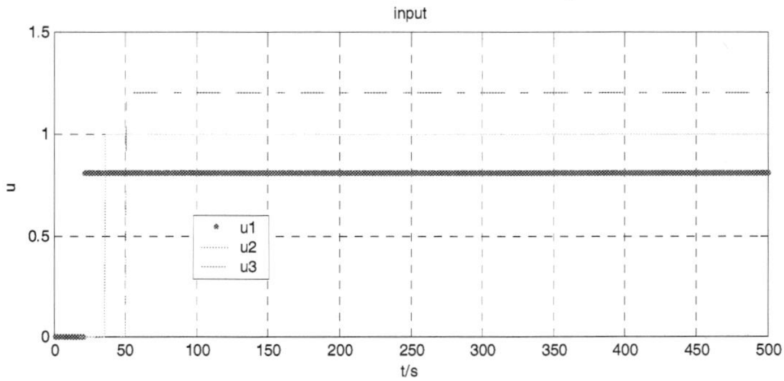

Fig. 2 System input signals

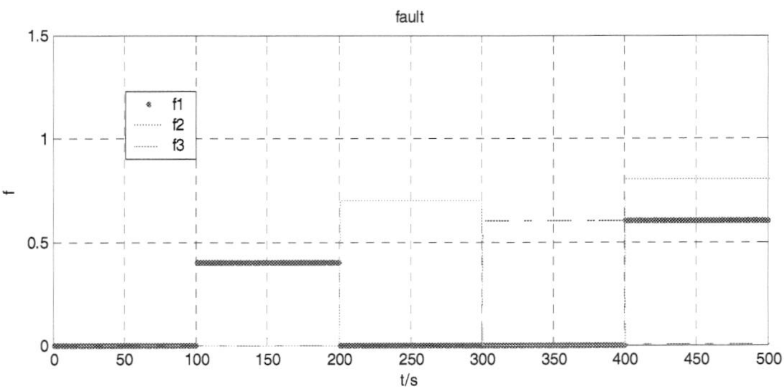

Fig. 3 Fault signals

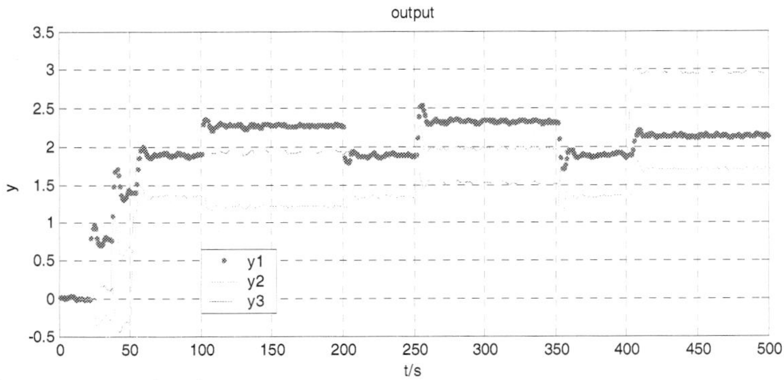

Fig. 4 System output signals

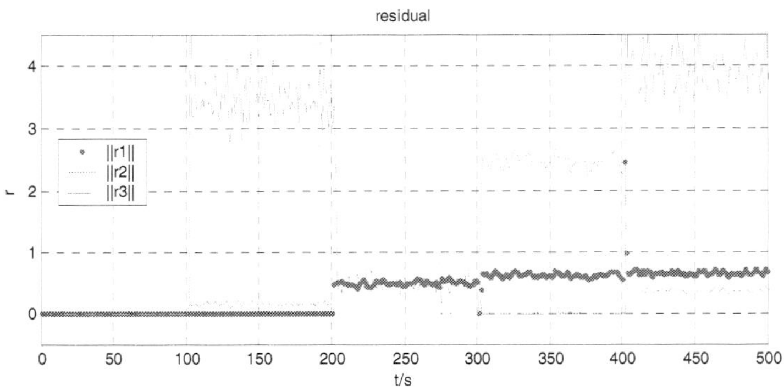

Fig. 5 Residual signals

There are 5 stages in the simulation process:

1. from time 0 second to time 100 second, the system works properly. $\|r_1\|$, $\|r_2\|$ and $\|r_3\|$ are near to zreo.
2. from time 100 second to time 200 second, the actuator 1 suffers from fault, while actuator 2 and actuator 3 work properly. $\|r_2\|$ and $\|r_3\|$ differ from zero while $\|r_1\|$ keeps zero nearby.
3. from time 200 second to time 300 second, the actuator 2 suffers from fault, while actuator 1 and actuator 3 work properly. $\|r_1\|$ and $\|r_3\|$ differ from zero while $\|r_2\|$ keeps zero nearby.
4. from time 300 second to time 400 second, the actuator 3 suffers from fault, while actuator 1 and actuator 2 work properly. $\|r_1\|$ and $\|r_2\|$ differ from zero while $\|r_3\|$ keeps zero nearby.
5. from time 400 second to time 500 second, the actuator 1 and actuator 2 suffer from fault at the same time, while actuator 3 works properly. $\|r_1\|$, $\|r_2\|$ and $\|r_3\|$ differ from zero simultaneously.

It can be concluded that when there is one actuator goes into fault, the above generalize residual set based on pariy space can detect the fault and isolate which actuator corrupted by the fault. However, when there are more than one actuators break into faults, the method can only detect the fault, while it have no idea that which actuators corrupted by the fault.

5. Conclusion

A fault detection and isolation scheme for discrete time-delay system has been proposed in this chapter. The scheme can not only detect the faults but also isolate (locate) the faults. To fulfill the FDI, a generalized residual set in form of parity space is designed by the recursion of the system equations. Each residual is sensitive to all but one actuator faults. The actuator with fault can be isolated from the normal ones exactly. A time delay operator is used to deal with the problem brought by the time-delay system. The effectiveness of the proposed method has been verified by a numerical example.
However, further studies are required which include the follow aspects:

1. To determine an optimal recursion step L . Such that the residuals can obtain a certain freedom to complete fault isolation, while the computation is minimized.
2. To extend the fault isolation result. The sensor faults and the actuator faults should be discerned.
3. To enhance the reliability and robust performance of the FDI system.

6. References

Chen, J.; Patton, R. J. (1999). *Robust Model-based Fault Diagnosis for Dynamic Systems,* Kluwer Academic Publishers

Chow, E. Y.; Willsky, A. S. (1984). Analytic redundancy and the design of robust fault detection systems. *IEEE Transaction on Automatic Control,* Vol. 29, Jul. 1984, 603-614

Dugard, L.; Verriest, E. I. (1997). *Stability and Control of Time-delay Systems. Lecture Notes in Control and Information Sciences,* Springer

Frank, P. M.; Ding, X. (1997). Survey of robust residual generation and evaluation methods in observer-based fault detection systems. *Journal of Process Control,* Vol. 7, Jun. 1997 pp. 403-424

Gertler , J. Li, B. (1991). Analytical redundancy methods in fault detection and isolation, *Proceedings of the IFACIIMACS Symposium SAFEPROCESS 9,* pp. 9-21, Baden-Baden 1991

KRATZ, F.; NUNINGER, W. & PLOIX, S. (1998). Fault detection for time-delay systems: a parity space approach, *Proceedings of the American Control Conference,* pp. 2009-2011, Philadelphia, Pennsylvania, 1998

Patton, R. J.; Frank, P. M. and Clark, R. N. (2000). *Issues of Fault Diagnosis for Dynamic Systems,* Springer

Venkatasubramanian, V.; Rengaswamy, R.; Yin, K. and Kavuri, Surya N. (2003). S A review of process fault detection and diagnosis: Part I: quantitative model-based methods. *Computers and Chemical Engineering,* Vol. 27, 2003 pp. 293-311

Yang, H.; Saif, M. (1998). Observer design and fault diagnosis for state retarded dynamical systems. *Automatica,* Vol. 34, Feb. 1998 pp. 217-227

Zhong, M.; Ye, H. & Wang, G. (2004). Multi-freedom design of fault detection filter for linear time-delay systems, *Proceedings of2004 8th Inernational Conference on Control, Automation, Robotics and Vision*, pp. 1630-1634, Kunming, China, 2004

7

Model-Based FDI Schemes For Robot Manipulators Using Soft Computing Techniques

Tolga YÜKSEL and Abdullah SEZGİN
Ondokuz Mayıs University
TURKEY

1. Introduction

While modern control methods were becoming widespread, in addition to demanded repeatability and accuracy specifications, reliability and detection and isolation of probable faults have become an obligation for automatic control systems. In the early 70's, first studies were appeared on this subject. While the first studies on fault detection and isolation (FDI) were implemented for supervisory of chemical processes, following studies were extended to systems like air and spacecrafts, automobiles, nuclear reactors, turbines and HVACs with high reliability mandatories after especially aircraft accidents with high mortality. In 1991, with extending and increasing studies, IFAC SAFEPROCESS comittee was founded and in 1993, this comittee issued some definitions about fault types, fault detection and isolation, fault diagnosis and fault tolerant control (FTC) (Isermann & Ballé, 1997).

Robots are accepted as an assistant subsystem or an individual part of a complex system in most applications. In addition to applications like serial product lines in which they can work harder, faster and with higher accuracy than humans, they are assigned to missions like waste treatment in nuclear reactors, data and sample collection, maintenance in space and underwater tasks which can be very risky for humans. As a consequence, a fault in one product line may cause a pause in all connected lines even in flexible automation systems or a developing and undetected fault may cause abortion of a whole space or underwater mission with big money costs, it may even cause harm to humans. With the increase of these events in real-life applications and with 90's, studies on robot reliability and fault detection and diagnosis in robotics have become common. In addition to these studies, NASA and US Army issued some standards on robots and on the reliability and fault possibilities of robotparts (Cavallaro & Walker, 1994).

This study is focused on model-based FDI schemes, how they can be applied to robot manipulators, how soft computing techniques can be used in these schemes and three different FDI schemes are proposed. Soft computing techniques which can overcome the difficulties of schemes using analytical methods for nonlinear systems are used as modelling, fault isolator and fault function approximator tools in the proposed schemes. In the following section, a literature overview on FDI for nonlinear systems and robot manipulators is given. In Section 3, defined faults and their physical causes are explained. In Section 4, 5 and 6, the proposed schemes are introduced and explained. Soft computing tools used in these schemes are introduced; furthermore, how they work and what their duties are in these schemes are explained in details. In Section 7, simulation implementations and results of these schemes for a two-link robot are given. In the last section, a comparison of these schemes according to some FDI specifications and future studies on these schemes are given.

Most studies in the literature are interested in sensor faults and locked and free-swinging joint (actuator) faults. This study is interested in abrupt partial actuator faults defined in Section 3 to contribute model-based FDI studies for robot manipulators. Furthermore, most studies using soft computing techniques for FDI are interested in how the parameters in their soft computing tools can be defined in terms of faults and how they can be updated according to faults like adaptive learning. In this study, these tools are used directly and without any modification to give appropriate outputs for appropriate inputs. From this point of view, this study can be accepted as a bridge between model-based FDI and data-based FDI methods. Furthermore, in this study, in addtion to soft computing tools, a hybrid soft computing tool M-ANFIS (multiple-ANFIS) which combines and utilizes benefits of neural network (NN) and fuzzy logic (FL) is used for modelling and function approximation. The two-link robot manipulator used in the simulations can be seen simple but accepted as a test platform for most studies. Besides, it has sufficient specifications for implementation of newly proposed schemes.

2. Literature overview of model-based FDI for nonlinear systems and robot manipulators

Studies and methods on fault detection and isolation can be divided into two main groups: model-based methods and data-based methods (Chen & Patton, 1999). Model-based methods are based on modelling the system and processing the difference signals between the model and the real system named as residuals. Data-based methods are based on processing the input and output signals of the system. Proposed schemes in this study are based on model-based FDI methods and studies on model-based FDI are examined in details. Information and surveys about data-based methods can be found in (Chen & Patton, 1999; Patton et al., 2000a; Venkatasubramanian et al., 2003). Model-based fault diagnosis (detection and isolation) is defined as detection, isolation and characterization of faults in components of a system from the comparison of the system's available measurements, with a priori information presented by the system's mathematical model (Chen & Patton, 1999). According to this definition, model-based FDI methods are formed of two steps. The first step is generation of difference signals called *residuals* between real and predicted or estimated output signals of the system. Discordance of these real and predicted or estimated output signals which means nonzero residual signals indicates a potential fault in the system. The second step is isolation of faults using these obtained difference signals according to a decision set.

Model-based FDI methods can be classified according to the method used for residual generation and the decision set is defined according to the specifications of each residual generation method. The point of classification can not only be the method but also linearity (linear-nonlinear-bilinear) type of the system that will be dealt with. The methods used for linear systems can be classified into three main titles. The first and mostly used methods are observer based methods. The main idea behind these methods is to estimate the output of the system using the measurements with Luenberger observes and deterministic adjustments or with Kalman filters and stochastic adjustments (Frank & Ding, 1997). Parity vector (relation) methods use definition of parallel or temporal redundancy which is named as defining a variable with two or more definitions and which can be obtained from measurements or analytical relations (Chow & Willsky, 1984). Parameter estimation methods are based on the principle that accepts sudden changes in parameters like friction, mass, viscosity, resistance etc. using system identification methods as a sign of faults (Isermann & Ballé, 1997; Moseler & Isermann, 2000). In addition to proposing new methods, all these methods are investigated for the robustness against disturbances and uncertainties. A detailed survey about these methods and studies can be found in (Chen & Patton, 1999; Patton et al., 2000; Frank & Ding, 1997).

Generally, two main approaches are adopted for FDI for nonlinear systems (Chen & Patton, 1999). First approach linearizes nonlinear models around one or multiple equilibrium(working) points and generates residuals insensitive to parameter changes in small equilibrium point neighbourhoods using robust techniques. This approach may give good and sufficient results only for systems with low level nonlinearities. But this approach is not suitable for nonlinear systems having highly nonlinear terms and wide working points. Second approach, as proposed for solving this problem, uses multiple nonlinear models for each working points. But this approach bring multiple FDI systems for each working points and this will be not practical for real-time implementations.

As mentioned below, to solve these problems, FDI methods which can deal with nonlinear systems directly must be proposed and developed. Therefore, most of the methods proposed for linear systems are adapted for nonlinear systems. One of the approaches is to use analytical or deterministic nonlinear observers (Chen & Patton, 1999; Patton et al. 2000; Frank & Ding, 1997; Adjallah et al.,1994; Garcia & Frank, 1997). Model of a nonlinear systems is accepted as below:

$$\begin{aligned} \dot{x}(t) &= g(x(t),u(t),f(t),d(t)) \\ y(t) &= h(x(t),u(t),f(t),d(t)) \end{aligned} \tag{1}$$

where $x(t)$ is the state vector, $y(t)$ is the output vector, $u(t)$ is the input vector, $f(t)$ is the fault vector, $d(t)$ is the disturbance vector and $g(\,.\,,\,.\,,\,.\,,\,.\,)$ and $h(\,.\,,\,.\,,\,.\,,\,.\,)$ are nonlinear functions. FDI problem is generating residuals using the observer form defined in (2):

$$\begin{aligned} \dot{\xi}(t) &= g_r(\xi(t),u(t),y(t)) \\ r(t) &= h_r(\xi(t),u(t),y(t)) \end{aligned} \tag{2}$$

and the residuals should prove (3):

$$\|r(t)\| \begin{cases} \approx 0 & f(t) = 0 \\ >> 0 & f(t) \neq 0 \end{cases} \tag{3}$$

Here, the aim is to design ξ appropriate for $g_r(\,.\,,\,.\,,\,.\,)$ and $h_r(\,.\,,\,.\,,\,.\,)$. $\xi(t)$ defines state estimator (Garcia & Frank, 1997). In the literature, there are some nonlinear observer desing approaches defined for certain type nonlinear systems (Frank & Ding, 1997; Adjallah et al., 1994; Garcia & Frank, 1997; Seliger & Frank, 1991; Yang & Saif, 1995; Kinnaert, 1999).

All approaches mentioned below use analytical methods but it is hard to obtain analytical models which nonlinear observers are based on. To overcome this problem, "universal approximator" soft computing modelling tools which can model nonlinear systems are preferred. Soft computing techniques involves NN, FL and genetic algorithms (GA). NN can be used to model multiple-input-multiple-output (MIMO) nonlinear systems by using nonlinear mapping capabilities in its hidden black-box structure (Haykin, 1999). Residual generation is implemented by comparing the real system outputs and estimated outputs by NN. Furthermore NNs are superior to analytical techniques on classification and a second NN can be used to isolate faults by evaluating (classifying) the residuals (Patton et al., 2000b; Marcu et al., 1998). Model-based FDI scheme with NNs is shown in Fig. 1:

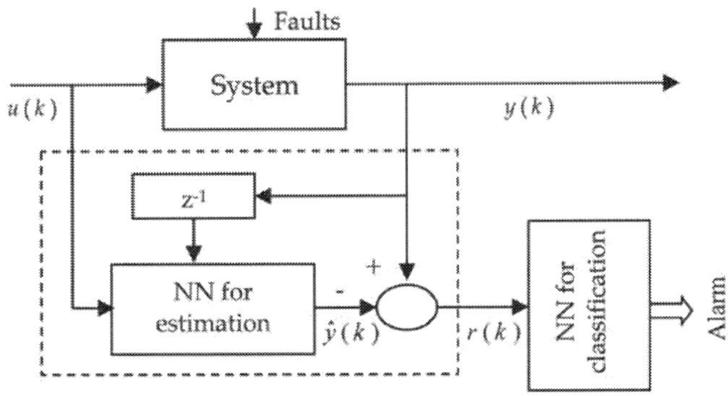

Fig. 1. Model based fault detection and isolation with two neural networks

In addition to these NN approaches, there are some approaches named *online approaches* using adaptive learning and defining NN parameters in terms of input signals anf fault functions (Zhang et al., 2002; Polycarpou & Helmicki, 1995; Polycarpou & Trunov, 2000)

NNs with self learning capabilities can be thought convenient for FDI but, as a black-box structure, keeping user experiences and interferences away from modelling is not a desired feature. To overcome this disadvantage, FL is used as modelling, observer, adaptive residual threshold selector tools for FDI for nonlinear systems (Dexter, 1995; Patton et al., 1998; Ballé, 1998). The main idea behind fuzzy observers is to define a nonlinear system using a set of locally linearized observers with Takagi-Sugeno fuzzy models. Local region definitions are generated using working points. Furthermore, FL is used for residual evaluation. The most important problem in residual evaluation is to set residual threshold. An adaptive threshold can prevent the FDI system from false and missed alarms and FL can be used as the adaptive threshold selector tool (Schneider & Frank, 1996).

FDI studies on robots continue on most of the robot types (mobile, flexible, kinematically redundant, parallel, mobile manipulator, humanoid, bipedal and multi-legged etc.). This study deals with the most common used serial, open chained and rigid robot manipulator and only a detailed review on this type of robots is given here. Studies on other type robots can be found in (Goel et al., 2000; Tinós & Terra, 2002). Most studies on FDI for robot manipulators are based on nonlinear observer approaches. Caccavale and Walker tried to adapt robot dynamics to a certain kind of nonlinear systems and used nonlinear observers which are convenient for these systems (Caccavale & Walker, 1997). Similarly, Schneider and Frank used a robust observer designed for nonlinear systems for robot dynamics and fuzzy logic for residual evaluation/fault isolation (Schneider & Frank, 1996). Leuschen et al. transferred analytical redundancy from linear to nonlinear systems, they defined nonlinear analytical redundancy term and they implemented these redundancies and FDI studies on a hydraulic robot manipulator platform and on a two-link IMI robot manipulator (Leuschen et al., 2005). De Luca and Mattone resembled robot dynamics to generalized notation of adaptive controllers and proposed an adaptive FDI scheme (De Luca & Mattone, 2004). Dixon et al. proposed a filter, they passed torque signals applied to a robot manipulator through this filter and they proposed an FDI technique robust to parametric uncertainties using the difference signals between these filtered signals and their predictions (Dixon et al. 2000). Abdul and Liu proposed an analytical method for the prediction of position and velocity signals of modular type robots and designed a fault tolerant controller using these predictions (Abdul & Liu, 2008). Brambilla et al. generated residuals using inverse robot model, proposed a sliding mode observer and isolated sensor faults using this observer (Brambilla, 2008). Chen and Saif resembled robot dynamics to systems with unknown inputs in state-space and implemented fault detection using output observers (Chen & Saif, 2008).

NNs as an approved tool for FDI for nonlinear systems are also used for robot manipulators. Naughton et al. used nonlinear observer proposed by Adjallah et al. for residual generation and NN for residual evaluation (Adjallah et al., 1994; Naughton et al., 1996). Vemuri and Polycarpou considered fault as a component of robot model function and used adaptive learning strategy of NNs to approximate fault function (Vemuri & Polycarpou, 1997). Terra and Tinós used some different types of NN structures for both residual generation and evaluation (Terra & Tinós, 2001). Lee et al. tried to use parameter identification methods for fault detection and ART type NNs for fault isolation on component and sensor type faults (Lee et al., 2003). Datta et al. tried to classify coefficients obtained from discrete wavelet transform (DWT) using a NN (Datta et al., 2007).

3. Faults defined for robot manipulators

Faults can be classified according to the part of the system, according to modelling or according to time characteristics (Chen & Patton, 1999). In this section, how faults defined for robot manipulators are classified according to the part of the system (Fantuzzi et al., 2003).

Generalized dynamics of robot manipulators are defined in (4) :

$$M(q)\ddot{q} + V(q,\dot{q}) + G(q) + F(\dot{q}) = \tau \tag{4}$$

In (4), n as the number of links, $q(t),\dot{q}(t),\ddot{q}(t) \in \Re^n$ are in order angular position, velocity and accelerations of each link, $M(q) \in \Re^{n\times n}$ is the positive defined inertia matrix, $V(q,\dot{q}) \in \Re^n$ is the Coriolis and centripedal vector, $G(q) \in \Re^n$ is the gravity vector, $F(\dot{q}) \in \Re^n$ is the friction vector and $\tau \in \Re^n$ is the vector of applied torques to joints. If the nonlinear terms in (4) except terms having angular acceleration are expressed as $N(q,\dot{q}) = V(q,\dot{q}) + G(q) + F(\dot{q})$

(4) becomes

$$M(q)\ddot{q} + N(q,\dot{q}) = \tau \qquad (5)$$

Faults are defined in (5) as follows:

a) **Actuator faults**: Motors and power transmission tools are charged as actuators for manipulators and faults occured in these actuators affects the ability of movement. These type of faults change dynamics as follows:

$$M(q)\ddot{q} + N(q,\dot{q}) + u(t-T)f(t) = \tau \qquad (6)$$

- Locked joint faults: These faults occur when the magnetic brake of the motor connected to a joint is locked and doesn't allow any movements (τ free - q_i fixed).
- Free-swinging joint faults: These faults occur when applied torque of the motor connected to a joint is zero because of a disconnected cable (τ zero - q_i free and under impact of other joints or gravity).
- Partial actuator faults: These faults occur when applied torque of the motor connected to a joint decreases (i.e. %20 decrease etc.) because of a fault at power electronics components.

b) **Component faults**: In robot manipulators, broken link, gear corrision or fixed or slipped chain can be component faults. Dynamics of the manipulator change as follows:

$$M(q)\ddot{q} + (N(q,\dot{q}) + f_N) = \tau \qquad (7)$$

c) **Sensor faults**: Optical encoders for joint positions, tachogenerators for joint velocities and tactile sensors for contact forces are the sensors used in robot manipulators. The faults defined for all sensors can be seen in these sensors (bias etc.). Dynamics of the manipulator change as follows:

$$y(t) = [q_1(t) \cdots q_n(t)\; \dot{q}_1(t) \cdots \dot{q}_n(t)]^T + \Delta f_{q,\dot{q}} \qquad (8)$$

In (6), (7) and (8), $u(t\text{-}T)$ is the delayed unit step function, T is the fault occurence instant, f_N is the component fault, $\Delta f_{q,\dot{q}}$ is the sensor fault.

4. FDI scheme with M-ANFIS and NN

The block diagram of the first scheme is shown in Fig. 2. The diagram includes not only the FDI block but also the control block to expose the whole system. To control robot, Computed Torque-PID (CT-PID) method is used and a brief introduction about this method is given in Section 6. CT-PID method needs exact robot model but it can cope with bounded uncertainties. M-ANFIS as the first block of FDI scheme is formed of multiple independent ANFIS structure and implements robot modelling. It takes instant torque values of each joint and unit time step delayed angular position and velocity values of each link. The model gives estimated values of instant angular position and velocity values of each link. Difference signals between real robot and robot model generate residuals. Fault detection alarm is accepted as overshooting of predefined thresholds by residuals obtained from healthy robot operations or simulations. Fault isolation process is implemented by classification of the generated residuals applied to a multilayer NN. In subsections, a review on M-ANFIS and NN with resilient propagation will be given.

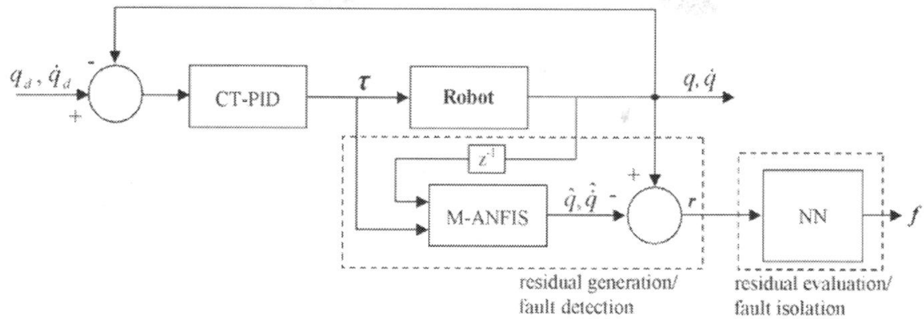

Fig. 2. The block diagram of the first FDI scheme

4.1 Residual generation with M-ANFIS

Difficulties in modelling nonlinear systems analytically, robustness obligations to uncertainty and disturbance directed researchers to soft computing techniques and structures with self, automatic learning and nonlinear mapping capabilities. Soft computing techniques involve NNs, FL, GA and hybrid structures of these tools. Despite NNs have lots of types and lots of learning algorithms, they behave like a black box due to their self-learning nature. Alike FL leaves all parametric adjustments to users, users' experiences become a parameter in modelling and performance of adjustments are dependent on the users. Jang et al. considered advantages of both structures and they decided to combine these advantages and proposed an adaptive network called ANFIS (Adaptive Neuro Fuzzy Inference System) which is functionally equivalent to fuzzy inference system (Jang et al., 1997). ANFIS has a 5-layered structure and a sample with 2 inputs is shown in Fig. 3. Functions of the layers are given below.

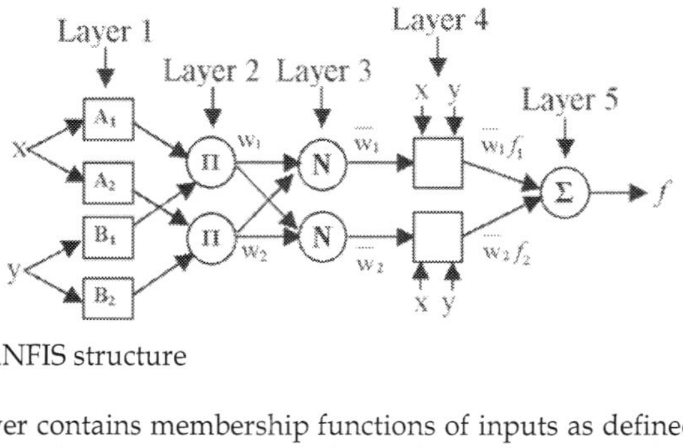

Fig. 3. General ANFIS structure

Layer 1: This layer contains membership functions of inputs as defined in (9) and all inputs are applied to these functions. Type and shape of the membership functions are defined by the user and generally, these functions are bell-shaped functions defined in (10):

$$L1x_i = \mu_{Ai}(x) \quad i = 1,2$$
$$L1y_i = \mu_{Bi}(y) \quad i = 1,2 \tag{9}$$

$$\mu_{Ai}(x) = \frac{1}{1+\left|\dfrac{x-c_{Ai}}{a_{Ai}}\right|^{2b_{Ai}}}, \quad \mu_{Bi}(y) = \frac{1}{1+\left|\dfrac{y-c_{Bi}}{a_{Bi}}\right|^{2b_{Bi}}} \quad i = 1,2 \tag{10}$$

where a_i is center, c_i is width and b_i is crossover gradient, (a_i, b_i, c_i) are parameters of defined function and named as premise parameters.

Layer 2: Each function value is multiplied by other values coming from other inputs due to defined rule base and the result values are named as firing strength of each rule:

$$L2_i = w_i = \mu_{Ai}(x).\mu_{Bi}(y) \quad i = 1,2 \tag{11}$$

Layer 3: Firing strengths are normalized:

$$L3_i = \overline{w}_i = \frac{w_i}{\sum_{j=1}^{2} w_j} \quad i = 1,2 \tag{12}$$

Layer 4: Normalized firing strengths are multiplied by a first order function of inputs:

$$L4_i = \overline{w}_i.f_i = \overline{w}_i(p_i x + q_i y + r_i) \quad i = 1,2 \tag{13}$$

where (p_i, q_i, r_i) are parameters of a first order function and these parameters are named as consequent parameters.

Layer 5: Values coming from all Layer 4 outputs are summed and output value is obtained.

$$L5 = \sum_{i=1}^{2} \overline{w}_i f_i \tag{14}$$

Points that use user experience in ANFIS like in FL are the choices of membership function types at Layer 1 and multiplication operation due to rule base arrangement at Layer 2. These dependencies are exposed in the demonstration of functional equivalence of ANFIS and fuzzy inference system under some circumstances in (Jang et al., 1997).

It is purposed in all network structures to adapt or update network parameters in order to give appropriate outputs against appropriate inputs. From this point of view, ANFIS updates its own parameters using learning algorithms like NNs. As a learning algorithm, backpropagation or hybrid learning expressed in forward and backward passes can be preferred. Table 1 explains hybrid learning. In forward pass, while premise parameters are fixed, inputs go forward until Layer 4 and consequent parameters are determined with least squares. In backward pass, consequent parameters are fixed, error is backpropagated until Layer 1 and premise parameters are determined with gradient descent.

Parameters	Forward pass	Backward pass
Premise	Fixed	Gradient descent
Consequent	LSE	Fixed

Table 1. Parameter updates for hybrid learning in two passes

Fig. 3 illustrates the main disadvantage of ANFIS, being multi-input-single-output (MISO). To model systems with multiple outputs (MIMO), multiple-ANFIS (M-ANFIS) which has independent parameters and outputs is used. A structure that considers correlations between outputs and uses some mutual parameters in order to decrease computational load coming from increasing parameter number exists and it is named as coactive-ANFIS (C-ANFIS) (Jang et al., 1997). In this study, M-ANFIS is preferred and M-ANFIS with 2 inputs and 2 outputs is illustrated in Fig. 4.

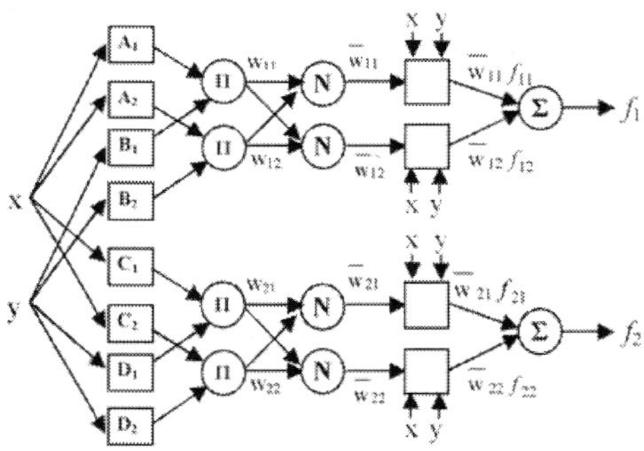

Fig. 4. M-ANFIS with 2 inputs-2 outputs

Robot model takes instant torque and unit time step delayed position and velocity signals belonging to each joint as inputs and generates instant predictions of position and velocity signals of each joint using an ANFIS for each signal. Residuals are generated from differences between real robot and robot model signals. Residual generation using M-ANFIS is shown in Fig. 5.

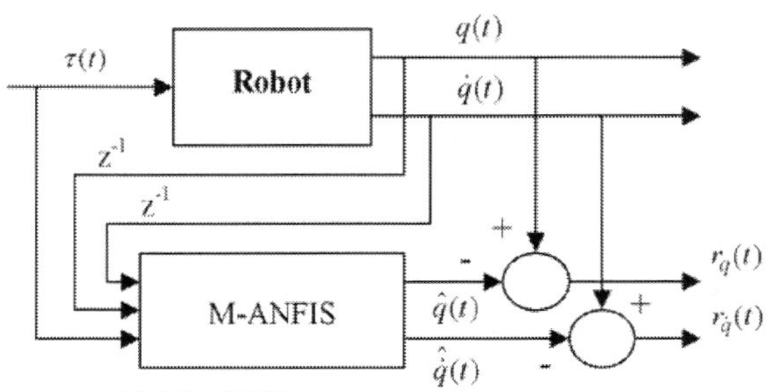

Fig. 5. Residual generation with M-ANFIS

If one of the residual signals overshoots the thresholds determined by simulations and implementations running under healthy conditions, it informs a fault alarm. In this study, in addition to generated residuals, signals called analytical redundancies are used for fault detection and isolation. These signals will be outlined in the following section.

4.2 Residual evaluation using NN with resilient propagation and analytical redundant signals

Residual generation is followed by residual evaluation to isolate faults. This operation is based on the fact that different types of faults show different residual characteristics and it is considered as a classification process. All techniques used for classification and pattern recognition can be used for this process. Multilayer feedforward NNs are very convenient in soft computing based classification tools (Haykin, 1999). (Haykin, 1999) gives all detailed information about NNs. Here, just a review on the used learning algorithm will be given.
NNs aim to update their defined parameters with learning algorithms to give appropriate outputs for appropriate inputs. The most common learning algorithm for this purpose is gradient descent algorithm and it is defined in (15):

$$
\begin{aligned}
w_{ij(n+1)} &= w_{ij}(n) + \Delta w_{ij}(n) \\
\Delta w_{ij}(n) &= -\eta . \frac{\partial E(n)}{\partial w_{ij}}
\end{aligned} \tag{15}
$$

where $E(n)$ as the n. step error function, w_{ij} as the weight from neuron i to neuron j and η as the learning rate parameter. The updates are dependent on the learning rate parameter η and it is known that if it is chosen too small, too many operation steps will be needed and if it is chosen too large, minimum value will not be reached and error value will oscillate around it. To avoid this problem and to accelerate converge, a momentum term with μ momentum parameter is added to (15):

$$
\Delta w_{ij}(n) = -\eta . \frac{\partial E(n)}{\partial w_{ij}} + \mu . \Delta w_{ij}(n-1) \tag{16}
$$

Despite the momentum term in (16), observations showed that this regulation is still dependent on the selected momentum parameter. And again to avoid these parameter dependencies, adaptive learning and momentum parameters are suggested.
These regulations neglect that weight updates are dependent not only the learning rate but also partial derivatives of $E(n)$ with respect to w_{ij}. Resilient Propagation (RP) learning algorithm removes this blurred adaptation from updates and performs updates directly according to the following steps (Riedmiller & Braun, 1993). Firstly, RP assigns Δ_{ij} update values to all weights. These values are updated as defined in (17) and (18):
(17) expresses that if the partial derivative of error with respect to weight changes its sign, the update value is too big and local minima is missed, it should be decreased by η^- factor taking values between 0 and 1 and if it remains with the same sign, it should be increased by η^+ factor taking values greater than 1. After the update value is calculated, the weight update is performed in (18):

$$\Delta_{ij}(n) = \begin{cases} \eta^{+}.\Delta_{ij}^{(n-1)} & \textit{if } \dfrac{\partial E(n-1)}{\partial w_{ij}}.\dfrac{\partial E(n)}{\partial w_{ij}} > 0 \\ \eta^{-}.\Delta_{ij}^{(n-1)} & \textit{if } \dfrac{\partial E(n-1)}{\partial w_{ij}}.\dfrac{\partial E(n)}{\partial w_{ij}} < 0 \\ \Delta_{ij}^{(n-1)} & \textit{else} \\ 0 < \eta^{-} < 1 < \eta^{+} \end{cases} \tag{17}$$

$$\Delta w_{ij}^{(n)} = \begin{cases} -\Delta_{ij}^{(n)} & if \dfrac{\partial E(n)}{\partial w_{ij}} > 0 \\ +\Delta_{ij}^{(n)} & if \dfrac{\partial E(n)}{\partial w_{ij}} < 0 \\ 0 & else \end{cases} \tag{18}$$

On the contrary of other learning algorithms, RP is more transparent, has a more powerful update process and is more efficient with respect to time, memory storage consumption and it is chosen as the learning algorithm in this study.
Simulation results in this study and in some other studies showed that it is hard to train a NN with sufficient fault isolation rates just using existing residual signals (Chen & Patton, 1999; Leuschen et al., 2005). To assist isolation process, some variant signals must be generated. This can be done by using the definition of analytical redundancy as defining one variable in two or more ways like derivation of position and integral of acceleration for velocity. It is clear that the derivative value of position residuals must be equivalent to velocity residuals mathematically and these derivatives can be applied to NN for isolation process and can be used for fault detection. It must be noted that these redundant signals are just used to help fault isolation (classification) process and these signals are not directly sensitive to a fault defined in (Leuschen et al., 2005). Definitions of analytical redundant signals and fault detection process in this study are given in (19) and (20), respectively. NN for fault isolation is shown in Fig. 6.

$$\begin{aligned} (r_q(t))' &= (q(t) - \hat{q}(t))' \\ &= \dot{q}(t) - \dot{\hat{q}}(t) \end{aligned} \tag{19}$$

$$\left.\begin{aligned} r_q(t) &> threshold\_up_{r_q} \\ r_q(t) &< threshold\_down_{r_q} \\ r_{\dot{q}}(t) &> threshold\_up_{r_{\dot{q}}} \\ r_{\dot{q}}(t) &< threshold\_down_{r_{\dot{q}}} \\ (r_q(t))' &> threshold\_up_{(r_q)'} \\ (r_q(t))' &< threshold\_down_{(r_q)'} \end{aligned}\right\} \Rightarrow \textit{fault alarm} \tag{20}$$

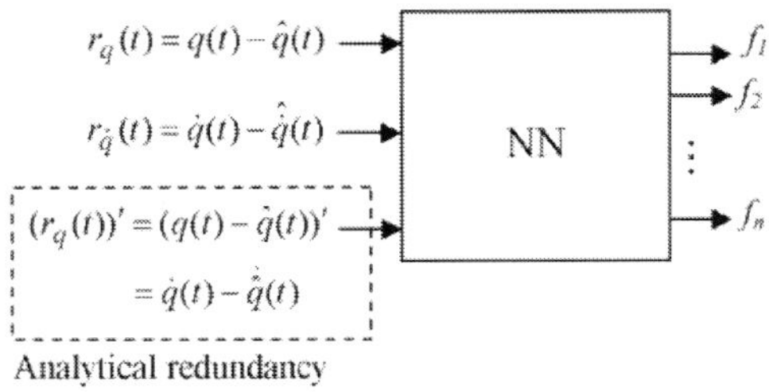

Fig. 6. Fault isolation with NN using residuals and analytical redundancy

5. FDI scheme using generalized observers with M-ANFIS

The design procedure of fault isolation using observes is based on defining relationships between faults and generated residuals. If the residual set can isolate all faults, it can be said that the residual set has the required isolation property. Two methods can be applied to residual sets that involve all residuals to verify this fault isolability property (Chen & Patton, 1999). For dedicated observer schemes (DOS), as the first method, each of the residuals must be sensitive to one fault and insensitive to others. Although this method sounds good, in practice, it is hard to design and to obtain robustness against modelling errors for dedicated observers using analytical approaches. For generalized observer schemes (GOS), as the second method, each of the residuals is sensitive to all but one fault and it is easier to design generalized observers using generated residuals. GOS approach can easily cope with uncertainties by modelling faulty systems one by one. Nevertheless, computational load coming from each system models arises real-time implementation problem. The block diagram of the second proposed model-based FDI scheme is shown in Fig. 7.

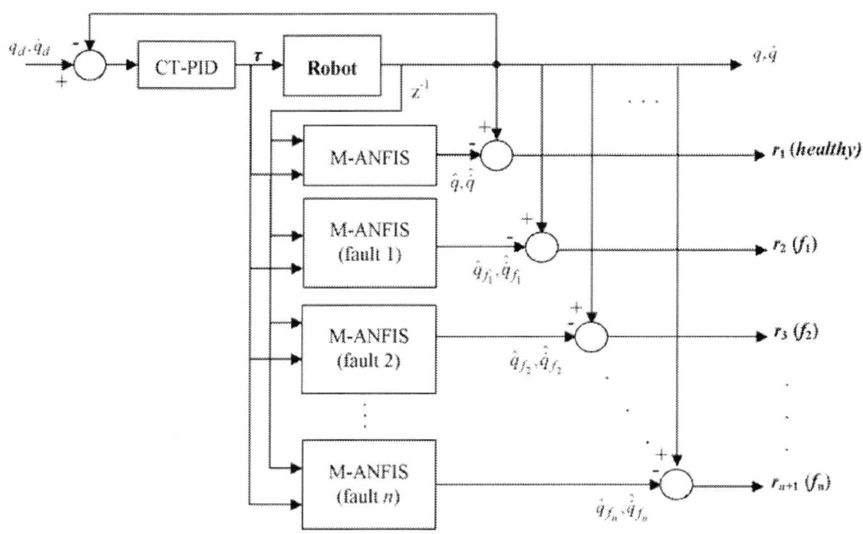

Fig. 7. The block diagram of the second FDI scheme

Again CT-PID is preferred as the robot control method. For the robot manipulator, $n+1$ systems, as one healthy and n faulty, are defined. Each system is modelled using M-ANFIS and generates residual sets (position and velocity residuals of each system) using the same definitions in Section 3.1. Using GOS approach, each of the residual sets overshoots thresholds of all other systems but itself determined by simulations or implementations. GOS combines fault detection and isolation process in one step and that makes the method more effective and attractive. But using models for each defined fault exposes heavy computational load. Table 2 illustrates residuals and residual evaluation process for the proposed scheme.

Faults	r_1	r_2	r_3	...	r_{n+1}
Healthy	0	1	1	1	1
f_1	1	0	1	1	1
f_2	1	1	0	1	1
...	1	1	1	0	1
f_n	1	1	1	1	0

Table 2. Residual evaluation for generalized observer scheme

6. Fault function approximator FDI scheme with M-ANFIS

Despite two proposed schemes are valid and effective for fault detection and isolation, due to their nature, they can be used only for predefined fault types and that makes them hard against partial actuator faults for nonlinear systems and robot manipulators. Furthermore, these schemes can not fully succeed in the design of fault tolerant controllers that use information coming from FDI.

These disadvantages makes fault function approximation more important and the third proposed scheme is focused on fault function approximation. The block diagram of the third scheme is shown in Fig. 8. Modelling part of the scheme is implemented using the same definitions in Section 3.1. In the design of this scheme, to approximate faults, firstly, a M-ANFIS that takes residuals and redundant signals as inputs is trained but the results are vain. As the second attempt, a M-ANFIS that uses instant torque, position and velocity signals as inputs is trained. Each ANFIS of M-ANFIS gives a fault function approximation of each actuator belonging to each joint. Activation of M-ANFIS and fault detection process are

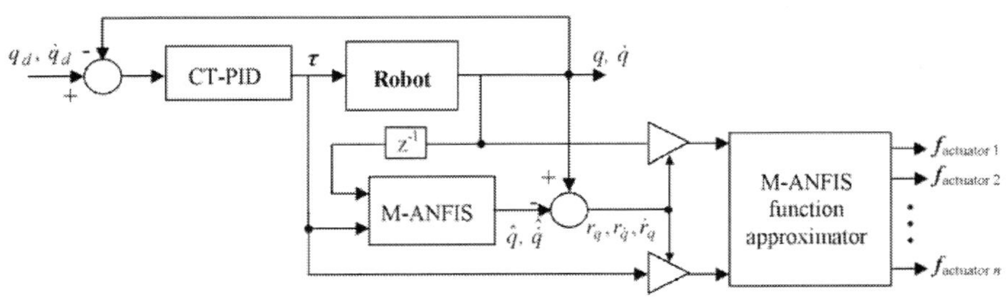

Fig. 8. The block diagram of the third FDI scheme

implemented by residuals and analytical redundant signals. M-ANFIS is designed to give one fault function, the other outputs will give zero value. That makes the fault isolation process unnecessary but simulation results showed that especially for small faults more than one output may give approximations. In this study to avoid this problem, bigger output is accepted as the real fault function approximation.

7. Simulation results

In this section, the proposed FDI schemes are simulated using MATLAB *Fuzzy Logic Toolbox* and *Neural Network Toolbox*. Robot manipulator used in the simulations is a two-link planar manipulator under gravity and masses of the links are defined at the end of the links (Lewis et al., 1993). The manipulator is shown in Fig. 9.

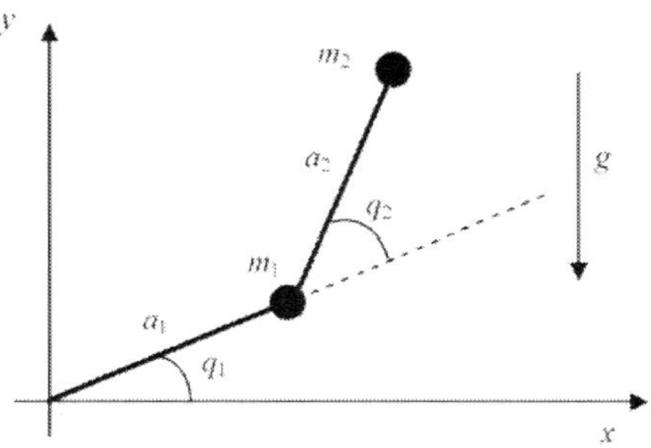

Fig. 9. Two-link manipulator under gravity

The dynamics of the manipulator and the generalized form of dynamics are given in (21) and (22), respectively:

$$\begin{aligned}
&M_{11} = (m_1 + m_2)a_1^2 + m_2 a_2^2 + 2m_2 a_1 a_2 \cos(q_2) \\
&M_{12} = M_{21} = m_2 a_2^2 + m_2 a_1 a_2 \cos(q_2),\ M_{22} = m_2 a_2^2 \\
&V_1 = -m_2 a_1 a_2 (2\dot{q}_1 \dot{q}_2 + \dot{q}_2^2)\sin(q_2) \\
&V_2 = m_2 a_1 a_2 \dot{q}_1^2 \sin(q_2) \\
&G_1 = (m_1 + m_2) g a_1 \cos(q_1) + m_2 g a_2 \cos(q_1 + q_2) \\
&G_2 = m_2 g a_2 \cos(q_1 + q_2) \\
&M(q) = \begin{bmatrix} M_{11} & M_{12} \\ M_{21} & M_{22} \end{bmatrix},\ V(q,\dot{q}) = \begin{bmatrix} V_1 \\ V_2 \end{bmatrix},\ G(q) = \begin{bmatrix} G_1 \\ G_2 \end{bmatrix}
\end{aligned} \tag{21}$$

$$\tau = M(q)\ddot{q} + V(q,\dot{q}) + G(q) = M(q)\ddot{q} + N(q,\dot{q}) \tag{22}$$

where $\tau \in \Re^2$ is the applied torques to joints, $M(q) \in \Re^{2\times 2}$ is the inertia matrix, $V(q,\dot{q}) \in \Re^2$ is the Coriolis/centripedal vector and $G(q) \in \Re^2$ is the gravity vector. Friction and disturbance terms are neglected. The link masses are $m_1 = m_2 = 1$ kg., the link lengths are $a_1 = a_2 = 1$ m. and the sampling frequency is 100 Hz.

Computed Torque control is a special application of feedback linearization of nonlinear systems to robotics. It has commonly PD or PID types and CT-PID is used in this study (Lewis et al., 1993). CT-PID is not very effective against uncertainties but in this study, the controller is not important because suggested FDI schemes use soft computing techniques and they use only the datas coming from the system. If another controller is used, only datas will change and soft computing tools will be trained again using these new datas. Equations of CT-PID are given in (23):

$$\begin{aligned} &\dot{\varepsilon} = e \quad , \quad e = q_d - q \\ &\tau = M(q)(\ddot{q}_d + K_d\dot{e} + K_p e + K_i \varepsilon) + N(q,\dot{q}) \end{aligned} \tag{23}$$

where e is the error, ε is the derivative of error. The gain matrices of the controller are K_d = 100 $I_{2\times2}$, K_p = 20 $I_{2\times2}$, K_i = 500 $I_{2\times2}$.

In this study, 4 different partial actuator faults are defined, examined and they are given in Table 3 with loss percentages. Expression of partial actuator faults defined in (6) with respect to time in (22) is given in (24).

$$M(q(t))\ddot{q}(t) + N(q(t),\dot{q}(t)) = \tau(t) - \alpha.u(t-T).\tau(t) \tag{24}$$

where $\alpha \in \Re^2$ is the partial actuator loss, $u(t) \in \Re^2$ is the unit step and T is the fault occurence instant.

Fault	Name
Actuator 1 %50 loss	f_1
Actuator 1 %30 loss	f_2
Actuator 2 %50 loss	f_3
Actuator 2 %30 loss	f_4

Table 3. Simulated faults

7.1 Case 1: FDI with the first scheme

As the first case, FDI scheme in Fig. 2 is implemented. Firstly, M-ANFIS is constructed as the model. M-ANFIS is formed of 4 independent ANFIS having instant torque and unit time step delayed position and velocity signals as common 6 inputs coming from 2 joints and each giving assigned instant position or velocity signal of each joint. Manipulator is simulated for 146 different joint trajectories defined in sin-, cos-shape with amplitudes varying between ±1 and 76 sampled datas from these simulations are used to train M-ANFIS. Each ANFIS has two bell-shaped membership functions for each input and hybrid learning is selected as the learning algorithm. To show the robustness of the model and residuals generated, time varying %5 dynamics uncertainty is added to (22) as accepted in (25) and the schemes are tested against different uncertainties added to robot dynamics.

$$\|M(q)\| \leq 1.05\|M(q)\|, \|N(q,\dot{q})\| \leq 1.05\|N(q,\dot{q})\| \tag{25}$$

Difference signals between this robot model and real robot generate residuals. Although residuals must be zero in healthy conditions, it may have very small values due to uncertainties. 12 trajectories of 146 trajectories are selected, up and down boundary thresholds of residuals generated from healthy 12 different trajectories are determined and any overshooting is defined as the fault alarm and fault detection part of the scheme.

A four-layered NN with 10-20-20-4 neurons using RP learning algorithm is constructed for residual evaluation/fault isolation. Four faults defined in Table 4 for the same 12 different trajectories are simulated, 4 residuals and 2 analytical redundant (position residual derivatives of each link) signals with 101 samples (6×4848) as 6 inputs and 4 outputs giving 0.9 value, each representing one fault in Table 4, are used to train NN. 7000 epochs and 0 error target are selected as training parameters. Training process is accomplished under 5 minutes and error value is under 10^{-3}.

Simulation tests showed that NN is confused by residuals during transition from healthy to faulty condition and this increases false alarm rate. To prevent this, 3.85 seconds delay is added before NN starts to evaluate residuals after fault detection alarm.

If one of the NN outputs overshoots 0.5, it is defined as fault isolation signal. Simulation studies showed that some faults belonging to some trajectories may cause two outputs (especially fault outputs of the same actuator) to overshoot 0.5 for short time durations, and these results increase false alarm rate. To regulate this, continuity is accepted as criteria and network output signals with time durations shorter than 0.35 s. are neglected.

With these specifications, simulation studies showed %89.58 fault isolation rate for defined faults and trajectories.

As an illustration of the proposed scheme, a trajectory is defined in (26), a %50 fault at actuator 1 (f_1) at t = 24 s. is simulated. Followed trajectories by joints, errors and applied torques to joints are given in Fig. 10, residuals and analytical redundant signals are given in Fig. 11 and NN outputs are given in Fig. 12 for 4-50 s. time interval:

$$\begin{aligned} q_{d1}(t) &= 0.8 * \cos(t/2) \\ q_{d2}(t) &= 0.8 * \sin(t/2) \end{aligned} \tag{26}$$

It can be seen in Fig. 12 that the outputs of f_1 and f_2 overshoots 0.5 but the time continuity rule neglects this time interval. And fault isolation instant t = 28.21 s. can be seen in Fig. 12.

7.2 Case 2: FDI with the second scheme

As the second case, scheme in Fig. 7 is implemented. Four actuator faults, as in the first case defined in Table 1, are considered and from this consideration, as can be seen in Fig. 7, 1 M-ANFIS for healthy and 4 M-ANFIS for faulty robot models are defined and each is formed of 4 independent ANFIS having instant torque and unit time step delayed position and velocity signals as common 6 inputs and each giving assigned instant position or velocity signal. As a result, 5 M-ANFIS with 20 independent ANFIS is constructed. To train, same 146 different joint trajectories defined in sin- and cos-shape with amplitudes varying between ±1 are simulated for healthy and faulty robot models and 76 sampled datas from these simulations are used.

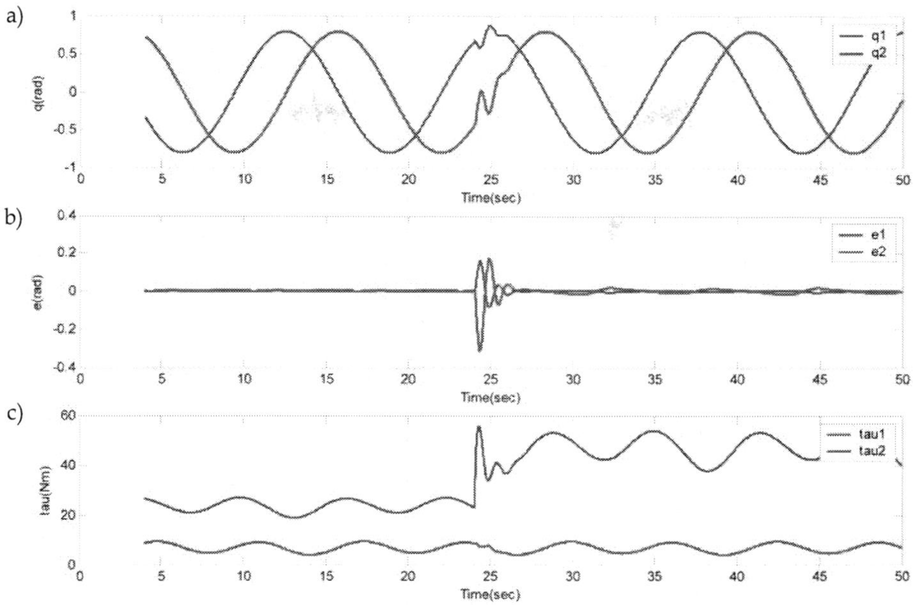

Fig. 10. Time histories of joint 1,2: a) Followed trajectories b) Errors c) Applied torques

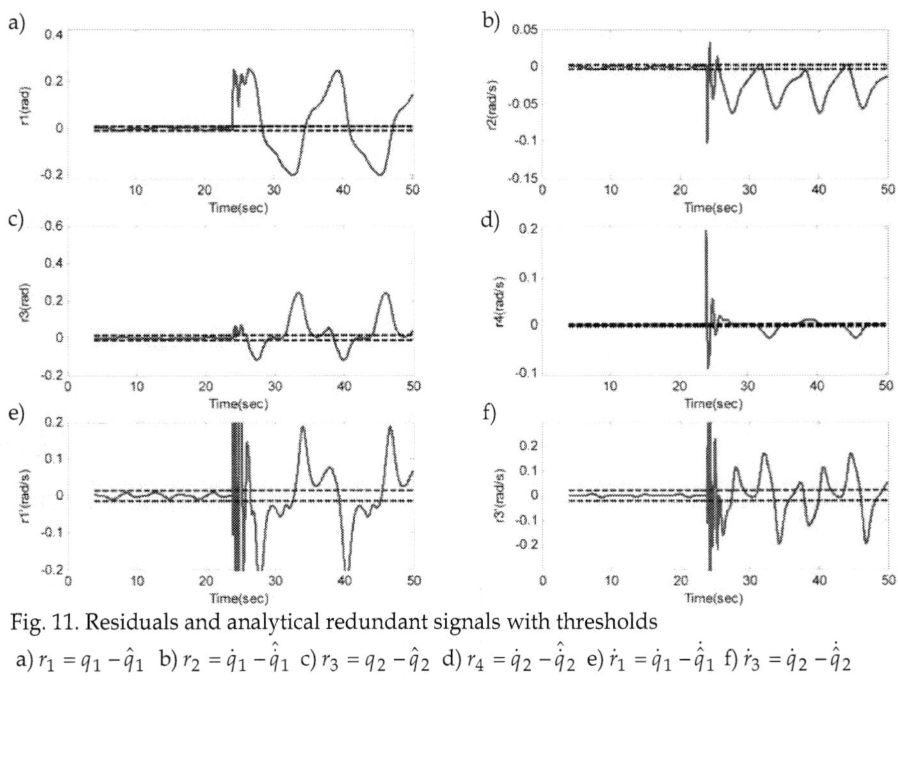

Fig. 11. Residuals and analytical redundant signals with thresholds

a) $r_1 = q_1 - \hat{q}_1$ b) $r_2 = \dot{q}_1 - \hat{\dot{q}}_1$ c) $r_3 = q_2 - \hat{q}_2$ d) $r_4 = \dot{q}_2 - \hat{\dot{q}}_2$ e) $\dot{r}_1 = \dot{q}_1 - \dot{\hat{q}}_1$ f) $\dot{r}_3 = \dot{q}_2 - \dot{\hat{q}}_2$

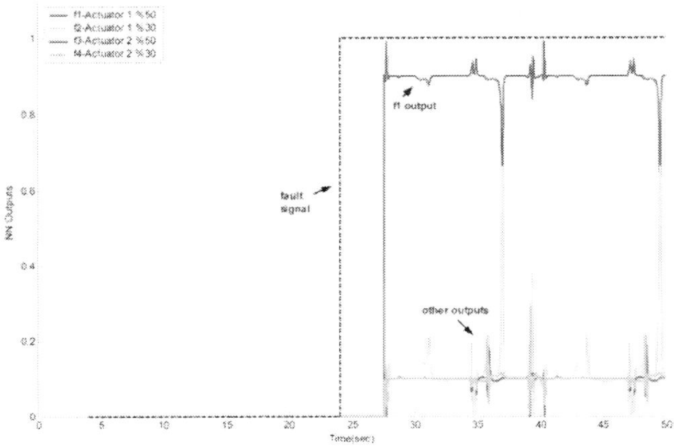

Fig. 12. Neural network outputs

The operation of GOS shows that it can detect and isolate faults at the same time instant. Although Table 2 implies that the residual of the defined fault or healthy condition must be zero, it may have very small values due to modelling errors. To define thresholds for each residual of defined conditions (1 healthy + 4 faulty), 80 trajectories are simulated, up and down thresholds are determined, overshooting of these thresholds for each residual is defined as another condition except belonging to itself. Analytical redundant signals are not defined for this scheme because the residual signals are sufficient and capable of isolating faults using GOS scheme.

Simulation studies for determining fault isolation rate showed that for some faults belonging to some trajectories, two residual sets coming from two models may fall between threshold intervals causing decrease of isolation rate. To increase this rate, residuals are evaluated in the sequence of *healthy*-f_1-f_2-f_3-f_4. Furthermore, again continuity is accepted as criteria and residuals falling between threshold intervals with time durations shorter than 0.8 s. are neglected.

With these specifications, simulation results showed %87.81 fault isolation rate for defined faults and trajectories. As an example of this scheme, the same trajectory in (26) and the same fault f_1 is accepted. Residuals of f_1 are given in Fig. 13 and isolation signals are given in Fig. 14. Fault isolation instant t = 24.93 s. can be seen in Fig. 14.

7.3 Case 3: FDI with the third scheme

As the third case, the scheme in Fig. 8 is implemented. Robot modelling, residual generation and fault detection processes are implemented in the same way with the first scheme. The scheme is simulated for the 80 trajectories used in the second scheme and residual thresholds are determined. To approximate the fault function, M-ANFIS is formed of two independent ANFIS (for each actuator) taking instant torque, position and velocity signals as inputs. Activation of M-ANFIS is realized by the residuals and analytical redundant

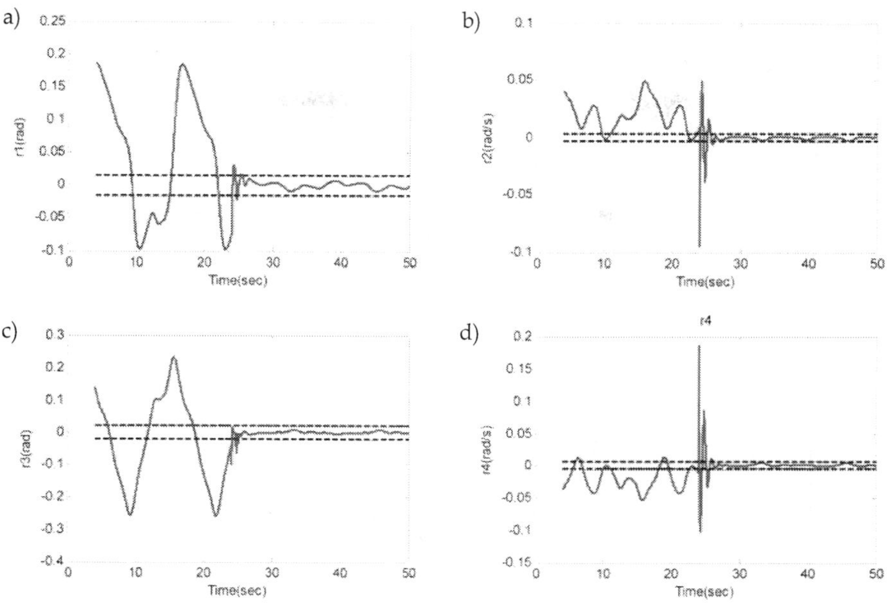

Fig. 13. Residuals of f_1 : a) $r_1 = q_1 - \hat{q}_{1f_1}$ b) $r_2 = q_1 - \hat{\dot{q}}_{1f_1}$ c) $r_3 = q_2 - \hat{q}_{2f_1}$ d) $r_4 = \dot{q}_2 - \hat{\dot{q}}_{2f_1}$

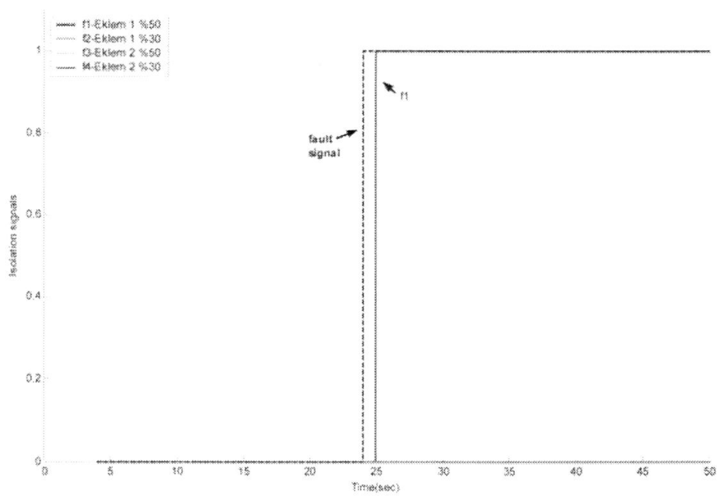

Fig. 14. Fault isolation signals of the second scheme

signals defined for the first scheme. For the training of M-ANFIS, to realize fault function approximation more correctly, 2 extra faults f_5-Actuator 1 %70 torque loss and f_6-Actuator 2 %70 torque loss are defined. Simulations are implemented for defined 6 faults and 101 sampled datas from these simulations are used for training.

Function approximator M-ANFIS is formed of two ANFIS and they are trained to give fault approximation of the faulty actuator. It is hard for the other ANFIS to give zero output because of uncertainties and that causes false alarms. This problem is solved by the rule "The bigger signal is the right signal". This rule is based on that ANFIS output torque of the faulty actuator is mostly bigger than other ANFIS output torques. This rule constitutes false alarms in the cases of low percentage actuator faults resulting close ANFIS outputs.

The second problem is the torque signals containing high frequency components coming from the change of the system region from stable to unstable in the cases of some very high percentage actuator faults. Sudden changes in short time intervals as the high frequency signals lead M-ANFIS to false alarms. These high frequency signals are removed by using a filter like sliding mode control with saturation but the problem still continues and these high frequency components are accepted as a part of the study.

In the simulation studies, like the first scheme, transition from healthy to faulty condition increased false alarms and to decrease these alarms, function approximator M-ANFIS is activated after 3.2 s. delay from fault detection instant.

Again for this scheme, 0.8 s. continuity test is added to the scheme to increase fault isolation rates.

With these specifications, simulation results showed %87.5 fault isolation rate for defined faults and trajectories. Fault function and aproximation signals are given in Fig. 15 and fault isolation signals are given in Fig. 16 for the same example defined for the other schemes. Fault isolation instant t = 28 s. can be seen in Fig. 16. In addition to approximating fault function, this approximation can give information if the fault function causes actuator saturation problem.

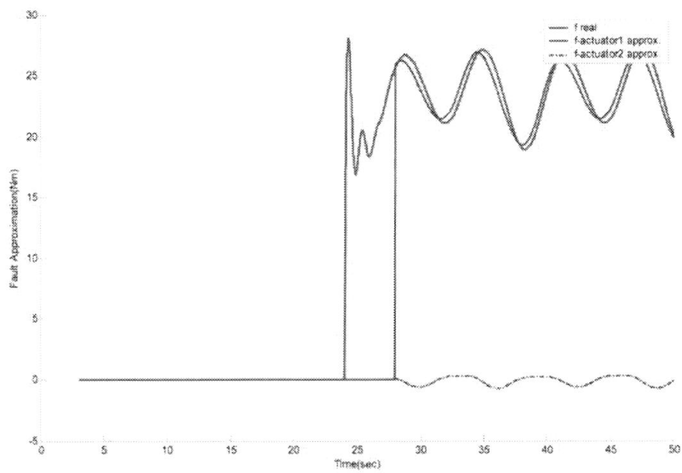

Fig. 15. Fault function and approximation signals

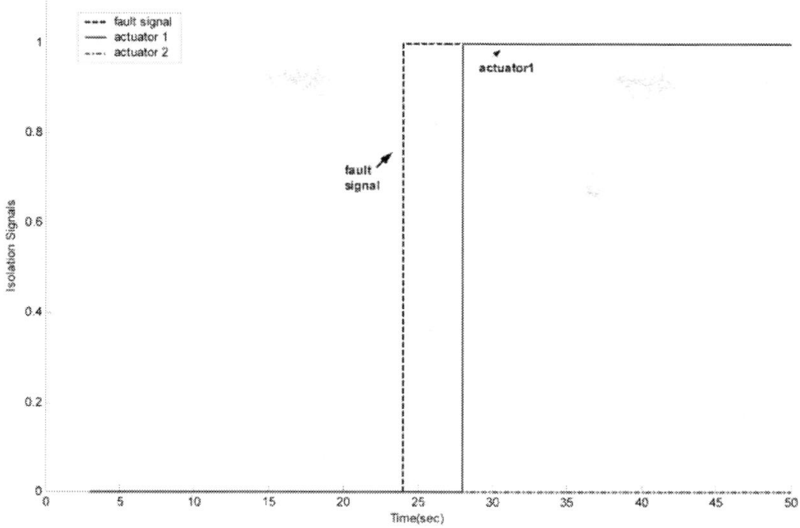

Fig. 16. Fault isolation signals of the third scheme

8. Comparison of the schemes and conclusions

In this study, three model-based fault detection and isolation schemes for robot manipulators using soft computing techniques are proposed. All schemes use multi-input-multi-output type of ANFIS, M-ANFIS structure for modelling. The first scheme uses overshooting of the residuals for fault detection and a NN for fault isolation and the second scheme uses generalized observer approach for fault detection and isolation. The third scheme differs from these schemes by approximating fault function and it uses M-ANFIS for approximation. A comparison of the schemes according to some important FDI specifications is given in Table 4.

All schemes have low false alarm rates, common and positive responses against robustness, and adding new trajectories and training. It is hard to define robustness of these schemes analytically but all schemes are tested against different uncertainties added to robot dynamics and all give similar isolation results. It is observed that new added position trajectories with lower values than ±1 peak values show same results. When new trajectories are added, residual thresholds should be calibrated, isolation structure should be trained and isolation delay and continuity time intervals should be revised if the same isolation results are desired.

One of the most demanded specification for FDI schemes is low or absent fault detection and isolation delays. None of the schemes need fault detection delay. The first and the third schemes have fault isolation delays arising from transition from healthy to faulty condition and these delays should be paid attention especially in real-time implementations. The second scheme realizes detection and isolation in the same method and doesn t need any delays.

Resembling of faulty condition residuals to healthy and other type residuals in short time intervals causes low isolation rates. To prevent this, continuity tests are added to all of the schemes.

Scheme	False Alarm Rate	Robustness	Fault Detection Delay	Fault Isolation Delay	Adding New Trajectories& Training
1. Scheme	low	reliable	no	yes	easy
2. Scheme	low	reliable	no	no	easy
3. Scheme	low	reliable	no	yes	easy
	Continuity Test	Computational Load	Fault Function Approximation	Usable For Fault Tolerance	
1. Scheme	yes	low	just defined faults	just defined faults	
2. Scheme	yes	very high	just defined faults	just defined faults	
3. Scheme	yes	low	yes	yes	

Table 4. Comparison of the schemes

The first and the third schemes can be considered to have low level, the second scheme to have very high level computational load (simulation durations as the proof). The main sources of the computational load in these schemes are soft computing structures. Especially, ANFIS and its big brother M-ANFIS bring high computational loads. The second scheme includes M-ANFIS for all models and that makes it hard for real-time implementations.

The main disadvantage of the first and second scheme is that both are functional for just defined faults. This is a limiting characteristic for fault function approximation and for fault tolerant controller (FTC) schemes as an extension of FDI schemes. Fault function approximation is important to give information about how fault function changes and to give the peak value of control signal to observe the saturation occurence. The third scheme can approximate all faults and that makes it the first choice for real-time implementation.

These three schemes, especially the third one, are suitable for real-time implementation. In the future studies, schemes involving fault tolerant controllers will be proposed. These FTC schemes will use gain scheduling nonlinear PID approach and all proposed schemes will be implemented on a real manipulator.

9. References

Abdul, S. & Liu, G. (2008). Decentralised Fault Tolerance and Fault Detection of Modular and Reconfigurable Robots with Joint Torque Sensing, *IEEE International Conference on Robotics and Automation*, pp. 3520-3526

Adjallah, K.; Maquin, D. & Ragot, J. (1994). Non-linear observer-based fault detection, *Third IEEE Conf. on Control Applications*, pp. 1115-1120

Ballé, P. (1998). Fuzzy Model-Based Symptom Generation and Fault Diagnosis for Nonlinear Processes, *IEEE International Conference On Fuzzy Systems*, Vol. 2, pp. 945-950

Brambilla, D.; Capisani, L. M.; Ferrara, A. & Pisu P. (2008). Fault Detection for Robot Manipulators via Second-Order Sliding Modes , *IEEE Transactions on Industrial Electronics*, Vol. 55, No. 11, pp. 3954-3963

Caccavale, F. & Walker, I. D. (1997). Observer-based fault detection for robot manipulators, *IEEE International Conference on Robotics and Automation*, pp. 2881-2887

Cavallaro, J. R. & Walker, I. D. (1994). A survey of NASA and military standards on fault tolerance and reliability applied to robotics, *AIAA*

Chen, J. & Patton, R. J. (1999). *Robust Model-Based Fault Diagnosis For Dynamic Systems*, Kluwer Academic Publishers

Chen, W. & Saif, M. (2008). Output Estimator Based Fault Detection for a Class of Nonlinear Systems with Unknown Inputs, *American Control Conference*, pp. 3307-3312

Chow, E. Y. & Willsky, A. S. (1984). Analytical redundancy and the design of robust failure detection system, *IEEE Transactions on Automatic Control*, Vol. 29 No. 7, pp. 603-614

Datta, A.; Mavroidis, C.; Krishnasamy, J. & Hosek, M. (2007). Neural network based fault diagnostics of industrial robots using wavelet multi-resolution analysis, *American Control Conference*, pp: 1858-1863

De Luca, A. & Mattone, R. (2004). An adapt-and-detect actuator FDI for robot manipulators, *IEEE International Conference on Robotics and Automation*, pp. 879-884

Dexter, A. L. (1995). Fuzzy Model-Based Fault Diagnosis, *IEE Proceedings On Cont. Theory and Applications*, Vol. 142, No. 6, pp. 545-550

Dixon, W. E.; Walker, I. D.; Dawson, D. M. & Hartranft, J. P. (2000). Fault detection for robot manipulators with parametric uncertainty: a prediction-error-based approach, *IEEE Transactions on Robotics and Automation*, Vol. 16, No. 6, pp. 689-699

Fantuzzi, C.; Secchi, C.; Visioli A. (2003). On The Fault Detection And Isolation Of Industrial Robot Manipulators, *7th International IFAC Symposium on Robot Control*, pp. 399-404

Frank, P. M. & Ding, X. (1997). Survey of robust residual generation and evaluation methods in observer-based fault detection systems, *J. Proc. Cont.*, Vol.7, No.6, pp. 403-424.

Garcia, E. A. & Frank, P. M. (1997). Deterministic nonlinear observer-based Approaches To Fault Diagnosis: A survey, *Control Eng. Practice*, Vol. 5, No. 5, pp. 663-670

Goel, P.; Dedeoglu, G.; Roumeliotis, S. I. & Sukhatme, G. S. (2000). Fault Detection and Identification In A Mobile Robot Using Multiple Model Estimation and Neural Network, *IEEE International Conference on Robotics and Automation*, pp. 2302-2309

Haykin, S. (1999). *Neural Networks: A Compherensive Foundation*, Prentice-Hall, 2. edition

Isermann, R.& Ballé, P. (1997). Trends in the application of model-based fault detection and diagnosis of technical processes, *Control Eng. Practice*, Vol. 5, No. 5, pp. 709-719

Jang, J.R.; Sun, C.T. & Mizutani, E. (1997). *Neuro-fuzzy And Soft Computing*, Prentice-Hall Inc.

Kinnaert, M. (1999) Robust fault detection based on observers for bilinear systems, *Automatica*, Vol. 35, pp. 1829-1842

Lee, I. S.; Kim, J. T.; Lee, J. W.; Lee, D. Y. & Kim K. Y. (2003). Model-based fault detection and isolation method using ART2 neural network, *International Journal Of Intelligent Systems*, Vol. 18, pp: 1087–1100

Leuschen, M. L.; Walker, I. D. & Cavallaro, J. R. (2005). Fault residual generation via nonlinear analytical redundancy, *IEEE Transactions on Control System Technology*, Vol. 13, No. 3, pp. 452-458

Lewis, F. L.; Abdallah, C. T. & Dawson D. M. (1993). *Control Of Robot Manipulators*, MacMillan Publishing

Marcu, T.; Mirea, L. & Frank, P. M. (1998). Neural Observer Schemes For Robust Detection And Isolation Of Process Faults, *UKACC International Conference on CONTROL '98*, pp. 958-963

Moseler, O. & Isermann, R. (2000). Application of model based fault detection to a brushless dc motor, *IEEE Transactions on Industrial Electronics*, Vol. 47, No. 5, pp. 1015-1020

Naughton, J. M.; Chen, Y.C. & Jiang J. (1996). A neural network application to fault diagnosis for robotic manipulator, *IEEE International Conference on Control Applications*, pp. 988-993

Patton, R. J.; Chen, J. & Lopez-Toribio, C. J. (1998). Fuzzy Observers for Nonlinear Dynamic Systems Fault Diagnosis , *37th IEEE Conference on Decision & Control*, pp. 84-89

Patton, R. J.; Frank, P. M. & Clark, R. N. (2000a). *Issues Of Fault Diagnosis For Dynamic Systems*, Springer-Verlag

Patton, R.J.; Uppal, F. J.; Lopez-Toribio, C. J. (2000b). Soft Computing Approaches To Fault Diagnosis For Dynamic Systems: A Survey, *IFAC-SAFEPROCESS*, pp. 298-311

Polycarpou, M. M. & Helmicki, A. J. (1995) Automated Fault Detection and Accommodation: A Learning Systems Approach, *IEEE Transactions On Systems, Man and Cybernetics*, Vol. 25, No. 11, pp. 1447-1458

Polycarpou, M. M. & Trunov, A.B. (2000). Learning Approach to Nonlinear Fault Diagnosis: Detectability Analysis , *IEEE Transactions On Automatic Control*, Vol. 45, No. 4, pp. 806-812

Riedmiller, M. & Braun, H. (1993). A direct adaptive method for faster backpropagation learning: the RPROP algorithm, *IEEE International Conference on Neural Networks*, Vol. 1, pp. 586-591

Schneider, H. & Frank, P. M. (1996). Observer based supervision and fault detection in robots using nonlinear and fuzzy logic residual evaluation, *IEEE Transactions on Control System Technology*, Vol. 4, No. 3, pp. 274-282

Seliger, R. & Frank, P. M. (1991). Fault Diagnosis By Disturbance Decoupled Nonlinear Observers, *30th Conf. on Decision and Control*, pp. 2248-2253

Terra, M. H. & Tinós, R. (2001). Fault detection and isolation in robotic manipulators via neural networks: a comparison among three architectures for residual analysis, *Journal of Robotic Systems*, Vol. 18, No. 7, pp. 357-374

Tinós R. & Terra M. H. (2002). Free-Swinging and Locked Joint Fault Detection and Isolation In Cooperative Manipulators , *E. Symp. on Artificial Neural Networks*, pp. 513-518

Vemuri, A. T. & Polycarpou, M. M. (1997). Neural-network-based robust fault diagnosis in robotic systems, *IEEE Transactions on Neural Networks*, Vol. 8, No. 6, pp. 1410-1420

Venkatasubramanian, V.; Rengaswamy, R.; Kavuri, S.N. & Yin K. (2003). A review of process fault detection and diagnosis part III: process history based methods, *Computers and Chemical Engineering*, Vol.27, pp. 327-346

Yang, H. & Saif, M. (1995). State Observation, Failure Detection And Isolation(FDI) In Bilinear Systems, *34th Conference on Decision & Control*, pp. 2391-2396

Zhang, X.; Polycarpou, M. M. & Parisini, T. (2002). A Robust Detection and Isolation Scheme for Abrupt and Incipient Faults in Nonlinear Systems, *IEEE Transactions On Automatic Control*, Vol. 47, No. 4, pp. 576-592

8

Condition Monitoring and Fault Detection of Electric Drives

Christian Kral[1] and Thomas G. Habetler[2]
[1]*Austrian Institute of Technology,* [2]*Georgia Institute of Technology*
[1]*Austria,* [2]*USA*

1. Introduction

An electric drive consists of an electric machine, which converts electrical power to mechanical power, power electronics to operate the machine and a unit to control the motion of the drive. These are the components of the drive. Parts of each of these components could fail and give rise to specific failure scenarios. The drive types investigated in this chapter are limited to asynchronous induction machine and permanent magnet synchronous machines, since these are the most common machine types in modern electric drive applications. Faults of power electronics are not discussed since most failures lead to the outage of the drive as the power electronics usually show no symptoms before failure.

The task of identifying and classifying drive failures from certain measured quantities is called fault detection. Under some conditions, fault detection may require certain safety protection actions. Example: A turn to turn short circuit in the stator winding of the machine is one example for a safety critical issue. If the short remains for a certain time, parts of the winding will be destroyed. This in turn could cause winding failures that lead to a larger short circuit current which may result in the failure and outage of the entire drive. In this sense, a safety critical issue is a time critical issue. If the failure cannot be detected within a certain time, the drive will be damaged and fails. It is thus highly demanded to accurately detect safety critical faults and to protect the drive (and the application) in this case.

1.1 Classification of Methods

Condition monitoring of electric drives is motivated by different intentions. According to these intentions methods can be classified:

- The operating conditions of the drive are stored in a specific format. From the reported state a maintenance action or a warning for the operator can be triggered. Such methods are *condition monitoring methods.*
- Some state quantities of the drive can be monitored to be fed back to the operating strategy of the drive. Therefore, these classes of methods may be called *condition based control* methods. An estimated operating temperature of, e.g., the winding of the stator or rotor of the machine can be used to control the machine such way that a certain temperature limit of the windings is not exceeded.

- A fault indicator can be used to trigger a protection switch to disconnect the drive from the power supply (Farag et al., 1996). In this case the purpose of the condition monitoring method is to protect the drive and the method is called a *protection method.*

Condition monitoring and fault detection of electric drives are very important tasks in order to maintain the reliability and safety of the drive. Additionally unexpected failures and expensive repair and downtime costs can be avoided or limited this way.

From a technical point of view condition monitoring and fault detection methods can be classified as passive (non-invasive) and active methods.

- *Passive methods* utilize measured quantities and derive certain fault or condition indicators. These methods are usually applied to mains supplied machines and have to no way of influencing the actual operating condition of the drive.
- *Active methods* do need any kind of actuator to control the condition of the drive. Such an actuator may be either the inverter of the drive or an additional power electronics device, connected in series to the feeder cables of the machine.

Another classification refers to physical domain for where the measured signals are acquired from.

- *Electrical monitoring* evaluates only electrical signals such as the currents, voltages, and - in some cases - the mechanical angle between the rotor and stator by means of encoder. For the detection of faults with electrical origin, usually, it is advantageous to evaluate electrical signals.
- *Magnetic monitoring* assesses the signatures determined from measurement coils. Such coils may either be embedded into the stator slots of the machine or are located externally as Rogowki coils (Henao et al., 2003).
- *Vibrational monitoring* is based on the evaluation of vibrational acceleration or velocity. Mechanical imbalances and faults are difficult to detect by electrical quantities. It is thus more reasonable to elaborate vibrational signals for the detection of faults with mechanical origin.
- For *thermal monitoring* the desired temperature can either be measured directly or estimated from state observers or models (Gao et al., 2008a; Kral et al., 2004a).
- Acoustic monitoring is based on the condition specific sound emissions of a drive (Gaylard et al., 1995). The quality of *acoustic monitoring*, however, is very much dependent on the background noise of the environment, the machine is operated at. It is therefore very difficult to implement such methods in a real industrial environment.
- For *mixed physical domain monitoring* methods signals from different domains are evaluate and processed.

In this chapter only electrical monitoring methods are investigated and discussed.

1.2 Structure of Monitoring Methods

Monitoring methods are usually based on the processing blocks illustrated in Fig. 1. In the first processing block the signals are measured. For currents and voltages usually current and voltage transducers are used. The rotor speed and the rotor angle, respectively, are either retrieved by an encoder or determined from the slotting harmonics of the currents in the frequency or time domain (Hurst & Habetler 1996; Kral et al., 2006). Alternative monitoring methods may measure other quantities such as temperatures, the vibrations of the housing, acoustic emissions or the flux in internal or external coils.

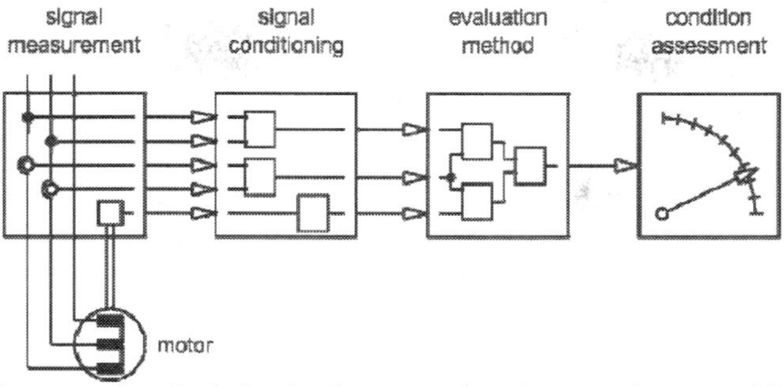

Fig. 1. Block diagram of methods for the detection of machine conditions and faults

Signal conditioning includes some initial processing of the measured quantities. Such processing includes the calculation of the phase voltages and the currents from the measured quantities, depending on whether the connection of the machine is either wye or delta. Additionally, for some methods the space phasors

$$\underline{v}_s = \frac{2}{3}\left(v_1 + e^{j2\pi/3} v_2 + e^{-j2\pi/3} v_3\right) \tag{1}$$

$$\underline{i}_s = \frac{2}{3}\left(i_1 + e^{j2\pi/3} i_2 + e^{-j2\pi/3} i_3\right) \tag{2}$$

are determined from the phase voltages v_1, v_2, v_3 and phase currents i_1, i_2, i_3.

The *evaluation method* incorporates knowledge on the computation of internal machine quantities or fault signatures from the measured signals. An internal quantity could be, e.g., the stator or rotor temperature of the machine. An example of a fault signature is the harmonic component of the current that indicates an electrical asymmetry of the squirrel cage of an induction machine. In this sense, physical or empirical models or practical findings are applied to indicate the condition of the machine.

If a sound knowledge based on the physical context of certain conditions exists, the *assessment* of the condition is straight forward. Example: The stator temperature of a machine is estimated by means of condition monitoring technique. The insulation of the stator winding is specified for a specific insulation class and maximum operating temperature. In this context it is clear that if the maximum operating temperature is exceeded, the machine may be damaged. If the stator temperature is integrated into the control of the load of the machine, the set values for the load can be modified such way, that maximum operating temperature is not exceeded. In another application one wants to predict how long the machine can be operated if it is operated above the maximum temperature. Even if the manufacturer of insulation materials provide life cycle curves which assess the total life time as a function of the operating temperature, it may still be difficult to forecast the remaining life time in an application. From this example the reader can suspect that it is very difficult the give a precise answer to the question: How long is it possible to operate the machine without failing, considering the given conditions or fault condition. The final question in this context is: when does the operator have to schedule maintenance for the machine. In the literature mostly condition monitoring and fault detection methods are investigated without having the focus on fault assessment (Kral et al., 2003). Only little effort has been spent on the prediction of the remaining life time of machines so far - because the subject is so complex and time and cost intense.

2. Faults in Permanent Magnet Synchronous Machines

The high power density and high efficiency of the permanent magnet (PM) machine has led to the use of this machine in applications in which the high reliability is a key feature, such as aerospace/aircraft actuators, automotive auxiliaries and traction, and other industrial and stand-by power generation applications. The switched reluctance motor drives are often used for these applications because of their inherent fault tolerance (Mecrow et al., 1996). While this is true, the switched reluctance machine has a lower power density compared to the PM machine, and therefore is often undesirable. If PM drives are to be considered for these applications, they have to be designed to be very reliable. Redundancy and conservative designs have been used for improving the reliability of these drives against the variety of faults that can occur (Bianchi et al., 1996). As an alternative to these expensive solutions, considerable diagnostic strategies and control schemes have been devised to ensure a fault tolerant drive.

Permanent magnet machines can be divided into two types: permanent magnet synchronous machines (PMSMs) (which have sinusoidal induced stator voltages), and brushless DC machines (BLDCMs) (which have trapezoidal induced stator voltages). PM machines have been very desirable since their torque density and efficiency is higher than any other machine type. However, like all machines, they can fail. Moreover, due to the presence of rotating permanent magnets, damage to the machine can progress even if the stator is disconnected from the line. Fault diagnosis, condition monitoring, and fault tolerant operation of PM machines is very important, if not critical.

2.1 Detecting Faults in Permanent Magnets

Rotor magnet defects can be detected by observing the motor current for the same frequencies as given by equation (7) below (Rajagopalan et al., 2004). They can also be detected by estimating the strength of the permanent magnet as demonstrated for PMSMs in (Le Roux et al., 2003). Estimating the instantaneous or RMS back-EMF gives a measure of the flux linkages resulting from the rotor excitation, making it possible to estimate the magnet strength. There are some examples in the literature of estimating the back-EMF, for example in (Wang et al., 2001).

One method for detecting magnet problems is by estimating the d-axis magnet flux linking the stator windings in the synchronous reference frame (Le Roux et al., 2003). This method is derived in the *rotor reference frame*. The instantaneous value of the d-axis permanent magnet flux linking the stator windings, $\lambda^{r}_{d(pm)}$, is estimated by, (averaging the current over an integer number of fundamental periods),

$$\hat{\lambda}^r_{d(pm)} = \frac{v^r_{qs} - r_s i^r_{qs}}{\omega_r} - L^r_{ds} i^r_{ds} \tag{3}$$

In eqn. (3) all quantities are instantaneous values and i^r_{ds} is the d-axis stator current, L^r_{ds} is the d-axis stator inductance, v^r_{qs} is the q-axis stator voltage, r_s is the stator resistance, and ω_r is the rotor speed. The average value of $\lambda^r_{d(pm)}$ is a direct measure of the magnet strength. In many applications, the rotor angle might not be available to determine the currents and voltages in the rotor rotating reference frame. For these reasons, the estimation is best done in the *synchronous reference frame*, by averaging $\hat{\lambda}^e_{d(pm)}$, which is given by,

$$\hat{\lambda}^e_{d(pm)} = \frac{v^e_{qs} - r_s i^e_{qs}}{\omega_e} - L^e_{ds} i^e_{ds} \,. \tag{4}$$

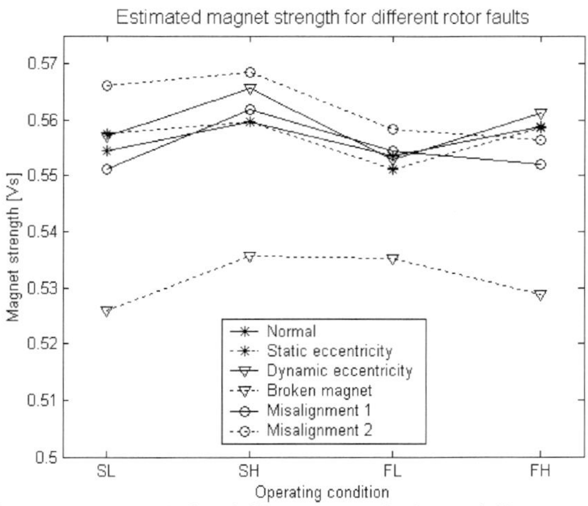

Fig. 2. Estimated magnet strength for different rotor faults at different operating conditions (SL: 640 rpm, 3 Nm; SH: 640 rpm, 6 Nm; FL: 1280 rpm, 4.4 Nm, FH: 1280 rpm, 8.7 Nm)

This method to estimate the magnet strength with (4) was implemented on the measurement data of a normal motor case and all the rotor fault cases. The results of this estimation on these motors operating at different speed and loads are shown in Fig. 2.

For BLDC machines, magnet fault can be found by estimating the mean value of the torque constant, K_t (Rajagopalan et al., 2004). For brushless dc machines (BLDC), this is simply the mean value of the added back-EMFs of the two conducting phases in any particular 60-degree rotor position region.

The method averages the supply voltage (V) and twice the DC-link current (I_{dc}) multiplied by the stator resistance (r_s) in every 60-degree region of rotor position during steady-state operation. However, this estimation neglects the variations in stator inductances. The estimation works better when the back-EMF (E) is used directly, but this is not available on a physical machine. However, the estimated magnet strength is independent of other faults such as the dynamic and static eccentricities, implying that this estimation could still be used to detect a decrease in the magnet strength. Thus, the estimation of the magnet strength is given by,

$$\hat{K} = \frac{V - 2r_s I_{dc}}{\omega_r} = \frac{E}{\omega_r}. \tag{5}$$

2.2 Detection of Rotor Eccentricity

Load unbalance, misalignment, improper mounting, and a bent rotor shaft can all cause rotor eccentricities. These eccentricities can be divided into static and dynamic eccentricities (Fig. 3). Static eccentricity is when the rotor is shifted from its normal position at the center of the stator and it rotates in that position. With dynamic eccentricity, the rotor is also shifted from the normal position, but now rotates around the center of the stator. These rotor faults cause problems such as vibration and noise due to unbalanced magnetic pull (UMP). It also causes dynamic problems by adding to torque pulsations.

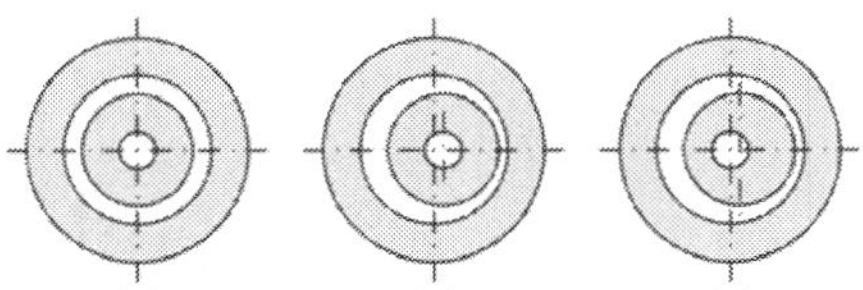

Fig. 3. (a) centric rotor, (b) rotor with static eccentricity, (c) rotor with dynamic eccentricity

Past research in the detection of rotor eccentricities in induction motors has shown that these faults affect certain frequency components in the stator current which can be monitored for use in diagnosing a fault (see section 3). These frequency components (Cameron et al., 1986) and are given by equation (20) in section 3.4 below. Setting k, c, and o in (20) to the appropriate integers yields the frequencies that have to be monitored to detect dynamic eccentricity. Setting $c = 0$, yields the principle slot harmonics, which are the frequencies that have to be monitored to detect static eccentricity. However (20) is of little practical use as the knowledge of stator slots is not available and BLDC machines do not have rotor slots.
Dynamic eccentricity in induction motors causes current components at frequencies of, (Dorrell et al., 1997),

$$f_{de} = f_e \pm \frac{2mf_e}{P}. \tag{6}$$

This equation is the same as equation (18) below, except with slip equal to zero. In (6), f_{de} is the dynamic eccentricity frequency, f_e is the fundamental frequency, and m is any integer. In the presence of dynamic and static eccentricity, current components at the rotating frequency sidebands of f_{de} can be monitored. This equation can be adapted for use in diagnosis of permanent magnet machines too (Le Roux et al., 2002). In the case of BLDC motor drives, there is no rotor slip. Also, there are no rotor bars or rotor windings and therefore there are no rotor slots. This means that $R = 0$ and $s = 0$ in (6). Thus, the only frequencies that can be used to detect dynamic eccentricity are those given in (6) and integer multiples of the supply frequency harmonics, thus $m.f_e$. The only frequencies that are influenced by static eccentricity are the integer multiples of the supply frequency harmonics (setting $R = s = c = 0$).

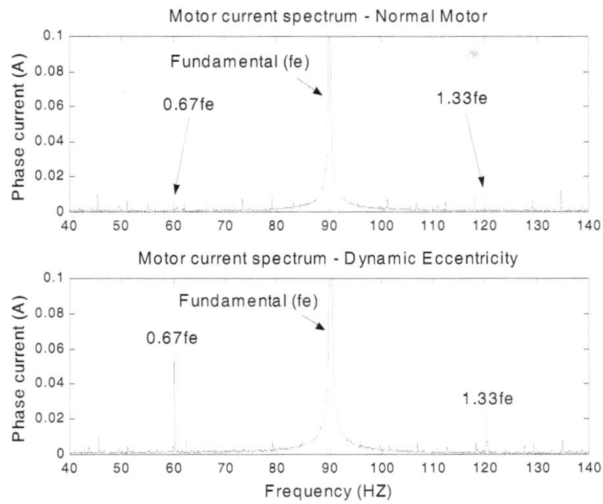

Fig. 4. Comparison of good motor current spectrum to a motor with a dynamic eccentricity

The harmonic spectrum of a good 6-pole PMSM and the same machine with a dynamic eccentricity are shown in Fig. 4. Note the presence of the sidebands of the fundamental at the rotating frequency.

2.3 Detection of Stator Faults

The different types of stator turn faults are depicted in Fig. 5. The first stage is small shorts between turns in the same winding. Among the five failure modes, turn-to-turn faults (stator turn fault) have been considered the most challenging one since the other types of failures are usually the consequence of turn faults. Furthermore, turn faults are very difficult to detect at their initial stages.

Conventional turn fault detection schemes merely monitor the negative sequence component of line currents (or the effective negative sequence impedance) and rely on mathematical models for symmetrical induction machines to account for the effect of unbalanced supply voltages on the negative sequence current (Kliman, 1996). However, neglecting inherent asymmetries can lead to misdetection, with catastrophic consequences. The issue of inherent asymmetries can be addressed by using a neural network-based approach (Tallam et al., 2000).

While this technique has been shown to give very good results for line-connected induction machines, it is often not suitable to PM synchronous machine applications since the current controller in the drive attempt to regulate the current to track the reference value. Since the current is regulated, the effect of the asymmetry from the fault is now reflected in the motor voltage. For this reason, voltage-based turn fault detection methods have been proposed, but require additional voltage sensors and cables.

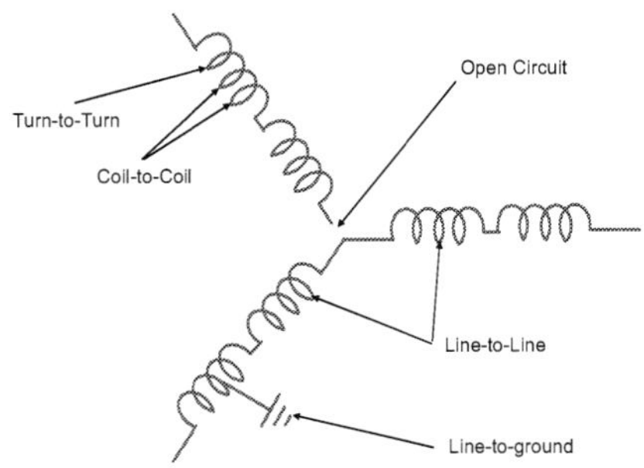

Fig. 5. Possible failure modes in wye-connected stator windings

By modeling a machine with a turn fault, it can be concluded that a bolted turn fault reduces the positive sequence components of the machine impedances and back-emf voltages, while increasing the negative sequence and coupling terms in the impedance matrix at the same time. The positive sequence current slightly increases under a stator turn fault condition in a mains-fed application where the power supply is a fixed voltage source (Lee et al., 2002). In a Current Controlled Voltage Sourced Inverter (CCVSI)-driven application, the inverter controls the line currents so as to follow their references by introducing negative sequence voltage and reducing positive sequence voltage under a stator turn fault condition (Lee et al., 2007). Since the inverter output voltages are produced according to the voltage references that are generated through the current controllers, the variations in the machine parameters will be reflected into the voltage references. This implies that for a given rotating speed and current references (or alternatively torque reference), the presence of a stator turn fault results in a reduced positive sequence component and an increased negative sequence component of the voltage references as compared to a machine without a turn fault. Thus, it can be concluded that the differences in positive and negative sequence components of the voltage references, for a given torque reference and rotating speed, under a stator turn fault and fault-free conditions can indicate the occurrence of a stator turn fault. The voltage references in the rotating and stationary reference frame until fault and no-fault conditions are shown in Fig. 6.

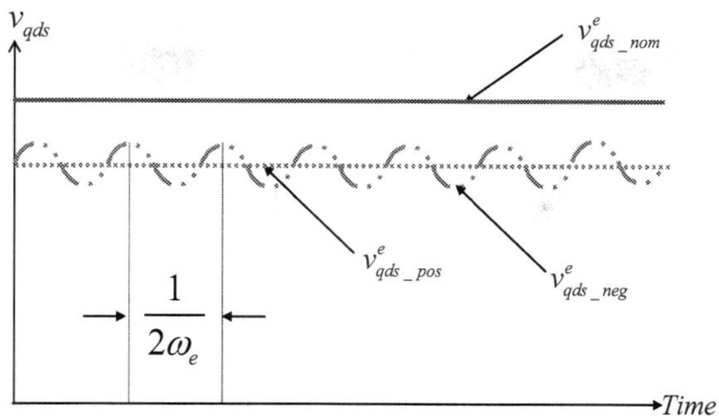

Fig. 6. Voltage references under fault-free and turn fault conditions

2.4 Fault Tolerant Operation

Turn faults are particularly problematic in Interior Permanent Magnet Synchronous Machine (IPMSM) drives in safety-critical applications. This is due to the fact that the rotating magnet can often not be stopped when a fault occurs, and therefore the fault current is allowed to flow until catastrophic or dangerous thermal damage is done to the machine. Therefore, PM machines not only require a reliable turn fault detection method, but also imperatively require a proper remedial action that can maintain the drive's uninterrupted operation.

The most desirable characteristic of a remedial action is to maintain the drive's uninterrupted operation without any degradation in the performance characteristics of the drive in the presence of a stator turn fault. Unfortunately, this is very difficult to achieve, and only redundancy-based approaches can solve this difficulty. But these approaches can be justified in specific applications. In transit applications such as traction drives, an uninterrupted operation during a short period of time, even with a limp operation, can prevent injury or death.

A simple stator turn fault-tolerant strategy for IPMSM drives that does not require any hardware modification to the standard drive configuration has been proposed (Lee et al., 2006). This strategy does not result in the complete loss of availability of the drive. Generally, the asymmetry in the stator voltages resulting from a stator turn fault has only a small effect on the overall stator voltage. Therefore the amplitude of the faulty phase voltage is almost the same as that of the complex stator voltage vector ($\tilde{v}_s^e$) in the synchronous rotating reference frame. Consequently, current in the faulted winding is given by,

$$\left|\tilde{i}_f\right| \approx \frac{\left|\tilde{v}_s^e\right|}{\left|\frac{R_f}{\mu} + r_s + j\omega_e\left[L_{ls} + \mu(L_1 - 3\tilde{L}_2)\right]\right|}. \tag{7}$$

Where R_f is the resistance of shorted section of the winding, μ is the fraction of the winding that is shorted, L_{ls} is the leakage inductance, and L_1 is the average self inductance of the winding, and L_2 is the first harmonic magnitude of the self inductance. This equation implies that an appropriate selection of q- and d-axis current combination for a given operating condition can reduce the stator voltage significantly; consequently, a significant reduction in i_f is achievable while maintaining the given operating condition.

2.5 Fault Detection under Non-Stationary Conditions (Rajagopalan et al., 2005)

Operating conditions of an electric motor usually change rapidly over time. Diagnostics of motor faults in such conditions is a challenging problem due to the need for application of sophisticated signal processing techniques that can process non-stationary signals. While some research has been reported in the detection of faults in induction motors operating under very slowly varying speed and load conditions (Yazici et al., 1999 and Kim et al., 2002), no research has been reported in the diagnosis of faults in motors operating under rapidly varying operating conditions.

The classical technique for characterizing the time evolution of non-stationary signals is the short-time Fourier transform (STFT), a linear time frequency representation (TFR) analysis technique. Though simple and rugged, the STFT still assumes that the non-stationary signal is slowly changing in the chosen time window. This and the choice of window length impair the use of STFT at low frequencies or when the signal is changing very fast dynamically. Beginning with the Wigner-Ville distribution a plethora of quadratic TFR techniques have been introduced in an attempt to improve the time-frequency detail over that achievable with the STFT (Marple et al., 1998). They have been extensively used in the area of mechanical engineering for detection of gear faults.

The use of quadratic TFRs has been presented as a solution for the diagnostics of rotor faults in electric motors operating under non-stationary load and speed conditions. Although the method could be applied to any motor, its application is limited in this discussion to BLDC motors. Four time-frequency representations have been considered - Wigner-Ville distribution (WVD) (Cohen, 1989), Choi-Williams Distribution (CWD) (Choi et al., 1989), and the Zhao-Atlas Marks Distribution (ZAM) (Zaho et al., 1990). The use of time-frequency distributions has proven suitable for detection of rotor faults in electric motors operating under continuous non-stationary conditions. However, the need to provide high frequency resolution along with good cross-term suppression leads to complicated kernels requiring large amounts of processing power.

Some of the commonly used TFRs have been implemented on a DSP platform to study their computational loads. It is observed that quadratic TFRs such as WVD and CWD are computationally more intensive than linear TFRs such as the spectrogram. The quadratic TFRs however provide much better frequency resolution and localization of energy with the ZAM distribution exhibiting the best performance. The CWD in particular is a good trade-off between the excellent frequency resolution of the WVD and the high cross-term suppression of the ZAM distribution. These distributions also do not depend on the size and type of the window as in the spectrogram. The increased computational load is the price paid if a better frequency resolution and good localization of energy is needed. In spite of the increased complexity involved, the computation time of a CWD is still in the order of a few tens of micro-seconds and hence is amenable to implementation in real-time. This computational time can be further decreased by paralleling several micro-programmed systems and using more optimized software routines. The results of Fig. 7 show that the fault frequency magnitudes can be clearly seen even in the case of a time-varying fundamental frequency for the machine.

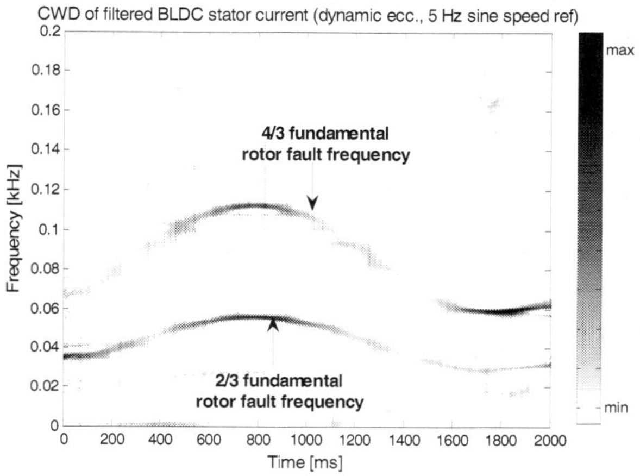

Fig. 7. CWD of a simulated BLDCM rotor fault.

3. Faults in Induction Machines

Many fault statistics have been published for induction machines, e.g., (Bell et al., 1985). In these statistics the failure causes are classified. In most statistics the majority of failures are due to bearing and stator faults. Since these statistics mostly refer to industrial applications the statistics may look different for, e.g., railway traction drives. Additionally, any statistic is inherently incomplete, since it very much depends on the boundary conditions and the integrity of the determined data. However, induction machine conditions are usually classified as:

- *Stator faults* are electrical faults in the stator winding of the machine. Stator faults may be turn to turn faults, interturn faults, phase to phase faults, and phase to ground faults.
- *Rotor faults* refer to electrical faults of the rotor of an induction machines. Since most machines have a squirrel cage rotor, these faults are also called *cage faults*.
- The determination of the *stator and rotor temperature* is of particular interest for the conditional monitoring and control of the machine. The temperature does not directly indicate a fault, but may be an indicator of disadvantageous operating conditions or severe stress to the machine.
- *Eccentricity faults* are caused by mechanical imbalances, misalignments, bent shafts, or non-uniform air gaps. These eccentricities may be caused by the machine, the mechanical coupling or the mechanical load or any combinations of these. To a certain extent eccentricity faults can be detected by means of electrical measurements over the reaction of magnetic field air gap on the stator currents.

- *Bearing faults* are due the failure of one of the bearings of the machine. The causes of failures are manifold and detection of the origin is very difficult. Particularly, the detection of bearing faults by means of electrical signals is a real challenge (Stack et al., 2006).

3.1 Detection of Stator Faults

In the case of a fully symmetric stator winding the stator magneto motive force (MMF) can be synthesized as a Fourier series, i.e., a fundamental and higher harmonic waves. A symmetric winding gives rise to a constant magnitude of the fundamental MMF wave. A fault of the stator winding leads to asymmetry of the stator magneto motive force (MMF) caused by the stator winding. This asymmetry causes the magnitude of the stator MMF not be constant any more. In this case the fundamental wave can be decomposed into a forward and backward traveling wave with constant magnitudes. The forward traveling wave is represented by the positive sequence component of the stator current $\underline{i}_{sp}$ and the additional backward traveling wave due to the stator fault is reflected by the negative sequence component of the stator current $\underline{i}_{sn}$. The sequence components are computed from the phase voltages and currents, however.
A negative sequence current could also be caused by a negative sequence voltage which is due to a supply asymmetry. For the reliable detection of stator turn and winding faults it is important not to confuse these faults with supply voltage imbalances. Nevertheless, the measured negative sequence current and voltage component are used to identify stator turn and winding faults (Tallam et al., 2007; Lee et al., 2003). Additionally, inherent winding asymmetries have to be taken into account for stator fault detection techniques.
The negative sequence voltage, $\underline{v}_{sn}$ is coupled to the positive and negative sequence current through impedances,

$$\underline{v}_{sn} = Z_{snp}\underline{i}_{sp} + Z_{snn}\underline{i}_{sn}\,. \tag{8}$$

In order to compensate inherent winding asymmetries the impedances Z_{snp} and Z_{snn} have to be identified. These impedances are (slightly) dependent on the operating conditions of the machine. The identification of the impedances may therefore be performed by means of neural networks or functional approximations (Tallam et al., 2000). This way supply voltage imbalances and inherent winding asymmetries are properly taken into account. For less sensitive methods, the impedances are assumed to be constant quantities (Kral et al., 2007). After a learning stage, where impedances are identified, the fault detection method can be applied in the regular operation of the machine.
For the identification of the impedances Z_{snp} and Z_{snn} a negative sequence voltage is required, since a symmetric voltage supply does not give rise to negative sequence currents in case of a symmetric machine. The negative sequence voltage can be generated by means of a single phase ohmic resistor which leads to asymmetric line to line voltages. If it is not possible to introduce a negative sequence voltage, the impedances cannot be indentified - except for inherent voltage asymmetries caused by asymmetric loads.
From the measured positive and negative sequence current, the computed negative sequence voltage (8) can be compared with the measured negative sequence voltage.

Alternatively the computed negative sequence current according to (8), incorporating the measured positive sequence current and the measured negative sequence voltage, can be compared with the measured negative sequence current. Any deviation of the computed from the measured value then indicates a stator fault.

In closed loop inverters the identification of the negative sequence component could be contained in the machine voltages and currents (Tallam et al., 2003). It is thus important to take both, the negative sequence voltage and current for the fault detection into account. Nevertheless, the artificially generation of negative sequence currents for the determination of the impedances applied in (8) is much easier since the reference currents can be superimposed with a negative sequence reference quantity.

An alternative method for the detection of stator faults in inverter fed drives is based on the statistical evaluation of the switching behavior of the current controller which is caused by an electrical asymmetry of the stator winding (Wolbank et al., 2003). The advantage of this method is that it easy to implement on an existing dive, since statistical evaluation has to be implemented by means of additional software routines.

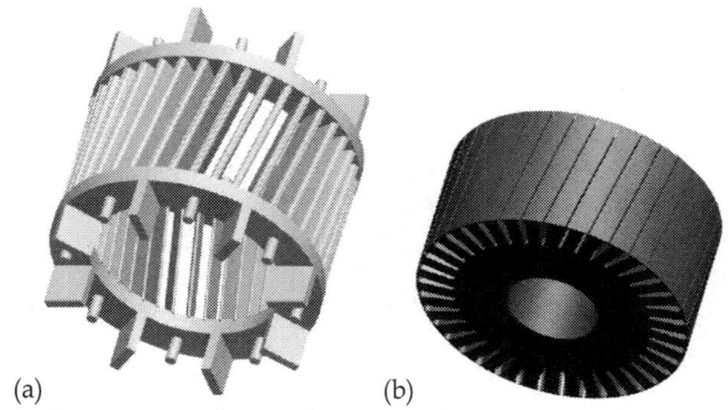

Fig. 8. The squirrel cage rotor of an induction machine consists of (a) rotor bars and end rings and (b) sheet iron

3.2 Detecting Faults in the Squirrel Cage

The squirrel cage of induction machine consists of R rotor bars and end rings on both ends which are placed in the sheet iron of the rotor as depicted in Fig. 8. The bars and end rings of small machines are usually die casted of aluminum or copper. For larger machines, the copper bars are insulated and manually fitted into the sheet iron. In this case the bars and end rings are either welded or hard-soldered.

In the symmetric cage of a squirrel cage induction machine the root mean square (RMS) values of the bar currents are equal - except for inherent asymmetries. Therefore, also end ring currents on both sides are equal. In case of a faulty rotor bar, which is physically reflected in an increase of the respective bar resistance, the current flow of this bar is diminished.

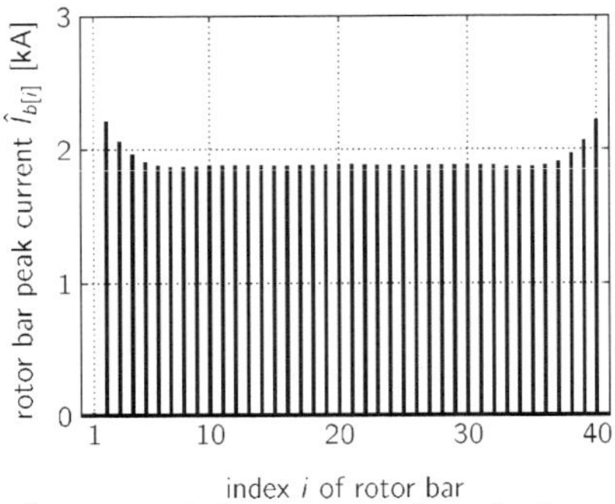

Fig. 9. RMS values of rotor bar currents for the case of one broken rotor bar (index 1); results refer to an 18.5 kW induction machine with 40 rotor bars

In Fig. 9 the RMS rotor bar currents of a squirrel cage with one broken rotor bar are depicted. Since bar number 1 is broken, the respective current is zero. Therefore, the bars adjacent to the fault location have to carry higher RMS bar currents than the rest of bars. The increased currents lead to an increased thermal stress of the respective bars. Due to this phenomenon the adjacent bars could also fail. This in turn leads to the spreading out of the fault from an initial fault location. Over time, the fault condition of the rotor gets worse and worse until the machine fails.

In a real machine a faulty bar leads to an additional phenomenon. Since the rotor bars are usually not insulated from the sheet iron, interbar currents occur. The original bar current is thus diverted to the iron parts of the rotor which in turn damages the sheets.

Apart from broken rotor bars, broken end ring or broken junctions of the bars and end ring segments may give also rise to an electrical asymmetry of the rotor. The entirety of these faults is also called *electrical rotor asymmetries*. These faults cause a distortion of the rotor MMF which can be interpreted as a forward and a backward traveling rotor MMF wave – with respect to the rotor reference frame. The backward traveling rotor MMF wave induces voltages in the stator winding at the so called *lower side band* frequency

$$f_{\mathrm{lsb}}^{r} = (1-2s)f_e, \tag{9}$$

where f_e is the fundamental stator supply frequency and s represents slip. The superscript r indicates that the frequency refers to a rotor fault. The interaction of this harmonic component with the flux causes a double slip frequency

$$f_t^r = 2sf_e \tag{10}$$

torque pulsation. Due to the finite inertia of the drive, rotor speed pulsations occur at the same frequency. The speed pulsations then induce additional side band harmonics in the stator currents, i.e.,

$$f_{\mathrm{lsb}[m]}^{r} = (1-2ms)f_e, \tag{11}$$

$$f^{r}_{\mathrm{usb}[m]} = (1 + 2ms) f_e \, . \tag{12}$$

where m is a positive integer order number. The index usb abbreviates the term *upper side band*, however. The entirety of current harmonics is also reflected in low frequency torque and instantaneous power pulsations at

$$f^{r}_{t[m]} = 2msf_e \, . \tag{13}$$

A typical stator current Fourier spectrum for steady state operation of an induction machine is depicted in Fig. 10. The equidistant displaced lower and upper side band currents are clearly arising in this figure.

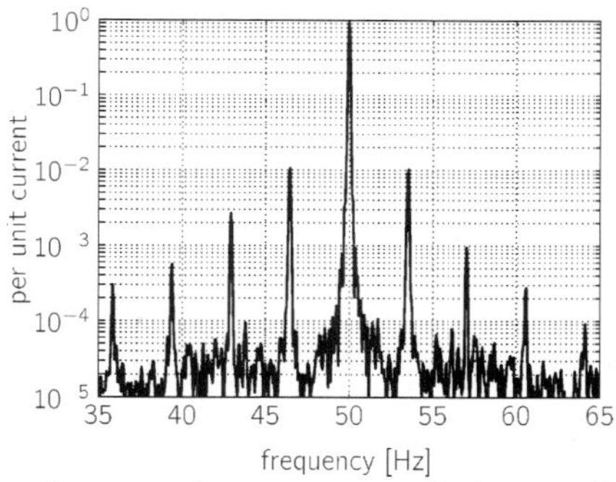

Fig. 10. Fourier spectrum of a per unit stator current during steady state operation; results refer to a 18.5 kW induction machine with one broken rotor bar out of 40 bars

A rotor fault can only be detected when the rotor currents lead to sufficient reactions on the air gap field. The magnitudes of the side band currents of stator current are proportional to load torque in steady state. The magnitudes of side band harmonics with low ordinal numbers m are larger than the magnitudes with higher ordinal numbers. When the mechanical load of the machine is relieved, slip decreases and the magnitudes of the side band currents decrease as well. In this case the side band harmonics merge with the fundamental according to (11) and (12). Therefore, a rotor fault cannot be detected when the machine is mechanically unloaded in steady state. For the reliable detection of rotor faults a minimum load torque of approximately 30% of the nominal torque is required.

The magnitudes of the lower and upper side band harmonics are also influenced by the total inertia of the drive. A large inertia leads to a dominant magnitude of the lower side band harmonic component (9) in the stator current, whereas a low inertia gives rise to a decreased lower side band and an increased higher side band component (Kral et al., 2008a).

Most rotor fault detection methods are based on the evaluation of one or more stator current signatures and are thus called *current signature analysis* (CSA) methods. For steady state operating conditions a conventional Fourier analysis can be used to identify electrical rotor asymmetries. For such applications, it is important to determine the magnitudes of the respective current harmonics with high precision. Since low frequency torque fluctuations

can also induce current sidebands as depicted in Fig. 10, it is important to exactly track the fault specific frequency components so that they do not get confused with load specific harmonics (Kral et al., 2005). From the magnitudes of the side band harmonics the extent of the fault can be estimated according to formulas presented in the literature, e.g., (Thomson & Ranking 1987; Bellini et al., 2000; Culbert & Rhodes 2007).
During transient operating conditions, load torque and thus slip are not constant. A conventional Fourier analysis is therefore not applicable in this case. The problem of time varying load torque can either be solved by means of compensating the modeled load torque (Schoen et al., 1995) or by mathematical transformations and algorithms (Wu et al., 2005). Other approaches use a short time Fourier analysis or multiple signal classification (MUSIC) algorithms to overcome the problem of identification of the non-stationary fault signatures (Benbouzid & Kliman 2003). The fault detection under transient operating conditions can also be performed by means of wavelet transforms (Antonino-Daviu et al., 2006).

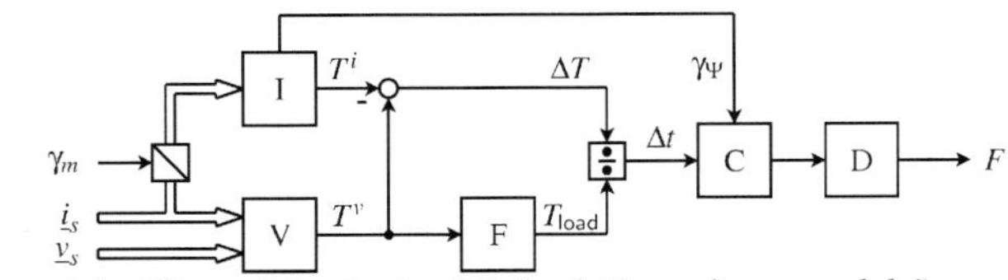

Fig. 11. Scheme of the Vienna Monitoring Method: V = voltage model, I = current model; F = low pass filter for determining the (average) load torque from the torque of the voltage model, C = spatial data clustering technique, D = discrete Fourier analysis

Apart from these methods, electrical rotor asymmetries can be detected by means of the Vienna Monitoring Method (VMM). A scheme of this method is depicted in Fig. 11. This method is a model based approach which compares the torque signals T^v and T^i of a voltage and current model of the machine, resptively; the superscript indicates the model reference. Both these mathematical models are the models of a symmetrical machine. The voltage model is based on the stator voltage equation in the stator reference frame, whereas the current model relies on the rotor voltage in the rotor reference frame. The input quantities of the voltage model are the stator voltage and current space phasor (1) and (2). Since the current model is utilized in the rotor reference frame the stator current space phasor has to be transformed by means of the mechanical angle γ_m - which is the electrical angle of the rotor with respect to the stator.
Applied to a fully symmetrical machine each model calculates the same torque and thus the torque difference

$$\Delta T = T^v - T^i \tag{14}$$

is zero. In case of a rotor fault, both models will calculate a double slip harmonic torque oscillation - as well as higher harmonics which are, however, not taken into account in the VMM. The double slip harmonic torque oscillations computed by the voltage and current model show different magnitudes and phase angles, since both models have a different model structure and different input signals. It turned out that the magnitude of the torque difference is directly proportional to the fault extent and the load torque (Wieser et al., 1999). If the torque difference is divided by the estimated load torque - which is determined from, e.g., the voltage model - the relative torque difference is obtained,

$$\Delta t = \frac{\Delta T}{T_{\text{load}}}. \tag{15}$$

The magnitude of the double slip frequency component of Δt is independent of the load torque and thus a measure for the electrical asymmetry of the rotor. Nevertheless, the frequency of the relative torque difference is depending on slip. In order to eliminate the time dependency of the relative torque difference, a spatial data clustering technique is applied. This technique investigates the relative torque difference versus the angle of the rotor flux space phasor, γ_Ψ ,which is determined by the current model in Fig. 11. Since the rotor flux with respect to the rotor reference frame rotates with slip frequency, the double slip frequency oscillation of the relative torque difference is mapped into a second harmonic component with respect to the rotor circumference. A clustering technique is used to average the wave forms of the obtained second harmonic signal. For this purpose, that rotor circumference is subdivided into n data segments. The data value of each segment is then derived by applying a recursive averaging algorithm. The data values represent an averaged and discretized wave of the second harmonic torque difference. After a certain measurement period a discrete Fourier analysis is applied to the data values. The magnitude of the second harmonic then represents the averaged magnitude of the double slip frequency component of the relative torque difference - which in turn serves as fault indicator for the VMM (Kral et al., 2008a).

A great advantage of the VMM is that it reliably detects rotor faults under almost any operation and boundary conditions. The VMM works

- independent of the load torque of the machine,
- under stationary and transient load conditions,
- under varying load and speed conditions,
- independent of the inertia of the drive,
- for mains supplied and inverter fed machines, independent of the applied control technique and structure.

3.3 Monitoring of Stator and Rotor Temperature

The stator temperature of a machine can be measured by means of sensors which are embedded in the winding, the winding heads or the sheet iron of the stator. This is a state of the art measurement technology. Measuring the rotor temperatures is much more difficult since the rotor is spinning and measuring lines can thus not be used. Even if there exist radio frequency (RF) transmission sensors such technology is almost impossible to be implemented in an industrial environment, because the machine to be monitored has to be dismounted and modified - which is highly undesired.

In some applications it may be undesired to even measure the stator temperatures. In this case it is then required to estimate the temperatures of the stator and rotor, respectively. Neither stator nor rotor temperature are precise terms, since temperature is a local quantity. Both the stator and the rotor have specific temperature distributions, depending on the actual operating conditions. It is for the symmetrical machine (and supply), however, useful to assume homogenous or averaged temperatures, for the conductors in the stator slots, each winding head as well as each rotor bar and both end rings. By means of a lumped parameter thermal model of the machine, these temperatures can be determined (Kral et al., 2008b) - even online. In order to parameterize such a model all geometric details of the stator, rotor, housing and cooling concept have to be known and modeled. A complex lumped parameter model is thus not applicable in practice if the only little is known about the machine.

For practical applications simplified lumped thermal parameter models can be used (Gao et al., 2008b). These models usually employ only a couple thermal time constants. The thermal time constants have to be determined during an identification test or are computed or measured in advance before the machine is put into operation. A lumped parameter thermal model can also be combined with any other resistance estimation method to a hybrid model (Kral et al., 2004a).

In practice, only the average temperature of the stator winding can be estimated. The average stator winding temperature increase is determined by estimating the stator resistance, considering the temperature dependence of the conductor according to,

$$R = R_{\mathrm{ref}}(1 + \alpha_{\mathrm{ref}}(T - T_{\mathrm{ref}})), \tag{16}$$

where R_{ref} is the resistance at temperature T_{ref}, and α_{ref} is the linear temperature coefficient at the reference temperature. In this equation T is the actual temperature and R is the respective actual resistance. One class of methods determines the average stator winding temperature by means of a direct current (DC) bias. Since a DC current does not cause any voltage drop across the stator inductance, the DC voltage represents the voltage drop across the stator resistance only. For symmetric machines it can be assumed that the winding temperatures are equal in all three phases. Therefore, the DC bias is injected only in one phase. For mains fed induction machines the bias could be applied by means of different devices: A resistor in parallel with a diode can be used to cause a voltage drop across this element which is different with respect to the positive and the negative half-wave (Lee & Habetler, 2003b). Other proposals provide the controlled bias by a soft starter, giving rise to a similar effect (Zhang et at., 2008), or use a zero sequence voltage to estimate the stator resistance of the machine (Jacobina et al., 2000).

Since the estimation of resistances is required for the tuning the control of inverter drives, such methods are often integrated into the digital signal processing (DSP) software of the drive. For this class of applications rotor and stator resistances may be focus of interest. In this context observers (Jeon et al., 2002), models (Gao et al., 2008a) and neural networks (Karanayil et al., 2007) are applied. In variable speed drives also signal injection methods can be used to determine the resistances since the inverter can be seen as active voltage source (Wu & Gao, 2006).

3.4 Detection of Rotor Eccentricity

Rotor eccentricities are usually classified by the terms *static* and *dynamic eccentricity*. A visualization of these classes is depicted in Fig. 3. For static eccentricity the axis of rotation of the rotor is displaced from geometric center of the stator. This eccentricity mode can be

caused by a misalignment of the bearings, the end shields or stator ovality. Since the air gap and the air gap field are non-uniform in this case a so called *unbalanced magnetic pull* arises. This pull acts in the direction of the minimum air gap. In the case of dynamic eccentricity the axis of rotation of rotor is aligned with center of the stator, but the axis of rotation is displaced from the rotor center. Dynamic eccentricity may be caused by a bent shaft, mechanical resonances, bearing wear and static eccentricity. An unbalanced magnetic pull is also present in this case (Dorrell 1996). In practice always mixed modes of eccentricity occur (Faiz & Ojaghi 2008).
Mixed static and dynamic eccentricity gives rise to harmonic components with frequencies

$$f_{i[m]}^{e} = f_e (1 \pm m \frac{1-s}{P}), \tag{17}$$

in the stator currents, where P is the number of pole pairs and superscript e indicates the reference to eccentricity related components and $m \in [1,2,3,...]$ is the ordinal number. Due to the interaction of current and flux, power and torque specific components arise at

$$f_{t[m]}^{e} = f_e m \frac{1-s}{P}. \tag{18}$$

The slotting of the rotor also gives rise to additional frequencies

$$f_{iR[m]}^{e} = f_e \left[(mR \pm c) \frac{1-s}{P} \pm o \right], \tag{19}$$

in the current, where $c \in [0,1]$ indicates static (zero) or dynamic (one) eccentricity, and o is an odd integer number, $o \in [1,3,5,\ldots]$.
For stationary operated induction machines, supplied by the mains, the eccentricity related harmonic components (17)) or (19)) can be determined by means of a conventional Fourier analysis (Thomson & Fenger 2001). Oscillating load effects require different approaches since a Fourier analysis is not applicable any more. An alternative method utilizes the negative sequence fault components of the stator current space phasor which is independent of the actual loading of the machine (Wu et al., 2007).
Another approach strictly avoids time domain frequency analysis and is based on the evaluation of (18) applied to instantaneous power (Kral et al., 2004b). A scheme of this method is depicted in Fig. 12. In order to evaluate a quantity independent of the actual rating of the machine, instantaneous power is divided by a reference power, which is, e.g., equal to the nominal power of the machine. A band pass filter extracts the frequency components (18) and a phase locked loop (PLL) is used to track the filtered signal. Since power is a real quantity, a Hilbert transform is applied to generate a complex phasor. The angle γ of the quantity associated to the output of the Hilbert transform is then used to employ a data clustering technique similar to that one used for the Vienna Monitoring Method in subsection 3.2. The data clustering averages the output of the filter in a spatial domain to eliminate any frequencies other than the fault frequency.

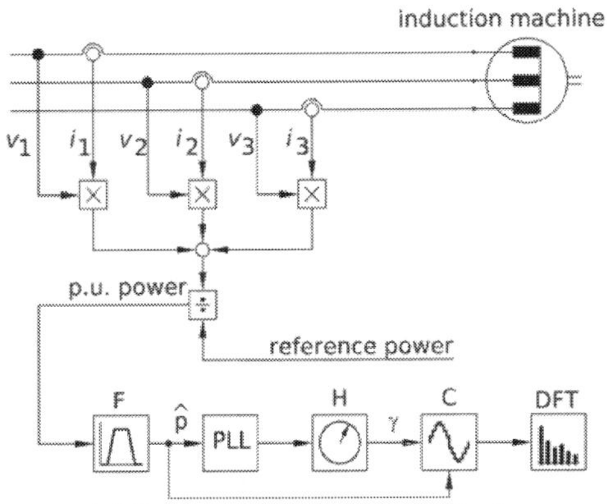

Fig. 12. Method for the eccentricity detection based on instantaneous power; F = band pass filter, PLL = phase locked loop, H = Hilbert transform, C = clustering technique, DFT = discrete Fourier analysis

Eccentricities have mechanical origins and therefore fault detection based on vibration signals leads to more direct - without the interaction of the magnetic field and the reaction on the stator currents - and thus more significant signatures. A fault signature according to (17) and (18) may also be caused by a mechanical mass imbalance of the rotating parts. This case was investigated in (Kral et al., 2004c) with a non-rigid mounting of the machine. It turned out that it is difficult to detect a mechanical problem by means of current signatures whereas vibrational signatures very clearly indicate the issue. The result of this investigation thus shows that the detection of mechanical problems by means of electrical signals is limited.

4. Conclusions

In this chapter state of the art fault detection methods for permanent magnet and induction machines are presented. The discussed methods rely on the evaluation of measured voltages, currents and speed, respectively. Each kind of fault gives rise to a specific fault pattern in either of the measured quantities. Advanced signal processing techniques and neural network methods are used to isolate and assess the fault severity accordingly.
The field of rotating electric machine monitoring and diagnostics has seen major advancements in the two decades. As a result of this, simple motor protection devices will soon be replaced with sophisticated monitoring devices which will provide early indication of impeding faults. The technology described in this chapter can be used to greatly improve factory and process reliability and availability by reducing unscheduled downtime. The potential cost savings are enormous. All of this can be achieved at little or no cost since all the technology here relies only on the use of data from sensors that are all ready installed. Virtually everything described here can be implemented with only software changes to existing microprocessor-based protection relays.

The future will be even more exciting as advances are made in sensorless monitoring of mechanical systems driven by electrical machines, motor bearings, and a wide variety of motor types in a myriad of applications.

5. References

Antonino-Daviu, J. A.; Riera-Guasp, M.; Folch, J. R. & Palomares, M. P. M. (2006). Validation of a new method for the diagnosis of rotor bar failures via wavelet transform in industrial induction machines, *IEEE Transactions on Industry Applications*, vol. 42, no. 4, pp. 990-996.

Bellini, A.; Fillipetti, F.; Franceschini, G.; Tassoni, C.; Passaglia, R.; Saottini, M. & Tontini, G. (2000). ENELs Experience with On-Line Diagnosis of Large Induction Motors Cage Failures, *Conference Record of the IEEE Industry Application Conference, IAS*.

Benbouzid, M. E. H. & Kliman, G. B. (2003). What stator current processing-based technique to use for induction motor rotor faults diagnosis?, *IEEE Transactions on Energy Conversion*, vol. 18, no. 2, pp. 238-244.

Bianchi, N., S. Bolognani and M. Zigliotto, (1996). Analysis of PM Synchronous Motor Drive Failures during Flux Weakening Operation, in Proceedings of the 27th Annual IEEE Power Electronics Specialist Conference - PESC96, Baveno, Italy, 23-27, pp. 1542-1548.

J.M. Cameron, W.T. Thomson and A.B. Dow. (1986). Vibration and Current Monitoring for Detecting Airgap Eccentricity in Large Induction Motors, IEE Proceedings, vol. 133, pt. B, no.3, pp. 155-163.

H. I. Choi and W. J. Williams, (1989). Improved time-frequency representation of multi-component signals using exponential kernels, *IEEE Trans. on Acoustics, Speech, and Signal Processing*, vol. 37, no. 6, pp. 862-871.

Cohen, L. (1989). Time-frequency distributions – A review, *Proceedings of the IEEE*, vol. 77, no. 7, , pp. 941-981.

Culbert, I. M. & Rhodes, W. (2007). Using Current Signature Analysis Technology to Reliably Detect Cage Winding Defects in Squirrel-Cage Induction Motors, *IEEE Transactions on Industry Applications* vol. 43, no. 2, pp. 422-428.

Dorrell, D. G. (1996). Calculation of Unbalanced Magnetic Pull in Small Cage Induction Motors with Scewed Rotors and Dynamic Rotor Eccentricity, *IEEE Transactions on Energy Conversion*, pp. 483-488.

Dorrell, D.G., Thomson, W.T., and Roach, S. (1997). Analysis of Airgap Flux, Current, and Vibration Signals as a Function of the Combination of Static and Dynamic Airgap Eccentricity in 3-Phase Induction Motors, *IEEE Transactions on Industry Applications*, vol. 33, no. 1, (January/February) pp. 24-34.

Faiz, J. & Ojaghi, M. (2009). Different indexes for eccentricity faults diagnosis in three-phase squirrel-cage induction motors: A review, *Mechatronics*, vol. 19, no. 1, pp. 2 - 13.

Farag, S. F.; Bartheld, R. G. & Habetler, T. (1996). An Integrated on-Line Motor Protection System, *IEEE Industry Applications Magazine*, vol. 2, no. 2, pp. 21-26.

Gao, Z.; Habetler, T. G.; Harley, R. G. & Colby, R. S. (2008a). A Sensorless Rotor Temperature Estimator for Induction Machines Based on a Current Harmonic Spectral Estimation Scheme, *IEEE Transactions on Industrial Electronics*, vol. 55, no. 1, pp. 407-416.

Gao, Z.; Colby, R. S.; Habetler, T. G. & Harley, R. G. (2008b). A Model Reduction Perspective on Thermal Models for Induction Machine Overload Relays, *IEEE Transactions on Industrial Electronics*, vol. 55, no. 10, (October 2008) pp. 3525-3534.

Gaylard, A.; Meyer, A. & Landy, C. (1995). Acoustic evaluation of faults in electrical machines, *Seventh International Conference on Electrical Machines and Drives, IEE (Conf. Publ. No. 412), pp.* 147-150.

Henao, H.; Demian, C. & Capolino, G. (2003). A frequency-domain detection of stator winding faults in induction machines using an external flux sensor, *IEEE Transactions on Industry Applications*, vol. 39, pp. 1272-1279.

Hurst, K. D. & Habetler, T. G. (1996). Sensorless speed measurement using current harmonic spectral estimation in induction machine drives, *IEEE Transactions on Power Electronics*, vol. 11, no. 1, pp. 66-73.

Jacobina, C. B.; Filho, J. E. C. & Lima, A. M. (2000). On-line estimation of the stator resistance of induction machines based on zero-sequence model, *IEEE Transactions on Power Electronics*, vol. 15, no. 2, pp. 346-353.

Jeon, S. H.; Oh, K. K. & Choi, J. Y. (2002). Flux observer with online tuning of stator and rotor resistances for induction motors, *IEEE Transactions on Industrial Electronics*, vol. 49, no. 3, pp. 653-664.

Karanayil, B.; Rahman, M. F. & Grantham, C. (2007). Online Stator and Rotor Resistance Estimation Scheme Using Artificial Neural Networks for Vector Controlled Speed Sensorless Induction Motor Drive, *IEEE Transactions on Industrial Electronics*, vol. 54, no. 1, pp. 167-176.

Kim, K. and Parlos, A. G. (2002). Induction motor fault diagnosis based on neuropredictors and wavelet signal processing, *IEEE/ASME Trans. Mechatronics*, vol. 7, no. 2, , pp. 201-219.

Kliman, G.B., W.J. Premerlani, R.A. Koegl, D. Hoeweler, A New Approach to On-Line Turn Fault Detection in ac Motors, *Conference Record of the IEEE-IAS Annual Meeting,* 1996, pp, pp. 687-693.

Kral, C.; Habetler, T. G.; Harley, R. G.; Pirker, F.; Pascoli, G.; Oberguggenberger, H. & Fenz, C. J. M. (2003). A Comparison of Rotor Fault Detection Techniques with Respect to the Assessment of Fault Severity, *Symposium on Diagnostics for Electric Machines, Power Electronics and Drives, SDEMPED*, pp. 265-270.

Kral, C.; Habetler, T. G.; Harley, R. G.; Pirker, F.; Pascoli, G.; Oberguggenberger, H. & Fenz, C. (2004a). Rotor Temperature Estimation of Squirrel-Cage Induction Motors by Means of a Combined Scheme of Parameter Estimation and a Thermal Equivalent Model, *IEEE Transactions on Industry Applications*, vol. 40, no. 4, pp, pp. 1049-1057, ISSN 0093-9994

Kral, C.; Habetler, T. G. & Harley, R. G. (2004b). Detection of Mechanical Imbalances of Induction Machines Without Spectral Analysis of Time-Domain Signals, *IEEE Transactions on Industry Applications*, vol. 40, no. 4, pp. 1101-1105, ISSN 0093-9994

Kral, C.; Fenz, C.; Plainer, M.; Pirker, F. & Pascoli, G. (2004c). Mechanical Imbalances---Test Bed, Measurement, Detection Technique, *International Conference on Electric Machines, ICEM.*

Kral, C.; Kapeller, H.; Pirker, F. & Pascoli, G. (2005). Discrimination of Rotor Faults and Low Frequency Load Torque Modulations of Squirrel Cage Induction Machines by Means of the Vienna Monitoring Method, *IEEE 36th Annual Power Electronics Specialists Conference, PESC, Recife, Brazil*, pp. 2861-2866.

Kral, C.; Gragger, J.; Pirker, F. & Pascoli, G. (2006). Sensorless Rotor Position Estimation through Current Signature Analysis Based on Rotor Saliency Harmonics, *International Symposium on Power Electronics, Electrical Drives, Automation and Motion, SPEEDAM 2006.*

Kral, C.; Kapeller, H. & Pirker, F. (2007). A Stator and Rotor Fault Detection Technique for Induction Machines in Traction Applications for Electric or Hybrid Electric Vehicles, *The World Electric Vehicle Association Journal, WEVA*, vol. 1, pp. 184-189.

Kral, C.; Pirker, F. & Pascoli, G. (2008a). The Impact of Inertia on Rotor Fault Effects - Theoretical Aspects of the Vienna Monitoring Method, *IEEE Transactions on power electronics*, vol. 23, no. 4, pp. 2136-2142.

Kral, C.; Haumer, A. & Bäuml, T. (2008b). Thermal Model and Behavior of a Totally-Enclosed-Water-Cooled Squirrel-Cage Induction Machine for Traction Applications, *IEEE Transactions on Industrial Electronics, Vol. 55, No. 10, October 2008 3555*, vol. 55, no. 10, pp. 3555-3565.

Le Roux, W., R.G. Harley and T.G. Habetler, (2002). Rotor Fault Analysis of a Permanent Magnet Synchronous Machine, *Proceedings of the 15th International Conference on Electrical Machine (ICEM02) on CD*, Brugge, Belgium, pp. 25-28.

Le Roux, W., R.G. Harley and T.G. Habetler, (2003). Detecting Rotor Faults in Permanent Magnet Synchronous Machines, *Proceedings of the 4th International Symposium on Diagnostics for Electric Machines, Power Electronics and Drives on CD*, Atlanta, Georgia, pp. 24-26.

Lee, S.-B., Habetler, T.G.; Harley, R.G.; and Gritter, D.J. (2002). "An evaluation of model-based stator resistance estimation for induction motor stator winding temperature," *IEEE Transactions on Energy Conversion,* vol. 17, no. 1, pp. 7-15.

Lee, S. B.; Tallam, R. M. & Habetler, T. G. (2003a). A robust, on-line turn-fault detection technique for induction machines based on monitoring the sequence component impedance matrix, *IEEE Transactions on Power Electronics,* Vol. 18, No. 3, pp. 865-872.

Lee, S. & Habetler, T. G. (2003b). An online stator winding resistance estimation technique for temperature monitoring of line-connected induction machines, *IEEE Transactions on Industry Applications*, vol. 39, no. 3, pp. 685-694.

Lee, Y.-K.; Habetler, T.G. (2006). A Stator Turn Fault Tolerant Strategy for Induction Motor Drives in Safety Critical Applications, *IEEE Power Electronics Specialists Conference,* pp. 1-7.

S. L. Marple Jr. (1998). Time-frequency signal analysis: issues and alternative methods, in *Proceedings of the IEEE-SP International Symposium on Time-Frequency and Time-Scale Analysis*, pp. 329-332.

B.C. Mecrow, A.G. Jack, J.A. Haylock and J. Coles, (1996). Fault-Tolerant Permanent Magnet Machine Drives, *IEE Proceedings-Electric Power Applications,* Vol. 143, No. 6, pp. 437-442.

Pal, S.K. (1991). Design Criteria for Brushless DC Motors with Hollow Rotor of Samarium Cobalt for Applications above 25000 RPM in Vacuum, *Proceedings of the 5th International Conference on Electrical Machines and Drives*, London, UK, pp. 115-120.

Rajagopalan, S., Le Roux, W., Habetler, T.G. and Harley, R.G. (2004). Diagnosis of potential rotor faults in brushless DC machines, Second International Conference on Power Electronics, Machines and Drives (PEMD 2004). vol. 2, pp. 668 - 673.

Rajagopalan, S.; Restrepo, J.A.; Aller, J.M.; Habetler, T.G.; Harley, R.G. (2005); Selecting time-frequency representations for detecting rotor faults in BLDC motors operating under rapidly varying operating conditions, *Annual Conference of the IEEE Industrial Electronics Society, IECON*, pp. 6-10, 2005, (Nov 2005) pp. 2579 - 2584.

Stack, J. R.; Habetler, T. G. & Harley, R. G. (2006), 'Fault-signature modeling and detection of inner-race bearing faults', *IEEE Transactions on Industry Applications* **42**(1), 61-68.

Tallam, R. M.; Habetler, T. G.; Harley, R. G.; Gritter, D. J. & Burton, B. H. (2000). Neural Network Based On-Line Stator Winding Turn Fault Detection for Induction Motors, *Conference Record of the IEEE Industry Application Conference, IAS*.

Tallam, R. M.; Habetler, T. G. & Harley, R. G. (2003). Stator winding turn-fault detection for closed-loop induction motor drives, *IEEE Transactions on Industry Applications*, vol. 39, no. 3, pp. 720-724.

Tallam, R. M.; Lee, S. B.; Stone, G. C.; Kliman, G. B.; Yoo, J.; Habetler, T. G. & Harley, R. G. (2007). A Survey of Methods for Detection of Stator-Related Faults in Induction Machines, *IEEE Transactions on Industry Applications*, vol. 43, no. 4, pp. 920-933.

Thomson, W. T. & Ranking, D. (1987). Case History of Rotor Winding Fault Diagnosis in Induction Motors, *Conference Proceedings of Condition Monitoring, University College Swanse*, 798-819.

Thomson, W. T. & Fenger, M. (2001). Current Signature Analysis to Detect Induction Motor Faults, *IEEE Industry Applications Magazine*, vol. 7, no. 4, pp. 26-34.

Wieser, R.; Kral, C.; Pirker, F. & Schagginger, M. (1999). On-Line Rotor Cage Monitoring of Inverter-Fed Induction Machines by Means of an Improved Method, *IEEE Transactions on Power Electronics*, vol. 14, no. 5, pp. 858-865.

Wolbank, T. M.; Loparo, K. A. & Wohrnschimmel, R. (2003). Inverter statistics for online detection of stator asymmetries in inverter-fed induction motors, *IEEE Transactions on Industry Applications*, no. 39, vol. 4, pp. 1102-1108.

Wu, L.; Habetler, T. & Harley, R. (2005). Separating Load Torque Oscillation and Rotor Fault Effects in Stator Current-Based Motor Condition Monitoring, *IEEE, International Electric Machines and Drives Conference*, 1889-1894.

Wu, Y. & Gao, H. (2006). Induction-motor stator and rotor winding temperature estimation using signal injection method, *IEEE Transactions on Industry Applications*, vol. 43, no. 4, pp. 1038-1044.

Wu, L.; Habetler, T. & Harley, R. (2007). A Review of Separating Mechanical Load Effects from Rotor Faults Detection in Induction Motors, *IEEE International Symposium on Diagnostics for Electric Machines, Power Electronics and Drives, The 6th, SDEMPED 2007, Cracow, Poland.*

B. Yazici and G. B. Kliman, (1999). An adaptive statistical time-frequency method for detection of broken bars and bearing faults in motors using stator current, *IEEE Trans. Industry Applications*, vol. 35, no. 2, pp. 442-452.

Zhang, P.; Lu, B. & Habetler, T. G. (2008). A Remote and Sensorless Stator Winding Resistance Estimation Method for Thermal Protection of Soft-Starter-Connected Induction Machines, *IEEE Transactions on Industrial Electronics*, vol. 55, no. 10, pp. 3611-3618.

Zhao Y.; Atlas, L. E.; Marks II, R. J. (1990). The use of cone-shaped kernels for generalized time-frequency representations of nonstationary signals, *IEEE Trans. on Acoustics, Speech, and Signal Processing*, vol. 38, no. 7, pp. 1084-1091.

9

Fault Detection in Crypto-Devices

K. BOUSSELAM, G. DI NATALE, M.-L. FLOTTES and B. ROUZEYRE
LIRMM (Université Montpellier II / CNRS UMR 5506)
Montpellier, France

This chapter presents a study on fault detection mechanisms involved in secure devices in order to prevent faults-based attacks. We explore the solutions based on the use of error detection codes (parity bits, CRC) and we discuss the strengths and the weaknesses of these solutions with regards to error and fault detection.

1. Introduction

Today's secure devices are mainly used for storage and processing of confidential data. Current products provide hardware and software secure solutions for civil and online identification, telecommunication, healthcare, banking, pay-TV, access control for restricted systems or areas, e-government... Tomorrow, they will include decision making capabilities for machine to machine applications. 20 billion of secure devices are forecasted in 2020 (4 billion in 2007) [Eurosmart, 2007].

Due to their applications, secure devices must be designed so that they can guarantee high levels of dependability and quality. But in addition to usual dependability features (reliability, availability, safety, robustness to environmental conditions), we also expect that they have the ability to protect information against unauthorized access and intentional misuse. The digital security mechanisms involved in such devices rely on various principles: secrecy of design and implementation, ciphering operations for encryption/decryption of confidential data, and hardware and software countermeasures for attack detection or tolerance.

Encryption is the process of transforming data in order to make it unreadable to anyone except those possessing the decryption key. Encryption can be symmetric or asymmetric. In symmetric encryption, a sender and a recipient share the same secret key, which is used for both encryption of plaintexts and decryptions of corresponding cipher texts. Conversely, the asymmetric algorithms use different keys for encryption and decryption. A sender S communicate his/her public key to all recipients R; messages encrypted with that public key by any R can only be decrypted by the sender S using his/her corresponding private key. The symmetric and asymmetric algorithms are generally public, while the secret and private keys are kept secret. Symmetric encryption is fast but senders and recipients need to define a "secure" key exchange process prior to start communication. Asymmetric encryption can be used for exchanging this secret key at the beginning of a communication, and then a symmetric-key algorithm using that secret key can be used for fast encryption during the remainder of the communication. The *section 2* details the Advanced Encryption Standard

(AES), a symmetric encryption process based on the Rijndael algorithm from Joan Daemen and Vincent Rijmen.

Due to their applications, secure devices are subject to attacks aiming to gather private information. Discovering the secret key of a symmetric cipher for instance allows decrypting the text encrypted with that key. Numerous types of attack rely on the hardware implementation of the cipher since cryptanalysis on recent algorithms is hopefully not practical. Most recent invasive attacks using probes or modifications of the chip are powerful but destroy the package, require time of specialists in laboratories and a proper budget. Non-invasive side-channel attacks use leakage information related to the processed data such as the operational timing, the power consumption of the chip, or the electromagnetic interferences of signals. Active but semi or non-invasive fault-based attacks rely on perturbation of the circuit behavior and use (expected) production of erroneous results for inferring secret information. *Section 3* gives an overview of implementation techniques and data analysis performed on the reference encryption standard AES for fault-based attacks.

Cipher algorithms are often integrated as coprocessors for better performance. As any other function implemented in hardware, these coprocessors need to be carefully tested in order to determine whether they are capable of performing the intended functions. So, test contributes to the dependability and the quality of the devices in the sense that it prevents insertion of failing hardware in dependable devices, and allows revealing faulty behaviours during the chip lifetime. Classically, targeted fault-set includes permanent (or intermittent) faults, which model physical defects due to manufacturing defaults or aging, and "natural" transient faults due to the environment (particle hit) (Reed et al.., 2003). However, test can also contribute to digital security by preventing maliciously injected transient faults to contribute in revealing confidential data. While the detection of permanent faults is generally performed during the execution of specific test modes after production or during maintenance times, transient fault detection must be concurrent to the mission mode of the circuit. Online test solutions are thus implemented in secure chips as part of protection mechanisms targeting fault-attacks. In this context, fault-set includes transient stuck-at faults on gate signals in the combination part of the cipher, and bit-flips on memory elements (Leveugle, 2007). In this context, potential faults are generally detected through identification of erroneous signal(s) on a function or sub-function output. *Section 4* presents the solutions from the literature for error-detection schemes in AES ciphers. These solutions are first compared using common evaluation criteria such as implementation cost and fault detection latency.

These solutions are then compared in terms of error detection capacity in *Section 5*. Experimental data shows the percentage of undetected errors for each scheme according to the number of erroneous bits injected in one or several bytes during execution of the AES encryption.

Finally, the quality of the protection mechanisms is also evaluated in terms of capacity to detect most frequent misbehavior. *Section 6* presents an analysis on the diffusion of errors in the responses of the AES cipher in the case of single transient stuck-at faults and discuss the choice of appropriate detection mechanisms.

2. Advanced Encryption Standard

Even if fault detection techniques described in this chapter are general enough to be implemented for secure devices implementing any cryptographic algorithm, we use the Advanced Encryption Standard (AES) as support example. This section details the AES which was adopted by the US government (FIPS-197, 2001).

The AES algorithm is a symmetric block cipher that can encrypt (encipher) and decrypt (decipher) information. Encryption converts data to an unintelligible form called cipher text; decrypting the cipher text converts the data back into its original form, called plaintext. Encryption and decryption are performed by means of the same cryptographic secret key. The AES algorithm is capable of using cryptographic keys of 128, 192, and 256 bits to encrypt and decrypt data in blocks of 128 bits. This section focuses on the encryption algorithm for 128-bits cryptographic keys (details on decryption and others key lengths are fully described in (FIPS-197, 2001)).

The basic unit for processing in the AES algorithm is a byte. Input, output and secret key bit sequences are internally processed on a two-dimensional array of bytes called the *State*. The State consists of four rows of four bytes. In the State array, denoted by the symbol s, each individual byte has two indices, with its row number r in the range $0 \le r < 3$ and its column number c in the range $0 \le c \le 3$.

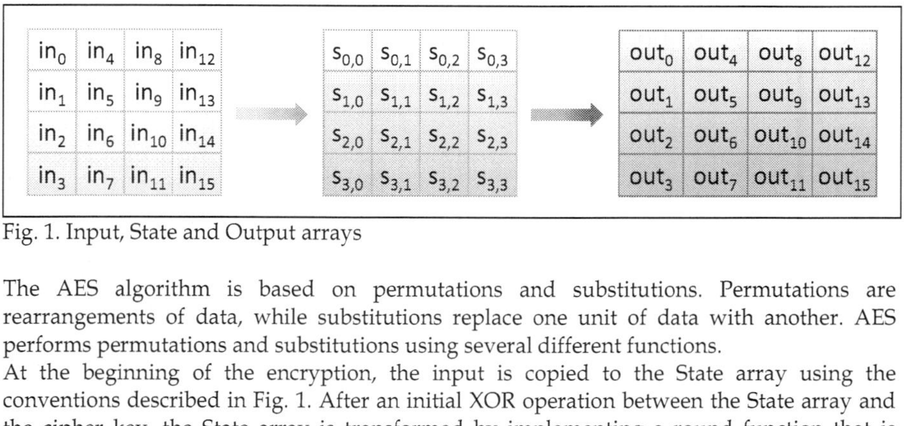

Fig. 1. Input, State and Output arrays

The AES algorithm is based on permutations and substitutions. Permutations are rearrangements of data, while substitutions replace one unit of data with another. AES performs permutations and substitutions using several different functions.

At the beginning of the encryption, the input is copied to the State array using the conventions described in Fig. 1. After an initial XOR operation between the State array and the cipher key, the State array is transformed by implementing a round function that is repeated 10 times (see Section 2.1), with the final round differing slightly from the first 9 rounds. The final State is then copied to the output as shown in Fig. 1.

The round function is parameterized using a Key Expansion function that generates a variation of the original cipher key at each round repetition (see Section 2.2).

Fig. 2 summarizes the AES algorithm.

2.1 Round

Round function is composed of 4 operations: SubBytes, ShiftRows, MixColumns, and AddRoundKey. These functions operate and modify the value of the State. As shown in Fig.

2, all rounds are identical with the exception of the final round, which does not include the MixColumns transformation.

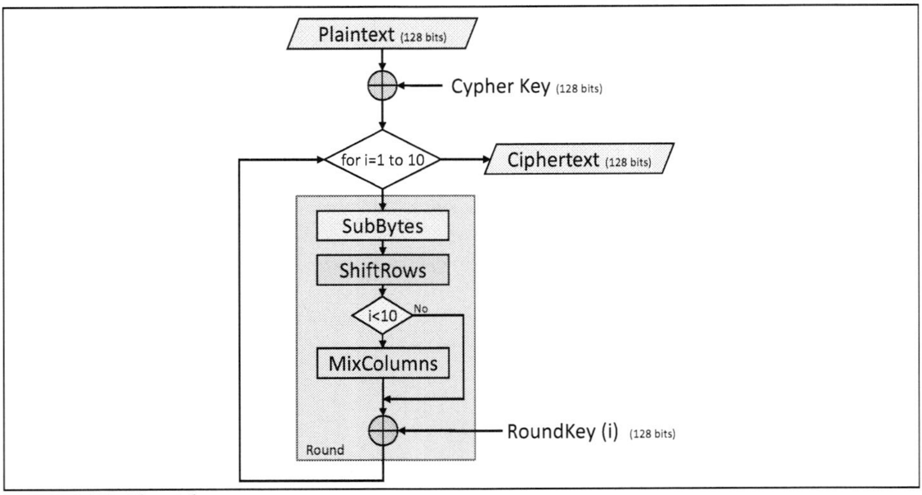

Fig. 2. AES Algorithm

A schematic view of the Round function is shown in Fig. 3. Next paragraphs describe in detail each of the 4 operations.

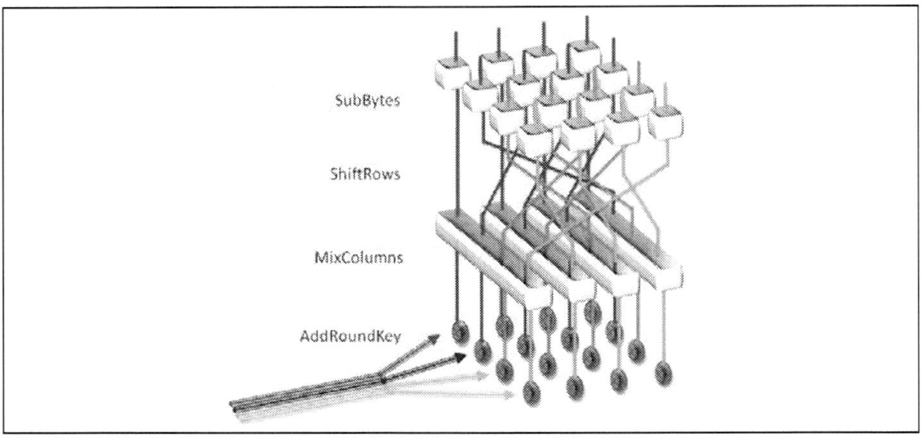

Fig. 3. AES Round

The **SubBytes** transformation is a non-linear byte substitution that operates independently on each byte of the State using a substitution table (S-box). The S-box is it is constructed by composing two transformations:

1. The multiplicative inverse in the finite field GF(2^8), where the element $(00000000)_2$ is mapped to itself;

2. The following affine transformation (over GF(2)):

$$b_i' = b_i \oplus b_{(i+4)\bmod 8} \oplus b_{(i+5)\bmod 8} \oplus b_{(i+6)\bmod 8} \oplus b_{(i+7)\bmod 8} \oplus c_i$$

for $0 \leq i \leq 7$, where b_i is the ith bit of the byte, and c_i is the ith bit of a byte c whose value is fixed by the standard and it is equal to {01100011}.

The S-box can be seen as a look-up table where for each 8-bits input there is a unique 8-bits output value. Several hardware implementations have been proposed in the literature. They can be classified in three major categories:

1. ROM-based (Zhang and Parhi, 2002);
2. Look-up table implementation (combinational logic): the S-box is written in any hardware description language as a look-up table and it is converted into a gate-level circuit using a logic synthesizer (Di Natale et al., 2007 A). This solution allows the best performances in terms of speed and dynamic power consumption;
3. Mathematical implementation: instead of describing the S-box as look-up table, the circuit is described as the sequence of operations in the finite field GF and then synthesized. This alternative can provide better solutions in terms of area occupation and leakage power consumption (Wolkerstorferm et al., 2002).

In order to provide an idea of the differences between look-up table based implementations and mathematical implementations, we implemented in VHDL the solutions proposed in (Di Natale et al., 2007 A) (look-up table) and in (Wolkerstorferm et al., 2002) (mathematical) and then we synthesized the two circuits. Table 1 shows the comparison of area, speed and power consumptions obtained using a 90nm technology library provided by ST Microelectronics, where a new input is provided to the circuit every 3 ns.

	Look-up table implementation	*Mathematical implementation*
Area [μm²]	1993	1122
Delay of critical path [ns]	1.25 (14 logic levels)	2.45 (25 logic levels)
Average Dynamic Power [mW]	0.136	0.907

Table. 1. Comparison of area, delay and power between look-up table implementation and Mathematical implementation (ASIC)

The **ShiftRows** operation changes the byte position in the State. It rotates each row with different offsets to obtain a new State as follows:

$$State = \begin{bmatrix} s_0 & s_4 & s_8 & s_{12} \\ s_5 & s_9 & s_{13} & s_1 \\ s_{10} & s_{14} & s_2 & s_6 \\ s_{15} & s_3 & s_7 & s_{11} \end{bmatrix}$$

The first row is unchanged; the second row is rotated one byte position to the left, the third row two byte positions, and the fourth row three byte positions. ShiftRows is a linear transformation. Hardware implementation of this transformation resumes to wiring.

The **MixColumns** operates column-wise altering all the bytes of the same column. It treats a column as a 3rd degree polynomial with coefficients in GF (2^8) and produces the new column by multiplying it with a constant polynomial. This operation is performed modulo a 4th degree polynomial with coefficients in GF (2^8).

Let s_i, s_{i+1}, s_{i+2}, and s_{i+3} be four consecutive bytes (column-wise), $i \in \{0,4,8,12\}$ in the State matrix before MixColumns. These four bytes are transformed as:

$$\begin{bmatrix} t_i \\ t_{i+1} \\ t_{i+2} \\ t_{i+3} \end{bmatrix} = \begin{bmatrix} 02 & 03 & 01 & 01 \\ 01 & 02 & 03 & 01 \\ 01 & 01 & 02 & 03 \\ 03 & 01 & 01 & 02 \end{bmatrix} \begin{bmatrix} s_i \\ s_{i+1} \\ s_{i+2} \\ s_{i+3} \end{bmatrix}$$

where the t_i's are the State bytes after the MixColumns operation. The MixColumns operation is typically implemented as XOR-trees.

The **AddRoundKey** adds the corresponding round key to the current State. In the $GF(2^8)$ field, addition is implemented as a bit-wise XOR operation between the two elements. The implementation of the AddRounkey operation resumes to a single layer of XOR gates.

2.2 Key Expansion

The AES algorithm takes the Cipher Key and performs a Key Expansion routine to generate a key schedule. The Key Expansion generates a total of 11 words (the initial key plus one roundkey for each of the ten rounds). Each generated key has 128 bits that are organized in a state array as defined for the data in Figure 1 (i.e., the jth byte of the key state word is composed of the bits from 8•j to 8•j+7).

We define K_i the key of ith round ($0 \leq i \leq 10$), where K_0 denotes the initial secret key. The generation of a new round key depends on the previous one. Fig. 4 shows the scheme of generation on a new key K_{i+1} starting from K_i. The transformation F_i is a non-linear function composed of 4 S-boxes (one for each byte), a rotation of bytes, and the addition of a round specific constant, defined by the standard. More details about the generation process are given in (FIPS-197, 2001).

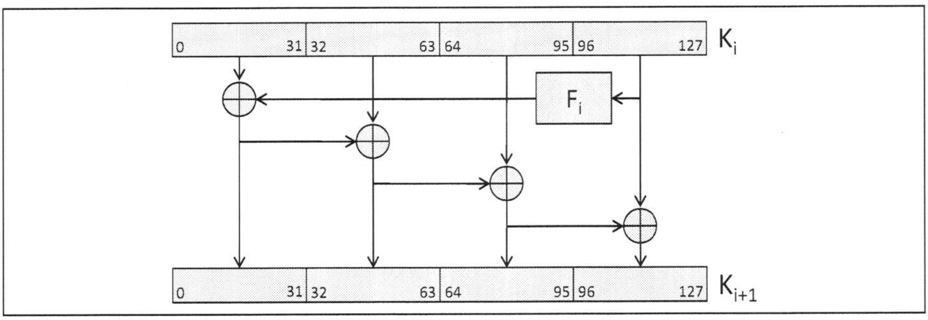

Fig. 4. AES Key Expansion Scheme

3. Fault attacks

As mentioned in the introduction, faults can be either permanent (due to fabrication defects) or transient (due to environmental conditions such as the exposure to cosmic radiations). For the particular case of crypto-coprocessors, transient faults can also be maliciously and intentionally injected in the circuit in order to retrieve the secret key (Blömer and Seifert,

2003). By comparing the result of a normal encryption with a faulty one, a hacker can deduce the key. These techniques are referred to as "Differential Fault Analysis" (DFA). Faults can be injected by different means such as temperature variation, clock frequency modification, exposure to radiations UV, X or visible light (Kim and Quisquater, 2007). The formers are quite efficient for software implementations of the crypto algorithm, while the latter (i.e., laser-based fault injection) is particularly well-suited for hardware implementations. The main advantage of laser-based fault injection is the localization of the fault (in the timing and the spatial domains).
In this context, it is thus of prime importance to be able to detect such faults. Nevertheless, independently of the mean used to inject the fault, the induced error must satisfy certain conditions in order to be successfully exploited. The following subsections quickly review the published DFA attacks on AES and analyze these conditions on the injected errors.

3.1 Attacks on 1 bit

One of the first cryptanalysis method using faults on AES has been published by (Giraud, 2005). The considered error is a single faulty bit on any of the 128 State bits at the input of the SubBytes operation during execution of the last encryption round (10th round). This error of multiplicity one (only one bit is affected)is equivalent to an error of multiplicity one appearing on the round key or on the input of the previous AddRoundKey operation. This error spreads, affects the output of the SubBytes operation and, since the last round does not include the Mixcolum operation, it also affects one (and only one) byte of the final Output array. The attack can be summarized as follows: let denote B_9 one of the output bytes of the last but one round (i.e., the one where the fault will be injected), e the 8-bit error mask (it contains only one asserted bit in the position where the error is considered), and BK_{10} the byte of the round key K_{10} at the same position of B_9 (among the 16 bytes). The correct output result should be $\{SubBytes(B_9)\oplus BK_{10}\}$. Because of the fault, the result is $\{SubBytes(B_9\oplus e)\oplus BK_{10}\}$. By adding these two results, it comes:

$$d = \{SubBytes(B_9)\oplus BK_{10}\}\oplus\{SubBytes(B_9\oplus e)\oplus BK_{10}\} = SubBytes(B_9)\oplus SubBytes(B_9\oplus e)$$

There are 8 possibilities for the faulty bit. For each possibility, the list of possible B_9 values is built up. The same process is iterated with other values of e. In average, the value of B_9 can be obtained with 3 faults. Then, by adding SubBytes(B_9) to the correct result, the byte BK_{10} of the key is obtained. In about 50 fault injections, the whole round key K_{10} can be fully determined. The primary key K_0 can be mathematically derived from K_{10}.

In (Blömer and Seifert, 2003) the authors proposed attacks based on the "safe error" principle, i.e. "the error affects the result or not". The considered fault is a stuck-at-0 affecting one bit of the key. If the result is faulty, the actual value of the key bit is 1.

In (Blömer and Krummel, 2006) the authors report another attack based on the injection of an error of multiplicity one which exploits collision effects, i.e. the fact that a two messages (one without, the other one with an error) will give the same result.

3.2 Attacks on 1 byte

In (Giraud, 2005), Giraud also proposes a more complex attack with the advantage of considering a less restrictive error model than errors on a single bit only. Here, errors of multiplicity x (x ≥1) affecting one byte are taken into account. The attack is composed of three steps:

1. an error is injected on the round key K_9 just before the last KeySchedule. This error affects 5 bytes of the next key K_{10}, four of them being on the same line of the Key State array. Using the same set of equations than in the previously mentioned attack, one byte of the last column of the last round key is obtained. By repeating this step 4 times, the 4 bytes of the last column of K_9 are retrieved.
2. an error is injected on the last column of K_8 before the penultimate KeySchedule. This spreads on 10 bytes. One of these faulty bytes gives information on the propagated error during the last KeySchedule. Since, the last column of K_9 is known, 3 bytes of K_8 can be definitely determined and it remains 130 possibilities of its fourth byte. This gives the penultimate column of K_9.
3. in the last step, an error is injected on the last column of the State array before round 10. Thus, all values which satisfy the observed differences on the output are determined using an exhaustive search on the 242 possibilities. Since, the last column before round 10 is known and also the two last columns of K_9, one can derive 14 bytes of K_{10}. The two last ones are determined using an exhaustive search.

Some other attacks relying on several fault injections on bytes are reported in (Chen and Yen, 2003), (Piret and Quisquater, 2003), and (Dusart et al., 2003).

The attack in (Blömer and Seifert, 2003), based on the "safe error" principle, can be extended to the case of attack on 1 byte. In this case all the input bits of an S-box must be stuck at 0. Then, by applying all 256 values at the input of the S-box, a collision appears: when the byte has the same value as the key, the result equals the one obtained with the stuck-at fault.

3.3 Attacks on several bytes on the same column

In (Moraidi et al., 2006) the authors propose a generalization of the attack on a single byte, focusing on 3 or 4 bytes in the first State array column. The faults can be injected at any step of the AES process. For the attack to be successful, it is necessary to know if the error affects 3 or 4 bytes. Furthermore, 6 and 1500 error injections have to be performed for 3 and 4 faulty bytes respectively.

3.4 Conclusions

In this paper, we do not discuss about the actual capabilities of injecting faults according to the hypothesis underlying those attacks. Nevertheless, from this overview, it can be concluded that:

- To our knowledge, errors affecting more than one byte (except (Moraidi et al., 2006)) cannot be exploited to perform an attack. Thus it's of prime importance to detect errors located within a byte. All error multiplicities (1 to 8) have to be considered.

- According to the variety of reported attacks, all time steps (rounds) of the AES algorithm are prone to fault injections. Thus, the data protection mechanism must span over the whole AES process.
- While errors affecting more than one byte are not exploitable by nowadays reported attacks, their detection is of interest since, firstly it helps in detecting an attack (for instance laser-based attacks need in practice many shots before succeeding in flipping bits within a single byte), and secondly the ingenuity of hackers may make efficient these attacks in the future.

4. Error Detection Methods

Concurrent fault detection for hardware implementations of secure devices is important not only to protect the system from random faults, but also to protect it the from an attacker who may maliciously inject faults in order to find the encryption secret key.
Several error detection solutions have been proposed in the literature, they rely on the use of some form of spatial or temporal redundancy. (Karri et al., 2002) propose a solution suitable for systems containing both encryption and decryption modules. The basic idea is to use the decryption module to decrypt the encrypted data (ciphertext), and to verify that the ciphertext is equal to the actual plaintext. (Maistri et al., 2007) propose a design solution that exploits temporal redundancy by a Double-Data-Rate mechanism in order to perform the computation twice without affecting adversely the overall throughput of the system. (Monnet et al., 2006) propose quasi-delay insensitive (QDI) asynchronous logic design to cope with fault attacks. Indeed, specific properties of asynchronous circuits make them inherently resistant against a large class of faults. Finally, several papers have been published on the use of codes to detect faults in secure devices. In particular, we focus the analysis on the architectures proposed in (Wu et al., 2004), (Bertoni et al., 2003), (Yen and Wu, 2006), and (Di Natale et al., 2007 B), as representative of the wide range of code-based solutions.
In a code-based fault detection scheme for a block of logic gates, the basic approach is that the output code of the block is predicted from the input data and this code is compared to the actual code calculated from the block output data of the block. The disadvantage of this technique is the use of the data bits to predict the code since corrupted bits may affect the code and the error may not be detected.
(Wu et al., 2004) propose the use of a parity bit for each of the 16 S-boxes that are XOR-merged in a single parity bit for the whole 128-bits data block. The single parity bit is used to check the correctness of the other blocks of the round. The datapath implementing the AES round with concurrent checking is shown in Fig. 5.
The principle is to compare the input parity of a round with the predicted output parity of the previous round, at each round. For that, the parity of the input values of a round is first determined ($P(x)$ in the figure). Then, the parity bit is modified according to the operations executed during the round. In particular, a parity bit of the word composed by the input (x_i) and the output (y_i) is predicted for each S-box. The 16 parity bits so calculated are then added to obtain a single parity bit ($P(x) \oplus P(y)$). This parity bit is added again to the input parity bit P(x) so that the parity P(y) is determined. P(y) is not affected by neither ShiftRows, which only changes the byte position in the state array, nor MixColumns operations (Wu et al., 2004). This parity bit is then modified by adding the parity of the key P(k) and stored in a

1-bit register. Errors are detected when this predicted parity bit differs from the input parity computed at the beginning of the next round.
The authors implemented this solution on a Xilinx Virtex 1000 FPGA. The hardware overhead is about 8% and the additional time delay is about 5%. Details about error detection capability are given in Section 5. It must be noted that this technique, while being effective when S-boxes are ROM-implemented, is quite costly for standard gates implementations.

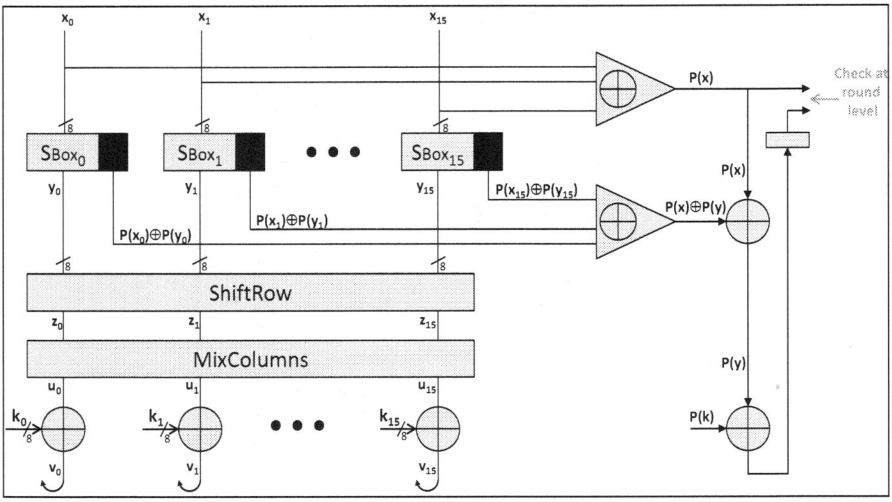

Fig. 5. (Wu et al., 2004) concurrent checking scheme

In (Bertoni et al., 2003) the authors propose to use 16 parity bits instead of a single bit. In particular, a parity bit is associated to each byte of the state (see Fig. 6).
Concerning the S-boxes, the authors propose a ROM-based implementation. The extra parity bit is stores in that ROM resulting in a 256x9 bits memory. Moreover, to detect input parity errors and some internal memory (data or decode) errors, authors propose the use of a 512×9 ROM, where the ninth bit is driven bit the input parity bit. They deliberately force all the ROM words corresponding to a wrong input address (i.e., S-boxes input with a wrong parity bit associated) with a dummy output value where the output parity is wrong, so that the detection mechanism will detect the fault.
As before, the parity bit associated to each byte does not change after the ShiftRows operation. On the contrary, the prediction of the output parity bits of the MixColumns is the most mathematically complex for this type of architecture, because it depends on the value of 4 bytes of the state (see details in (Bertoni et al., 2003)). For AddRoundKey operation, the prediction of the output parity bit easily consists in adding the current parity bits with the parity bits of the corresponding round key.
(Yen and Wu, 2006) propose the use of a systematic (n+1,n) Cyclic Redundancy Check (CRC) over $GF(2^8)$ to detect errors during encryption, where $n\in\{4,8,16\}$ is the number of bytes contained in the message (see Fig. 7). The generator polynomial is $g(x)=1+x$. The CRC

byte can be associated either to each column of the state (i.e., to 4 bytes and so they use CRC(5,4)), or to two columns (i.e., 89 bytes, so CRC(9,8)), or to the whole State (CRC(17,16)). This solution allows a very high error detection level at the cost of high area overhead (in the order of a thousand additional ports).

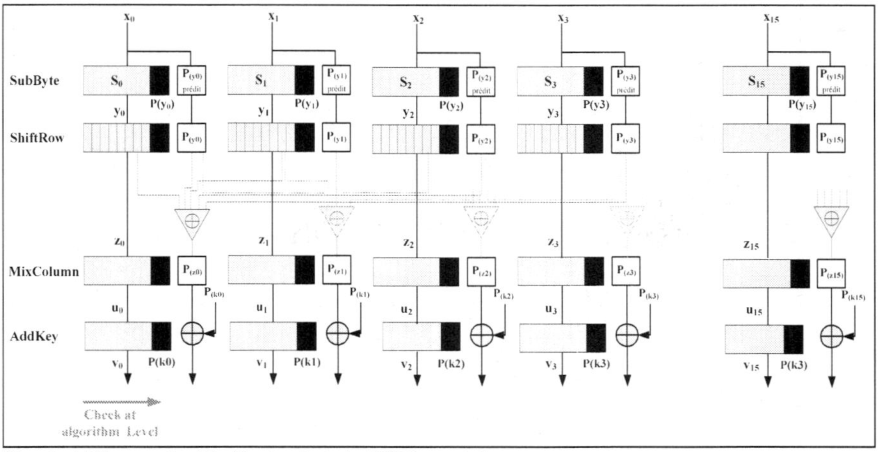

Fig. 6. AES round with (Bertoni et al., 2003) model

Since S-boxes represent the largest part of the circuit, (Di Natale et al., 2007 B) propose a solution that focuses on S-boxes only. They propose the use of two parity bits per S-box, one parity bit for the input byte of the S-box and one for its output byte. This double parity per S-box was also proposed in (Wu et al., 2004), but the two parity bits are now independently generated by dedicated prediction logics.

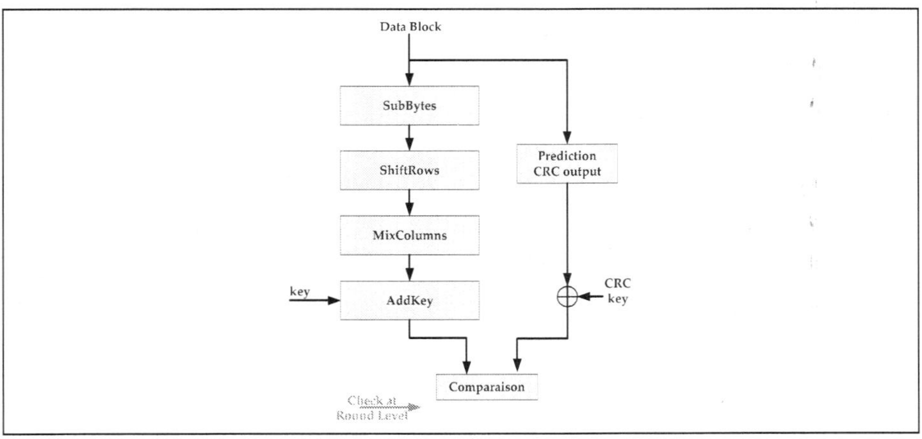

Fig. 7. AES round with (Yen and Wu, 2006) model

Then, the actual output parity is compared with the predicted output parity bit, and the actual input parity bit is compared to the predicted one (Fig. 8). When the S-box and the prediction circuits are synthesized as combinational logic, the area overhead is 38.33% with respect to the original S-box. This double parity checking allows additional detection of 27% of errors of even multiplicity compared to the solution of (Wu et al., 2004), besides all odd multiplicity of errors.

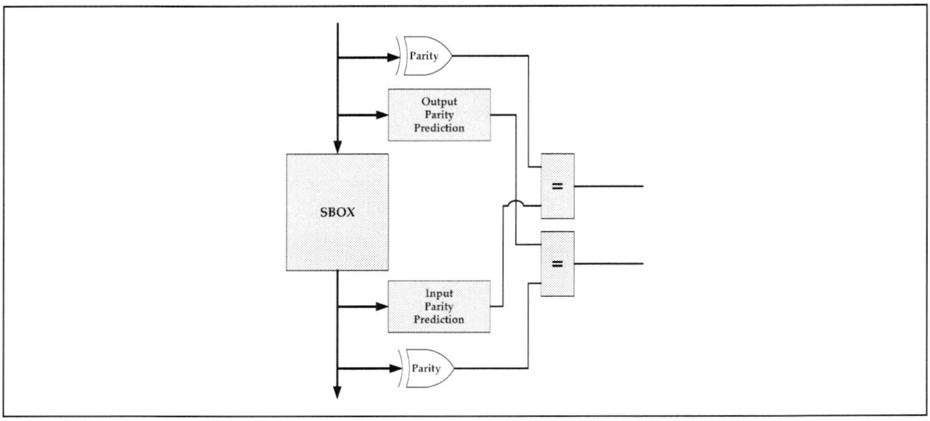

Fig. 8. SubBytes transformation with (Di Natale et al., 2007 B) model

Concerning error detection flags, the predicted and actual parity bits can be compared after each operation, each round, or after execution of the whole encryption. These solutions are equivalent from the detection capability point of view. They only differ in terms of hardware overhead and detection latency. Multiplicity of checkpoints decreases the error detection latency but increases the hardware overhead.

5. Error Detection

The error detection schemes detailed in the preceding section are now compared in terms of error detection capabilities.

Concerning the experimental setup, we assume single transient stuck-at faults, which are supposed to occur during the execution of one operation in one round (Fig. 9). S-boxes are implemented using random logic. Concerning the scheme presented in (Yen and Wu, 2006), we used a CRC (5,4) i.e. 4 CRC bytes, one for each column of the State array (32 bits). Since the error detection scheme presented in (Di Natale et al, 2007 B) only addresses errors on the S-boxes outputs, this protection scheme is completed with the solution proposed in (Bertoni et al., 2003): a faulty parity bit is affected to an S-box operation when either faulty input or output parities are detected. This S-box parity bit is subsequently used in the following prediction operations.

We first analyzed the detection capabilities of these four techniques with respect to errors affecting a single byte. Errors have been exhaustively injected in every byte, every operation and at every round. Table 2 reports the error multiplicity (one to eight faulty bits), the number of simulated errors for each error multiplicity (e.g. $16 * 40 * \binom{2}{8} = 17920$ error instances for 2 faulty bits among 8 bits, affecting one byte over 16 in the State array, after one of the 40 operations executed during the encryption), and the number of the undetected errors for each technique under evaluation.

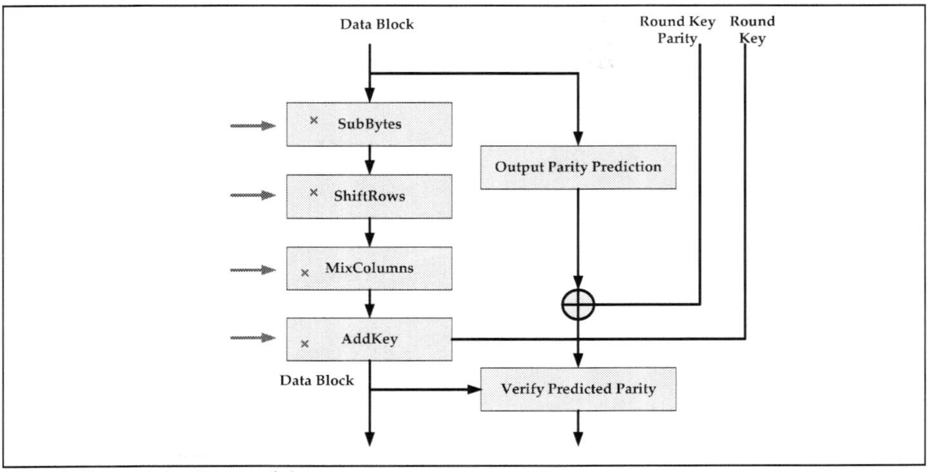

Fig. 9. Errors injection model

As expected, errors with odd multiplicity are more easily detected with detection schemes based on parity codes (Wu et al., 2004), while the CRC-based scheme proposed in (Yen and Wu, 2006) outperforms other techniques by detecting even-multiplicity errors too.

Output errors multiplicity	# of possible errors	Number and percentage of the undetected errors			
		(Wu et al., 2004) (1 parity bit)	(Bertoni et al., 2003) (16 parity bits)	(Yen and Wu, 2006) (4 CRC bytes)	(Di Natale et al., 2007 B) (16 parity bits + S-box protection)
1	5120	125 (2.44%)	0 (0%)	0 (0%)	0 (0%)
2	17920	2644 (14.75%)	16912 (94.38%)	0 (0%)	14666 (81.84%)
3	35840	841 (2.35%)	0 (0%)	0 (0%)	0 (0%)
4	44800	6614 (14.76%)	39760 (88.75%)	0 (0%)	34095 (76.10%)
5	35840	898 (2.51%)	0 (0%)	0 (0%)	0 (0%)
6	17920	2620 (14.62%)	14896 (83.13%)	0 (0%)	12642 (70.55%)
7	5120	110 (2.15%)	0 (0%)	0 (0%)	0 (0%)
8	640	98 (15.31%)	496 (77.50%)	0 (0%)	171 (26.72%)

Table 2. Detection capability for errors in a single byte

In the next experiments we injected random errors affecting any of the 128 state bits, with error multiplicity ranging from 1 to 64 faulty bits. Note that if errors affecting several bytes are not easily exploitable during DFA, their detection is of prime interest for detecting the attack itself. For each error multiplicity, 1000 randomly-chosen injection positions have been selected among randomly-chosen State array bits after execution of a randomly-chosen encryption operation. Simulation results are reported in Table 3.

All the techniques detect errors with multiplicity larger than 6 bits, except the detection scheme in (Wu et al., 2004) that provides only one parity bit for the whole State array.

Error multiplicity	Number and percentage of the undetected errors			
	(Di Natale et al., 2007 B) (16 parity bits + S-box protection)	(Yen and Wu, 2006) (4 CRC bytes)	(Bertoni et al., 2003) (16 parity bits)	(Wu et al., 2004) (1 parity bit)
1	0 (0.00%)	0 (0.00%)	0 (0.00%)	975 (2.44%)
2	1883 (4.71%)	549 (1.37%)	2164 (5.41%)	5992 (14.98%)
3	0 (0.00%)	2 (0.01%)	0 (0.00%)	1037 (2.59%)
4	262 (0.66%)	70 (0.18%)	320 (0.80%)	6069 (15.17%)
5	0 (0.00%)	1 (0.00%)	0 (0.00%)	981 (2.45%)
6	61 (0.15%)	0 (0.00%)	80 (0.20%)	6050 (15.13%)
7	0 (0.00%)	0 (0.00%)	0 (0.00%)	995 (2.49%)
8	0 (0.00%)	0 (0.00%)	0 (0.00%)	5956 (14.89%)
9	0 (0.00%)	0 (0.00%)	0 (0.00%)	1022 (2.56%)
10	0 (0.00%)	0 (0.00%)	0 (0.00%)	5933 (14.83%)
11	0 (0.00%)	0 (0.00%)	0 (0.00%)	991 (2.48%)
12	0 (0.00%)	0 (0.00%)	0 (0.00%)	6021 (15.05%)
13	0 (0.00%)	0 (0.00%)	0 (0.00%)	1001 (2.50%)
14	0 (0.00%)	0 (0.00%)	0 (0.00%)	6021 (15.05%)
15	0 (0.00%)	0 (0.00%)	0 (0.00%)	1018 (2.55%)
16	0 (0.00%)	0 (0.00%)	0 (0.00%)	5997 (14.99%)
....				
63	0 (0.00%)	0 (0.00%)	0 (0.00%)	1019 (2.55%)
64	0 (0.00%)	0 (0.00%)	0 (0.00%)	6008 (15.02%)

Table 3. Detection capability for random errors in 1 to 64 bits of a State

6. Faults and Error multiplicity

In the previously mentioned papers, the authors present the error detection capability of the methods (for instance, parity based architectures can detect all odd multiplicity errors). However, none of the papers correlatesthe error multiplicity, that appear at the output of round operations, with detected transient stuck-at faults affecting signals in logic blocks during execution of these operations.

In this section we evaluate the effectiveness of the solutions described in Section 4 with respect to fault detection. We focused this study on the S-box only, for the following reasons:

- it implements a 8 input bits function so it is possible to perform an exhaustive study;
- most of the attacks are performed on this block;
- Shiftrow module is anyway out of concern since it resumes to wiring;

- AddRoundKey resumes to a single layer of XOR gates. Thus, under the selected fault model, only error of multiplicity 1 may occur when a fault affects a XOR gate signal;
- MixCloumns operates on 32 input bits so an exhaustive study cannot be performed.

Moreover, since the error propagation on the output of the SubBytes operation strongly depends on the implementation (netlist) of the S-box, we implemented different versions of the S-box using different synthesis parameters and implementation styles. In particular, we implemented the following designs using the AMS 0.35μm technology library:

- Sbox1: description in VHDL as combinational look-up table, synthesis with Cadence©, 553 cells
- Sbox2: description in VHDL as combinational look-up table, synthesis with Design Compiler (Synopsys©) with "-map_effort high" option, 477 cells
- Sbox3: description in VHDL as combinational look-up table, synthesis with Design Compiler with "-map_effort medium" option, 482 cells
- Sbox4: description in VHDL as combinational look-up table, synthesis with Design Compiler with "-map_effort low" option, 474 cells
- Sbox5: mathematical description in VHDL of two blocks: the inversion in $GF(2^8)$ described as combinational look-up table plus the affine transformation, synthesis with Design Compiler with "-map_effort high" option, 481 cells
- Sbox6: mathematical description in VHDL, by using the decomposition of calculations in $GF(2^4)$ as described in (Wolkerstorferm et al., 2002), synthesis with Design Compiler with "-map_effort high" option, 193 cells

For each S-box implementation we performed exhaustive fault simulation, i.e. we applied all the possible input values (256 values) and we fault simulated the behavior of the device for each possible stuck-at in the circuit. The fault simulation produced a fault dictionary composed by all the possible couples C formed by { input value / fault } and, associated to each of them, the number of erroneous bits at the output of the S-box.

Table 4 reports the number of couples C leading to error multiplicities ranging from 0 to 8 for each S-box implementation. The first column (error multiplicity = 0) represents all the cases where the fault is not excited by a specific input value, or its effect is not propagated to the output. For all the other cases, the cell reports, besides the number of couples C, the percentage with respect to the overall number of couples and, in bold, the percentage with respect to the number of couples that lead to at least 1 error at the output of the S-box.

It is possible to note that, for instance, among the 11% of fault simulations for which a fault injection results in erroneous data on the Sbox1 output (89% of experiments result in 0 error), about 78% of the cases result in only one erroneous output bit, justifying therefore the use of code-based solution that exploits simple parity bit. Likewise, the 14 couple [fault/vector] providing 8 output errors correspond to faults at the input of the Sbox. It might not be necessary to protect the Sbox with detection schemes able to detect 8 error bits, if these errors are detected at the input of the Sbox.

Obviously, different S-box implementations lead the series of fault injection to different profiles in terms of error multiplicity. The presented implementations differ in terms of power consumption, performance and area according to chosen synthesis parameters. But the synthesis tool generates also quite different implementations whether the initial

VHDL model is described as a look-up table or a mathematical expression.When starting from look-up table, the number of errors at the output of the S-box is concentrated around 1 or 2 erroneous bits. On the contrary, mathematical-based architectures (Sbox5 and Sbox6)

are composed of several blocks that operate in cascade, and an error in a particular element is spread over several output bits. Therefore this type of implementation is first more sensitive to faults (in the case of Sbox6, 37% of couples generates an error at the output, while only 11% for Sbox1), and second it generates a higher number of output errors (the highest average is between 4 and 5).
Such experiments must be conducted for selecting appropriate fault detection schemes.

	0	1	2	3	4	5	6	7	8
Sbox1	450382 90%	36551 7% / **78%**	6448 1% / **14%**	2033 0% / **4%**	953 0% / **2%**	474 0% / **1%**	235 0% / **1%**	62 0% / **0%**	14 0% / **0%**
Sbox2	441836 89%	23664 5% / **43%**	16744 3% / **30%**	8547 2% / **15%**	3922 1% / **7%**	1631 0% / **3%**	637 0% / **1%**	147 0% / **0%**	24 0% / **0%**
Sbox3	438510 89%	23849 5% / **44%**	16609 4% / **30%**	8212 2% / **15%**	3801 1% / **7%**	1544 0% / **3%**	619 0% / **1%**	145 0% / **0%**	23 0% / **0%**
Sbox4	441855 89%	23658 5% / **43%**	16745 3% / **30%**	8534 2% / **15%**	3924 1% / **7%**	1629 0% / **3%**	636 0% / **1%**	147 0% / **0%**	24 0% / **0%**
Sbox5	418618 86%	7311 2% / **11%**	8973 2% / **13%**	4808 1% / **7%**	9949 2% / **15%**	27511 6% / **41%**	8486 2% / **13%**	212 0% / **0%**	20 0% / **0%**
Sbox6	144592 63%	9392 4% / **11%**	8515 4% / **10%**	10443 5% / **12%**	24581 11% / **29%**	21571 9% / **26%**	4473 2% / **5%**	3401 1% / **4%**	2152 1% / **3%**

Table 4. Fault vs error multiplicity on S-boxes output

7. Conclusions

This chapter presented a study on mechanisms involved for detecting faults-based attacks on crypto-processors. Using the example of a standard in symmetric cryptographic, fault-based attacks were discussed with respect to their requirements in terms of error multiplicity (spatial and timing characteristics). We presented countermeasures to fault-based attacks that consist in detecting errors on the ciphered information. We analyzed more precisely some error detection schemes based on code-redundancy, with respect to their cost and ability to detect errors occurring at run time. Analysis on error detection has been conducted according to the error multiplicity in order to check the ability of the protection schemes to detect exploitable errors. We also analyzed error detection schemes performances in terms of transient fault detection (natural or maliciously injected).
Current attacks using laser-based fault injection require errors localized on very few bytes, without need of large precision as for the attack launch instant. Error detection schemes can be used for preventing these attacks or detecting natural transient faults. Dedicated parity and CRC -based solutions exploit typical features of the cipher for optimization of cost factors (area, latency) and improvement of their error detection capacity. The correlation between the transient faults affecting the combinational parts of the circuit and the errors produced on sub-function outputs strongly depends on the implementation of the cipher. Experiments show that for some particular implementations, most of the faults results in only one or two erroneous output bits, while there is no internal faults affecting all the output bits. Such analysis on ciphering operations can justify simple and non expensive code-redundancy solutions.

8. References

Bertoni, G.; Breveglieri, L.; Koren I.; Maistri, P. & Piuri, V. (2003), Error Analysis and Detection Procedures for a Hardware Implementation of the Advanced Encryption Standard, *IEEE Trans. on Computers*, Vol. 52., No.4, April 2003

Blömer J. & Krummel, V. (2006), *Fault Based Collision Attacks on AES, Proceedings of FDTC 2006*, pp. 106-120.

Blömer, J. & Seifert, J.P. (2003), Fault Based Cryptanalysis of the Advanced Encryption Standard (AES), *Proceedings of CHESS 2003*, pp 162-181.

Chen, C.N. & Yen, S.M. (2003), Differential Fault Analysis on AES Key Schedule and Some Countermeasures, *Proceedings of Australasian Conference on Information Security and Privacy 2003*, LNCS 2727, Springer-Verlag, pp. 118–129.

Di Natale, G.; Flottes M.L. & Rouzeyre, B. (2007 B), An On-Line Fault Detection Scheme for SBoxes in Secure Circuits, *Proceedings of 13th IEEE International On-Line Testing Symposium, IOLTS 2007*, pp. 57-62.

Di Natale, G.; Flottes, M.L. & Rouzeyre, B. (2007 A), A Novel Parity Bit Scheme for SBox in AES Circuits, *Proceedings of IEEE Design and Diagnostics of Electronic Circuits and Systems, DDECS 2007*, April 2007, pp. 1–5, DOI 10.1109/DDECS.2007.4295295

Dusart, P.; Letourneux, G. & Vivolo, O. (2003), Differential Fault Analysis on A.E.S., *Applied Cryptography and Network Security*, Springer Ed., Vol. 2846/2003, pp 293-306.

Eurosmart (2007), Vision paper 2020, http://www.eurosmart.com/index.php/publications/vision-paper-2020.html

FIPS-197 (2001), Advanced Encryption Standard (AES), *Federal Information Processing Standards Publication 197*, http://csrc.nist.gov/publications/, November 26, 2001

Giraud, C. (2005), DFA on AES, *Proceedings of 4th International Conference on AES 2005*, Springer publisher, pp 27-41.

Karri, R.; Wu, K.; Mishra, P. & Kim, Y. (2002), Concurrent Error Detection Schemes for Fault-Based Side-Channel Cryptanalysis of Symmetric Block Ciphers, *IEEE Transactions on Computer-Aided Design of Integrated Circuits and Systems*, Vol. 21, N. 12, December 2002, pp. 1509-1517

Kim, C.H. & Quisquater, J.-J. (2007), Faults, Injection Methods, and Fault Attacks, *IEEE Design & Test of Computers*, Nov.-Dec. 2007, Vol.24, Issue 6, pp.544-545.

Leveugle, R. (2007), Early Analysis of Fault-Based Attack Effects in Secure Circuits, IEEE Transactions on Computers, Vol. 56, N. 10, October 2007, pp. 1431-1434

Maistri, P.; Vanhauwaert, P. & Leveugle, R. (2007), A Novel Double-Data-Rate AES Architecture Resistant against Fault Injection, Workshop on Fault Diagnosis and Tolerance in Cryptography, 2007, DOI 10.1109/FDTC.2007.8, pp. 54-61

Monnet, Y.; Renaudin, M. & Leveugle, R. (2006), Designing Resistant Circuits against Malicious Faults Injection Using Asynchronous Logic, *IEEE Transactions on Computers*, Vol. 55, N. 9, September 2006, pp. 1104-1115

Moradi, A.; Manzuri Shalmani, M.T. & Salmasizadeh, M. (2006), A Generalized Method of Differential Fault Attack Against AES Cryptosystem, *Proceedings of Cryptographic Hardware and Embedded Systems*, CHESS 2006, pp 91-100.

Piret, G. & Quisquater, J.J. (2003), A Differential Fault Attack Technique against SPN Structures, with Application to the AES and KHAZAD, *Proceedings of Chess 2003*, pp 77-88.

Reed, R.A.; Kinnison, J.; Pickel, J.C.; Buchner, S.; Marshall, P.W.; Kniffin, S. & LaBel, K.A. (2003), Single-event effects ground testing and on-orbit rate prediction methods: the past, present, and future, *IEEE Transactions on Nuclear Science,* Vol. 50, pp. 622-634, June 2003

Wolkerstorferm, J.; Oswald, E. & Lamberge, M. (2002), An ASIC Implementation of the AES SBoxes, *Proceedings of The Cryptographer's Track at the RSA Conference on Topics in Cryptology 2002, Lecture Notes In Computer Science,* Vol. 2271; pp. 67-78

Wu, K; Karri, R.; Kuznetsov, G. & Goessel, M. (2004), Low Cost Concurrent Error Detection for the Advanced Encryption Standard, *Proceedings of IEEE International Test Conference, 2004.* pp 1242- 1248

Yen, C.H. & Wu, B.F. (2006), Simple Error Detection Methods for Hardware Implementation of Advanced Encryption Standard, *IEEE Trans. on Computers,* June 2006, Vol. 55, No.6, pp 720-731

Zhang, X. & Parhi, K.K. (2002), Implementation approaches for the Advanced Encryption Standard algorithm, *IEEE Circuits and Systems Magazine,* Volume 2, Issue 4, Fourth Quarter 2002, pp. 24-46, DOI 10.1109/MCAS.2002.1173133

10

Sensor and Actuator/Surface Failure Detection Based on the Spectral Norm of an Innovation Matrix

Chingiz Hajiyev
Istanbul Technical University
Turkey

1. Introduction

The problem of changes detection in dynamical properties of signals and systems appears in many problems of signal processing, navigation and control (Basseville & Benveniste, 1986; Benveniste *et al.*, 1987; Gadzhiev, 1992; Chen & Patton, 1999; Chan *et al.*, 1999; Hajiyev & Caliskan, 2003; Vaswani, 2004; Tykierko, 2008; Li & Jaimoukha, 2009). Abnormal measurements, sudden shifts appearing in the measuring channel, faultiness of measuring devices, changes in statistical characteristics of noises of an object or of measurements, malfunctions in the computer, and also a sharp change in the trajectory of a monitoring process, etc. should be enumerated among these changes. In real situations of exploiting an object, the problem occurs of operative detection of such changes in order to subsequently correct estimators or to make timely decisions on the necessity and character of control actions with respect to the process of technical exploitation of the object. Under this process, different methods of control and diagnostics are used.
Many fault detection methods have been developed to detect and identify sensor and actuator faults by using analytical redundancy (Zhang & Li, 1997; Rago *et al.*, 1998; Maybeck, 1999; Larson *et al.*, 2002; Lee & Lyou, 2002). In (Larson *et al.*, 2002) an analytical redundancy-based approach for detecting and isolating sensor, actuator, and component (i.e., plant) faults in complex dynamical systems, such as aircraft and spacecraft is developed. The method is based on the use of constrained Kalman filters, which are able to detect and isolate such faults by exploiting functional relationships that exist among various subsets of available actuator input and sensor output data.
A statistical change detection technique based on a modification of the standard generalized likelihood ratio (GLR) statistic is used to detect faults in real time. The GLR test requires the statistical characteristics of the system to be known before and after the fault occurs. As this information is usually not available after the fault, the method has limited applications in practice. An integrated robust fault detection and isolation (FDI) and fault tolerant control (FTC) scheme for a fault in actuators or sensors of linear stochastic systems subjected to unknown inputs (disturbances) is presented in (Lee & Lyou, 2002). The FDI modules is constructed using banks of robust two-stage Kalman filters, which simultaneously estimate the state and the fault bias, and generate residual

sets decoupled from unknown disturbances. All elements of residual sets are evaluated by using a hypothesis statistical test, and the fault is declared according to the prepared decision logic. In this work it is assumed that single fault occurs at a time and the fault treated is of random bias type. The diagnostic method presented in the article is valid only for the control surface FDI.

In (Zhang & Li, 1997; Rago *et al.*, 1998) the algorithms for detection and diagnosis of multiple failures in a dynamic system are described. They are based on the Interacting Multiple-Model (IMM) estimation algorithm, which is one of the most cost-effective adaptive estimation techniques for systems involving structural as well as parametric changes. The proposed algorithms provide an integrated framework for fault detection, diagnosis, and state estimation. In (Maybeck, 1999) Multiple model adaptive estimation (MMAE) methods have been incorporated into the design of a flight control system for the variable in-flight stability test aircraft (VISTA) F-16, providing it with the capability to detect and compensate for sensor/actuator failures. The algorithm consists of a "front end" estimator for the control system, composed of a bank of parallel Kalman filters, each matched to a specific hypothesis about the failure status of the system (fully functional or a failure in any one sensor or actuator), and a means of blending the filter outputs through a probability-weighted average. In methods described in (Zhang & Li, 1997; Rago *et al.*, 1998; Maybeck, 1999), the faults are assumed to be known, and the Kalman filters are designed for the known sensor/actuator faults. As the approach requires several parallel Kalman filters, and the faults should be known, it can be used in limited applications.

In the references (Napolitano *et al.*, 1993; Raza, *et al.*, 1994; Napolitano, *et al.*, 1996; Borairi & Wang, 1998; Alessandri, 2003) the neural network based methods to detect sensor/actuator failures are developed and discussed. In the reference (Napolitano *et al.*, 1993) a neural network is proposed as an approach to the task of failure detection following damage to an aerodynamic surface of an aircraft flight control system. This structure, used for state estimation purpose, can be designed and trained on line in flight and generates a residual signal indicating the damage as soon as it occurs. In (Raza *et al.*, 1994) the problem of detecting control surface failures of a high performance aircraft is considered. The detection model is developed using a linear dynamic model of an F/A-18 aircraft. Two parallel models detect the existence of a surface failure, whereas the isolation and magnitude of any one of the possible failure modes is estimated by a decision algorithm using either neural networks or fuzzy logic. The reference (Napolitano *et al.*, 1996) describes a study related to the testing and validation of a neural-network based approach for the problem of actuator failure detection and identification following battle damage to an aircraft control surface. Online learning neural architectures, trained with the Extended Back-Propagation algorithm, have been tested under nonlinear conditions in the presence of sensor noise. In (Borairi & Wang,1998) an approach for the fault detection and diagnosis of the actuators and sensors in non-linear systems is presented. First, a known non-linear system is considered, where an adaptive diagnostic model incorporating the estimate of the fault is constructed. Further, unknown nonlinear systems are studied and a feed forward neural network trained to estimate the system under healthy conditions. Genetic algorithms is proposed as a means of optimizing the weighting connections of neural network and to assist the diagnosis of the fault. In (Alessandri, 2003) a neural network based method to detect faults in nonlinear systems is proposed. Fault diagnosis is accomplished by means of a bank of estimators, which provide estimates of parameters that describe actuator, plant, and sensor faults. The problem of designing such estimators for general nonlinear systems is solved by searching for optimal estimation functions. These functions are approximated by feed forward neural networks and the problem is reduced to find the optimal neural weights. The methods based on artificial neural networks and genetic algorithms do not have physical bases. Therefore according to the different data corresponding to the same event the model gives different solutions. Thus, the model should continuously be trained by using the new data.

The reference (Perhinschi *et al.*, 2002) focuses on specific issues relative to real-time on-line estimation of aircraft aerodynamic parameters at nominal and post-actuator failure flight conditions. A specific parameter identification (PID) method, based on Fourier Transform, has been applied to an approximated mathematical model of the NASA IFCS F-15 aircraft. The direct evaluation of stability and control derivatives versus the estimation of the coefficients of the state space system matrices evaluation has been considered. This method may not produce good results when the number of the stability and control derivatives is high.

In this direction of studies, it is necessary to mention the theory of diagnostics of a dynamic system by the innovation sequence of the Kalman filter (Mehra & Peschon, 1971; Willsky, 1976; Basseville & Benveniste, 1986; Gadzhiev, 1992, 1993; Hajiyev & Caliskan, 2003, 2005). The advantages of these methods are as follows: they provide the monitoring of the correctness of the result obtained by current working input actions; they do not require a priori information on the values of changes in the statistical characteristics of the innovation sequence in the case of fault; they allow one to solve the fault detection problem in real time; they require small computational expenditures for their realizations since they do not increase, in contrast to the most algorithmic methods, the dimension of the initial problem.

As is known (Mehra & Peschon, 1971), in the case where a system is normally operated, the normalized innovation sequence in the Kalman filter compatible with the model of dynamics is the white Gaussian noise with zero mean and identity covariance matrix. The faults appearing in the system of estimations lead to the changes in these statistical characteristics of the normalized innovation sequence. Therefore, in this case, the fault detection problem is reduced to the problem of fastest detection of the deviation of these characteristics from nominal.

In (Hajiyev & Caliskan, 2005) the sensor and control surface/actuator failures that affect the mean of the innovation sequence have been considered. The methods of testing the correspondence between the innovation sequence and the white noise and of revealing the change of its expectation are based on the classical statistical methods and are considered in detail in the literature (Mehra & Peschon, 1971; Hajiyev & Caliskan, 2003, 2005) therefore, it shall not be concentrated on testing these characteristics.

Testing, in real time, the covariance matrix of the innovation sequence of the Kalman filter turns out to be very complicated and not well developed, since there are difficulties in the determination of the confidence domain for a random matrix. Moreover, the existing methods of high-dimensional statistical analysis (Anderson, 1984; Kendall & Stuart, 1969) usually lead to asymptotic distributions; this sharply diminishes the operativeness of these methods. The method of testing the covariance matrix of the innovation sequence proposed in (Gadzhiev, 1992) on the basis of using the statistics of the ratio of two quadratic forms, whose matrices are reversed sample and theoretical covariance matrices, is free from the above-mentioned shortcoming. Nevertheless, the results obtained in (Gadzhiev, 1992) are valid only in the case where the reversed matrices which enter the expression of the monitoring statistics are nonsingular.

In practice, therefore, one makes use of a scalar measure of this matrix such as the trace, sum of the matrix elements, generalized variance (determinant), eigenvalues of a matrix, etc., each characterizing one or another geometrical parameter of the correlation ellipsoid.

The algorithm for testing the trace of the covariance matrix of the innovation sequence is presented in (Mehra & Peschon, 1971). But the trace testing algorithm ignores the off-diagonal elements of the covariance matrix. Therefore this algorithm cannot detect very small changes, in the measurement channel (Hajiyev & Caliskan, 2003).

In (Gadzhiev, 1993) a confidence range has been constructed for the generalized variance of the Wishart matrix using Chebyshev inequality. However as it is known (Krinetsky *et al.*, 1979), the Chebyshev inequality gives the extended confidence range for the random variables. Therefore in this case the miss-failure probability increases.

Most of fault detection tests are based on the statistical properties of the eigenvalues of the sample covariance matrix (Bienvenu & Kopp, 1983; Wax & Kailath, 1985). In (Wu *et al.*, 1995) an algorithm based on the geometrical location of these eigenvalues has been proposed. In (Grouffaud *et al.*, 1996) a new kind of test based on an analytic expression of the ordered eigenvalues profile, obtained under noise only hypothesis. Strategy in this work consists in looking for a break in profile by comparing observed profile and noise only one. The decision is taken by comparing the error of prediction with the threshold, which is obtained by solving the integral equation. Unfortunately, the distributions entering in this equation are not analytically known, hence it is difficult to determine the threshold and perform the proposed algorithm.

There exists some interesting results on the distribution of eigenvalues, characteristic function of eigenvalues, and distribution and moments of the smallest eigenvalue of Wishart distributed matrices (Malik, 2003; Zanella *et al.*, 2008; Edelman, 1991; Everson & Stephen, 2000). But application of mentioned works to fault detection problem of multidimensional dynamic systems turns out to be very complicated since there are difficulties in determining the confidence domain (or intervals) for the eigenvalues of random matrix.

In this study, an approach to detect the aircraft sensor and actuator/surface failures based on the spectral norm of an innovation matrix is proposed. A real-time detection of sensor and actuator/surface failures affecting the mean and variance of the innovation process applied to F-16 fighter flight dynamic is examined. A decision approach to isolate the sensor and actuator/surface failures based on the Adaptive Extended Kalman Filter insensitive to sensor failures is proposed.

The structure of this chapter is as follow. In Section 2, the failure detection problem in multidimensional dynamic systems using spectral norm of the innovation matrix of the Kalman filter is formulated. The upper confidence bound of the spectral norm of a Gaussian random matrix that consists of normally distributed random variables with zero mean is found and a new failure detection approach based on the properties of the spectral norm of the innovation matrix is proposed in this Section. In Section 3 the AFTI/F-16 aircraft model description is given and the Extended Kalman filter (EKF) for the F-16 nonlinear dynamic model estimation is designed. In Section 4 an adaptive EKF for the F-16 aircraft state estimation which is insensitive to sensor failures is designed and a decision approach to isolate the sensor and actuator/surface failure is proposed. In Section 5 some simulations are carried out for the sensor and actuator/surface failure detection problem in the AFTI/F-16 aircraft flight control system. The changes that affect the mathematical expectation and variance of the innovation sequence have been considered. Simulation results of adaptive EKF insensitive to sensor failures are given in this section too. Section 6 gives a brief summary of the obtained results and conclusions.

2. Failure Detection Using Spectral Norm of the Innovation Matrix

In diagnosing some dynamic systems, of special interest now are the methods of dynamic diagnosis that take into account influence of failures on system dynamics, in particular, revealing failures based on the analysis of the innovation sequence. Let us consider the linear dynamic system described by the equation of state

$$x(k+1) = \Phi(k+1,k)x(k) + G(k+1,k)w(k) \tag{1}$$

and the equation of measurements

$$z(k) = H(k)x(k) + V(k), \tag{2}$$

where $x(k)$ is an N - dimensional vector of system state; $\Phi(k+1,k))$ is the $N \times N$ transition matrix of the system; $w(k)$ is a random N - dimensional vector of disturbances (system noise); $G(k+1,k)$ is the $N \times N$ transition matrix of system noise; $z(k)$ is the n - dimensional vector of measurements; $H(k)$ is the $n \times N$ matrix of measurements of the system; and $V(k)$ is a random n - dimensional vector of measurement noise. Assume that random vectors $w(k)$ and $V(k)$ are a Gaussian white noise. Their mean values and covariance are determined by the expressions

$$\begin{aligned} &E[w(k)] = 0; E[V(k)] = 0; \\ &E\left[w(k)w^T(j)\right] = Q(k)\delta(kj); \\ &E\left[V(k)V^T(j)\right] = R(k)\delta(kj). \end{aligned} \tag{3}$$

Here E is the operator of statistical averaging; T is the sign of transposition; and $\delta(kj)$ is the Kronecker delta symbol. Note that $\{w(k)\}$ and $\{V(k)\}$ are assumed mutually uncorrelated.

Estimate of the state vector $\hat{x}(k/k)$ and covariance matrix of estimation errors $P(k/k)$ can be found using the optimum linear discrete Kalman filter (Sage and Melsa, 1971):

$$
\begin{aligned}
&\hat{x}(k/k) = \hat{x}(k/k-1) + K(k)\nu(k);\\
&\nu(k) = z(k) - H(k)\hat{x}(k/k-1);\\
&K(k) = P(k/k-1)H^T(k)\left[H(k)P(k/k-1)H^T(k) + R(k)\right]^{-1};\\
&P(k/k) = \left[I - K(k)H(k)\right]P(k/k-1);\\
&P(k/k-1) = \Phi(k,k-1)P(k-1/k-1)\Phi^T(k,k-1) +\\
&+ G(k,k-1)Q(k-1)G^T(k,k-1),
\end{aligned}
\tag{4}
$$

Here $K(k)$ is the gain matrix of the Kalman filter; $\nu(k)$ is the innovation sequence; I is a unit matrix; $P(k/k-1)$ is the covariance matrix of extrapolation errors and $P(k-1/k-1)$ is the covariance matrix of estimation errors at the previous step.
If there are no faults in the estimation system, then the normalized innovation sequence

$$\tilde{\nu}(k) = \left[H(k)P(k/k-1)H^T(k) + R(k)\right]^{-1/2}\nu(k), \tag{5}$$

in the Kalman filter (4) coordinated with the model dynamics is a Gaussian white noise with zero mean and a unit covariance matrix (Mehra and Peschon, 1971)

$$E\left[\tilde{\nu}(k)\right] = 0; E\left[\tilde{\nu}(k)\tilde{\nu}^T(j)\right] = P_{\tilde{\nu}} = I\delta(kj). \tag{6}$$

Failures that change system dynamics due to abrupt changes or shifts in components of the state vector, faults in computer, abnormal measurements, sudden shifts appearing in the channel of measurement, divergence of the estimation algorithm, and also such faults as a decrease in device accuracy, noise increase, etc. will result in changes of the above characteristics of the sequence of $\tilde{\nu}(k)$. Of interest is development of an on-line method of a simultaneous check of mathematical expectation and variance of the normalized innovation sequence (5) that does not require a priori information on the values of their changes in case of failure and allows one to detect on-line faults in the estimation system. To do this, two hypotheses are introduced:

γ_0 : the Kalman filter operates normally;
γ_1 : a failure takes place.

To reveal a failure, let us construct a matrix whose columns are vectors of innovation of the Kalman filter (Hajiyev, 2007). The following definitions are introduced.

Definition 1. By the innovation matrix of the Kalman filter a rectangular $n \times m$ matrix (n is the dimension of the innovation vector; $n \geq 2; m \geq 2$) is mentioned, whose columns are the innovation vectors $\nu(k)$ that correspond to m different instants of time.

Definition 2. The innovation matrix composed of the normalized innovation vectors $\tilde{\nu}(k)$ is referred as the normalized innovation matrix of the Kalman filter.

Hereinafter, to check the innovation sequence, the normalized innovation matrices A that consist of a finite number of normalized innovation vectors will be used. For a real-time check, it is expedient, at the instant of time $k(k \geq m)$, to construct the matrix $A(k)$ from a finite number m $(m \geq 2)$ of sequential innovation vectors:

$$A(k) = [\underbrace{\cdots \tilde{v}(k-2), \tilde{v}(k-1), \tilde{v}(k)}_{m}]. \tag{7}$$

To verify the hypothesis γ_0 and γ_1, a spectral norm of the matrix (7) below will be used.

2.1 Deriving the Upper Confidence Bound of the Spectral Norm of a Random Matrix

As is generally known (Horn and Jonson, 1986), the spectral norm $\|\cdot\|_2$ of a real matrix $A(k)$ is determined by the formula

$$\|A(k)\|_2 \equiv \max\left\{\left(\lambda_i\left[A^T(k)A(k)\right]\right)^{1/2}\right\}, \tag{8}$$

where $\lambda_i\left[A^T(k)A(k)\right]$ are eigenvalues of the matrix $A^T(k)A(k)$. Square roots of the eigenvalues of the matrix $A^T(k)A(k)$, i.e., the quantities $\left(\lambda_i\left[A^T(k)A(k)\right]\right)^{1/2}$, are called the singular values of the matrix $A(k)$. Therefore, spectral norm of the matrix $A(k)$ is equal to its maximum singular value. Since the matrix $A^T(k)A(k)$ is Hermitian, $\left[A^T(k)A(k)\right]^T = A^T(k)\left[A^T(k)\right]^T = A^T(k)A(k)$, and is positive definite, i.e., for any nonzero vector $x(k)$ the relation is true

$$x^T(k)A^T(k)A(k)x(k) = (A(k)x(k), A(k)x(k)) \geq 0, \tag{9}$$

(parentheses designate here a scalar product), the singular values are real and positive. For the same reasons, computation of singular values and, consequently, also of spectral norm, is more simpler than deriving eigenvalues for an arbitrary matrix. This explains the choice of spectral norm of matrices as a controllable scalar measure in solving some diagnosis problems. Of interest is here deriving the upper bound of the spectral norm of random matrices.

In the present study, based on the calculation of a respective vector and matrix norms, an analytical expression is found for the upper bound of spectral norm of a random matrix $A(k) \in R^{n \times m}$ composed of normally distributed random variables with zero mathematical expectation. The results of the above analysis are applied to the case of dynamic diagnosis of the Kalman filter in innovation sequence.

Let the Euclidean norm (or the 2-norm) of the vector $x \in R^n$ and the spectral norm (or the 2-norm) of the matrix $A(k) \in R^{n \times m}$ be determined by the expressions

$$\|x\|_2 = \sqrt{x_1^2 + \ldots + x_n^2}\,;$$
$$\|A\|_2 = \sqrt{\lambda_{max}\left[A^T A\right]} = \sigma_{max}[A], \tag{10}$$

where $\lambda_{max}[.]$ and $\sigma_{max}[.]$ are the maximum eigenvalue and maximum singular value of the respective matrix.
The Frobenius norm of the matrix $A = \left[a_{ij}\right] \in R^{n \times m}$ is determined as follows:

$$\|A\|_F = \sqrt{tr\left(A^T A\right)} = \sqrt{\sum_{i=1}^{n}\sum_{j=1}^{m} a_{ij}^2}, \tag{11}$$

where $tr(\cdot)$ is the trace of the matrix. The Frobenius norm and the 2-norm are related as follows (Chan *et al.*,1999):

$$\|Ax\|_2 \le \|A\|_F \|x\|_2, \tag{12}$$

where $A(k) \in R^{n \times m}$ and $x \in R^m$. Since $x \ne 0$, let us present expression (12) in the equivalent form

$$\frac{\|Ax\|_2}{\|x\|_2} \le \|A\|_F. \tag{13}$$

Inequality (13) is true for all $A(k) \in R^{n \times m}$, $x \in R^m$, $x \ne 0$, including the maximum value of the left-hand side of the inequality, i.e.,

$$\max_{x \ne 0} \frac{\|Ax\|_2}{\|x\|_2} \le \|A\|_F. \tag{14}$$

As is generally known (Horn and Jonson, 1986), the matrix norm associated with the respective vector norm is the relation

$$\|A\| = \max_{x \ne 0} \frac{\|Ax\|}{\|x\|}. \tag{15}$$

If the Euclidean norm $\|x\|_2$ is selected as the vector norm, then the respective matrix norm is the maximum singular value of the matrix *A*, i.e.,

$$\|A\| = \sigma_{max}[A]. \tag{16}$$

Regarding (14) - (16), the following inequality can be written,

$$\|A\|_2 = \sigma_{max}[A] \le \|A\|_F . \tag{17}$$

Let the matrix $A(k) \in R^{n \times m}$ be composed of normally distributed random variables with zero mathematical expectation and the mean-square deviation σ, i.e.,

$$a_{ij} \in N(0,\sigma). \tag{18}$$

Consequently, the quantity $\sum_{i=1}^{n}\sum_{j=1}^{m} a_{ij}^2 \Big/ \sigma^2$ will have the χ^2-distribution with $k = nm - 1$ degrees of freedom (Rao, 1965). It is easy to establish a relation between $\sigma_{max}[A]$ and χ^2:

$$\sigma_{max}[A] \le \|A\|_F , \tag{19}$$

where $\|A\|_F = \sigma\sqrt{\chi^2}$.

Specifying the significance level α, the following condition can be used

$$P\left\{\chi^2 \le \chi^2_{\beta,(nm-1)}\right\} = \beta , \tag{20}$$

where $\beta = 1 - \alpha$ is the confidence probability, and tables of the χ^2-distribution to derive $\chi^2_{\beta,(nm-1)}$, which is a confidence boundary (quantile) of the χ^2-distribution with $nm - 1$ degrees of freedom. Substituting $\chi^2_{\beta,(nm-1)}$ into (19), it is obtained finally

$$\sigma_{max}[A] \le \sigma\sqrt{\chi^2_{\beta,(nm-1)}} . \tag{21}$$

Formula (21) determines the upper confidence bound for spectral norm of a random matrix *A*. Thus, an analytical expression, convenient for practical calculations, is found for the upper bound of spectral norm of a random matrix, composed from normally distributed random variables with zero mathematical expectation. The obtained result may be used in applied statistical problems, in particular, to check statistical compatibility of data of statistical simulation with the results of field tests, and in health monitoring and diagnosis of multidimensional technical systems.

2.2 Failure Detection Using Spectral Norm of the Innovation Matrix

Since the random normalized innovation matrix (7) consists of normally distributed stochastic elements with zero mathematical expectation and a finite variance $a_{ij} \in N(0,1)$, inequality (21) may be used for solution of the above diagnosis problem. Expression (21) characterizes the relation between the mean-square value σ of the elements of a random matrix A and its spectral norm and may be used to derive the upper confidence boundary of the spectral norm of matrix (7). In this case, one may state that if elements a_{ij} of the controlled normalized innovation matrix of the Kalman filter obey the distribution $N(0,1)$, then inequality (21) should be fulfilled. Non-fulfillment of inequality (21) points to the shift of zero mean of the elements a_{ij}, a change of the unit variance or to difference of $\{a_{ij}\}$ from white noise. If the case $\sigma = 1$ is considered, then inequality (21) can be written in a more simple form:

$$\sigma_{max}[A] \le \sqrt{\chi^2_{\beta,(nm-1)}} \,. \qquad (22)$$

As is seen from expression (22), in the case being considered, the upper confidence boundary of the spectral norm of the normalized innovation matrix of the Kalman filter is determined by the dimension of the innovation vector (or dimension of the measurement vector), the number of sequential innovation vectors used, and the confidence probability selected.

In view of that stated above, in solving the diagnosis problem posed, the decision rule concerning the hypotheses introduced has the form

$$\begin{gathered} \gamma_0 : \sigma_{max}[A] \le \sqrt{\chi^2_{\beta,(nm-1)}} \,, \quad \forall k \ge m; \\ \gamma_1 : \exists k \ge m, \sigma_{max}[A] > \sqrt{\chi^2_{\beta,(nm-1)}} \,. \end{gathered} \qquad (23)$$

The boundary for the spectral norm of the normalized innovation matrix found is quite simple and allows one to check on-line simultaneously mathematical expectation and variance of the innovation sequence. Under operating conditions, the algorithm proposed can be reduced to the following sequence of calculations performed at each step of measurements.

1. Using expressions (4), calculate the Kalman estimate for the vector of system state and the value of the vector of the normalized innovation sequence at the current step k.
2. Compose the normalized innovation matrix for the Kalman filter according to (7) for the given $n \ge 2, m \ge 2$ and $k \ge m$.
3. Determine eigenvalues of the matrix $A^T(k)A(k)$ as solutions of the equation $\det(A - \lambda I) = 0$ and the spectral norm

$$\|A(k)\|_2 = \sigma_{max}[A(k)] = \max\{\lambda_i [A^T(k)A(k)]^{1/2}\}. \qquad (24)$$

4. Check realization of inequality (22) and make decision on detection of a failure in the Kalman filter based on the decision rule (23).

5. Repeat the sequence of calculations, beginning with step 1, for the next instant of time $k+1$.

Qualitative characteristics of the proposed algorithm of failure revealing are probabilities of a correct detection and a false alarm. These characteristics are calculated in a usual way with the use of the table of the χ^2-distribution (Grishin and Kazarinov, 1985). Deriving the required characteristics involves a large volume of mathematical simulation for a justified choice of m number of innovation vectors $v(k)$, that correspond to m different instants of time, from which the matrix of innovation A is composed. A too large m smoothes effects caused by the system failure, and a too small m, increases probability of a false alarm.

3. Design of the EKF for the F-16 Aircraft State Estimation

3.1 F-16 Aircraft Model Description

The technique for failure detection is applied to an unstable multi-input multi-output model of an AFTI/F-16 fighter. The fighter is stabilized by means of a linear quadratic optimal controller. The control gain brings all the eigenvalues that are outside the unit circle, inside the unit circle. It also keeps the mechanical limits on the deflections of control surfaces. The model of the fighter is as follows (Lyshevski, 1997):

$$x(k+1) = Ax(k) + Bu(k) + F(x(k)) + Gw(k) \tag{25}$$

where $x(k)$ is the 9-dimensional state vector of the aircraft, A is the transition matrix of order 9x9 of the aircraft, B is the control distribution matrix of order 9x6 of the aircraft, $u(k)$ is the 6-dimensional control input vector, $F(x(k))$ is the 9-dimensional vector of nonlinear elements of system, $w(k)$ is the random 9-dimensional vector of system noises with zero mean and the covariance matrix $E\left[w(k)w^T(j)\right] = Q(k)\delta(kj)$, G is the transition matrix of the system noises.

The aircraft state variables are:

$$x = [v, \alpha, q, \theta, \beta, p, r, \phi, \psi]^T,$$

where, v is the forward velocity, α is the angle of attack, q is the pitch rate, θ is the pitch angle, β is the side-slip angle, p is the roll rate, r is the yaw rate, ϕ is the roll angle, and ψ is the yaw angle.

The fighter has six control surfaces and hence six control inputs are:

$$u = [\delta_{HR}, \delta_{HL}, \delta_{FR}, \delta_{FL}, \delta_C, \delta_R],$$

where δ_{HR} and δ_{HL} are the deflections of the right and left horizontal stabilizers, δ_{FR} and δ_{FL} are the deflections of the right and left flaps, δ_C and δ_R are the canard and rudder deflections. The following hard bounds (mechanical limits) on the deflections of control surfaces are assumed: $|\delta_{HR}, \delta_{HL}| \le 0.44$ rad, $|\delta_{FR}, \delta_{FL}| \le 0.35$ rad, $|\delta_C| \le 0.47$ rad and $|\delta_R| \le 0.52$ rad. A,B and F(x) for the sampling period of 0.03 s. are:

$$A=\begin{bmatrix} 0.9995 & 0.2457 & -0.0273 & -0.2885 & -0.0391 & -0.0075 & -0.002 & 0 & 0 \\ -0.0001 & 0.9663 & 0.0291 & 0 & 0.0011 & 0.0017 & 0.0003 & 0 & 0 \\ 0 & 0.1135 & 0.9765 & 0 & 0.0007 & 0.0002 & 0.0012 & 0 & 0 \\ 0 & 0.0017 & 0.0296 & 1 & 0 & 0 & 0 & 0 & 0 \\ -0.0001 & 0.0048 & 0.0011 & 0.0288 & 0.977 & 0.0037 & -0.0268 & 0 & 0 \\ 0.0001 & 0.0165 & 0.0005 & -0.0198 & -1.3515 & 0.8977 & 0.025 & 0 & 0 \\ 0 & -0.0268 & 0.0015 & 0.0041 & 0.2716 & -0.0003 & 0.9811 & 0 & 0 \\ 0 & 0.0003 & 0 & -0.0002 & -0.0207 & 0.0285 & 0.0003 & 1 & 0 \\ 0 & -0.0004 & 0 & 0 & 0.0041 & 0 & 0.0297 & 0 & 1 \end{bmatrix}$$

$$B=\begin{bmatrix} 0.071 & 0.0067 & -0.011 & -0.0117 & -0.0004 & -0.0001 \\ -0.009 & -0.0088 & -0.0091 & -0.009 & 0 & 0 \\ -0.2818 & -0.2819 & -0.0746 & -0.0746 & 0 & 0 \\ -0.0042 & -0.0042 & -0.0011 & -0.0011 & 0 & 0 \\ -0.001 & 0.0007 & -0.0006 & 0.0005 & 0.0123 & 0.0018 \\ -0.0822 & 0.0819 & -0.0881 & 0.0879 & 0.0122 & 0.025 \\ 0.0921 & -0.0924 & 0.0233 & -0.0232 & 0.02 & -0.0132 \\ -0.0013 & 0.013 & -0.013 & 0.013 & 0.0002 & 0.0004 \\ 0.0014 & -0.0014 & 0.0003 & -0.0003 & 0.0003 & -0.0002 \end{bmatrix}$$

$$F(x)=\begin{bmatrix} F_V \\ F_\alpha \\ F_q \\ F_\theta \\ F_\beta \\ F_p \\ F_r \\ F_\phi \\ F_\psi \end{bmatrix} = \begin{bmatrix} 0 \\ -0.03p\cos\alpha\tan\beta - 0.03r\sin\alpha\tan\beta \\ 0.028pr - 0.00018p^2 + 0.00018r^2 \\ 0.03q\cos\phi - 0.03r\sin\phi \\ 0.03p\sin\alpha - 0.03r\cos\alpha \\ 0.00026qp - 0.017qr \\ -0.025qp - 0.00026qr \\ 0.03q\tan\theta\sin\phi + 0.03r\tan\theta\cos\phi \\ 0.03q\cos^{-1}\theta\sin\phi + 0.03r\cos^{-1}\theta\cos\phi \end{bmatrix}$$

Below the Extended Kalman Filter (EKF) to estimate the F-16 aircraft motion is designed.

3.2 Deriving of the EKF

Let us define the estimated vector as:

$$x^T(k)=[v(k),\alpha(k),q(k),\theta(k),\beta(k),p(k),r(k),\phi(k),\psi(k)]$$

and apply the Kalman filter to estimate this vector. The nonlinear mathematic model for the longitudinal and lateral F-16 aircraft motion is given in (25).

The measurement equations can be written as:

$$z(k) = Hx(k) + V(k), \tag{26}$$

where H is the measurement matrix, which is 9×9 unit matrix, $V(k)$ is the measurement noise and its mean and correlation matrix respectively are:

$$E[V(k)] = 0; E[V(k)V^T(j)] = R(k)\delta(kj).$$

By using quasi-linearization method let us linearize the equation (25):

$$\begin{aligned} x(k) &= A\hat{x}(k-1) + B\hat{u}(k-1) + F(\hat{x}(k-1)) + A[x(k-1) - \hat{x}(k-1)] + \\ &F_x(k-1)[x(k-1) - \hat{x}(k-1)] + B[u(k-1) - \hat{u}(k-1)] + Gw(k-1) \end{aligned} \tag{27}$$

where $F_x = \left[\frac{\partial F}{\partial x}\right]_{\hat{x}(k-1)}$.

Among the procedures of estimation theory, the Bayes procedure has the most accuracy because it is based on both the experimental data in likelihood function and a priori data expressed by a priori density of the estimated parameters. The more data, the more accuracy yields. Moreover, the Bayes procedure does not require the system to be linear and stationary, and produces a solution for the filtering when the initial conditions of the state vector are unknown (Gadzhiev, 1996). Therefore, the Bayes procedure to filter the state vector of the aircraft motion is preferred. A posteriori distribution density of the state vector is given by the Bayes formula:

$$P\left[{x(k)}/{Z^k}\right] = P\left[{x(k)}/{Z^{k-1}, z(k)}\right] = \frac{P\left[{x(k)}/{Z^{k-1}}\right]}{P\left[{z(k)}/{Z^{k-1}}\right]} P\left[{z(k)}/{x(k), Z^{k-1}}\right], \tag{28}$$

where $Z^k = \{z(1), z(2), z(3), \ldots, z(k)\}$; $Z^{k-1} = \{z(1), z(2), \ldots, z(k-1)\}$.

When the probability density functions in (28) are substituted and the conditional mathematical expectation of the a posteriori probability density function is taken as the optimum estimation value, the following recursive EKF algorithm for the state vector estimation of the F-16 aircraft motion is obtained as (Caliskan and Hajiyev, 2003):

Equation of the estimation value

$$\hat{x}(k) = \hat{x}(k/k-1) + K(k)v(k) \tag{29}$$

Equation of the extrapolation value

$$\hat{x}(k/k-1) = A\hat{x}(k-1) + B\hat{u}(k-1) + F(\hat{x}(k-1)) \tag{30}$$

The innovation sequence

$$v(k) = z(k) - H\left[A\hat{x}(k-1) + B\hat{u}(k-1) + F(\hat{x}(k-1))\right] \tag{31}$$

The gain matrix of filter

$$K(k) = P(k)H^T(k)R^{-1}(k) \tag{32}$$

The covariance matrix of estimation errors

$$P(k) = M(k) - M(k)H^T\left[HM(k)H^T + R(k)\right]^{-1}HM(k) \tag{33}$$

The covariance matrix of extrapolation errors

$$\begin{aligned} M(k) &= AP(k-1)A^T + BD_u(k-1)B^T \\ &+ F_x(k-1)P(k-1)F_x^T(k-1) + GQ(k-1)G^T \end{aligned} \tag{34}$$

where D_u is the covariance matrix of the control input error, $Q(k-1)$ is the covariance matrix of system noise.

4. Adaptive EKF Insensitive to Sensor Failures

An adaptive EKF for the F-16 aircraft state estimation may be designed in order to isolate the detected sensor and actuator/surface failures. The following approach for the solution of the filtration problem is proposed for this case (Hajiyev, 2006). In the case of normal operation of measurement system, the filter works according to the conventional EKF algorithm (29)-(34). But if the condition of the operation of the measurement system does not correspond to the models, used in the synthesis of the filter, then the gain coefficient (32) of the discrepancy automatically changes due to the change in the covariance matrix of the innovation sequence according to the rule

$$P_v(k) = HM(k)H^T + S(k)R(k) \tag{35}$$

in which weight coefficient $S(k)$ is calculated from the discrepancy (31) analysis results. In this case the filter gain coefficient (32) can be written in the form of

$$K(k) = M(k)H^T\left[HM(k)H^T + S(k)R(k)\right]^{-1} \tag{36}$$

According to the proposed approach the gain coefficient (32) is changed when the following condition is valid

$$\begin{aligned} &tr\left\{v(k)v^T(k)\right\} \geq tr\left\{E\left[v(k)v^T(k)\right]\right\} = \\ &tr\left\{E\left[H\left(x(k) - \hat{x}(k/k-1)\right) + v(k)\right] \times \left[H\left(x(k) - \hat{x}(k/k-1)\right) + v(k)\right]^T\right\} = \\ &tr\left\{HM(k)H^T + R(k)\right\} \end{aligned} \tag{37}$$

where $tr(.)$ is the trace of matrix. When a significant change in the conditions of the operation of the measurement system occurs, the prediction of observations in (31), $H\hat{x}(k/k-1)$, will considerably differ from the observation results $z(k)$. Consequently, the sum of the discrepancy squares on the left side of (37) will characterize the real filtration error, while the right side determines the theoretical accuracy of the innovation sequence, obtained on the basis of a priori information. If condition (38) is met, then the real filtration error exceeds the theoretical error. Therefore, it is necessary to correct the filter gain matrix (32). In this case by substituting (35) in (37) the following equation can be obtained;

$$tr\left\{v(k)v^T(k)\right\} = tr\left\{HM(k)H^T\right\} + S(k)tr\left\{R(k)\right\} \tag{38}$$

Hence taking the expression $tr\left\{v(k)v^T(k)\right\} = v^T(k)v(k)$ into consideration, the following formula for the weighting factor $S(k)$ is obtained:

$$S(k) = \frac{v^T(k)v(k) - tr\left\{HM(k)H^T\right\}}{tr\left\{R(k)\right\}} \tag{39}$$

Using (35), (36) and (39) in the estimation algorithm (29) -(34) gives the possibility to accomplish the adaptation of the filter to the change of measurement system operation conditions. If the left side of the expression (37) is greater than the right side, the value of coefficient $S(k)$ will increase. This corresponds to the beginning of the adaptation of filter. Consequently, both the covariance matrix of innovation sequence $P_v(k)$ (35) and the filter gain matrix $K(k)$ (32) increase, and that cause to the strengthening of the corrective influence of discrepancy in (29) which makes the estimation value $\hat{x}(k)$ approach to the actual value $x(k)$. This will lead to the decrease of discrepancy $v(k)$ and coefficient $S(k)$, weakening of the corrective influence of discrepancy, etc.

The final expressions of the proposed adaptive filtration algorithm with the filter gain correction insensitive to measurement faults can be written in the following form:

$$\hat{x}(k) = \hat{x}(k/k-1) + K(k)v(k)$$

$$
\begin{aligned}
\hat{x}(k/k-1) &= A\hat{x}(k-1) + B\hat{u}(k-1) + F(\hat{x}(k-1)) \\
v(k) &= z(k) - H\left[A\hat{x}(k-1) + B\hat{u}(k-1) + F(\hat{x}(k-1))\right] \\
K(k) &= P(k)H^{T}(k)\left[S(k)R(k)\right]^{-1} \\
P(k) &= M(k) - M(k)H^{T}\left[P_{v}(k)\right]^{-1}HM(k) \\
P_{v}(k) &= HM(k)H^{T} + S(k)R(k) \\
S(k) &= \frac{v^{T}(k)v(k) - tr\left\{HM(k)H^{T}\right\}}{tr\left\{R(k)\right\}} \\
M(k) &= AP(k-1)A^{T} + BD_{u}(k-1)B^{T} \\
&\quad + F_{x}(k-1)P(k-1)F_{x}^{T}(k-1) + GQ(k-1)G^{T}
\end{aligned}
\tag{40}
$$

where $P_v(k)$ is the covariance matrix of the innovation sequence, and $S(k)$ is the weighting factor. The other filter parameters in (40) are same with the ones given in the expressions (29)-(34).

In contrast to the EKF algorithm (29)-(34), in which the filter gain $K(k)$ changes by program, in the proposed algorithm the current measurements have larger weight, since the coefficients of matrix $K(k)$ are corrected by the results of each observation. This algorithm is adapted to the measurement system operation conditions by the approximation of the theoretical covariance matrix $P_v(k)$ to the real covariance matrix of the innovation sequence, due to the change in the weighting factor $S(k)$. Mentioned change is accomplished because of regarding the matrix $v(k)v^T(k)$, which characterizes the real filtration error. Proposed adaptive EKF for the F-16 aircraft state estimation will ensure the guaranteed adaptation of the filter to the change of the measurement system operation conditions, consequently it will become insensitive to sensor failures.

The designed adaptive EKF (40) is not an optimum filter, unlike the EKF (29)-(34), because of the $S(k)$ factor. Even in the absence of a failure, the estimation error could be larger than that of the conventional filtration algorithm (29)-(34). Therefore, adaptive algorithm is operated only when the measurements are faulty or in order to isolate the detected sensor and actuator/surface failures. In all other cases procedure is run optimally with regular EKF (29)-(31).

4. Simulation Results of Failure Detection and Adaptive EKF Algorithms

The technique for failure detection is applied to multi-input multi-output model of an AFTI/F-16 fighter (25). The measurements are processed using Kalman filter (29)-(32) that allows us to determine the estimate of the state vector of F-16 aircraft and the covariance matrix of the estimate errors at each kth step.

If there are no faults in the estimation system, then the normalized innovation sequence

$$\tilde{v}(k) = \left[HM(k)H^T + R(k)\right]^{-1/2} v(k), \tag{41}$$

of the EKF (29)-(32) is a Gaussian white noise with zero mean and a unit covariance matrix. Sensor and control surface/actuator failures will result with changes in the above characteristics of the sequence of $\tilde{v}(k)$. To verify the hypothesis γ_0 and γ_1, let us use a spectral norm of the matrix constructed as (7).

In the simulations, $m = 10; n = 9; \beta = 0.997$ are taken, and the threshold value $\sqrt{\chi^2_{\beta,(nm-1)}}$ is found as 11.4. Decisions as to reveal a failure in the system are made based on the rule (23). The results of simulations are shown in Figs. 1–16.

4.1 A Sensor Failure (Shift in the Pitch Rate Gyroscope)

Shift in the pitch rate gyroscope is simulated at iteration 30 as follows;

$$z_q(k) = q(k) + V_q(k) + 3, \ (k \geq 30). \tag{42}$$

The graph of the spectral norm $\sigma_{max}[A]$ is shown in Figure 1 when a shift occurs in the pitch rate gyroscope.

As seen in Figure 1, until the sensor failure occurs $\sigma_{max}[A]$ is lower than the threshold. When a failure occurs in the pitch rate gyroscope, $\sigma_{max}[A]$ grows rapidly, and after 1 iteration it exceeds the threshold. Hence γ_1 hypothesis is judged to be true. This failure causes a change in the mean of the innovation sequence. The innovation sequences in case of a shift in the pitch rate gyroscope are shown in Figures 2-4.

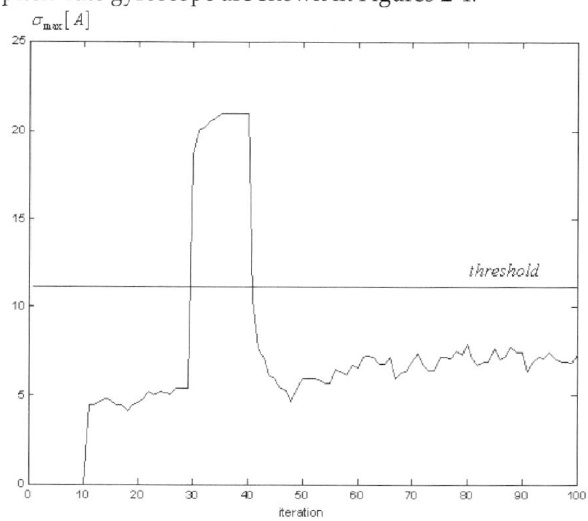

Fig. 1. Behavior of the spectral norm $\sigma_{max}[A]$ in case of a shift in the pitch rate gyroscope

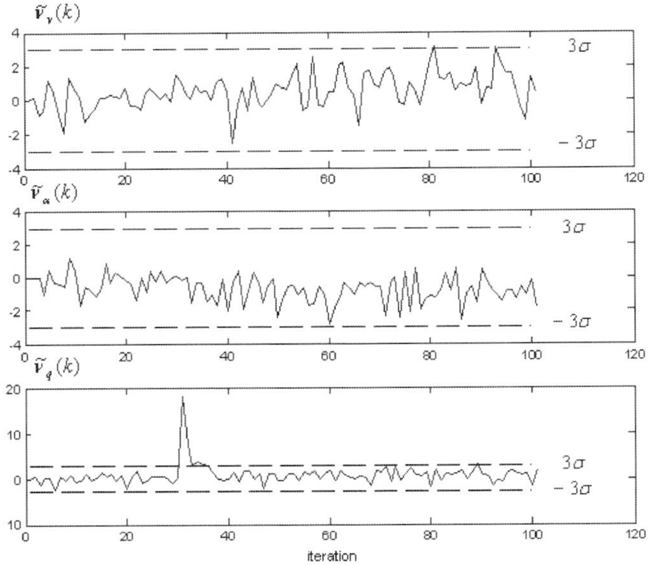

Fig. 2. Normalized Innovation Sequences $\tilde{\nu}_v(k), \tilde{\nu}_\alpha(k), \tilde{\nu}_q(k)$ in the case of a shift in the pitch rate gyroscope

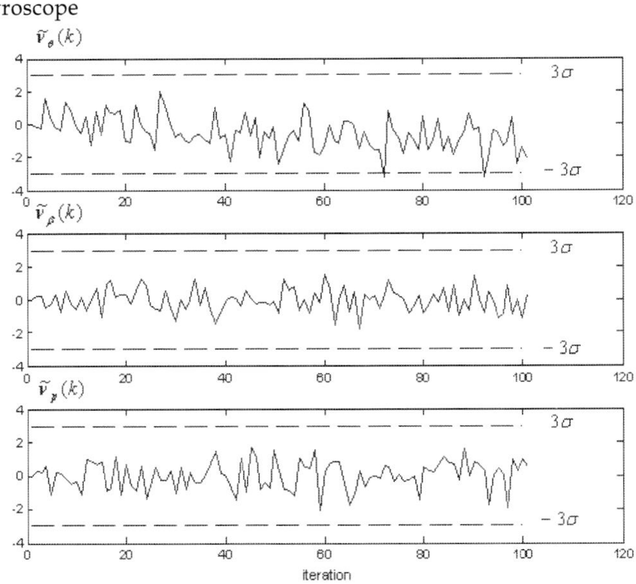

Fig. 3. Normalized Innovation Sequences $\tilde{\nu}_\theta(k), \tilde{\nu}_\beta(k), \tilde{\nu}_p(k)$ in the case of a shift in the pitch rate gyroscope

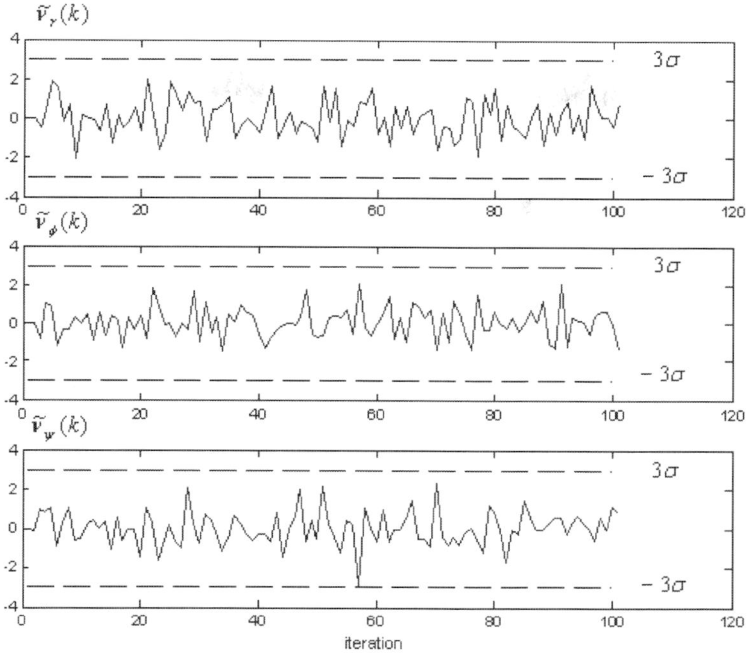

Fig. 4. Normalized Innovation Sequences $\tilde{v}_r(k)$, $\tilde{v}_\phi(k)$, $\tilde{v}_\psi(k)$ in the case of a shift in the pitch rate gyroscope

4.2 A Sensor Failure (the Noise Variance in the Pitch Rate Gyroscope is Changed)

The noise variance in the pitch rate gyroscope is changed at iteration 30 as follows;

$$z_q(k) = q(k) + 3V_q(k), \ (k \geqslant 30). \quad (43)$$

Figure 5 shows that the value of $\sigma_{max}[A]$ sharply increases after the 30th step and intersects its admissible bound at the step $k = 42$. As a result, based on the decision rule (23), estimation system failure is noted. This failure causes a change in the variance of the innovation sequence. The innovation sequences in case of changes in the noise variance of the pitch rate gyroscope are shown in Figures 6-8.

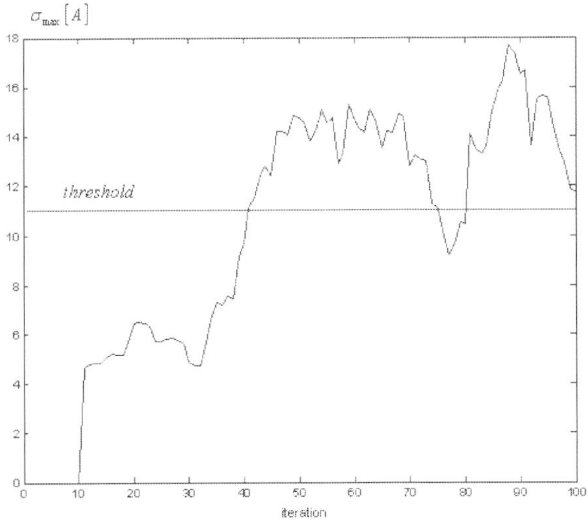

Fig. 5. Behavior of the spectral norm $\sigma_{max}[A]$ in case of changes in the noise variance of the pitch rate gyroscope

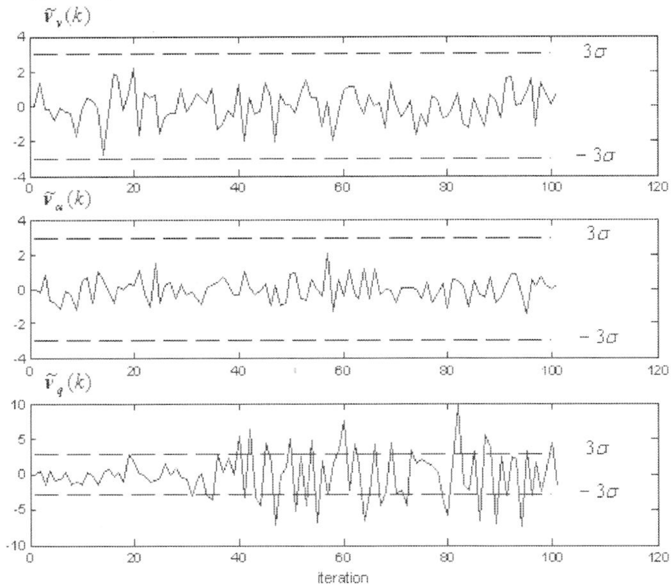

Fig. 6. Normalized Innovation Sequences $\tilde{v}_v(k), \tilde{v}_\alpha(k)$, $\tilde{v}_q(k)$ in case of changes in the noise variance of the pitch rate gyroscope

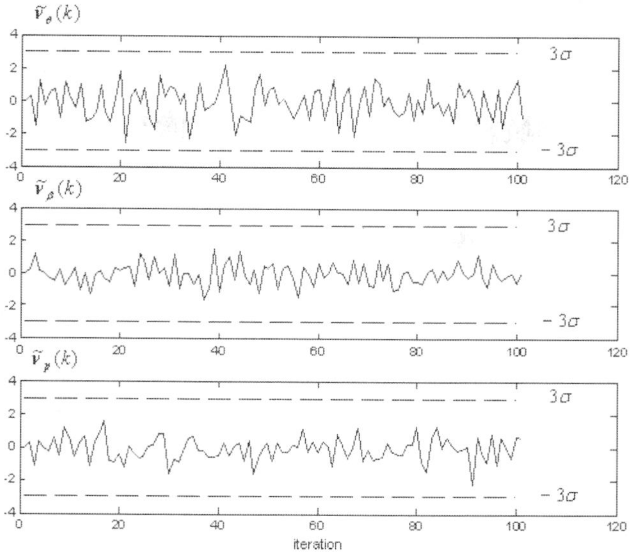

Fig. 7. Normalized Innovation Sequences $\tilde{v}_\theta(k)$, $\tilde{v}_\beta(k)$, $\tilde{v}_p(k)$ in case of changes in the noise variance of the pitch rate gyroscope

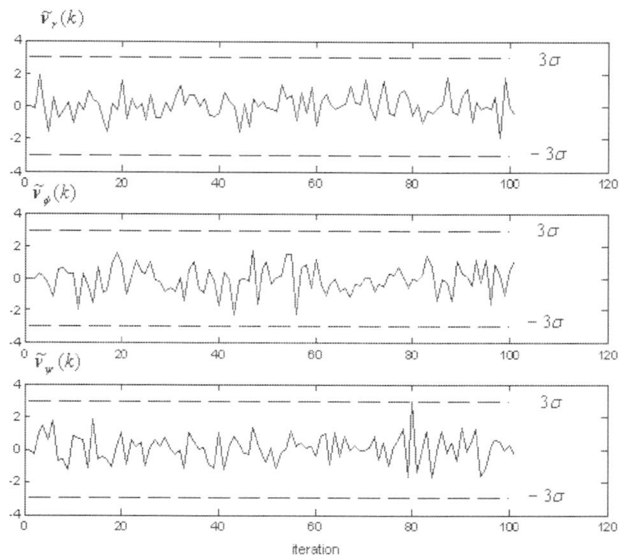

Fig. 8. Normalized Innovation Sequences $\tilde{v}_r(k)$, $\tilde{v}_\phi(k)$, $\tilde{v}_\psi(k)$ in case of changes in the noise variance of the pitch rate gyroscope

4.3 The Actuator Motor Failure

Two kinds of failures can occur in an actuator: actuator motor failure, and control surface failure. For simulation of the actuator motor failure, the control input δ_{HR} (deflection of the right horizontal stabilizer) has been changed to $\delta_{HR}(k) = \delta_{HR}(k) + 1^{\circ}$, $(k \geq 30)$ at iteration 30. The graph of the spectral norm $\sigma_{max}[A]$ is shown in Figure 9 when a shift occurs in the actuator motor at the step 30. This failure causes a change in the mean of the innovation sequence. As seen in Figure 9, until the actuator failure occurs, spectral norm $\sigma_{max}[A]$ is lower than the threshold. When a failure occurs in the actuator $\sigma_{max}[A]$ grows rapidly, and after 3 steps it exceeds the threshold. Hence γ_1 hypothesis is judged to be true. The innovation sequences in the case of actuator motor failure are shown in Figures 10-12.

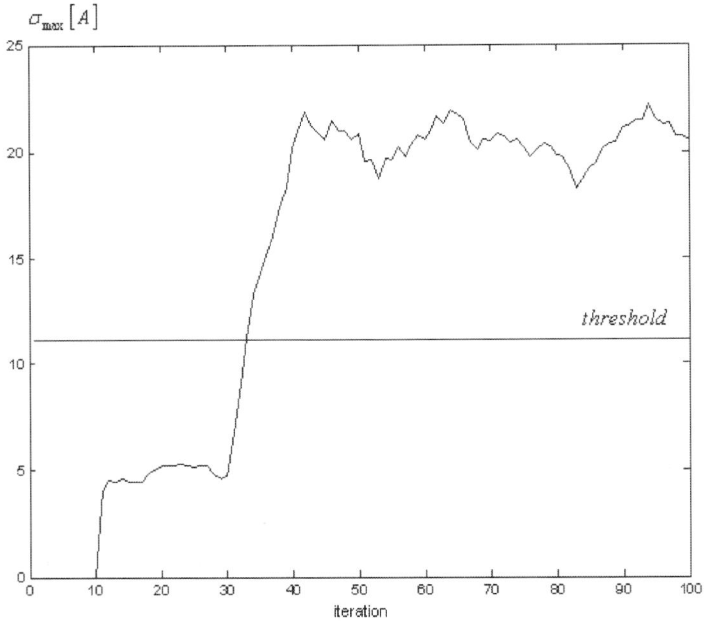

Fig. 9. Behavior of the spectral norm $\sigma_{max}[A]$ in case of actuator motor failure

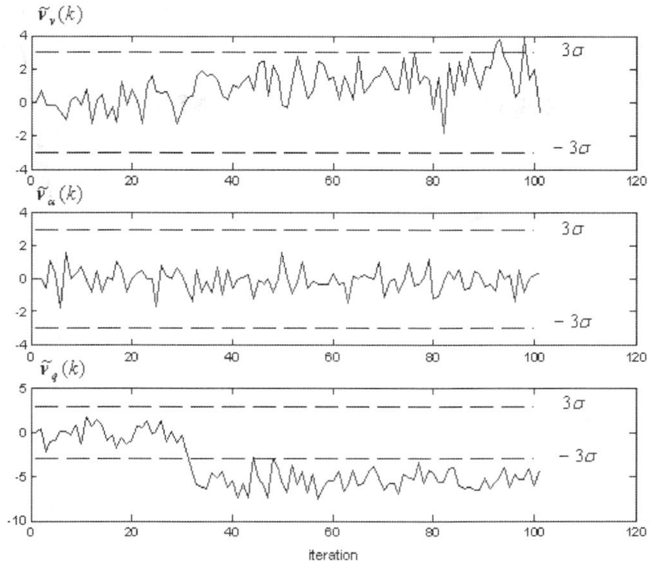

Fig. 10. Normalized Innovation Sequences $\tilde{v}_v(k), \tilde{v}_\alpha(k), \tilde{v}_q(k)$ in case of actuator motor failure

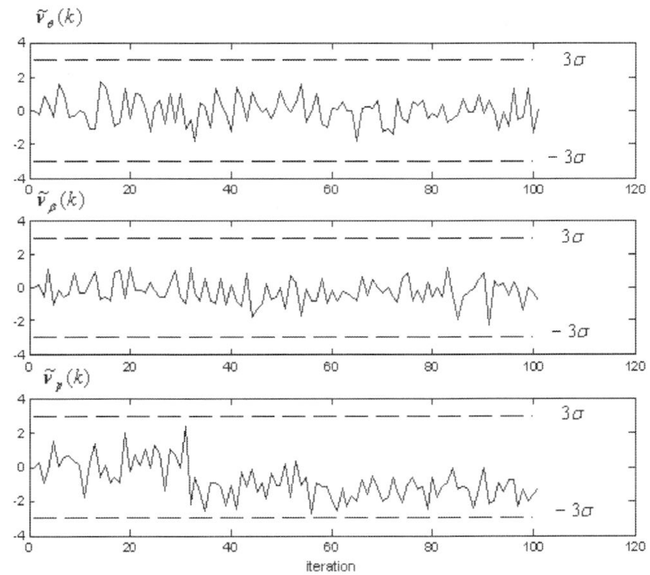

Fig. 11. Normalized Innovation Sequences $\tilde{v}_\theta(k)$, $\tilde{v}_\beta(k), \tilde{v}_p(k)$ in case of actuator motor failure

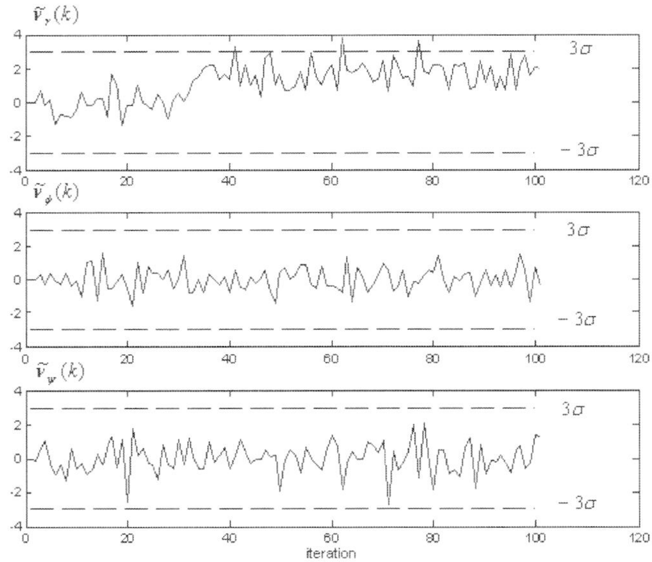

Fig. 12. Normalized Innovation Sequences $\tilde{v}_r(k)$, $\tilde{v}_\phi(k)$, $\tilde{v}_\psi(k)$ in case of actuator motor failure

4.4 The Control Surface Failure

The proposed failure detection algorithm is used below to detect the control surface failures. The control derivatives corresponding to the first control surface (right horizontal stabilizer) has been changed as follows at iteration 30;

$$B(i,1) = B(i,1) + 0.08; \; i = \overline{1,9}, \; (k \geq 30) \tag{44}$$

The graph of the spectral norm $\sigma_{max}[A]$ is shown in Figure 13 when a shift occurs in the control surface. As seen in Figure 13, until the control surface failure occurs, $\sigma_{max}[A]$ is lower than the threshold. When a fault occurs in the control surface $\sigma_{max}[A]$ grows rapidly, and after 28 iterations it exceeds the threshold. Hence γ_1 hypothesis is judged to be true. This failure causes a change in the mean of the innovation sequence. The innovation sequences in case of control surface failure are shown in Figures 14-16.

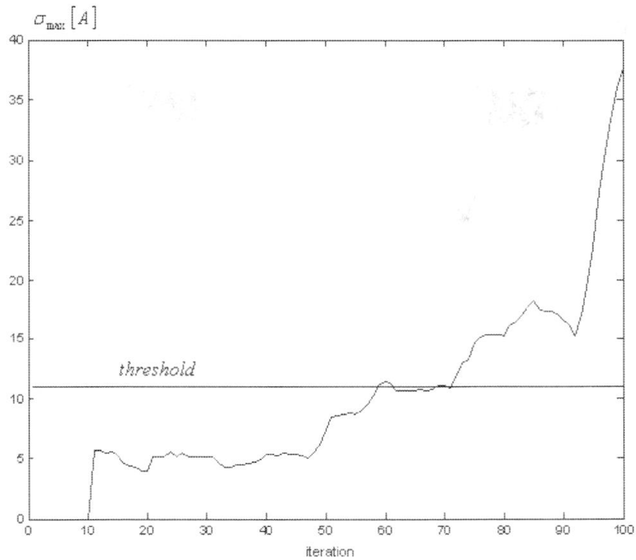

Fig. 13. Behavior of the spectral norm $\sigma_{\max}[A]$ in case of control surface failure

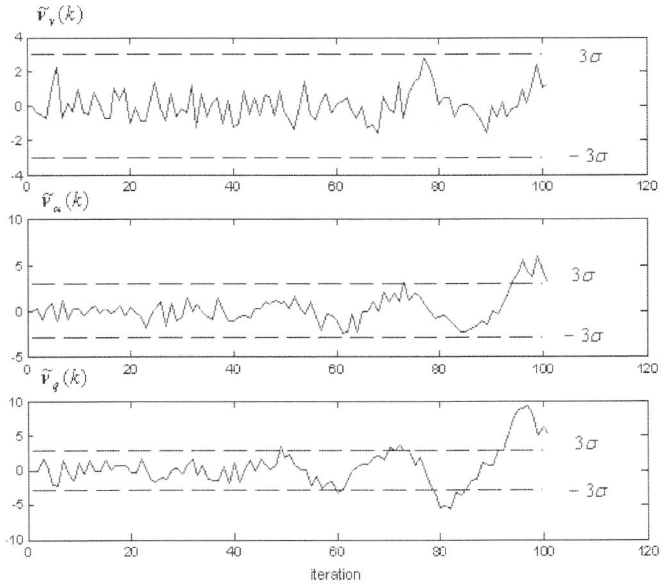

Fig. 14. Normalized Innovation Sequences $\tilde{\nu}_v(k)$, $\tilde{\nu}_\alpha(k)$, $\tilde{\nu}_q(k)$ in case of control surface failure

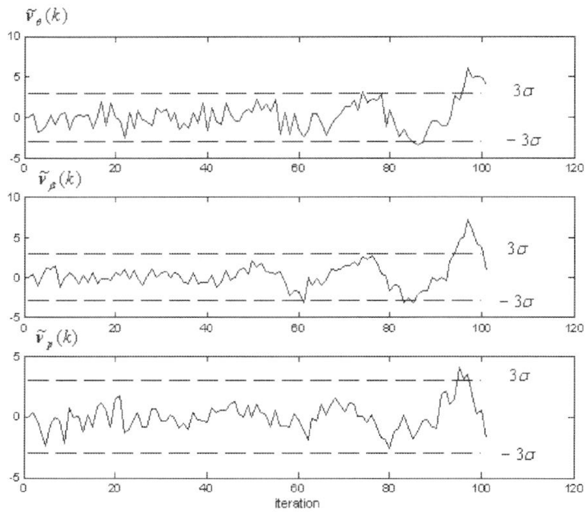

Fig. 15. Normalized Innovation Sequences $\tilde{v}_\theta(k)$, $\tilde{v}_\beta(k)$, $\tilde{v}_p(k)$ in case of control surface failure

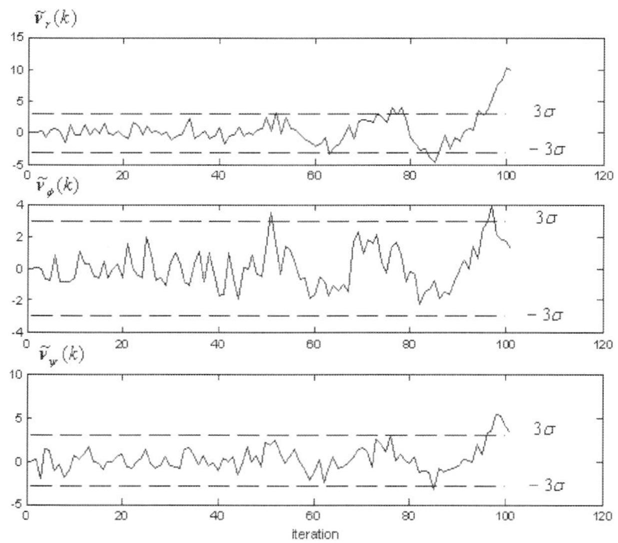

Fig. 16. Normalized Innovation Sequences $\tilde{v}_r(k)$, $\tilde{v}_\phi(k)$, $\tilde{v}_\psi(k)$ in case of control surface failure

The simulations show that both sensor and actuator/surface failures affect the spectral norm of the normalized innovation matrix. The simulation results justify the obtained theoretical calculations and show the practical applicability of the proposed failure detection algorithm.

The introduction of the developed fault detection algorithms does not distort the estimation results of the filter and has no influence on their accuracy.

In real situations of exploiting an object, the proposed algorithm enables operative detection of faults such as: abnormal measurements, sudden shifts appearing in the measuring channel, faultiness of the measuring devices, changes in the statistical characteristics of the noises of an object or of measurements, reduction in the actuator/surface effectiveness, friction between moving parts of the control surfaces, partial loss of a control surface (break off of a part of control surface), malfunctions in the computer, and also a sharp change in the trajectory of a monitoring process, etc in order to subsequently correct estimators or to make timely decisions on the necessity and the character of the control actions with respect to the process of technical exploitation of the object.

4.5 Simulation Results of Adaptive EKF Insensitive to Sensor Failures

Simulation of the proposed adaptive EKF for the F-16 aircraft state estimation is performed. The measurements were processed using adaptive EKF (40) insensitive to sensor failures. To verify the hypothesis γ_0 and γ_1 in cases of the sensor and control surface/actuator failures, the spectral norm of the matrix constructed as (7) is used. Decisions as to reveal a failure in the system are made based on the rule (23). The results of simulations are shown in Figures 17-21.

Behavior of the spectral norm $\sigma_{max}[A]$ in case of changes in the noise variance of the pitch rate gyroscope (sensor failure), when the adaptive EKF is used, is given in Figure 17. The noise variance of the pitch rate gyroscope has been changed corresponding to (43) at iteration 30.

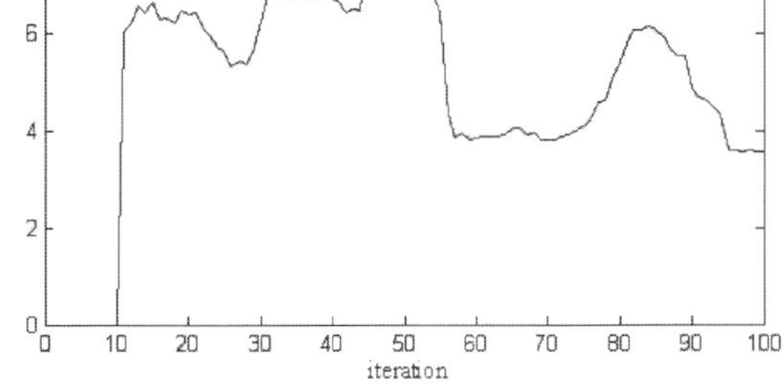

Fig. 17. Behavior of the spectral norm $\sigma_{max}[A]$ in case of changes in the noise variance of the pitch rate gyroscope (adaptive EKF was used)

As seen in Figure 17, in spite of the sensor failure, in all iterations $\sigma_{max}[A]$ is lower than the threshold. Consequently, via the decision rule (23) γ_0 hypothesis is judged to be true. The normalized innovation sequence $\tilde{v}_q(k)$ in case of changes in the noise variance of the pitch rate gyroscope, when adaptive EKF is used, is shown in Figure 18. The results presented in the Figures 17 and 18 show that the adaptive EKF (40) is insensitive to sensor failures. Behavior of the weighting factor $S(k)$ of the adaptive filter is given in Figure 19.

Behavior of the spectral norm $\sigma_{max}[A]$ in case of control surface failure, when adaptive EKF insensitive to sensor failures is used, is presented in Figure 20. The control derivatives corresponding to the first control surface (right horizontal stabilizer) has been changed corresponding to (44) at iteration 30. As seen in Figure 20, until the control surface failure occurs, $\sigma_{max}[A]$ is lower than the threshold. When a fault occurs in the control surface, $\sigma_{max}[A]$ grows rapidly, and after 26 iterations it exceeds the threshold. Hence γ_1 hypothesis is judged to be true. The normalized innovation sequence $\tilde{v}_q(k)$ in case of control surface failure, when adaptive EKF is used, is given in Figure 21.

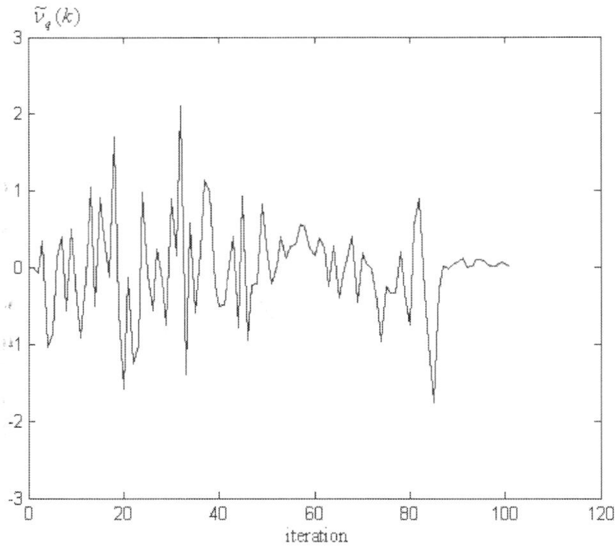

Fig. 18. Normalized Innovation Sequences $\tilde{v}_q(k)$ in case of changes in the noise variance of the pitch rate gyroscope (adaptive EKF was used)

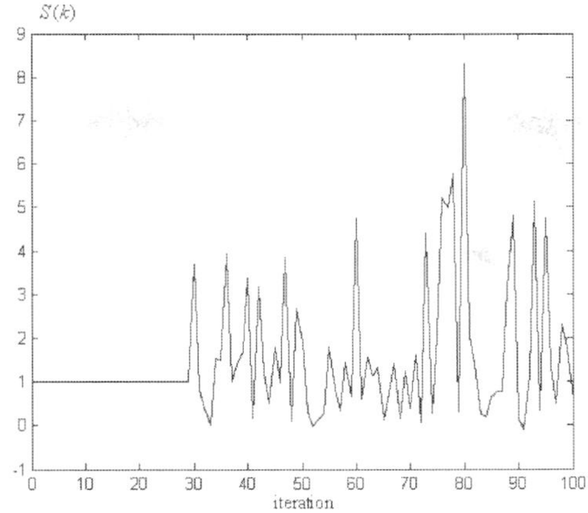

Fig. 19. Behavior of the weighting factor S(k)

The obtained simulation results show that the proposed adaptive EKF for the F-16 aircraft state estimation can isolate the detected sensor and actuator/surface failures. This filter is insensitive to sensor failures but sensitive to actuator/surface failures. When a regular EKF is used, the decision statistics changes regardless to the failure in the sensors or in the actuators/surfaces. On the other hand if the adaptive EKF insensitive to sensor failures is used, it is easy to distinguish the sensor and actuator failures.
The further fault isolation - finding in which component the fault has occurred (determining the location of the fault) - can be performed via the innovation approach based fault isolation methods described in (Hajiyev & Caliskan, 2003; Hajiyev & Caliskan, 2005; Hajiyev, 2009).

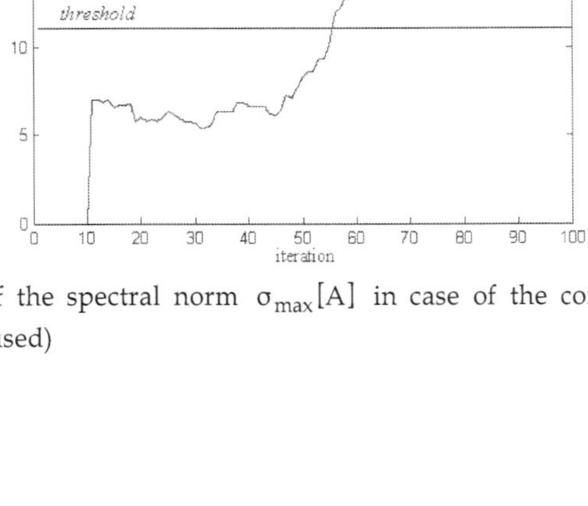

Fig. 20. Behavior of the spectral norm $\sigma_{max}[A]$ in case of the control surface failure (adaptive EKF was used)

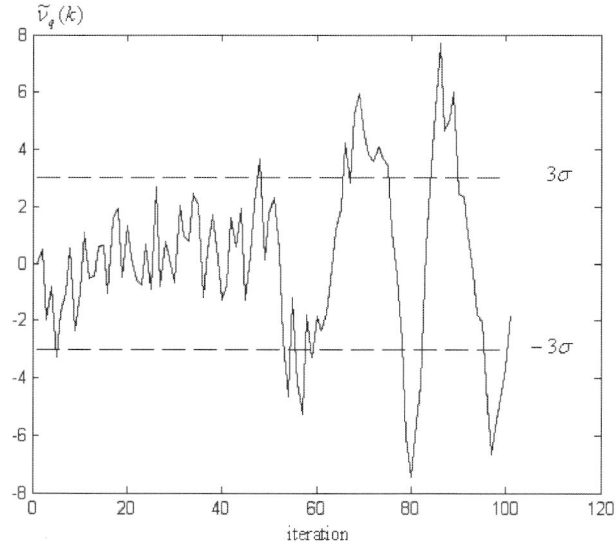

Fig. 21. Normalized Innovation Sequence $\tilde{v}_q(k)$ in case of the control surface failure (adaptive EKF was used)

4.6 About Choosing the Number of the Innovation Vectors (Samples)

Note that, in this case, the inertia (the delay) of the failure detection depends on the number of the innovation vectors (samples) m, that correspond to m different instants of time, from which the matrix of innovation A is composed and with an increase in this number, this characteristic worsens. On the other hand, a very small value of m leads to frequent false faults. Furthermore, the estimates of the eigenvalues of the matrix $A^T(k)A(k)$ and consequently, the singular values of the matrix $A(k)$ and the spectral norm of the matrix $A(k)$ will be biased for small sample sizes in general. Less unlikely larger number of the samples m causes to the biasness of the estimates. However, larger number of the samples reduces the ability of the algorithm to correctly trace high-frequency changes of the trajectory, e.g. turns (Mohamed & Schwarz, 1999). Therefore, the trade-off between the biasness and the frequent false faults on the one hand and the tractability of the estimates and bad inertia characteristic of the fault detection on the other hand should be taken into account according to the application at hand. In addition, the proper choice of the number of innovation vectors m, depends significantly on the motional dynamics. Since the number of samples m, is chosen empirically, there is no theoretically justified choice of it at present. Deriving the required correct detection and the false alarm characteristics involves mathematical simulation for a justified choice of the number of innovation vectors m. For this purpose simulations of the failure detection algorithm are performed for the different number of samples m. During simulations four kinds of scenario are considered:

1. In the simulations, m, n, and β are taken as $m = 6; n = 9; \beta = 0.95$, and the threshold value $\sqrt{\chi^2_{\beta,(nm-1)}}$ is found as 8.1. The noise variance of the pitch rate gyroscope is changed at iteration 30.
2. In the simulations, m, n, and β are taken as $m = 13; n = 9; \beta = 0.95$, and the threshold value $\sqrt{\chi^2_{\beta,(nm-1)}}$ is found as 11.91. The noise variance of the pitch rate gyroscope is changed at iteration 30.
3. In the simulations, m, n, and β are taken as $m = 30; n = 9; \beta = 0.95$, and the threshold value $\sqrt{\chi^2_{\beta,(nm-1)}}$ is found as 17.55. The noise variance of the pitch rate gyroscope is changed at iteration 60.

Decisions as to reveal a failure in the system are made based on the rule (23). The results of the simulations are shown in Figures 22–24. Graphs show the behavior of the statistic $\sigma_{max}[A]$ and its admissible bound in case of changes in the noise variance of the pitch rate gyroscope.

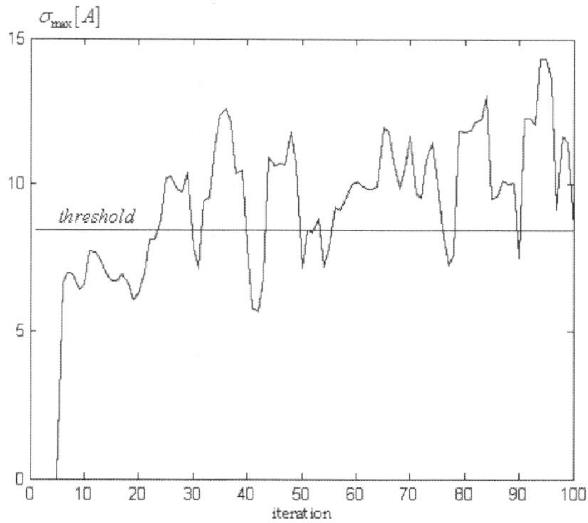

Fig. 22. Behavior of the spectral norm $\sigma_{max}[A]$ in case of changes in the noise variance of the pitch rate gyroscope at iteration 30 ($m = 6$)

As it is seen from Figure 22, in the first scenario ($m = 6$),the value of the statistic $\sigma_{max}[A]$ exceeds its admissible bound until the 30th step and via the decision rule (23) the false failure in the system is detected. When a failure occurs in the pitch rate gyroscope at the iteration 30, $\sigma_{max}[A]$ grows rapidly, and after 3 iteration (0.09 s after fault occurs) it exceeds the threshold. Hence γ_1 hypothesis is judged to be true. Figure 23 show that in the second scenario ($m = 13$), $\sigma_{max}[A]$ is lower than the threshold until the pitch rate gyroscope fault occurs. When a fault occurs in the pitch rate gyroscope at the 30th step,

$\sigma_{max}[A]$ grows abruptly and at the step $\tau = 37$ (0.21 s after fault occurs) it exceeds its admissible bound and the inequality (22) becomes not fulfilled. As a result, on the base of decision rule (23) failure in the system is detected. As seen from Figure 24, in the third scenario ($m = 30$), the statistic $\sigma_{max}[A]$ is lower than the threshold until the pitch rate gyroscope fault occurs. After the 60th step (after the fault occurs), the value of the statistic $\sigma_{max}[A]$ increases, and at the step $\tau = 84$ (0.72 s after fault occurs) it exceeds the threshold. Hence γ_1 hypothesis is judged to be true.

In the practical implementations the usage of the values between 10 and 20 are recommended for the number of samples m.

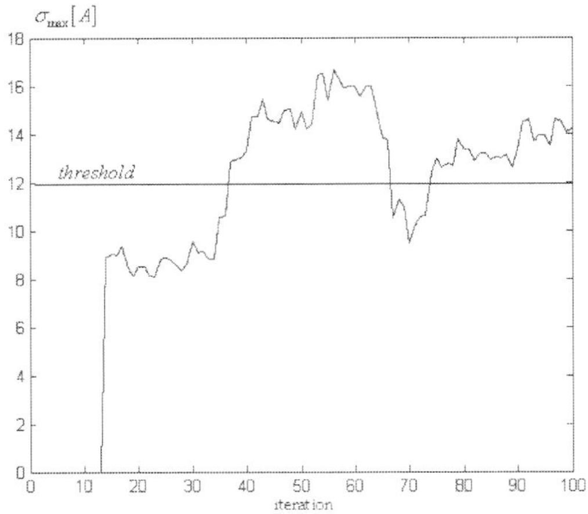

Fig. 23. Behavior of the spectral norm $\sigma_{max}[A]$ in case of changes in the noise variance of the pitch rate gyroscope at iteration 30 ($m = 13$)

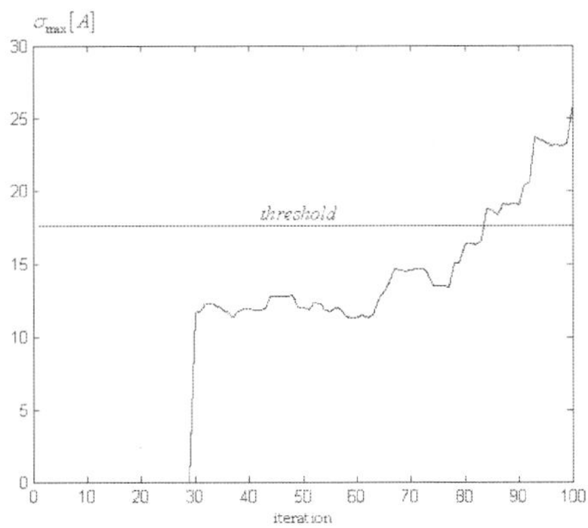

Fig. 24. Behaviour of the spectral norm $\sigma_{max}[A]$ in case of changes in noise variance in the pitch rate gyroscope at iteration 60 ($m = 30$)

5. Conclusions and Discussions

An approach to detect the aircraft sensor and actuator/surface failures based on the spectral norm of an innovation matrix is proposed. An upper confidence bound for the spectral norm of a random innovation matrix $A(k) \in R^{n \times m}$ that consists of normally distributed random variables with zero mathematical expectation is found. The outlined approach allows check of the mathematical expectation and the variance of the innovation sequence simultaneously online and does not require a priori information on the quantitative changes of its statistical characteristics in case of failure.

The suggested approach to the failure detection is used for the sensor and actuator/surface failure detection problem in the AFTI/F-16 aircraft flight control system. An extended Kalman filter has been developed for nonlinear flight dynamic estimation of an F-16 fighter. Failures in the sensors and actuators/surfaces affect the characteristics of the innovation sequence of the EKF. The failures that affect the mean and the variance of the innovation sequence have been considered. The following failures, that affect the characteristics of the innovation sequence of the EKF are examined:

a) Shift in the measurement noise of the pitch rate gyroscope (sensor failure);
b) Changes in the noise variance of the pitch rate gyroscope (sensor failure);
c) Shift in the control input, corresponding to the deflection of the right horizontal stabilizer (actuator motor failure);
d) Changes in the control derivatives corresponding to the right horizontal stabilizer (control surface failure).

The theoretical results are confirmed by the simulations carried out on a nonlinear dynamic model of the F-16 aircraft. The obtained simulation results have confirmed the practical possibility of the diagnostics of the flight control system using the introduced spectral norm of the innovation matrix. The introduction of the developed failure detection algorithm does not distort the estimation results of the filter and has no influence on their accuracy.

An adaptive EKF for the F-16 aircraft state estimation insensitive to sensor failures is designed and a decision approach to isolate the sensor and actuator/surface failure is proposed. When a regular EKF is used, the decision statistics changes regardless to the failure in the sensors or in the actuators/surfaces. On the other hand, if the adaptive EKF insensitive to sensor failures is used, it is easy to distinguish the sensor and actuator failures.

It is shown that the inertia of the fault detection depends on the number of samples m and with an increase in this number, this characteristic worsens. On the other hand, a very small value of m leads to frequent false faults. Some recommendations for the choice of the number of samples m in the practical implementations are given in this study.

The presented failure detection method has the following disadvantages: this method is of a statistical approach and a particular statistics must be accumulated, and the method has no ability to determine the value of the fault (fault identification).

The future work is to investigate the integrated sensor and actuator/surface failure detection, isolation and identification, and reconfigurable control problems together for the innovation approach based active fault-tolerant flight control system design.

6. References

Alessandri, A. (2003). Fault diagnosis for nonlinear systems using a bank of neural estimators. *Computers in Industry,* 52(3), pp. 271-289.

Anderson, T.W. (1984). *An Introduction to Multivariate Statistical Analysis,* 2nd edition. John- Wiley & Sons, Inc., New York, USA.

Basseville, M. & Benveniste, A. (Eds.) (1986). *Detection of Abrupt Changes in Signals and Dynamical Systems.* Springer-Verlag.

Benveniste, A., Basseville, M. & Moustakides, G.V. (1987). The asymptotic local approach to change detection and model validation. *IEEE Transactions on Automatic Control,* AC-32, pp. 583-592.

Bienvenu, G. & Kopp, L. (1983). Optimality of high resolution array processing using the eigensystem approach. *IEEE Transactions on Acoustics, Speech and Signal Processing,* 31, pp. 1235-1248.

Borairi, M. & Wang, H. (1998). Actuator and sensor fault diagnosis of non-linear dynamic systems via genetic neural networks and adaptive parameter estimation technique, In: *Proceedings of the IEEE Conference on Control Applications.* V.1, pp. 278-282.

Caliskan, F. & Hajiyev, Ch.(2003). Actuator Failure Detection and Reconfigurable Control for F-16 Aircraft Model. In: *Preprints of the 3rd IFAC Workshop "Automatic Şystems for Building the Infrastructure in Developing Countries"(DECOM-TT 2003),* Istanbul, Turkey, pp. 231-236.

Chan, C.W., et al.(1999). On-line fault detection and isolation of nonlinear systems. In: *Proc. Amer. Control Conf. (ACC' 99),* San Diego, California, pp. 3980–3984.

Chen, J. & Patton, R.J. (1999). *Robust Model - Based Fault Diagnosis for Dynamic Systems.* Kluwer Academic Publishers, USA.

Edelman, A. (1991). The distribution and moments of the smallest eigenvalue of a random matrix of Wishart type. *Linear Algebra and its Applications,* 159, pp.55-80.

Everson, R. & Stephen, R. (2000). Inferring the eigenvalues of covariance matrices from limited noisy data. *IEEE Transactions on Signal Processing,* 48, pp. 2083-2091.

Grouffaud, J., Larzabal, P. & Clergeot, H. (1996). Some properties of ordered eigenvalues of a Wishart matrix: application in detection test and model order selection, *Proceedings of the IEEE International Conference on Acoustics, Speech, and Signal Processing, ICASSP,* Atlanta, GA, USA, 5, pp. 2463-2466.

Gadzhiev, Ch. M. (1992). Dynamic systems diagnosis based on Kalman filter updating sequences. *Automation and Remote Control,* 1, pp. 147-150.

Gadzhiyev, Ch. M. (1993). Checking multivariate model fit from the generalized Wishart-statistic variance. *Measurement Techniques,* 36(12), pp. 1316-1319.

Gadzhiev, Ch. M. (1996). *The Information Provision of Offshore Platform Supervision and Control* (in Russian), Elm, Baku.

Grishin, Yu. P. & Kazarinov, Yu. M. (1985). *Fault-Tolerant Dynamic Systems* (in Russian), Radio i Svyaz, Moscow.

Hajiyev, Ch. & Caliskan, F. (2003). *Fault Diagnosis and Reconfiguration in Flight Control Systems.* Kluwer Academic Publishers, Boston.

Hajiyev, Ch. & Caliskan, F. (2005). Sensor and control surface/actuator failure detection and isolation applied to F-16 flight dynamics. *Aircraft Engineering and Aerospace Technology: An International Journal,* 7, pp. 152-160.

Hajiyev, Ch. (2006). Adaptive filtration algorithm with the filter gain correction applied to integrated INS/Radar altimeter, *Proceedings of the 5th International Conference on Advanced Engineering Design,* Prague, Czech Republic.

Hajiyev, Ch. (2007). Sensor and Control Surface/Actuator Failure Detection Based on the Spectral Norm of an Innovation Matrix. In: *Preprints of the 17th IFAC Symposium on Automatic Control in Aerospace,* Toulouse, France, Paper No.22.

Hajiyev, Ch. (2009). Innovation approach based sensor FDI in LEO satellite attitude determination and control system. In the book: *Kalman Filter: Recent Advances and Applications,* I-Tech Education and Publishing KG, Vienna, Austria, pp. 347-374.

Horn, R. & Jonson, Ch. (1986). *Matrix Analysis,* Cambridge University Press, Cambridge, UK.

Kendall, M.G. & Stuart, A. (1969). *The Advanced Theory of Statistics.* Griffin, London, UK.

Krinetsky, Ye.I., *et al.,* (1979). *Flight Testing of Rocket and Space Apparatus* (in Russian), Mashinostroyeniye, Moscow.

Larson, E.C., et al. (2002). Model-based sensor and actuator fault detection and isolation. In: *Proceedings of the American Control Conference,* V. 5, pp. 4215-4219.

Lee, J.H. & Lyou, J. (2002). Fault diagnosis and fault tolerant control of linear stochastic systems with unknown inputs. *Systems Science,* 27(3), pp. 59-76.

Li, Z. & Jaimoukha, I.M. (2009). Observer-based fault detection and isolation filter design for linear time-invariant systems, *International Journal of Control,* 82, pp. 171-182.

Lyshevski, S.E. (1997). State-space identification of nonlinear flight dynamics. In: *Proceedings of the Conference on Control Applications,* Hartford, Connecticut, pp. 496-498.

Malik, R.K. (2003). The pseude-Wishart distribution and its application to MIMO systems. *IEEE Transactions on Information Theory,* 49, pp. 2761-2769.

Maybeck, P.S. (1999). Multiple model adaptive algorithms for detecting and compensating sensor and actuator/surface failures in aircraft flight control systems. *International Journal of Robust and Nonlinear Control,* 9(14), pp.1051-1070.

Mehra, R.K. & Peschon, J. (1971). An innovations approach to fault detection and diagnosis in dynamic systems. *Automatica,* 7, pp. 637-640.

Mohamed, A.H. & Schwarz, K.P. (1999). Adaptive Kalman filtering for INS/GPS. *Journal of Geodezy,* 73, pp. 193-203.

Napolitano, M.R., Chen, Ch. & Naylo, S.(1993). Aircraft failure detection and identification using neural networks. *Journal of Guidance, Control, and Dynamics,* 16(6), pp. 999-1009.

Napolitano, M.R., et al. (1996). Online learning neural architectures and cross-correlation analysis for actuator failure detection and identification. *International Journal of Control,* 63(3), pp. 433-455.

Perhinschi, M.G., et al. (2002). On-line parameter estimation issues for the NASA IFCS F-15 fault tolerant systems. In: *Proceedings of the American Control Conference,* V.1, pp. 191-196.

Rago, C., et al. (1998). Failure Detection and Identification and Fault Tolerant Control using the IMM-KF with applications to the Eagle-Eye UAV. *Proceedings of the IEEE Conference on Decision and Control,* V 4, pp. 4208-4213.

Rao, S.R. (1965). *Linear Statistical Inference and Its Applications,* New York, John Wiley & Sons, Inc.

Raza, H., Ioannou, P. & Youssef, H.M. (1994). Surface failure detection for an F/A-18 aircraft using neural networks and fuzzy logic. In: *Proceedings of the IEEE International Conference on Neural Networks*, V 5, pp. 3363-3368.

Sage, A.P. & Melsa, J.L. (1971). *Estimation Theory with Applications in Communication and Control*. McGraw-Hill, New York.

Tykierko, M. (2008). Using invariants to change detection in dynamical system with chaos. *Physica D: Nonlinear Phenomena*, 237, pp. 6-13.

Vaswani, N. (2004). Change detection in partially observed nonlinear dynamic systems with unknown change parameters. *Proceedings of the American Control Conference (ACC'2004)*, pp. 5387-5393.

Wax, M. & Kailath, T. (1985). Detection of signals by information theoretic criteria. *IEEE Transactions on Acoustics, Speech and Signal Processing*, 33, pp. 387-392.

Willsky, A.S.(1976). A survey of design methods for failure detection in dynamic systems. *Automatica*, 12(6), pp. 601-611.

Wu, H.T., Yang, J.F. & Chen, F.K. (1995). Source number estimators using transformed Gerschgorin Radii. *IEEE Transactions on Signal Processing*, 43, pp. 1325-1333.

Zanella, A., Chiani, M. & Win, M.Z. (2008). A general framework for the distribution of the eigenvalues of Wishart matrices, *Proceedings of the ICC-2008 conference*, pp. 1271-1276.

Zhang, Y. & Li, X.R. (1997). Detection and diagnosis of sensor and actuator failures using Interacting Multiple-Model estimator, *Proceedings of the IEEE Conference on Decision and Control*, V.5, pp. 4475-4480.

11

Fault Detection, Isolation, and Control of Drive By Wire Systems

Sohel Anwar
Indiana University Purdue University Indianapolis
USA

1. Introduction

Fault tolerance is a property of a system that continues operating properly in the event of failure of some of its parts. It provides the ability of a system to provide a service complying with the specification in spite of faults. If operating quality decreases at all, the decrease is proportional to the severity of the failure, as compared to a naively-designed system in which even a small failure can cause total breakdown.

Fault-tolerant design of a drive-by-wire (DBW) system provides key-method for achieving safe and reliable systems (Isermann et al, 2002; Shibahata, 2005). It is generally required in the design of all mechanical, electrical and software components of drive-by-wire systems. Extensive analysis is performed during the design and testing phase, combined with quality control methods during manufacturing. This design method reduces faulty behaviour dramatically and forms the basis for safe and reliable systems. The design method identify potential hazards and associated avoidance requirements, translates safety requirements into engineering requirements, provides design assessment and trade-off to support the ongoing design, assesses the relative compliance of the design to the requirements and document the findings, directs and monitors specialized safety testing, and monitors and reviews test and field issues for safety trends.

It is noted that certain faults and failures cannot be avoided totally. Components in a system can fail even after with a fault tolerant design. These faults should be tolerated by additional measures and compensated in such a way that they do not cause the overall system (or certain critical functions) to fail. The most obvious way to reach this goal is to implement duplication in order to avoid single points of failure.

Duplication can offer fault-tolerance in three different ways:

- Replication: Providing multiple identical instances of the same system, directing tasks or requests to all of them in parallel, and choosing the correct result on the basis of a quorum.
- Redundancy: Providing multiple identical instances of the same system and switching to one of the remaining instances in case of a failure (fall-back or backup).

- Diversity: Providing multiple different implementations of the same specification, and using them like replicated systems to cope with errors in a specific implementation.

For redundancy, ideally it is desirable to have as much redundancy as possible. Instead of just two, or three, sensors one would perhaps want ten to be even safer. But the design is limited by cost and weight. Diversity in redundant components is desirable. There is also redundancy in software, where different programming teams work on solving the same problem using different methods. This method gives higher levels of safety than just duplicating the same code when running multiple redundant systems. Otherwise the same error is likely to happen in the backup system as well. There are three types of hardware redundancies as described below:

1. Static hardware redundancy

Three or more parallel modules are used that have the same input signal and are all active. A voter compares their output signals and decides by majority which is the correct one. In a commonly used triple-redundant modular architecture one fault can be masked without the use of special error detection methods. The output is parsed through a voter that examines the values. Decision is made based on what the majority thinks is a correct view of the reality. There can be a single module that fails, sending faulty output values, and system still have a correct view. For a system having n redundant modules, system will tolerate (n-1)/2 faults. n must be odd so that there exists no tie. Static redundancy is based on the voting of the outputs of a number of modules to mask the effects of a fault within these units. The simplest form of this arrangement consists of three modules and a voter (TMR).

2. Dynamic hardware redundancy

It requires fewer modules at the cost of more information processing. A minimal configuration consists of two modules, where one module is in operation, and the second module takes over in case of an error. When the second module is continuously operating this method is called 'hot standby', which has the advantage of shorter downtime of the system. However since it is operating all the time aging of the module becomes a disadvantage. In a 'cold-standby' configuration, the backup system is normally out of function. It only becomes operational in case of an erroneous primary system. Both configurations need for error-detection methods observing if a module becomes faulty. Methods range from simple ones such as limit value and plausibility checks, parity checking and watchdog timers, to sophisticated signal-model-based or process-model-based methods. In dynamic redundancy fewer modules are used but we have to do more information processing. The system has to detect if a module is malfunctioning and reconfigure the system so that the module is shutdown and the backup module is brought online. Dynamic redundancy based on fault detection rather than fault masking, achieved by using two modules and some sort of comparison on their outputs that can detect possible faults. It offers lower component count but is not suitable for real-time applications.

3. Hybrid hardware redundancy

In this configuration, a combination of voting, fault-detection, and module switching techniques is used. It is a hybrid of static and dynamic redundancies.

In order to bring down the cost, the total number of redundant components must be reduced without compromising the fault tolerance. One possible solution to this problem is to utilize analytical redundancy or model based fault detection, isolation, and accommodation. Model-based Fault Detection and Isolation explicitly use a mathematical model of the system. It is motivated by the conviction that utilizing deeper knowledge of the system results in more reliable diagnostic decisions. The main idea is "analytical redundancy" which makes comparison of measurement data with known mathematical model of the physical process. It is superior to "hardware redundancy" generated by installing multiple sensors for the same measured variable. They offer simplicity, flexibility in the structure, less hardware, less weight, and cost.

2. Analytical Redundancy

Model-based Fault Detection and Isolation is achieved by implementing more complex failure detection algorithms that take careful account of system dynamics; utilizing such algorithms, one may be able to reduce requirements for costly hardware redundancy. Analytical redundancy based FDI (failure detection and isolation) uses a model of the dynamic system to generate the redundancy required for failure detection. In many systems, all of the states cannot be measured because of cost, weight and size considerations, therefore, FDI schemes for such systems must extract the redundant information from dissimilar sensors, using the differential equations that relate their outputs. In addition to taking hardware issues into consideration, the designer should consider the issue of computational complexity. Most model-based FDI methods rely on analytical redundancy. In contrast to physical redundancy, when measurements from different sensors are compared, now sensory measurements are compared to analytically obtained values of the respective variable and the resulting differences are called residuals. The deviation of residuals from the ideal value of zero is the combined result of noise, modelling errors and faults. A logical pattern is generated showing which residuals can be considered normal and which ones indicate a fault. Such a pattern is called the signature of the failure. The final step of the procedure is the analysis of the logical patterns obtained from the residuals, with the aim of isolating the failures that cause them. Such analysis may be performed by comparison to a set of patterns known to belong to sample failures or by the use of some more complex logical procedure.

3. Case Study: Analytical Redundancy Based Fault Detection, Isolation, and Accommodation of a Steer By Wire System

Proposed Methodology

It is clear that a major challenge for steer by wire (SBW) system is to reduce the cost and the number of hardware components (e.g. sensors) while maintaining safety. Analytical redundancy method has been investigated by several applications before (Gertler, 1992; Dong & Hongyue, 1996; Suzuki et al, 1999; Venkateswaran et al, 2002; Anwar & Chen, 2006). However, most of these investigations focused on hardware redundancy based fault detection and simulation on aircraft applications as opposed to hardware experiment of analytical redundancies for automobiles with SBW system. The objective of this research is to study the feasibility of an analytical redundancy method for a SBW system through hardware-based experiment instead of software simulation.

The analytical redundancy method discussed in this paper makes the following assumptions: The probability of two sensors failing simultaneously is much less than that for only one sensor failing. Similarly, the probability of three sensors failing simultaneously is much less than that for two-sensor-fault.
It simulation studies, it has been shown that the accuracy of the Fault Detection and Isolation Accommodation (FDIA) algorithm won't be sacrificed by the reduction of the number of physical sensors since analytical redundancy will provide additional information (Anwar & Chen, 2006).
The research methodology in this paper involves several parts. These parts are:

- Build of a simplified SBW system hardware-in-loop (HIL) bench with necessary hardware components.
- Development of control models for SBW system. These models involve:
  - A modified 4th order vehicle model
  - A nonlinear observer model based on the vehicle model
  - A long range prediction model with different prediction horizons
  - A FDIA algorithm for sensor fault detection based on analytical redundancy.

Completion of the build of the HIL bench for the SBW system (illustrated in Figure 1) involves:
• A steering wheel module with three angular sensors and one electric motor for haptic feedback.
• A rapid prototyping controller from dSpace, MicroAutoBox Model 1401, which receives input signals from all angular sensors on the steering wheel, feedback signals from all angular sensors on the pinion of the rack and pinion assembly, sends commands to motor drivers and servo motors. And, all necessary wiring and cable devices between controllers, drivers, motors, and sensors.
• A set of motors: the motor drivers, Parker OEM770T, the servo motors, Parker BE342KJ-K10N, and a set of gears and a gear box converting the servo motor's rotation into rack's linear motion.
• A rack and pinion assembly with fixture. The rack and pinion assembly is that of a Volkswagen SI-9281-9-2, which provides lateral movement of the steering system. Three angular sensors situated on the pinion provide the feedback signal to controller. And two springs with spring coefficient 2000 N/in are attached a both ends of the rack to simulated the road loads for the SBW HIL bench.
For real vehicles, the road wheel friction force created at the contact patch of road and tire can be transmitted to rack via front wheels. In building the hardware of the SBW system in this paper (Figure 1), there are two springs that replace front wheels and these two springs also provide simulated force caused by front wheels contacting road surface. The rack was made for power steering wheel system (with pinion on top). In a close-loop control system, there must be a feedback from the road wheel system (rack/pinion). Thus, the pinion is taken as the feedback signal source for close-loop control (Figure 2).

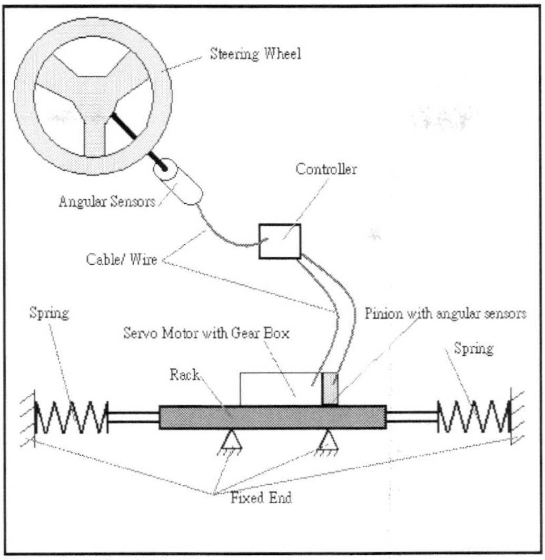

Fig. 1. Loop bench for simplified SBW system

Developing the overall analytical redundancy based FDIA and control model and performing the validation tests of FDIA algorithms is the main work in this research. Before modelling the FDIA algorithms as a part of fault tolerant control, it is necessary to understand the complete control model considered in this paper (Figure 3). The complete model shows that the fault tolerant control is made possible by FDIA. θ (pinion angle) from vehicle model and from predictor are checked by FDIA logic to isolate any sensor fault on the pinion. The vehicle model converts motor's current into pinion angle via the nonlinear observer. Ideally, we wished to make the steering wheel rotation and the pinion angle rotation simultaneously. But, due to the dynamic model used in the observer and usage of low pass filter to reduce noise in the feedback signals, a delay is introduced in the after the pinion angle is reconstructed using the observer. In order to reduce this delay, a long range predictor is used in the observer model that reduces the delay between steering wheel's angle and pinion angle (Anwar & Chen, 2006). And, the observer in the overall model provides the information about other state variables inside the controller as well (Hasan & Anwar, 2008; Nise, 1994).

Lastly, by taking the advantage of Matlab/Simulink and the code-compiling functions provided by dSpace, the steering system model, the modified vehicle model, the nonlinear observer, the predictor, and the FDIA algorithm were implemented in real-time environment which was then downloaded into the dSpace MicroAutoBox controller.

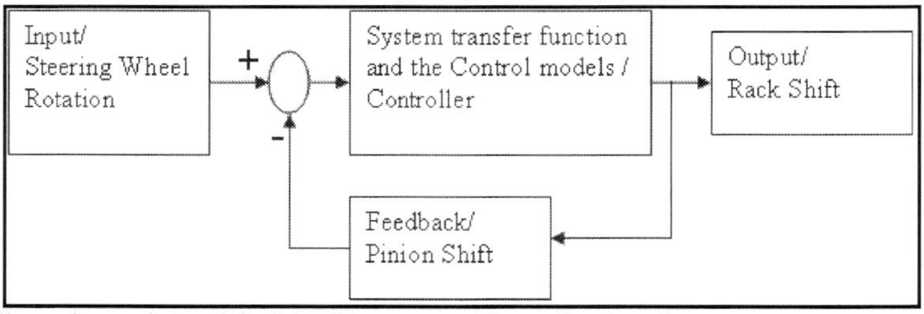

Fig. 2. Flow chart of simplified SBW system with mechanism/component

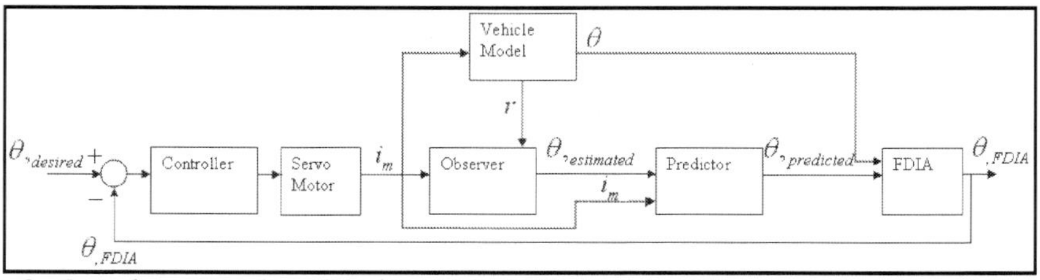

Fig. 3. Complete control model of simplified SBW system

3.2 SBW Hardware-In-Loop Bench Design and Construction

In building the loop bench, there are several components and parts (Table 1) used in machine constructing. The overall road wheel portion (including rack and pinion) of the HIL bench is shown in Figure 4.

Name of Component/Part	Quantity	Description
Servo Motor	1	Parker BE342KJ-K10N, with a gear box coupled with the rack.
Drive	1	Parker OEM770T, selected for the servo motor.
Rack	1	Volkswagen SI-9281-9-2, with a pinion on top
Angular Sensor	3	Model: MH22B, coupled with the pinion
Steering Wheel System	1	The system has been already built before, with angular signal as output.
Controller	1	DSpace MicroAutoBox 1401

Table 1. List of components/parts for building the loop bench

For the design and construction of mechanical components of the bench, each component was been modelled by using Pro/E Wildfire 3.0 and tested by using ANSYS 14.0 to check the maximum deformation within the material (steel).

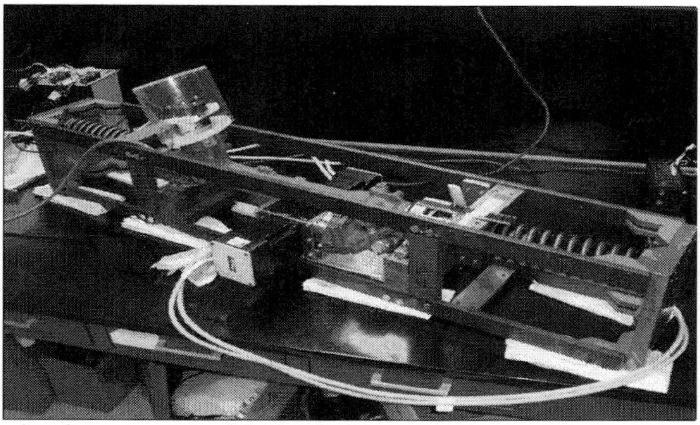

Fig. 4. Completed construction of the hardware

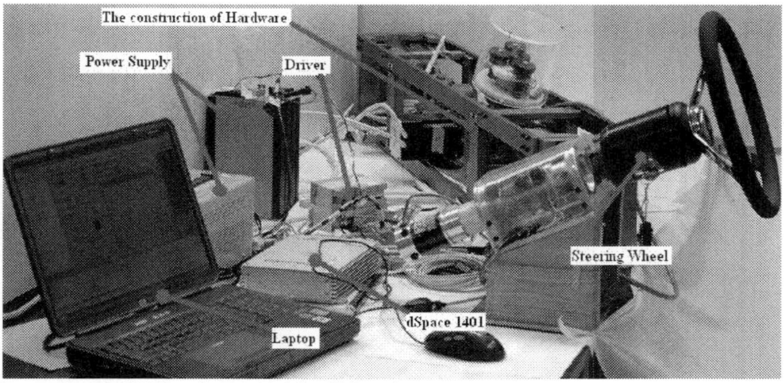

Fig. 5. Complete SBW hardware with all hardware components

Once the construction of hardware was finished, all the electrical and electronic components of the HIL bench (listed in Table 1) were then connected using the designed wiring diagram. The complete SBW system with the mechanical components, driver, controller, laptop computer, and power supply is represented in Figure 5.

4. Observer, Predictor, and FDIA Algorithm

A majority of the theoretical developments for analytical redundancy has been described in references (Anwar & Chen, 2006; Hasan & Anwar, 2008). A summary of the observer, predictor, and FDIA algorithms are presented below.
The overall vehicle and steering system model are given by (Anwar, 2005):

$$
\begin{aligned}
&\dot{x} = Ax + Bi_m \\
&y = Cx \\
&x = \begin{bmatrix} \beta \\ r \\ \theta \\ \dot{\theta} \end{bmatrix}; A = \begin{bmatrix} \frac{-C_{\alpha,f} - C_{\alpha,r}}{mV} & -1 + \frac{C_{\alpha,r} b - C_{\alpha,f} a}{mV^2} & \frac{C_{\alpha,f}}{mV} & 0 \\ \frac{C_{\alpha,r} b - C_{\alpha,f} a}{I_z} & \frac{-C_{\alpha,f} a^2 - C_{\alpha,r} b^2}{I_z V} & \frac{C_{\alpha,f} a}{I_z} & 0 \\ 0 & 0 & 0 & 1 \\ \frac{(t_p + t_m) C_{\alpha,f}}{J} & \frac{a(t_p + t_m) C_{\alpha,f}}{JV} & \frac{-(t_p + t_m) C_{\alpha,f}}{J} & \frac{-b}{J} \end{bmatrix} \\
&B = \begin{bmatrix} 0 \\ 0 \\ 0 \\ \frac{k_m}{J} \end{bmatrix}; E = \begin{bmatrix} 0 \\ 0 \\ 0 \\ -\frac{1}{J} \end{bmatrix}; C = \begin{bmatrix} 0 & 0 & 1 & 0 \end{bmatrix}
\end{aligned}
\tag{1}
$$

Whrere, β is the vehicle body side slip angle, r is the vehicle yaw rate at the center of gravity, θ is the steering angle at road wheel, $C_{\alpha,f}$ & $C_{\alpha,r}$ are cornering coefficients for the front & rear wheels respectively, a & b are distance between front & rear axles to the vehicle center of gravity respectively, m is vehicle mass, V is vehicle nominal velociy, I_z is vehicle inertia in yaw direction, J is wheel ineria, t_m & t_p are mechanical and pneumatic trails repectively, k_m is the motor torque constant. The details of the above model equations can be found in (Hasan, 2007) and is based on simplified single track vehicle model (or bicycle model) as illustrated in Figure 6.

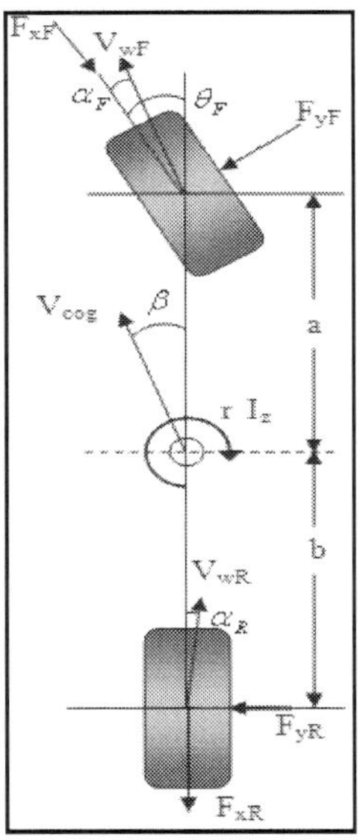

Fig. 6. Linear vehicle model simplified as bicycle model

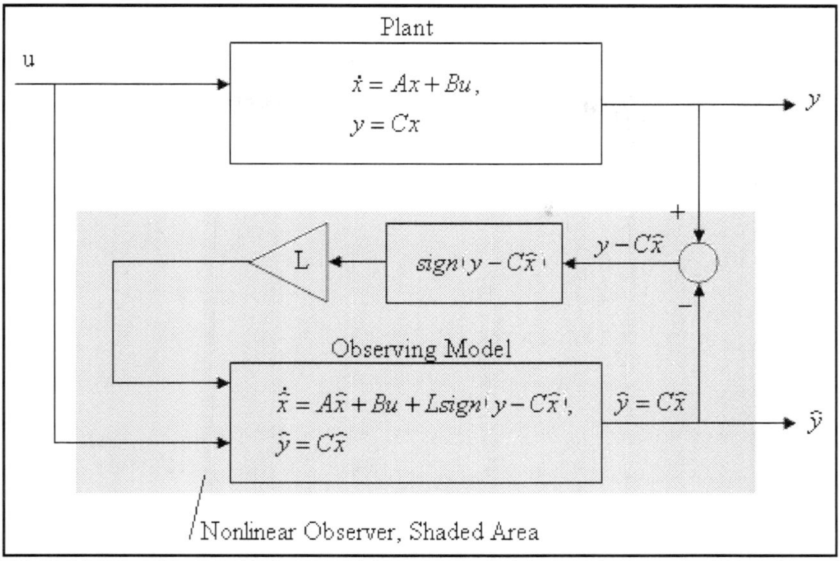

Fig. 7. Nonlinear observer (sliding mode) in state space form

A sliding mode observer was designed based on the above vehicle model. The details of the sliding mode observer developed are described in (Hasan, 2007; Utkin et al, 1999; Stotsky, & Hu, 1997) and hence skipped here. A schematic of this observer is given in Figure 7.

A long range prediction algorithm was designed based on Diophantine equation (Clarke et al, 1987). This algorithm was then used to predict the future output (e.g. steering angle at road wheel) which is then used in the FDIA algorithm. This objective of this predictor is to reduce the latency in fault detection due to computational delays. A detailed account of this prediction algorithm can be found in (Hasan & Anwar, 2008). Only prediction equation is stated here for reference.

$$\theta(t+j) = E_j(z^{-1})A(z^{-1})\Delta U(t+j-1) + F_j(z^{-1})\theta(t) \tag{2}$$

Where, t is current time, z^{-1} is the backward shift operator in z-domain, E_j and F_j are polynomials in Diophantine equation which are uniquely defined given the system characteristic polynomial $A(z^{-1})$, ΔU is the system input.

The fault detection, isolation, and accommodation algorithms (FDIA) are based on a majority voting scheme and are given in details in (Hasan, 2007; Niu, 2009). Only the fundamental algorithm in flow chart format is given here for brevity (Figures 8 and 9).

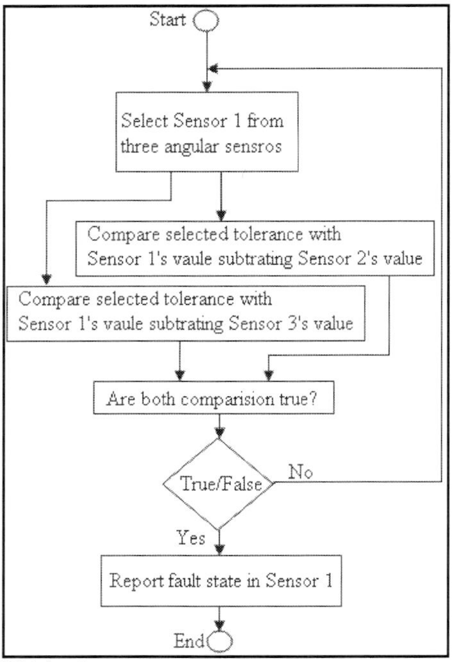

Fig. 8. Single-sensor FDIA logic

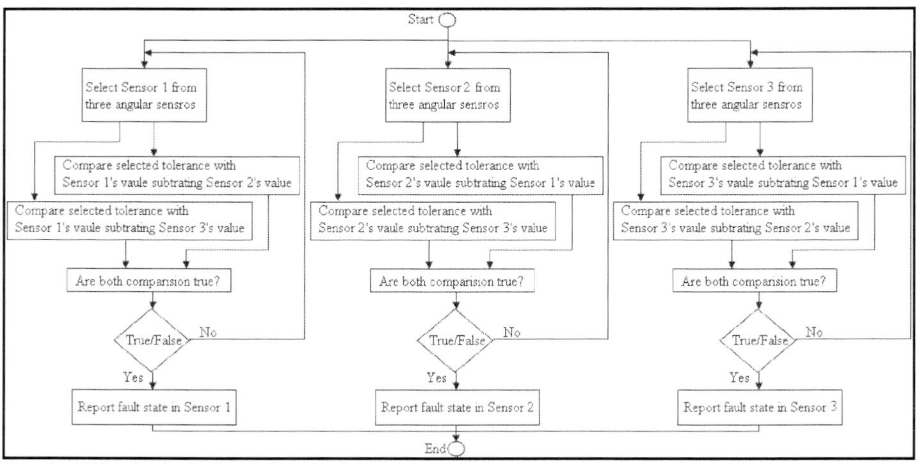

Fig. 9. FDIA logic in detecting fault state of three sensors

5. Experimental Results

Before we illustrate the FDIA algorithms used in this work, it is necessary to describe the types of Faults which are tested in this research. Commonly, faults can be classified in three main types: transient fault (Figure 10), hard fault (Figure 11), and soft fault (Figures 12 - 13).

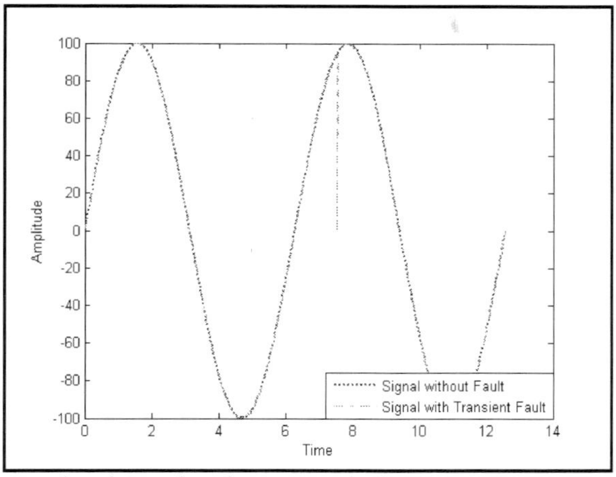

Fig. 10. Normal signal and signal with transient fault

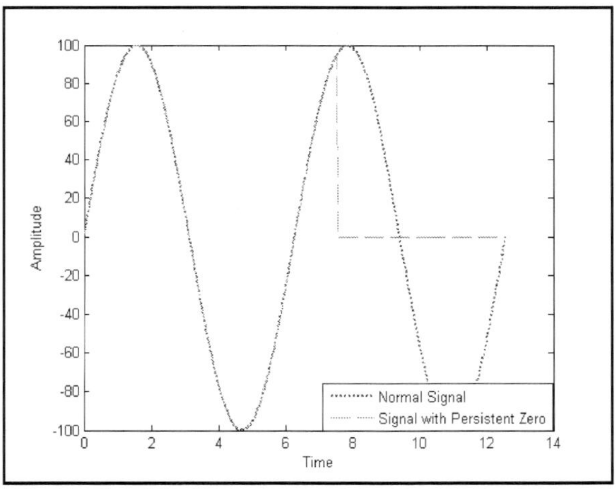

Fig. 11. Persistent fault, a hard fault

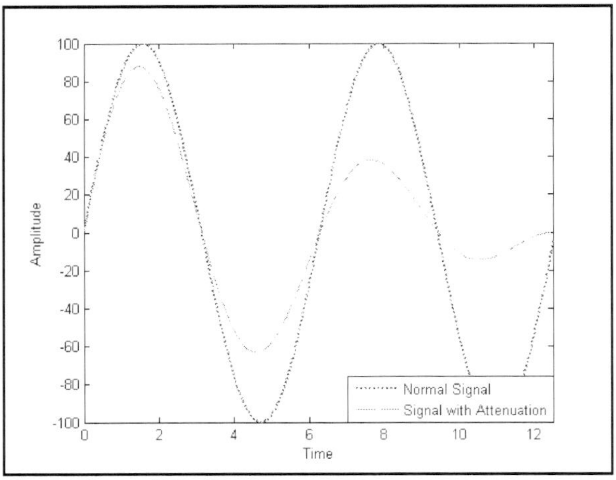

Fig. 12. Attenuating fault, a soft fault

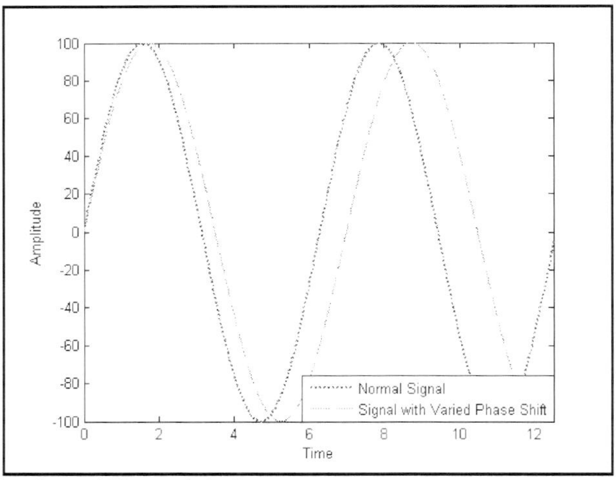

Fig. 13. Variable phase shift, a soft fault

5.1 Servo Motor Control Modeling by Using Matlab/Simulink

In order to build the control program for the experiment with the SBW system, it is first necessary to develop the basic functions to control the angular rotation of motor via feedback information from steering wheel's angular sensors. Servo motor control function (Figure 14) is the combination of the controller block and the servo motor block in Figure 3, and involves a counter, a PID controller, and a command block.

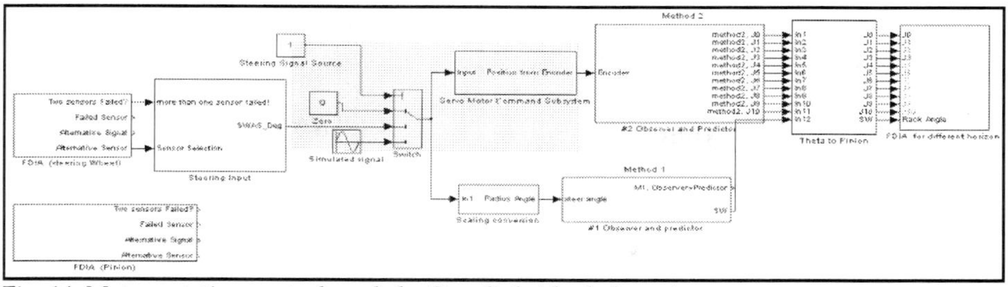

Fig. 14. Motor rotation control made by Simulink (shaded area)

5.2 Vehicle Parameters

The vehicle parameters used in a previous work (Hasan & Anwar, 2008) has been adopted for the modified vehicle model here and is modeled in Matlab/Simulink. These vehicle parameters are given in Table 2.

Parameter	Value
$C_{a,f}$	22000 (N/rad)
$C_{a,r}$	55000 (N/rad)
m	1400 (kg)
V	15 (m/s)
b	1.5 (m)
a	1 (m)
I_z	4000 (kg-m2)
t_p	0.025 (m)
t_m	0.03 (m)
J	20 Kg-m2

Table 2. Parameter values used for the modified vehicle model

A validated sliding mode observer model was developed using the HIL bench. The overall observer model along with the predictor is shon in Figure 15 will all the input/output relationships. It is noted that if the steering model in Figure 15 is $G(s)=\theta(s)/i_m(s)$, then the expression for motor current reading is be given by $i_m(s) = G^{-1}(s)\theta(s)$.

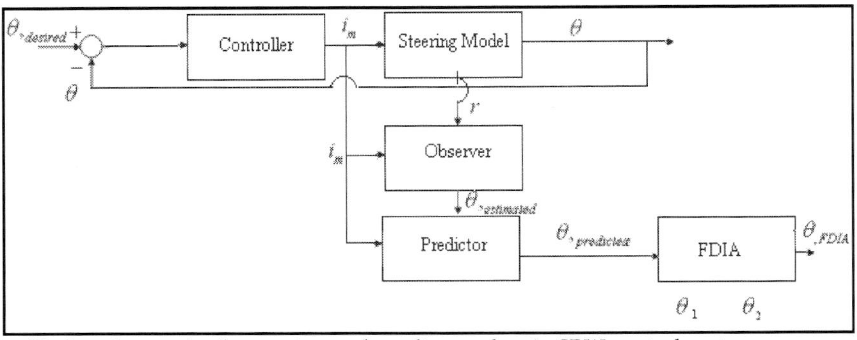

Fig. 15. Steering angle observation and predictor taken in SBW control system

Usage of ControlDesk program provides a convenient connection between modelling control program using Matlab/Simulink and using dSpace D1401 as a prototyping controller. After the control model built by using Matlab/Simulink is downloaded into D1401, ControlDesk program can help users to view and record various model parameters and inputs/output data. Additionally, other functions such as testing hardware with any control model, reporting state, etc. can also be performed. A number of interfaces were built by using ControlDesk to test steering function to move the rack, to test FDIA logic with different fault types, and to test the detection speed in detecting fault of predicted with different prediction horizon.

The purpose of adding a predictor in a control system is to reduce the delay between input and system's response. If we remove the FDIA and Predictor, then the system's feedback in a close-loop control system becomes $\theta_{estimated}$. Similarly, if the predictor hasn't been removed, then $\theta_{predicted}$ can be equal to $\theta_{estimated}$ when selecting the horizon as zero ($J = 0$).

Now, Figure 16 and Figure 17 represent the delay between $\theta_{desired}$ and $\theta_{estimated}$. In Figure 15, it is the same as the delay between $\theta_{desired}$ and $\theta_{predicted}$ for J equal to zero because $\theta_{estimated}$ is equal to $\theta_{predicted}$ for $J = 0$.

From these figures, it is noted that the controller with servo motor in the SBW System causes a 40 millisecond delay between $\theta_{desired}$ and $\theta_{estimated}$.

When the SBW System with Predictor (Figure 3) is used with different number of J, it is noted that selecting a higher number of J makes $\theta_{predicted}$ closer to $\theta_{desired}$ (Figure 17). Thus, using long range predictor with higher number of horizon can shorten the delay time between steering angle input and predicted steering angle output.

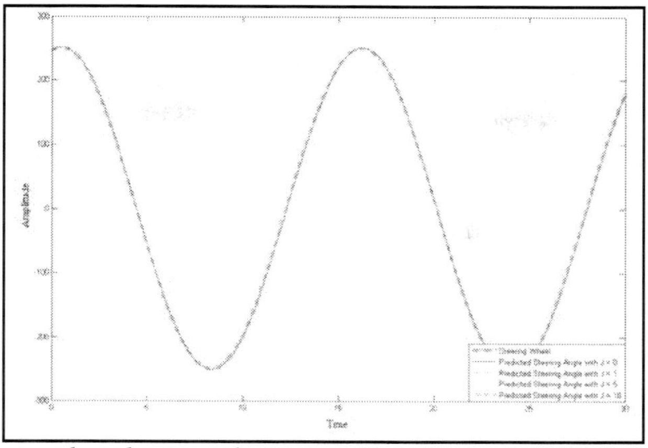

Fig. 16. Steering angle and estimated steering angle

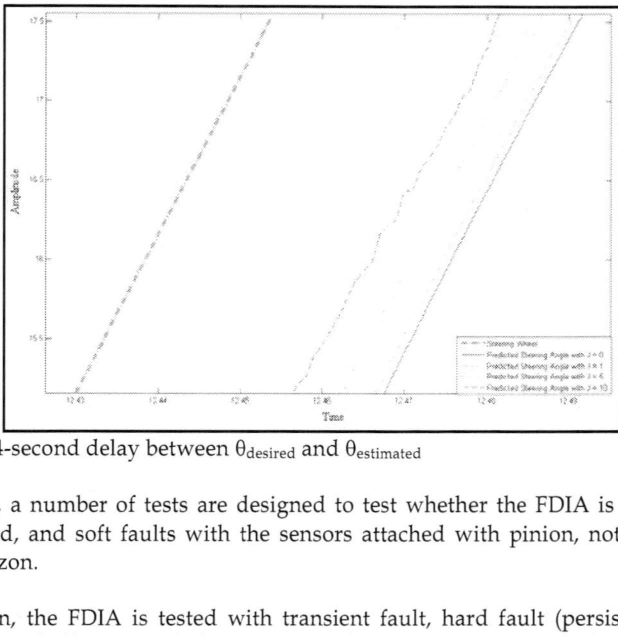

Fig. 17. A 0.04-second delay between $\theta_{desired}$ and $\theta_{estimated}$

In this paper, a number of tests are designed to test whether the FDIA is able to detect transient, hard, and soft faults with the sensors attached with pinion, not $\theta_{predicted}$ with different horizon.

In this section, the FDIA is tested with transient fault, hard fault (persistent zero and constant phase shift), and soft fault (increased amplitude, attenuated amplitude and variable phase shift). With appropriate designed FDIA using Matlab/Simulink and using Control Desk program, the FDIA interface built with Control Desk program is able to report the times of transient fault, the fallen sensor, and the state of morn than one sensor fallen.

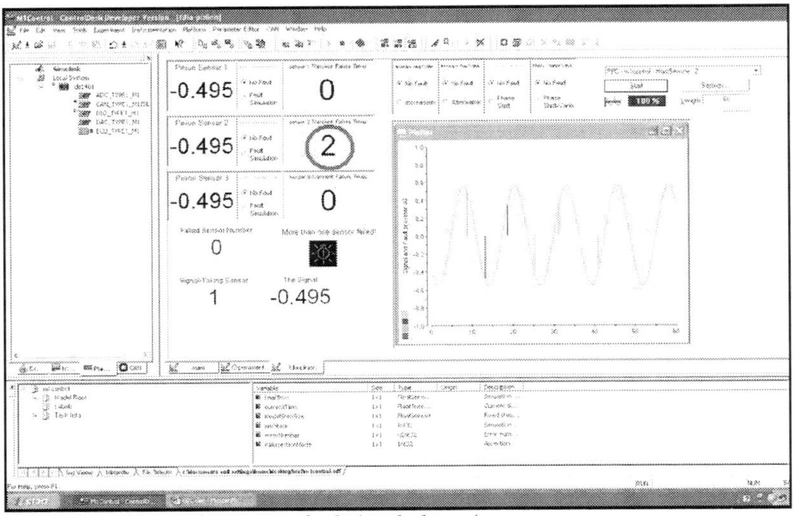

Fig. 18. Reported times of transient fault (circled area)

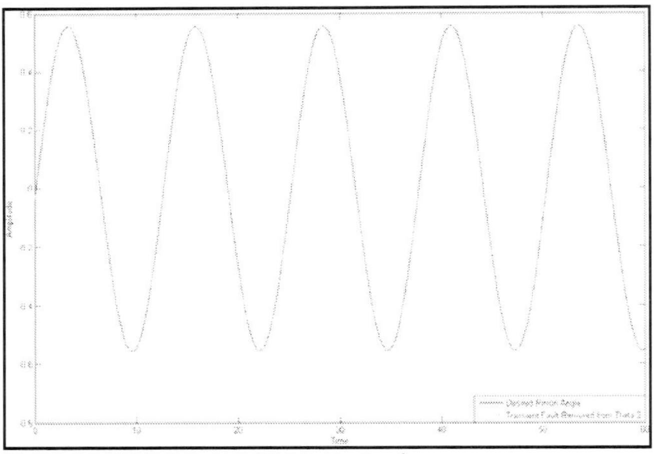
Fig. 19. Normal signal after transient zero is removed

To test transient fault and hard fault for the SBW System with mechanism, three manually operated switched are wired between the angular sensors coupled with pinion and D1401. Also, to test soft fault, Simulink blocks with signal-increasing, signal-attenuating and phase-shifting functions are added into the SBW System.

After manually operated switches and Simulink blocks are made, the ability of detecting fault using FDIA algorithm with transient, hard, and soft faults respectively was tested.

In testing transient fault, the transient-fault signal is made by quickly turning off and then turning on a manually operated switch with a selected sensor coupled with pinion. Thus,

a transient signal can be created for testing purpose. After doing experiment, the FDIA interface (Figure 18) represents the detection of transient fault with θ_2; it shows that the FDIA interface is able to record the times of transient fault. Certainly, the interface keeps the times of transient fault (Figure 18) after transient fault is removed (Figure 19).
Next, we test persistent zero with FDIA interface. The persistent-zero signal is created by turning off the manually operated switch with a selected sensor on pinion. Then, the interface immediately reports a fault state of the selected sensor (Figure 20). Later, once the switch is turned on (Figure 20), the fault state is updated as fault-out (Figure 21).

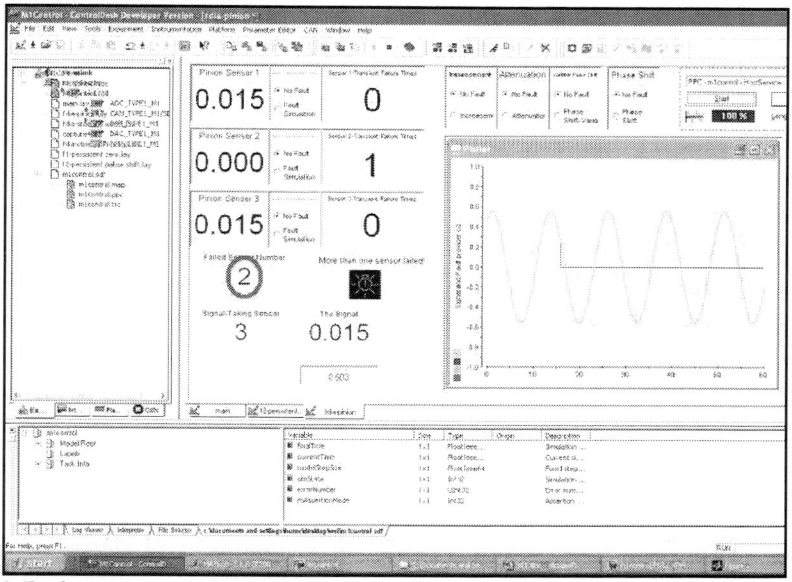

Fig. 20. Fault state reported by FDIA interface for persistent zero with θ_2(circled area)

To test the attenuating-amplitude fault with FDIA interface, the ControlDesk interface creates a decreasing-amplitude fault with a selected θ, and fault detection from FDIA interface is then observed (Figure 22). After the test was completed, it was noted that attenuating-amplitude fault is removed, and then it was observed that the FDIA interface updated the fault state (Figure 23).

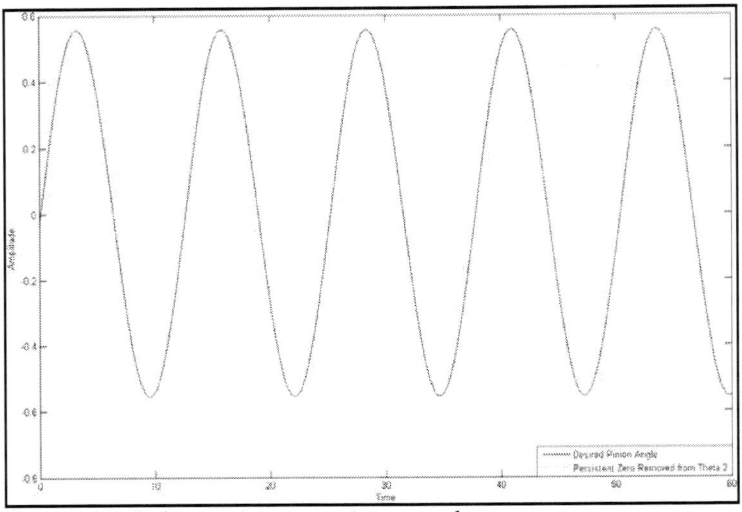

Fig. 21. Normal signal after persistent zero is removed

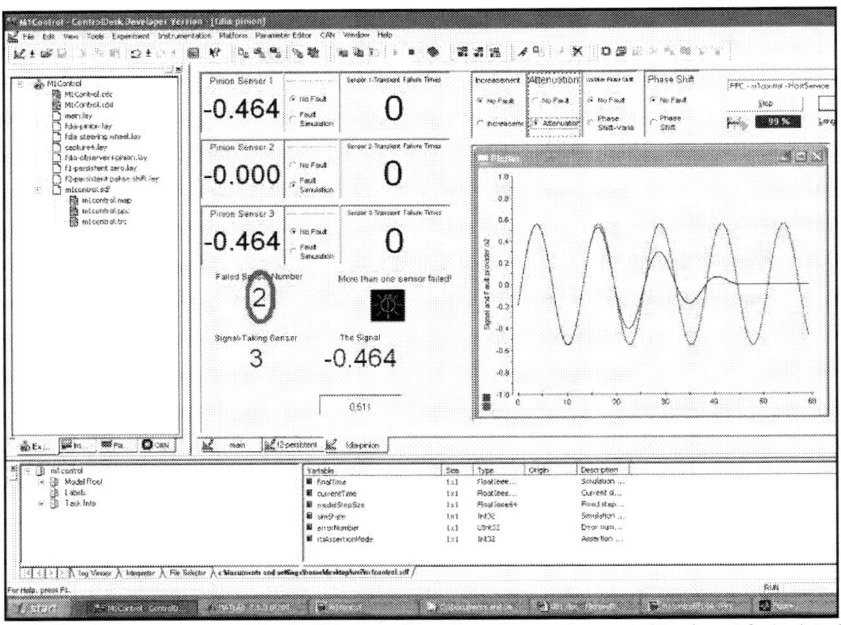

Fig. 22. Fault state reported by FDIA interface for attenuating amplitude with θ_2 (circled area)

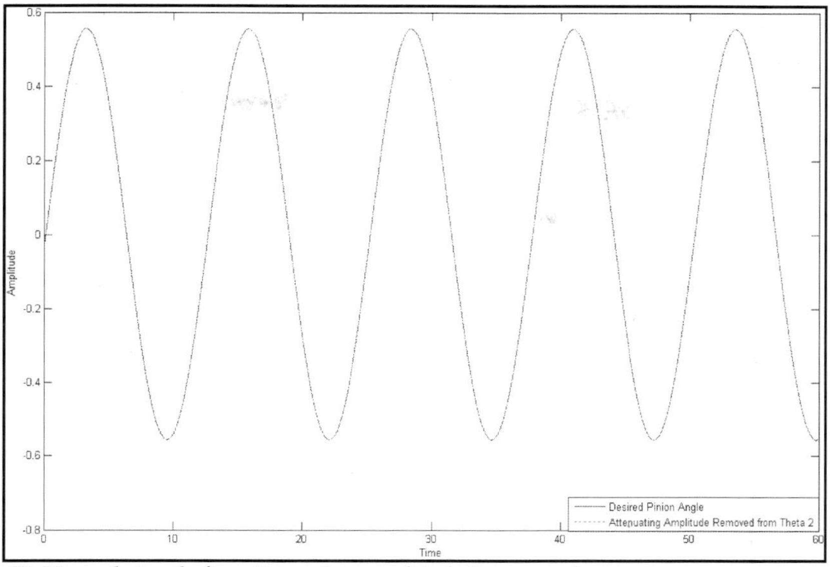

Fig. 23. Normal signal after attenuating amplitude is removed

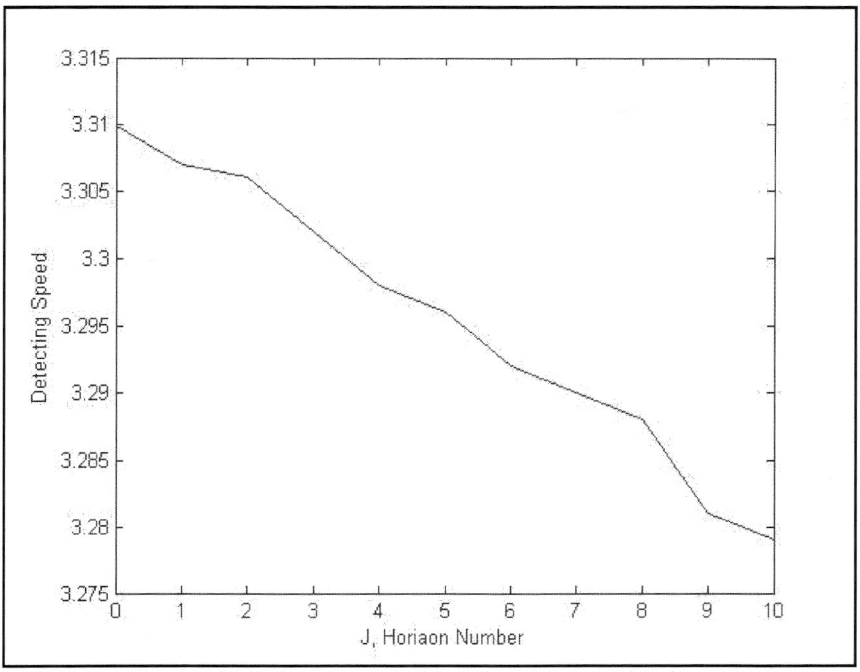

Fig. 24. Horizon versus fault-detecting speed for attenuating amplitude

For this fault case, the impact of the predictor with various prediction horizons is studied. Here the fault detection times are recorded for each prediction horizon selection. After completing the experiment with all selected prediction horizon, the result for attenuating-type fault is shown in Figure 24. It is clear that selecting higher number of horizon makes the FDIA report the detecting speed faster. Again, this result is similar to the ones tested in software-simulation experiment (Hasan & Anwar, 2008).

Similarly, to test variable-phase-shift fault with FDIA interface, the interface creates a variable-phase-shift fault with a selected θ, and fault detection from FDIA interface is then observed (Figure 25). After the test was completed, it is noted that the fault is removed, and then it was observed that the FDIA interface updated fault state. (Figure 26).

Finally, in verifying the performance with varying-phase-shifting-type fault, the result shows that selecting a higher number for prediction horizon makes the FDIA report the detecting speed slower (Figure 27). However, this result is different that the once as tested in (Hasan, 2007) using software-simulation.

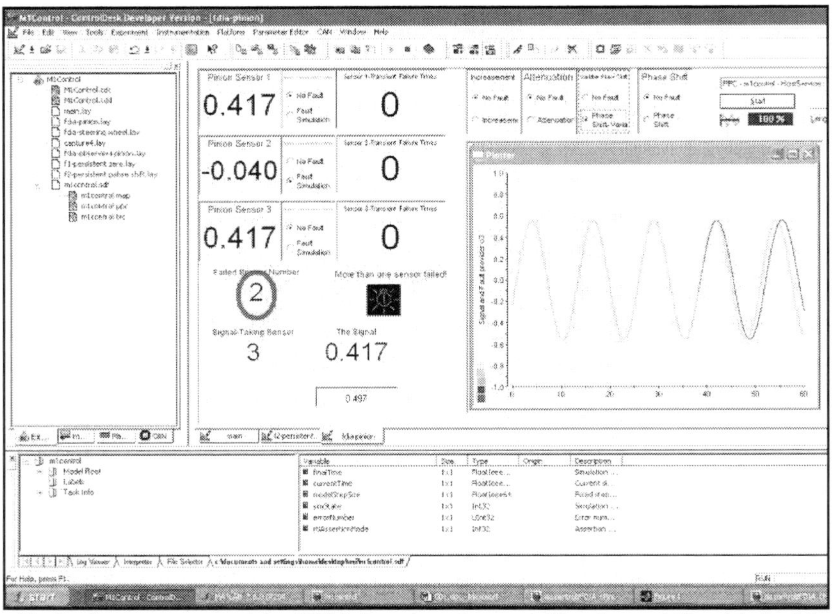

Fig. 25. Fault state reported by FDIA interface for variable phase shift with θ_2 (circled area)

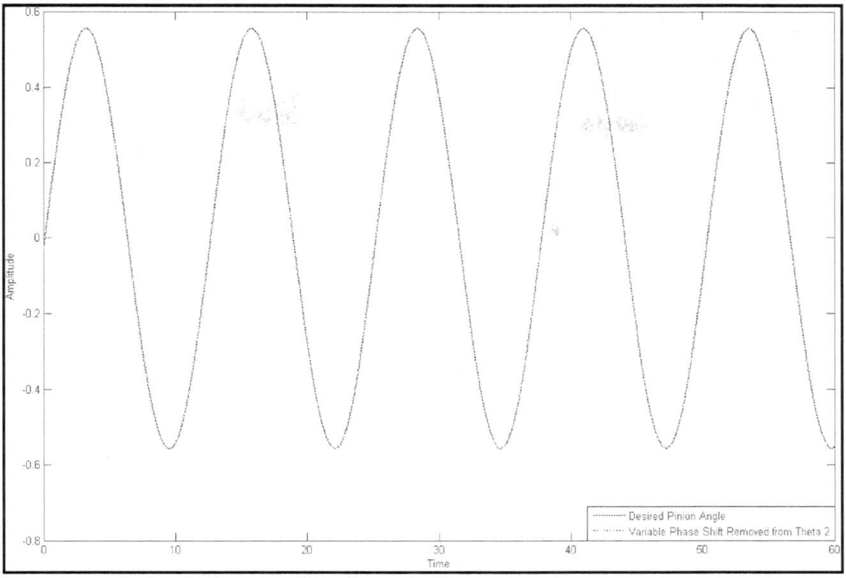

Fig. 26. Normal signal after variable phase shift is removed

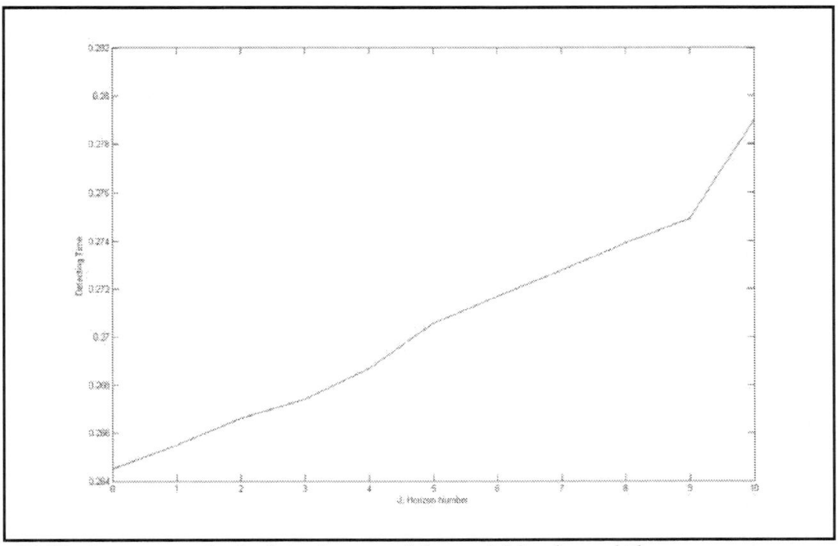

Fig. 27. Horizon versus fault-detecting speed for variable phase shift.

6. Summary and Conclusion

Although automobiles with Steer-by-Wire Control applications are still limited to only prototypes without any near-term commercialization scenario, the potential has been proven by recent research work based on simulation and prototype-based experiments. In this paper, we developed analytical redundancy algorithms using a sliding mode observer and long-range predictor that have been validated on a steer by wire hardware in loop bench. It has also been shown that overall robustness of the SBW system is not sacrificed through the usage of analytical redundancy for sensors along with the designed FDIA algorithm. It is also shown that the faults can be detected faster using the developed analytical redundancy based algorithms for amplifying-type and attenuating-type faults as shown in results section.

However, the fault detection speed for different prediction horizons for the varying phase-shift type faults, the developed analytical redundancy based FDIA algorithms developed for this paper doesn't work as good as in software-simulation. A close look at the result of detecting varying-phase-shift faults reveals that selecting longer prediction horizon for the predictor would reduce the delay time between input and system's output. But, this also means that if there is a fault with gradually changing phase, the system with longer horizon w\would not be able to react or isolate the fault quickly. The reason for the inability of detecting varying-phase-shifting fault faster is that phase of system's response is changed when the input's phase is changed (Figure 28).

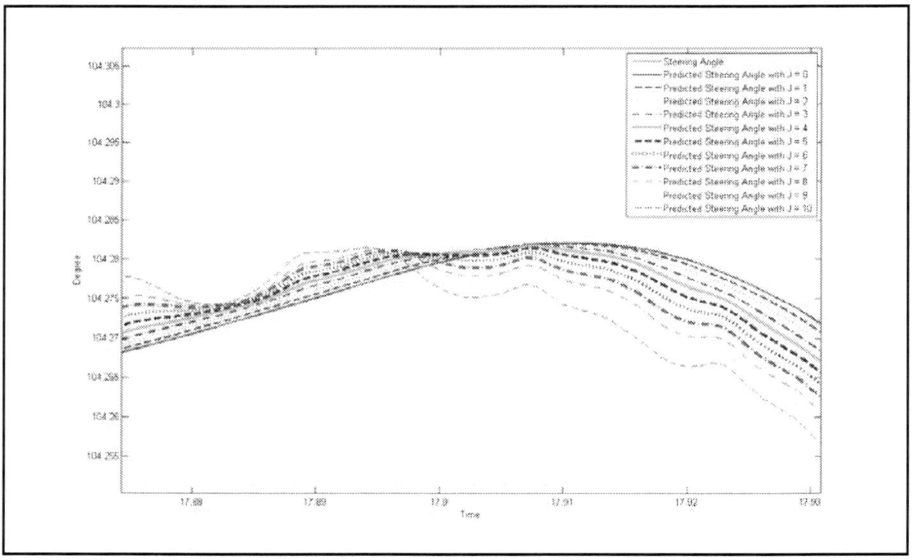

Fig. 28. Phase change occurring when variable phase shift happens with different horizon

Therefore, in order to make possible improvements in solving this problem, an enhanced FDIA algorithm is needed such as tracking the phase for certain duration, etc.

7. References

Anwar, S. (2005). Generalized Predictive Control of Yaw Dynamics of a Hybrid Brake-By-Wire Equipped Vehicle. *Mechatronics*, Vol. 15, November, 2005, pp. 1089-1108.

Anwar, S. & Chen, L. (2006). Analytical Redundancy Based Fault Tolerant Control of a Steer-By-Wire System, *American Society of Mechanical Engineers (ASME), Dynamic System and Control Division (Publication) DSC*, November, 2006, Chicago, IL.

Clarke, D. W., Mohtadi, C., & Tuffs, P. S. (1987). Generalized Predictive Control - Part I: The Basic Algorithm, *Automatica*, Vol. 23, No. 2, 1987, pp. 137-148.

Dong, Y. & Hongyue, Z. (1996). Optimal Design of Robust Analytical Redundancy for a Redundant Strapdown Inertial Navigation System, *Control Engineering Practice*, Vol. 4, No. 12, 1996, pp. 1747-1752.

Gertler, J. (1992). Analytical Redundancy methods in Fault Detection and Isolation, *IFAC Fault Detection, Supervision and Safety for Technical Processes*, Vol. 06, pp. 09-21, 1992, Baden - Baden, Germany.

Hasan, M. S. (2007). Sliding Mode Observer and Long Range Predictive Based Fault Tolerant Control of a Steer-By-Wire Equipped Vehicle, *M.S. Thesis*, Indiana University Purdue University Indianapolis, 2007, pp. 16-39.

Hasan, M. S. & Anwar, S. (2008), Sliding Mode Observer and Long Range Predictive Based Fault Tolerant Control of a Steer-By-Wire Equipped Vehicle, *SAE World Congress and Exposition, Paper No.* 2008-01-0903, April, 2008.

Isermann, R., Schwarz, R. & Stolzl, S. (2002). Fault-Tolerant Drive-By-Wire Systems, *IEEE Control Systems Magazine*, Vol. 22, No. 5, October 2002, pp. 64-81.

Nise, N. S. (1994). *Control Systems Engineering*, 2nd Edition, Addison-Wesley, 1994.

Niu, W. (2009). Fault Tolerant Control Of A Steer By Wire System Using Nonlinear Observer And Predictive Method On Hardware In Loop Bench, *M.S. Thesis*, Indiana University Purdue University Indianapolis, 2009.

Shibahata, Y. (2005). Progress and Future Direction of Chassis Control Technology, *Annual Reviews in Control*, Vol. 29, No. 1, 2005, pp. 151-158.

Stotsky, A. & Hu, X. (1997). Control of Car-like Robots using Sliding Mode Observers for Steering Angle Estimation, *Proceedings of the 36th Conference on Decision & Control*, December, 1997, San Diego, CA, USA.

Suzuki, H., Kawahara, T., Matsumoto, S., Ikeda, Y., Nakagawa, H., & Matsuda, R. (1999). Fault Diagnosis of Space Vehicle Guidance and Control Systems Using Analytical Redundancy, *Space Technology*, Vol. 19, No. 3-4, 1999, pp. 173-178.

Utkin, V., Guldner, J., & Shi, J. (1999). *Sliding Mode Control in Electromechanical Systems*, Taylor & Francis, 1999.

Venkateswaran, N., Siva, M. S., & Goel, P. S. (2002). Analytical Redundancy Based Fault Detection of Gyroscopes in Spacecraft Applications, *Acta Astronautica*, Vol. 50, No. 9, 2002, pp. 535-545.

12

Wavelet Based Diagnosis and Protection of Electric Motors

M. Abdesh Shafiel Kafiey Khan and M. Azizur Rahman
Memorial University of Newfoundland
St. John's, NL, Canada

1. Introduction

Electric machines are electromagnetic energy conversion devices that convert one form of energy into another form. Electric machines have been playing important roles in the developments of modern industrial technology for over a century. Better understanding of energy conversion principles coupled with evolution of new and improved materials has contributed to the developments of machine designs. The applications of electric machines are increasing rapidly with increased technological advancements. The advances of modern digital computers and recent developments in power electronics and semiconductor devices have made revolutionary contributions on the design and control of electric machines. The direct current (dc), induction, and synchronous machines are the three major electric machines that serve industrial, commercial and household needs. The time-stepping finite element analysis has helped in further developments and design optimization of electric machines. Thus, new electric machines such as brushless dc (BLDC) motor, switched reluctance motor, permanent magnet hysteresis motor, permanent magnet synchronous motor, self-excited induction generator, and doubly fed induction generator are developed and implemented for household and industry applications (Rahman, 1980; Slemon, 1992).
The electric machines comes in many sizes and forms, and fulfill their function either independently or as part of a highly complex process in which all elements must function smoothly so that production can be maintained. The function of an individual electric machine is normally seen as separable from the rest of the electromechanical system. Not withstanding their high reliability, electric machines face various stresses including faults during different operating conditions. Hence the condition monitoring, faults diagnostic, and protection become necessary in order to avoid catastrophic failures of electric machines. The use of comprehensive monitoring schemes for the continuous assessment of electrical machines is becoming increasingly important. It is possible to provide adequate warning of imminent/incipient failures using new condition monitoring techniques. It is also possible to schedule future preventive maintenance and repair work in addition to present maintenance needs. This can result in minimum downtime and optimum maintenance schedules. Faults diagnosis allows a machine operator to have the necessary spare parts before the machine is stripped down, which also reduce machine outage times. If faults diagnosis is integrated into the maintenance policy, the usual maintenance at specified

intervals can be replaced by condition centered maintenance. This can also eliminate unnecessary maintenance (Vas, 1996). It is important to stress the fact that protection system for electric machines is basically designed to act only when a fault has occurred in order to initiate some remedial action. Virtually all electric machine protection systems embody some form of protective devices. In a typical machine, they are used in some or all of the following schemes

- Earth fault protection,
- over current protection,
- differential current protection,
- under and overvoltage protection,
- negative phase sequence protection,
- field failure protection,
- reverse power protection,
- over speed protection,
- excessive vibration protection,
- thermal overload protection, etc.

The executive action of a protection system is the disconnection of the piece of machinery of a plant from the supply. Such action is acceptable if the machine is readily dissociated from the process it is involved with, or if it exists substantially in isolation. However, if the machine is vital to the operation of a process, then an unscheduled shutdown of the complete process may occur. The losses involved may then be significantly greater than those resulting simply from the loss of output during a schedule shutdown. The capital cost of an individual electric machine is more often not small compared with the capital costs involved in a plant shutdown. Maintenance is most effective when it is planned to service many items in the course of a single outage. Therefore, condition monitoring of an electric machine is not necessarily aimed at the machine itself, but also at the wider health of the process involving the machines (Tavner et al., 2008).

2. State of the Art Faults Diagnostic and Condition Monitoring Technologies

Intensive research has been conducted to develop and implement new and reliable techniques for condition monitoring, faults diagnostic and protection of electric machines. The modern techniques are based on the application of advanced digital signal processing tools on stator currents. The signal processing tools include discrete Fourier transform (DFT), fast Fourier transform (FFT), wavelet transform (WT), and other high level frequency spectra analysis tools. The model reference adaptive system and artificial intelligence have also been used for faults diagnostic and protection of electric machines. In these non-invasive faults diagnostic techniques, either stator current or vibration signals of electric machines is used to detect a fault. The new faults diagnostic techniques for electric machines can be broadly classified into three categories: (a) artificial intelligence based techniques, (b) standard digital signal processing based techniques, (c) advanced digital signal processing based techniques. The artificial intelligence (AI) based techniques include the applications of expert systems (ES), genetic algorithm (GA), support vector machine (SVM), neural networks (NN), fuzzy logic (FL), and neuro-fuzzy. The standard digital signal processing based techniques include the applications of discrete Fourier transform (DFT), fast Fourier transform (FFT), short time Fourier transform (STFT), and higher order spectra (HOS) such as bi-spectrum, tri-spectrum, etc. The advanced digital signal processing based techniques include the applications of continuous wavelet transform (CWT), discrete wavelet transform (DWT), wavelet packet transform (WPT), and wavelet neural network (WNN). The use of partial discharge (PD), and measurements of stator temperature and negative sequence impedance have also been documented in the literatures for faults diagnostic of electric machines. A state of the art review of various forms of condition monitoring and faults diagnostic techniques for squirrel cage and wound rotor induction motors, permanent magnet synchronous motors, interior permanent magnet (IPM) motors, separately excited synchronous generators, etc, are given in the following subsections.

2.1 Application of artificial intelligence

The artificial intelligence (AI) is the study of system conditions through the use of computational models. The AI tools are of great practical significance in engineering to solve various complex problems, which require human intelligence. Recently, significant efforts have been made on the use of artificial intelligence tools to develop condition monitoring and faults diagnostic techniques for electric machines. Filippetti et al., (1988), have outlined an expert system (ES) based on the knowledge representation for faults diagnostic of induction motors. The knowledge based ES uses instantaneous line currents, line voltages, and rotor speed as input variables. Leith et al., (1988), have presented an on-line real time ES for diagnosing faults in induction motors. The knowledge base consists of a failure tree, an observation tree, and a case tree. The proposed ES require theoretical and practical studies of fault mechanisms, and case histories of fault analyses. This method is vulnerable to uncertainty and is not quite suitable from computational point of view.

Pöyhönen et al. (2002), have implemented support vector machine (SVM) based faults diagnostic and classification technique for an inverter fed squirrel cage induction motor. The magnetic field analysis is used to get virtual data of the healthy and faulty operating conditions of the induction motor. The power spectra of stator current are used as inputs to the SVM based classifier to distinguish healthy condition from normal unfaulted condition. However, the technique may fail if two separate classes get equal amount of votes. In addition, it did not consider the possible redundancy from pair-wise outputs of the classifier. Chow et al. (1991), have implemented a three-layer feed forward neural network for condition monitoring of induction motors in real time. The stator inter-turn and motor bearing faults are investigated at constant load torque. The stator currents and rotor speed are used as inputs during the off-line training of the network. The network is implemented in real time using a digital signal processor board. The network showed satisfactory performances with higher number of hidden nodes. However, the technique is not quite accurate due to the dynamic nature of machine parameters. In addition, it requires large number of training data set in order to cover all the operating conditions including the faulted and unfaulted conditions of the motor. Lasurt et al. (2000), have implemented a fuzzy logic based condition monitoring and faults diagnostic technique for induction motors. The proposed technique implemented the higher order statistical (HOS) analysis of the machine vibration signal. The fuzzy logic procedures are then applied to the HOS signatures in order to enable diagnosis of a machine fault. Park et al. (2004), have presented an adaptive neuro-fuzzy inference system (ANFIS) based faults diagnostic technique for an inverter fed pulse width modulated induction motor drive system. The proposed technique involves data acquisition and feature extraction of fault currents, and then use of adaptive neuro fuzzy inference system (ANFIS) for faults diagnostic. The technique used the mean values of direct and quadrature axis phase currents as the input pattern for the ANFIS.

2.2 Application of standard digital signal processing techniques

The standard signal processing tools such as DFT, FFT, STFT, etc are widely used for condition monitoring and faults diagnostic of electric machines in last fifteen years. These techniques have traditionally been applied separately in time and frequency domains. The time domain analysis focuses particularly on statistical characteristics of vibration, temperature, and stator currents, which include peak level, standard deviation, kurtosis, skewness, and crest factor of the diagnostic signal. The frequency domain approach uses Fourier methods to transform the time domain signal in the frequency domain, where further analysis is carried out. The use of either domain implicitly excludes the direct use of information present in other domain (Yang et al., 2003).

Yang et al. (2003), have used vibration analysis based on bi-spectra and wavelet transform for the diagnostic of induction motor rolling element bearing faults. The singular value decomposition (SVD) technique is applied to extract the most significant features from vibration signatures. The features are used as inputs to an artificial neural network to identify the type of fault. Roux et al. (2003), have investigated two condition monitoring techniques in order to detect rotor faults in surface mounted type permanent magnet synchronous motor. The first technique is based on harmonic spectra analysis of stator voltage and current in the natural reference frame. The second technique is based on spectra analysis of *d-q* axis voltage the rotor reference frame. The static and dynamic eccentricity, broken magnets, and rotor misalignments are investigated. The harmonic spectra analysis method based on fast Fourier transform (FFT) was able to differentiate all faults except the static eccentricity from the normal case. Therefore, the harmonics of the negative sequence component of the stator current were used for the detection of static eccentricity, and it successfully distinguished the static eccentricity from the normal case. However, the main disadvantage of FFT based harmonic spectra analysis is the impact of side lobe leakage due to windowing of finite data sets.

Zanardelli et al. (2007), have implemented the short time Fourier transform (STFT) of the torque producing current component in order to diagnose faults in surface mounted permanent magnet synchronous motor. The field oriented *d-q* axis currents are used in this analysis. The energy of the STFT coefficients is used to detect a fault in the motor, and linear discriminator analysis is used to classify faults. However, the STFT based technique uses stationary and periodic basis functions. But fault currents are often non-stationary and non-periodic. As a result, the performances of the technique are limited due to the constraint on the window size. In addition, the performances of the proposed technique have not been investigated in real time. Schoen et al. (1995), have implemented motor current spectral analysis (MCSA) for diagnostic of rolling element bearing in induction motors. The vibration and current frequencies are modeled in order to detect incipient bearing failures. Yazici et al. (1999), have developed an adaptive statistical time-frequency approach for diagnostic of broken bars and bearing faults in induction motors. The proposed technique has four stages such as preprocessing, training, testing, and post processing. In the preprocessing stage, analog current data are filtered by low pass circuit in order to prevent aliasing in the frequency domain. Next, the time-frequency spectra of the digital data are computed and used as inputs of a neural network. In the training stage, fault frequencies are determined, and a window with frequency components near to the estimated frequencies is applied to form a feature vector. Next, feature vectors are segmented into homogenous sections along the time axis in time-frequency space. Segmentation is performed by a statistical method, which divided the time-frequency spectra into statistically homogenous regions along the time axis. However, the proposed technique shows strong interactions between machine operating conditions and machine variables.

2.3 Application of advanced digital signal processing techniques

Majority of the signal processing based faults diagnostic techniques involve the analyses of vibrations signal or stator currents in either time or frequency domains assuming stationary and periodic nature of fault currents. Thus these techniques are not fully suitable for localizing and identifying short duration dynamic phenomena. Therefore, the applications of advanced signal processing techniques are required, which include signal modeling, filtering, and time-frequency analysis. Among the latter, the wavelet transform algorithms are the recent mathematical tools adopted and implemented for faults diagnostic and protection of electric machines (Dalpiaz & Rivola, 1997).

Zanardelli et al. (2005), have developed a failure prognosis technique for surface mounted permanent magnet synchronous motor drives based on undecimated discrete wavelet transform (UDWT) of torque current components. The energy of the UDWT coefficients of normal unfaulted and faulted currents is used to detect a fault in the motor. The linear discriminator analysis is used to classify faults. The same authors (Zanardelli et. al., 2002) have made a comparative analysis of wavelet based faults diagnostic and protection techniques in electric machines. Khan & Rahman (2009), have developed and implemented a novel fault diagnostic and protection technique for interior permanent magnet (IPM) synchronous motors using wavelet packet transform (WPT) and artificial neural network (ANN). In the proposed technique, the line currents of different faulted and normal conditions of an IPM motor are preprocessed by the WPT. The second level wavelet packet transformed coefficients of line currents are used as inputs of a three-layer feed forward neural network. The proposed protection technique is successfully simulated and experimentally tested on a line-fed and an inverter-fed IPM motors. [a]Khan & Rahman (2008), have developed a wavelet transform based diagnostic and protection technique for inverter faults of IPM motor drives. The proposed technique is implemented in real time for a voltage source inverter fed IPM motor. The WPT coefficients of motor currents are used as inputs of a three-layer wavelet neural network (WNN) for detecting inverter faults in the drive system. A feature vector based on the energy of WPT coefficients is used to classify different faulted conditions. [b]Khan & Rahman (2008), have developed and implemented a WNN based diagnostic and protection algorithm for inverter faults in vector controlled induction motor drive system. The proposed technique is tested on-line for a laboratory 1-hp induction motor drive using the digital signal processor board ds1102. [a]Khan et al. (2007), have developed and implemented a novel wavelet power based faults diagnostic and protection algorithm for separately excited synchronous generator. The proposed algorithm is based on the comparison of instantaneous wavelet power of terminal voltage and current of a synchronous generator for different faulted and normal (unfaulted) conditions. The wavelet power of second level high frequency details (dd^2) of fault currents and voltages using a selected mother wavelet show distinctive features between different faulted and normal conditions. The proposed technique is tested on-line on a laboratory 1.6 kW three-phase synchronous generator. The WPT based diagnostic and protection technique is implemented in ([b]Khan et al., 2007) for three-phase induction motors. The WPT coefficients at second level of resolution using a selected mother wavelet are compared with a fault threshold in order to detect a fault in induction motor. The proposed technique is tested on both squirrel cage and wound rotor induction motors. The proposed protection technique initiated a trip signal almost at the instant or within one cycle of fault current in all cases of investigated faults. [c]Khan & Rahman (2007), have developed and implemented a hybrid WPT and ANN based faults diagnostic and protection technique for three-phase IPM motors. The proposed technique is compared with discrete Fourier transform (DFT) based protection algorithm at dynamic operating conditions. The proposed technique showed better performances than the DFT based technique. [d]Khan et al. (2007), have developed and implemented a wavelet power based diagnostic and protection technique for stator faults in synchronous generators. The stator phase unbalance, line to ground (L-G), line-to-line (L-L), and turn-to-turn faults are investigated to evaluate the performances of the proposed technique.

Kim et al. (2002), have developed a model based faults diagnostic technique based on recurrent dynamic neural networks and multi-resolution signal decomposition for predicting transient response and extracting features of fault currents in induction motors, respectively. The transient model is used to generate residual fault current. Then, the wavelet packet transform based decomposition algorithm is implemented on residuals in order to generate decoupled fault indicators. The wavelet transform is applied in (Chow et al., 2004) for extracting vibration spectra, which contain features of critical frequencies for faults diagnostic in induction motors running at different speeds. The wavelet basis functions are matched with related signals through careful selection of basis function parameters. An on-line fault detection approach based on the continuous wavelet transform of vibration signals for detecting bearing faults in induction motors has been reported in (Luo et al., 2003). Douglas et al. (2004), have developed a new faults diagnostic algorithm based on the signature analysis of starting currents of induction motors. The proposed algorithm estimated the amplitude, frequency, and phase of a single sinusoid signal of the non-stationary fault current waveforms. The DWT is applied to residual current vector in order to discriminate a healthy motor from the damaged motor. However, the proposed technique is used to detect passive faults rather than incipient failures in induction motors. Toliyat et al. (2003), have implemented WPT for detecting defects in railroad track. The energy of WPT coefficients is used to detect a fault. The experimental results showed deviation of energy of the DWT coefficients in the faulted motor from the healthy motor. Yen et al. (2000), have outlined a systematic procedure for selecting best WPT features, which exploit specific differences among interesting signals. In this method the signal is first decomposed via the wavelet packet transform (WPT) in order to extract the time-frequency information. Several feature components, which contain little discriminator information are discarded with the help of a statistic based feature selection criterion. Zhengjia et al. (1996), have implemented WPT for condition monitoring and faults diagnostic of turbo generators. The proposed method successfully diagnosed weak defects and looseness in ball bearings of inside the bearing terminal of a 50MW turbine generator.

There have been many condition monitoring and faults diagnostic techniques for electric machines. The artificial intelligence, motor current signature analysis, and finally time-frequency analysis based on short time Fourier transform or wavelet transforms are widely used in last twenty years. The wavelet transform is a relatively new technique for condition monitoring and faults diagnostic of electric machines. It replaces other condition monitoring techniques because of its better frequency resolution and time localization properties. It is free from any learning or training of the experimental data covering all operating conditions of the motor. In addition, desirable basis function related to a specific application can be chosen in wavelet transform based faults diagnostic technique whereas in Fourier transform based diagnostic technique the basis functions are fixed to sinusoid or cosine function.

3. DFT, STFT, and NN Based Faults Diagnostic and Protection

The discrete Fourier transform (DFT), short time Fourier transform (STFT), and neural network are implemented for faults diagnostic and protection of three-phase interior permanent magnet (IPM) motors. At the beginning the discrete Fourier transform (DFT) is implemented using the fast Fourier transform (FFT) based algorithm to estimate the spectra of fault current in order to differentiate normal conditions from abnormal conditions. The short time Fourier transform (STFT) based faults diagnostic algorithm is implemented after the DFT based technique. Finally a pattern recognition technique based on three-layer feed forward neural network is implemented for faults diagnostic of laboratory 1hp IPM motor. A short analysis and real time implementation of each type are given for diagnostic and protection of faults in electric motors.

3.1 Application of Discrete Fourier Transform (DFT)

The DFT analysis and synthesis equations can be expressed as

$$X[k] = \sum_{n=0}^{N-1} x[n] W_N^{kn}, \quad 0 \le k \le N-1 \tag{1}$$

$$x[n] = \frac{1}{N} \sum_{k=0}^{N-1} X[k] W_N^{-kn}, \quad 0 \le n \le N-1 \tag{2}$$

where,

$$W_N = e^{-j\frac{2\pi}{N}} \tag{3}$$

For the case of non-periodic and non-stationary fault current signals, the windowed DFT is normally used. If a signal is sampled with sampling interval of Δt such that there are $N/\Delta t$ samples per cycle, then the DFT basis function coefficients can be calculated as (Khan, 2006)

$$S_k = \frac{2}{N} \sum_{n=1}^{N-1} x[n] \sin\left(\frac{2\pi kn}{N}\right) \tag{4}$$

$$C_k = \frac{2}{N}\sum_{n=1}^{N-1} x[n]\cos\left(\frac{2\pi kn}{N}\right) \tag{5}$$

The Fourier harmonic coefficients can be calculated as

$$F_k = \sqrt{S_k^2 + C_k^2} \tag{6}$$

where F_k is the kth harmonic Fourier coefficient, and $x[n]$ is the sampled sequence of the continuous signal $x(t)$. Three types of electrical faults such as single phasing, stator winding line to ground (L-G), and stator winding line to line (L-L) faults are investigated. The DFT is implemented to determine the spectra of stator current of different faulted and normal unfaulted conditions.

Figure 1 shows the DFT based spectra of stator current of different unfaulted and faulted conditions of a laboratory 4-pole 1-hp IPM motor. The fundamental harmonic (30Hz) is found significant in all the operating conditions of the IPM motor of Figs. 1(*a*)-1(*d*). The fundamental spectrum varied between faulted and normal unfaulted (healthy) conditions and also within disturbances. The normalized magnitude of fundamental spectrum varied between 0.5 and 0.6 for the case of unfaulted operating conditions of the motor. It showed peak values of greater than 0.69 and less than 0.3 for the cases of disturbances. Based on the observations it can be asserted that the DFT based spectra analysis of fault currents can detect and classify possible disturbances in IPM motors. However, the DFT based technique is not suitable for non-stationary and non-periodic signals. In addition, faults cannot be localizes in time domain using DFT based technique.

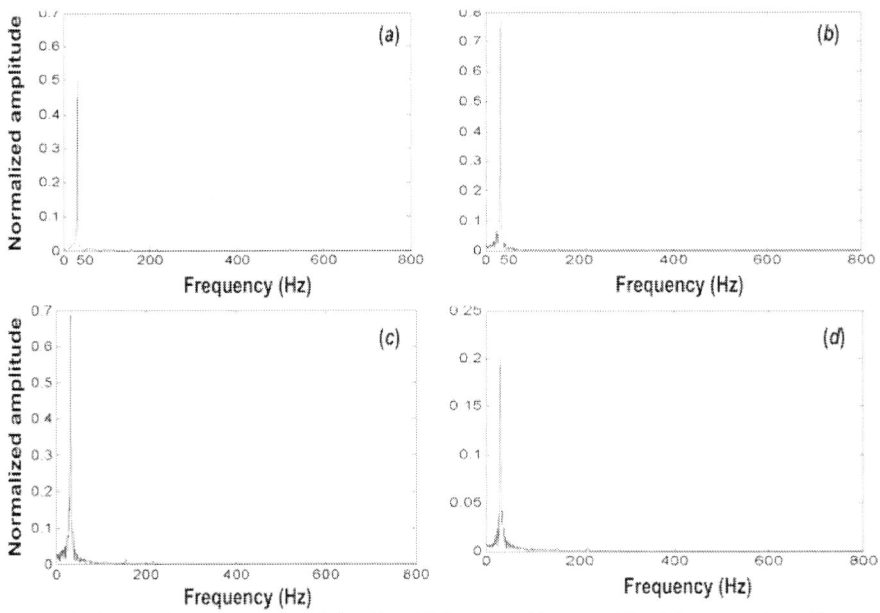

Fig. 1. The DFT of stator current of the 1hp IPM motor: (*a*) normal healthy operation, (*b*) stator winding line to ground (L-G) fault, (*c*) stator winding line to line (L-L) fault, (*d*) single phasing.

3.2 Application of Short Time Fourier Transform (STFT)

The short time Fourier transform (STFT) is an extension of fast Fourier transform (FFT), which is able to analyze non-stationary and non-periodic signals. In the STFT, the discrete signal is divided into segments, and each segment is analyzed using the FFT. The results of the STFT are intuitive and easy to correlate with the original signal. The tiling of the STFT is shown in Fig. 2. The tiling shows how the spectrum of a signal changes with time in the STFT. In the implementation of the STFT, a design tradeoff is normally made between time and frequency resolution. This is due to the uncertainty principle, which limits the lower bound of the time-bandwidth product. Figure 3 shows the block diagram of the STFT algorithm. In Fig. 3, nfft is the length of the FFT, noverlap is the number of overlap samples, and window is a weighting vector applied to the FFT input. The spectrogram is the graphical way to display the output of the STFT.

The STFT based spectrogram of stator current of the 1hp IPM motor for the case of normal unfaulted and faulted conditions are shown in Figs. 4(*a*)-4(*d*). In the detailed analysis, a 512-point FFT with 475 overlap samples between data segments is used to estimate the frequencies of discrete signal. A 500-point Kaiser window (Mathworks, 2007) is applied in each data segment. The analysis generated 257 frequency points in 141 time-axis values. The energy concentration of the fundamental harmonic (30 Hz) of the 4-pole IPM motor is uniform over the entire time axis of the spectrogram for the case of healthy operating condition of the IPM motor of Fig. 4(*a*). However, the concentrations of energy of the fundamental, third, fifth, and seventh harmonics are different than those of the healthy motor during the inception and clearing of faults of Figs. 4(*b*)-4(*d*). Based on the analysis performed, it can be asserted that the STFT based algorithm can detect faults in the IPM motors in both time and frequency domains. However, the STFT based technique does not provide good energy resolution for a specific point of the data signal as the length of the window is fixed in each of the data segment of the discrete signal. In addition, the frequency analysis is performed using the sinusoidal basis functions.

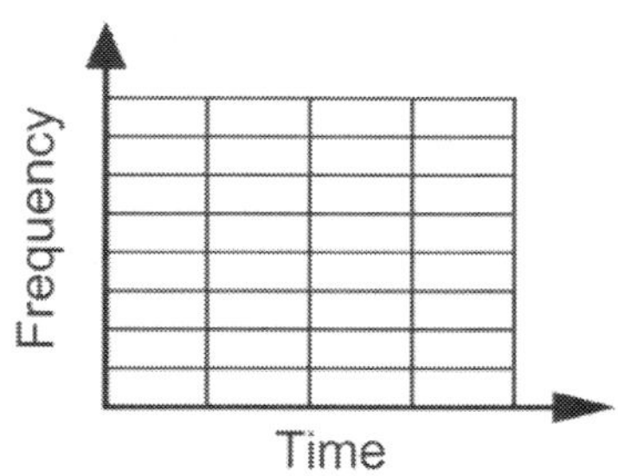

Fig. 2. Time-frequency distribution of the STFT.

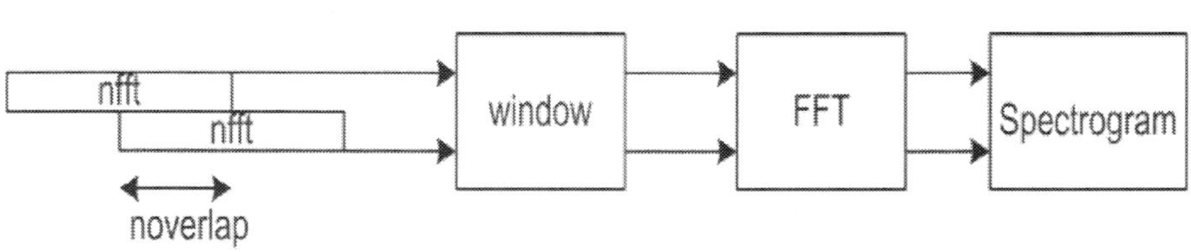

Fig. 3. Block diagram of the STFT algorithm.

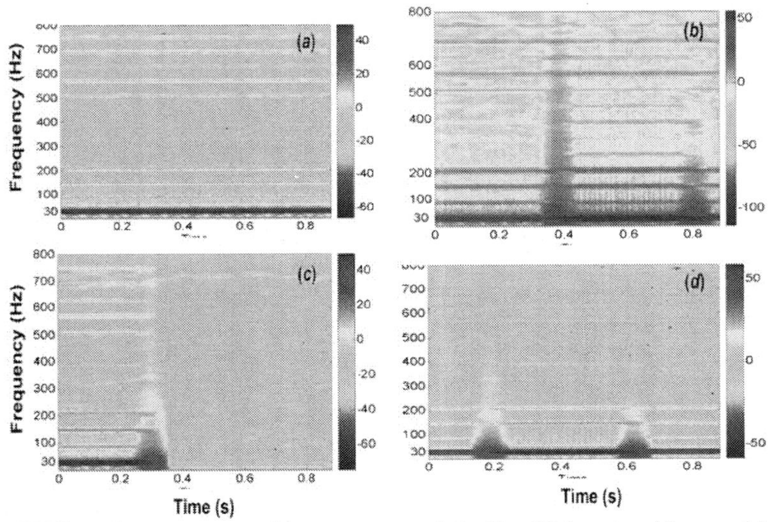

Fig. 4. STFT based spectrogram of stator current of the 1hp IPM motor: (*a*) normal healthy operation, (*b*) stator winding line to ground (L-G) fault, (*c*) single phasing, (*d*) stator winding line to line (L-L) fault.

3.3 Application of artificial neural network (NN)

The primary step in developing and implementing artificial neural network (NN) based faults diagnostic and protection technique for electric motors is to choose a suitable network structure. A three-layer feed forward network with three neurons in the hidden layer and one neuron in the output layer is chosen in this work for the NN based faults diagnostic of IPM motors. The numbers of neurons in the hidden layer are selected by trial and error, which ensured stability and higher convergence rate. The Nguyen-Widrow initialization algorithm (Mathworks, 2007) is used in order to initialize weights and biases of the network. The activation function *log-sigmoid* (Mathworks, 2007) is used in both hidden and output layers of the network.

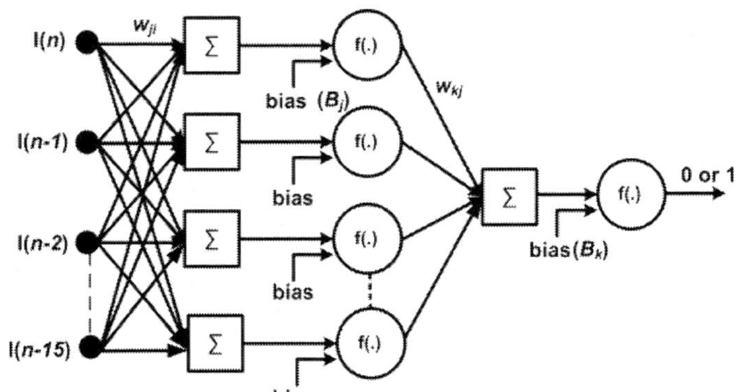

Fig. 5. Specific structure of a three-layer feed forward neural network (FFNN).

The specific structure of a three-layer feed forward neural network is shown in Fig. 5. The network is trained off-line in a supervised manner with the back propagation function *traingdm* (Mathworks, 2007), which updates weight and bias values of the hidden and output layers according to the gradient descent with momentum. In the NN based faults diagnostic algorithm for an IPM motor, the stator currents are used as inputs to the neural network. The discrete data of normal unfaulted and faulted conditions are used to train the network so that it can differentiate normal conditions from the abnormal conditions. In order to generate the realizable training patterns for the NN based faults diagnostic, samples of the squared summation of three-phase stator currents are compared with a predefined threshold to convert it a binary value of either 1 or 0, depending on whether the value is greater or smaller than the threshold, respectively. In this way each training pattern became a different combination of 1 and 0. It is expected that the starting current and fault currents data would not have same training pattern. The elements of the target vector for the case of normal unfaulted (both no load and full load) and starting current samples are chosen equal to binary '0'. On the other hand, the elements of the target vector are equal to binary '1' for the case of fault current samples. After training the network with one set of training pattern, which includes the samples of the normal unfaulted and faulted currents, and starting currents, the network is tested off-line in the MATLAB environment with the different set of testing pattern. Figure 6 shows the off-line test results of NN based faults diagnostic and protection algorithm of the 1hp IPM motor.

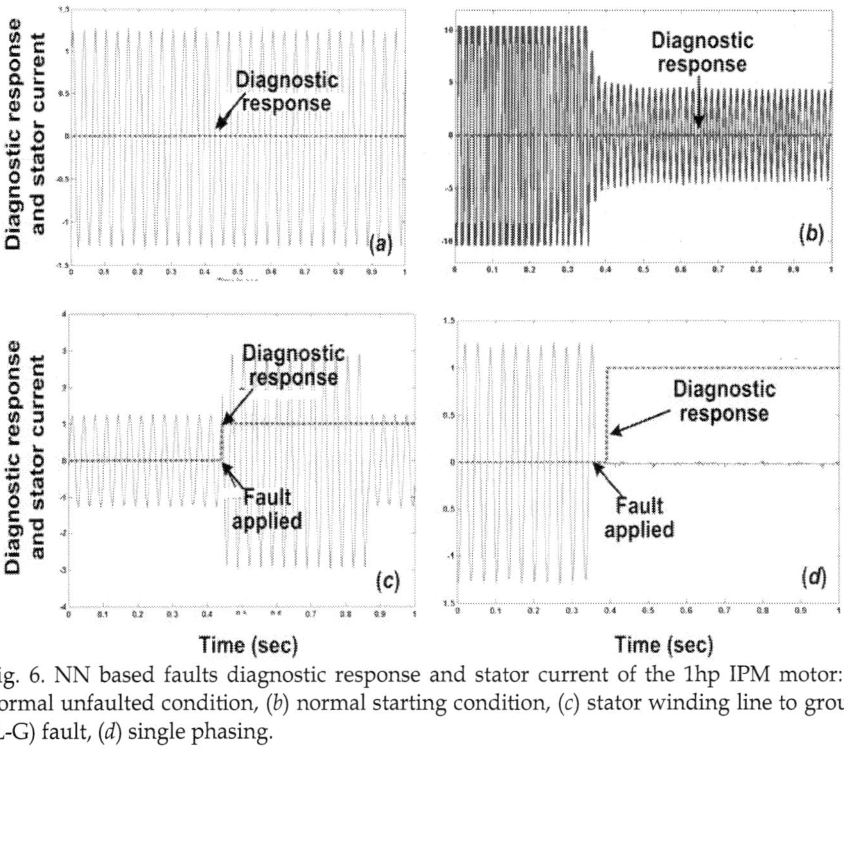

Fig. 6. NN based faults diagnostic response and stator current of the 1hp IPM motor: (*a*) normal unfaulted condition, (*b*) normal starting condition, (*c*) stator winding line to ground (L-G) fault, (*d*) single phasing.

Figures 6(*a*)–6(*b*) show the NN based faults diagnostic response and stator current of the 1hp IPM motor for the case of normal (unfaulted) and starting conditions. The neural network (NN) based diagnostic algorithm identified these as normal conditions and did not generate any trip signal. The stator current with the associated trip signal for the case of stator winding line to ground (L-G) fault and single phasing are shown in Figs. 6(*c*)–6(*d*). The algorithm identified these properly and initiated a trip signal almost at the instant of the fault occurrence. However, the technique needs a large number of data files to train the network effectively. In addition, number of hidden layers may have to be increased to improve the accuracy. Therefore, more memory may be needed to accommodate the weights and biases of the new layers, and as a result, many trials are required to determine the learning rate, so as to improve the functionality of the NN based diagnostic algorithm.

4. Wavelets and Wavelet Transforms

The wavelet transforms analyze a signal simultaneously in time and frequency domains. The wavelet transforms are very useful in analyzing non-stationary, non-periodic, intermittent, and transient signals. Therefore, a number of wavelet based techniques are developed and implemented for signal manipulation and interrogation. The wavelet transforms are applied in the investigation of diverse physical phenomena such as climate analysis, financial market analysis, heart monitoring, condition monitoring and protection of rotating machines, de noising of seismic signal and astronomical images, characterization of crack surface and turbulent intermittency, compression of video image and medical signal records, etc. The wavelets are little waves of short duration. These have finite energy and decay quickly in time. The wavelets also have oscillating feature, which comes along with the location in time and frequency. These basic features make wavelets highly adequate for signal representation. The wavelet functions of orthogonal type have a companion function, which is known as the scaling function. It is responsible for generating basis functions, which are required during the decomposition or reconstruction of a signal. Figures 7(*a*) and 7(*b*) show the *Daubechies* ('*db*3') wavelet function and its scaling function, respectively. In certain application, it is necessary to use real and symmetric wavelets. One way to get the symmetric wavelets is to construct two sets of bi-orthogonal wavelets, which are wavelet function $\psi_{m,n}(t)$ and it's dual $\hat{\psi}_{m,n}(t)$. The first set is used during the decomposition, and the other one is used during the reconstruction process. Figures 8(*a*)–8(*d*) show the *spline* bi-orthogonal ('*bior*2.6') wavelet functions and their scaling functions during the decomposition and reconstruction of a signal.

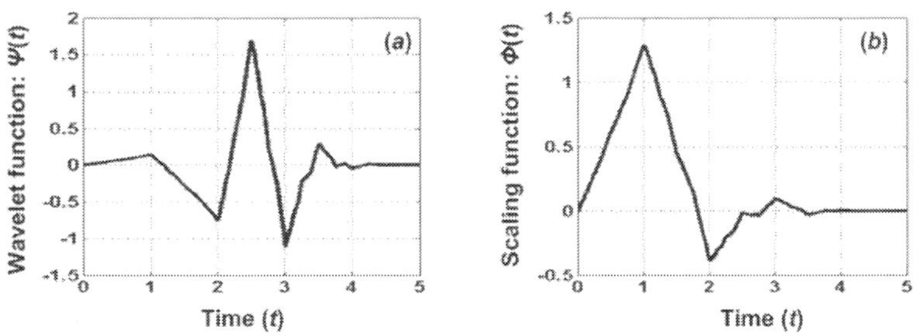

Fig. 7. The *Daubechies* ('*db*3') wavelet: (*a*) mother wavelet function and (*b*) scaling function.

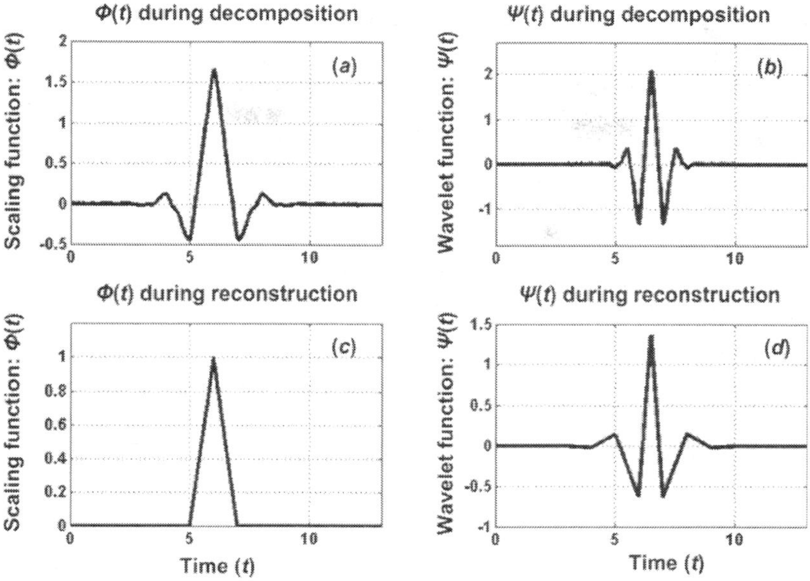

Fig. 8. The *Spline* bi-orthogonal (*'bior*2.6') wavelet: (*a*) scaling function during decomposition, (*b*) mother wavelet function during decomposition, (*c*) scaling function during reconstruction, and (*d*) mother wavelet function during reconstruction.

The wavelet transforms use little wavelike functions, which are known as wavelets. Wavelets are used to transform a signal under investigation into another representation of a more useful form. From the mathematical point of view, the wavelet transform is a convolution of the wavelet function with the signal. The wavelet function is manipulated in two ways: it is moved to various locations on the signal, and it is stretched or squeezed. If a wavelet function matches the shape of a signal well at a specific scale and location, then a large transformation value will be generated. On the other hand, if the wavelet function and signal do not correlate well, then a low value of transformation will be generated (Addison, 2002). The wavelet transform can be applied to both continuous and discrete signals. In the following subsections, different forms of wavelet transforms and their mathematical formulations are briefly presented.

4.1 Continuous wavelet transform

The wavelet transform of a continuous signal $x(t)$ with respect to the wavelet function $\psi(t)$ can be defined as

$$T(a,b) = w(a)\int_{-\infty}^{\infty} x(t)\psi^{*}\left(\frac{t-b}{a}\right)dt \qquad (7)$$

where $w(a)$ is the weighting function, a and b are the dilation and translation parameters, respectively. The asterisk indicates that the complex conjugate of the wavelet function is used in the transformation. The wavelet transform can be thought of as the cross correlation of a signal with a set of wavelets of various widths. Typically, $w(a)$ is set to $1/\sqrt{a}$ for the reason of energy conservation. It ensures that wavelets at each scale have identical energy. If one sets $w(a) = 1/\sqrt{a}$, then the wavelet transform of the continuous signal $x(t)$ can be rewritten as

$$T(a,b) = \frac{1}{\sqrt{a}} \int_{-\infty}^{\infty} x(t)\psi^*\left(\frac{t-b}{a}\right) dt. \tag{8}$$

The equation (8) is known as the continuous wavelet transform (CWT) of the signal $x(t)$. It contains both dilated and translated wavelets $\psi((t-b)/a)$, and the continuous signal $x(t)$. The signal $x(t)$ may be a beating heart, an audio signal, a financial index, the gearbox vibration signal, a spatial signal such as crack profile or land surface heights. The normalized wavelet function can be written more compactly as

$$\psi_{a,b}(t) = \frac{1}{\sqrt{a}} \psi\left(\frac{t-b}{a}\right) \tag{9}$$

where the normalization is in the sense of wavelet energy. Now the transform integral of equation (9) can be rewritten as

$$T(a,b) = \int_{-\infty}^{\infty} x(t)\psi_{a,b}^*(t)\, dt \tag{10}$$

The dilation and contraction of the mother wavelet function is governed by the dilation parameter a, which is the distance between center of the wavelet function and its crossing on the time axis. The movement of the wavelet function along the time axis is governed by the translation parameter b. Figure 9 shows the stretching (double) and squeezing (half) of the Mexican Hat wavelet function on the time axis. Figure 10 shows the translation of the Mexican Hat wavelet function on the time axis from b_1 via b_2 to b_3.

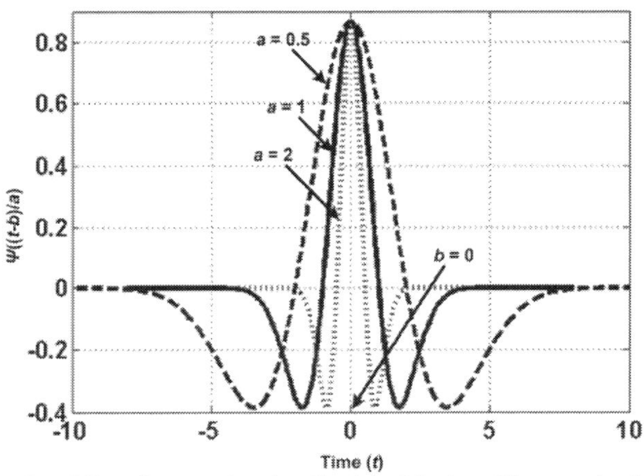

Fig. 9. Stretching (a = 0.5) and squeezing (a = 2) of the Mexican Hat wavelet function.

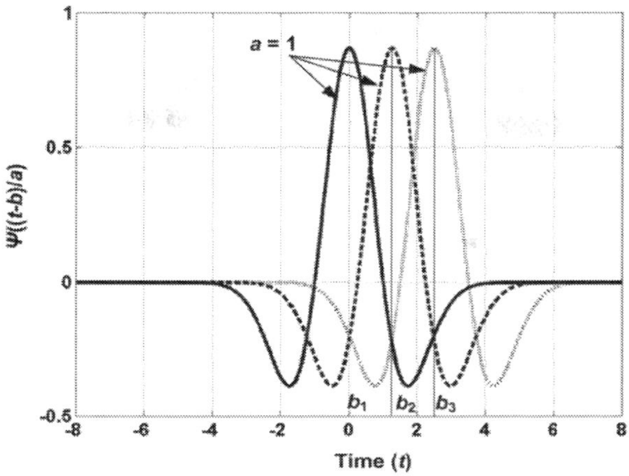

Fig. 10. Translation (b_1 via b_2 to b_3) of the Mexican Hat wavelet function.

4.2 Discrete wavelet transform

In the continuous wavelet transform (CWT), the mother wavelet is dilated and translated continuously over a real continuous number system $(\mathbb{R})$. Therefore, it can generate substantial redundant information. The mother wavelet can be dilated and translated discretely by replacing $a = a_0^m$ and $b = nb_0a_0^m$, where a_0 and b_0 are the fixed constants with $a_0 > 1$, $b_0 > 0$, and $m,n \in \mathbb{N}$. Here $\mathbb{N}$ is the set of positive integers. Then, the discretized mother wavelet function can be defined as (Addison, 2002)

$$\psi_{m,n}(t) = a_0^{-m/2}\,\psi\left(\frac{t - nb_0\,a_0^m}{a_0^m}\right) \tag{11}$$

and the corresponding discrete wavelet transform (DWT) can be defined as

$$DWT_{\psi}x(m,n) = \int_{-\infty}^{\infty} x(t)\psi_{m,n}^{*}(t)\,dt \tag{12}$$

In the DWT, the family of dilated wavelets constitutes an orthonormal basis by careful selections of a_0 and b_0. There are several implications of the orthonormal basis. The orthonormality ensures no information redundancy among the decomposed signals. With the optimal choices of a_0 and b_0, there exists an elegant algorithm known as the multiresolution signal decomposition. It decomposes a signal into various scales with different time and frequency resolutions. In the DWT, the procedure starts with passing the discrete signal x[n] of length N through a digital low pass filter with impulse response g[n] and a digital high pass filter with impulse response h[n]. The low pass and high pass filters are called scaling and wavelet filters, respectively. The outputs from the low pass filter are approximation coefficients of the discrete signal at first level of resolution of the DWT. The outputs from the high pass filter are detail coefficients of the discrete signal at first level of resolution of the DWT. The output of these filters consists of N wavelet coefficients. This constitutes first level of decomposition of the discrete signal and can be mathematically expressed as

$$a^1[n]=\sum_{k=0}^{N-1} g[k]x[n-k] \tag{13}$$

$$d^1[n]=\sum_{k=0}^{N-1} h[k]x[n-k] \tag{14}$$

The approximation coefficients (a^1) at first level of resolution are used as inputs for another pair of wavelet filters (identical with the first pair) after being down sampled by two. The filters at second level of resolution generate sets of approximations and details coefficients of length $N/2$. This constitutes second level of decomposition of the discrete signal and can be mathematically expressed as

$$a^2[n]=\sum_{k=0}^{N/2-1} g[k]a^1[2n-k] \tag{15}$$

$$d^2[n]=\sum_{k=0}^{N/2-1} h[k]a^1[2n-k] \tag{16}$$

Figure 11(a) shows the two-level decomposition of a discrete signal of the discrete wavelet transform. It uses the high pass filters (H) and the low pass filters (G) in the decomposition process.

4.3 Wavelet packet transform

The wavelet packets are alternative bases, which can be formed from the linear combinations of usual wavelet functions. These bases inherit properties such as orthonormality and time-frequency localization from their corresponding wavelet functions. A wavelet packet function is a function of three indices j, k and n, and is defined as

$$W_{j,k}^{n}(t)=2^{j/2}W^{n}(2^{j}t-k) \tag{17}$$

where the integers j and k are the indices for scale and translation operations, respectively. The index n is defined as the modulation or oscillation parameter. The first two wavelet packet functions are the scaling function and mother wavelet function, and these are defined as

$$W_{0,0}^{0}(t)=\varphi(t) \tag{18}$$

$$W_{0,0}^{1}(t)=\psi(t). \tag{19}$$

The wavelet packet functions for n = 2,3,... can be computed as

$$W_{0,0}^{2n}(t)=\sqrt{2}\sum_{k} g(k)\,W_{1,k}^{n}(2t-k) \tag{20}$$

$$W_{0,0}^{2n+1}(t)=\sqrt{2}\sum_{k}h(k)\,W_{1,k}^{n}(2t-k) \quad (21)$$

where $g(k)$ and $h(k)$ are the quadrature mirror filters associated with the predefined scaling and mother wavelet functions. To measure specific time-frequency information of a signal, one simply takes the inner product of the signal with a particular basis function. The wavelet packet decomposition (WPD) involves applying both high pass and low pass filters to a discrete signal, and then recursively to each intermediate signal. The procedure is illustrated in Fig. 11(*b*) up to the second level of resolution. The first level of decomposition of the discrete signal $x[n]$ of length N in the wavelet packet transform (WPT) generates two-frequency sub-bands, which are the approximation coefficients $a_n^1=\left[a_0^1\ a_1^1\ a_2^1 \ldots a_{N-1}^1\right]$ and detail coefficients $d_n^1=\left[d_0^1\ d_1^1\ d_2^1 \ldots d_{N-1}^1\right]$. The second level of decomposition generates four-frequency sub-bands using same set of filters of the first level of resolution. These are defined as

$$aa^2[n]=\sum_{k=0}^{N/2-1} g[k]a^1[2n-k] \quad (22)$$

$$ad^2[n]=\sum_{k=0}^{N/2-1} h[k]a^1[2n-k] \quad (23)$$

$$da^2[n]=\sum_{k=0}^{N/2-1} g[k]d^1[2n-k] \quad (24)$$

$$dd^2[n]=\sum_{k=0}^{N/2-1} h[k]d^1[2n-k] \quad (25)$$

The frequency sub-band aa^2 is defined as second level low frequency approximations of original signal. The frequency sub-band ad^2 is defined as second level low frequency details of original signal. The frequency sub-band da^2 is defined as second level high frequency approximations of original signal. The frequency sub-band dd^2 is defined as second level high frequency details of original signal.

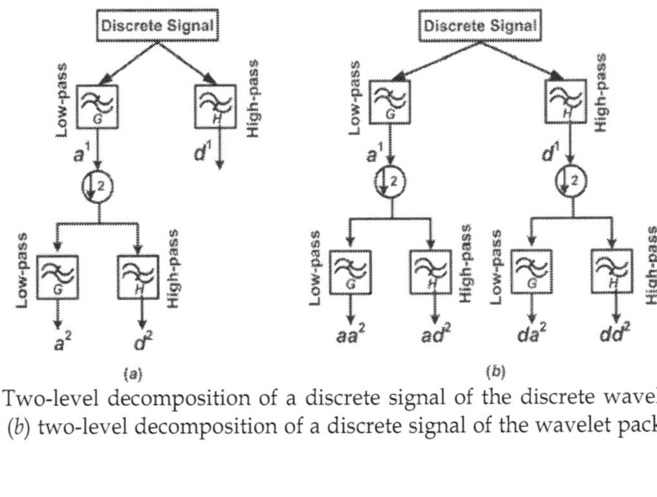

Fig. 11. (*a*) Two-level decomposition of a discrete signal of the discrete wavelet transform (DWT) and (*b*) two-level decomposition of a discrete signal of the wavelet packet transform (WPT).

5. Wavelet Transform Based Faults Diagnostic and Protection

One of the primary tasks of wavelet based faults diagnostic technique for electric machines is to develop experimental setup and to collect stator current of faulted and normal unfaulted conditions. The collected data are to be employed for selecting optimum mother wavelet and optimal levels of resolution, and for off-line testing of the proposed technique. In the data acquisition setup, the current transformers (CTs) are connected in series with motor terminals to collect different faulted and normal unfaulted stator currents. The CTs are rated at 200/5 A (rms) and 15 V (max.). The data acquisition instrument consists of DSP controller board ds1102, which includes a floating-point digital signal processor TMS320C31. The digital data are acquired through on-board three-channel analog-to-digital (A/D) converters. The data are collected at the sampling rate of 8 kHz, and stored in a personal computer through dSPACE TRACE module. Then, these data are converted to ASCII format for further processing. The wavelet based faults diagnostic and protection technique is tested on laboratory prototype electric machines. These include 1hp and 5hp interior permanent magnet (IPM) motors, 1hp squirrel cage induction motor (IM), 1.5hp wound rotor induction motor (IM), and 1.6 kW separately excited synchronous generator. Electric machines may experience different type of faults. The majority of these faults are stator faults such as turn-to-turn fault, which appears as phase-to-phase or phase-to-ground faults later, loss of a phase or field faults, and rotor faults such as static eccentricity, dynamic eccentricity, broken bars, and defects in buried permanent magnets or field windings. Faults such as stator inter-turn, loss of supply (single phasing), line to ground (L-G), and line-to-line (L-L) faults are considered in this work.

5.1 Feature extraction using WPT coefficients

The wavelet packet transform (WPT) is suitable for detection of high frequency components superimposed on the fundamental frequency. In addition, a feature can be extracted by the existence of details or approximations coefficients of a signal at any level of resolution of the wavelet packet tree, and such feature can be used to identify the type of a fault. The collected data of different faulted and unfaulted conditions of an IPM motor are decomposed up to the second level of resolution of the WPT using the selected mother wavelet '*db3*'. The minimum description length data criterion (Hamid et al., 2002) is used for the selection of optimum mother wavelet from a set of orthogonal and non orthogonal wavelet functions. Figures 12–13 show the second level WPT coefficients of normal and fault currents of an inverter-fed 1-hp IPM motor. The second level WPT coefficients of stator current for the case of faulted condition in Fig. 13 are larger than those of unfaulted condition in Fig. 12 at the inception of fault occurrence. Therefore, these feature coefficients can be used for faults diagnostic and protection of IPM motors. A feature vector F is defined using the de noised second level WPT components of stator currents. The feature vector F is defined as (Khan & Rahman, 2009)

$$F = \left[W_{aa^2} \, W_{ad^2} \, W_{da^2} \, W_{dd^2} \right] \tag{26}$$

$$W_{aa^2} = \sqrt{\sum_{n=1}^{N} aa^2(n) / N} \tag{27}$$

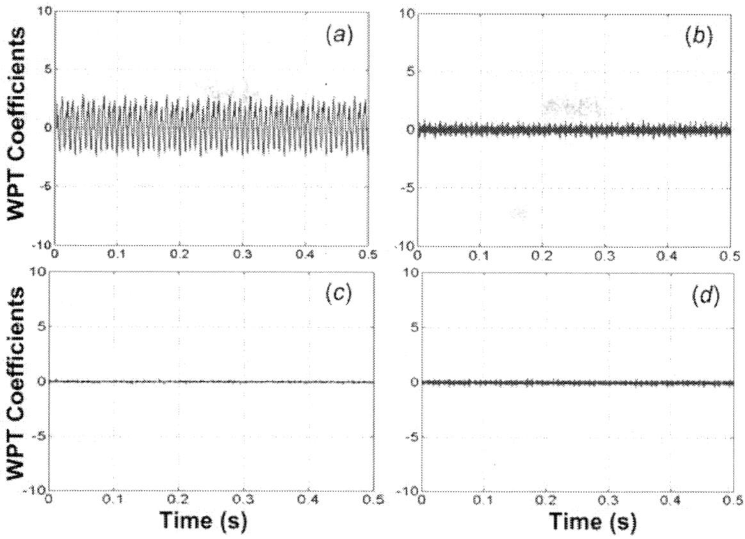

Fig. 12. Second level WPT coefficients of normal current: (*a*) low frequency approximations (aa^2), (*b*) low frequency details (ad^2), (*c*) high frequency approximations (da^2), and (*d*) high frequency details (dd^2).

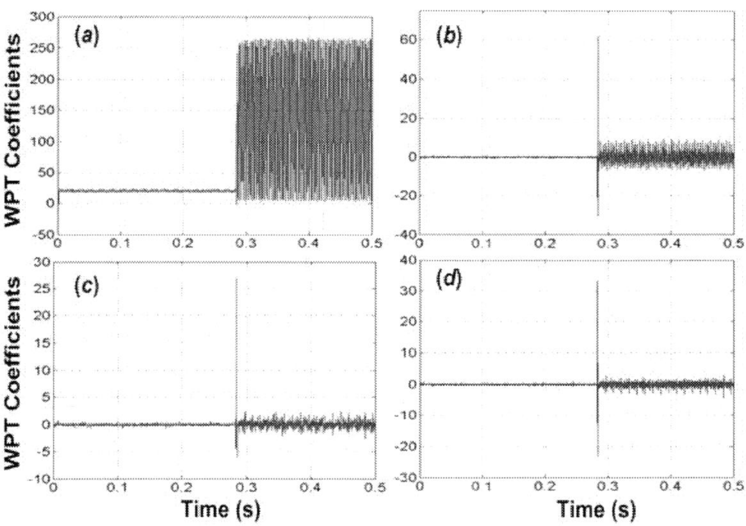

Fig. 13. Second level WPT coefficients of fault current: (*a*) low frequency approximations (aa^2), (*b*) low frequency details (ad^2), (*c*) high frequency approximations (da^2), and (*d*) high frequency details (dd^2).

$$W_{ad^2} = \sqrt{\sum_{n=1}^{N} ad^2(n) / N} \tag{28}$$

$$W_{da^2} = \sqrt{\sum_{n=1}^{N} da^2(n) / N} \tag{29}$$

$$W_{dd^2} = \sqrt{\sum_{n=1}^{N} dd^2(n) / N} \tag{30}$$

where N denotes total number of coefficients in a certain node of wavelet packet tree. Table-I shows the comparisons of feature vector calculated using equations (26)-(30) between faulted and normal conditions of an IPM motor. The feature vectors clearly differentiate faulted conditions from normal conditions and also within faulted conditions.

Type of faults	W_{aa^2}	W_{ad^2}	W_{da^2}	W_{dd^2}
Normal	17.31	0.2757	0.0479	0.1222
Inter-turn	30.33	0.3654	0.1125	0.1854
L-G	34.34	0.4164	0.1278	0.2039
L-L	114.75	0.7929	0.2588	0.3607
Single phasing	21.88	0.7485	0.1036	0.2671

Table 1. Feature vector

5.2 Feature extraction based on signature analysis of WPT coefficients

The signature analysis technique is used for feature extraction of fault currents of the proposed wavelet based faults diagnostic technique for induction motors. The discrete signal of stator current is decomposed up to the second level of resolution of the wavelet packet tree using the selected mother wavelet '*db*3'. The second level high frequency details (dd^2) of stator currents are used to analyse the signatures of various faults in an induction motor as most of the fault current signals contain high frequency components superimposed on the fundamental frequency. The 2nd level high frequency details (dd^2) of stator currents of different faulted and normal unfaulted conditions are given in Figs. 14(*a*)–14(*d*). The details (dd^2) of stator current of Figs. 14(*c*)–14(*d*) showed high density of color strips between the faulted region as compared to those of normal currents (loaded or unloaded) of Figs. 14(*a*)–14(*b*). Therefore, the significant features for faults detection can be extracted based on the density of WPT coefficients (dd^2) of stator currents.

5.3 Wavelet based faults diagnostic algorithm

The new faults diagnostic and protection algorithm is developed by combining the features of wavelet packet transform (WPT) coefficients and neural network (NN) algorithm. A three-layer feed forward neural network with four inputs and one output is used.

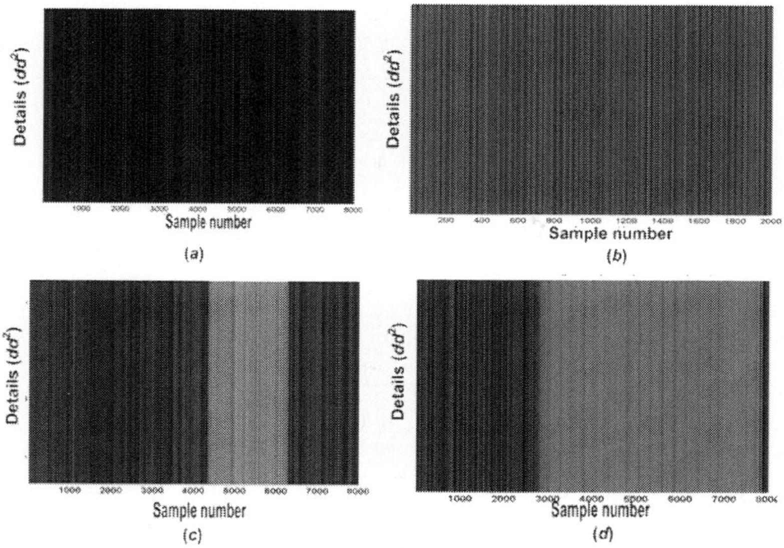

Fig. 14. Second level high frequency details (dd^2) of stator current of the 1hp induction motor: (a) unloaded current, (b) full load current, (c) stator winding phase to ground fault current, and (d) single phasing current.

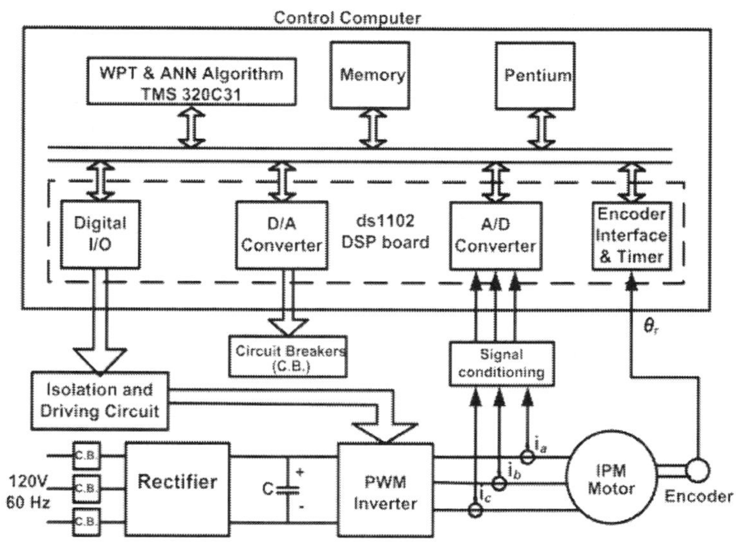

Fig. 15. Experimental setup of the proposed WPT and NN based faults diagnostic and protection algorithm for inverter fed IPM motor using the DSP board ds1102.

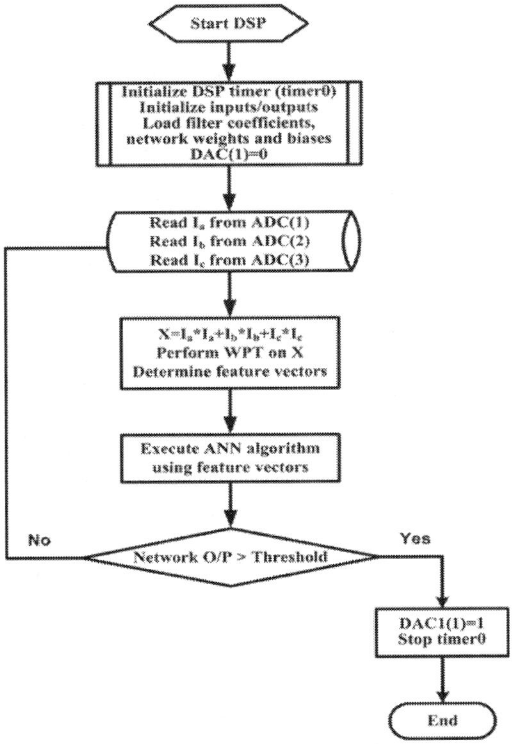

Fig. 16. Flow chart of the new WPT and NN based faults diagnostic and protection algorithm for electric motors.

In the proposed faults diagnostic and protection technique, the inputs of the NN are feature vectors of second level WPT coefficients of faulted and normal currents. The outputs are binary values of 0 or 1 to indicate whether the measured current is a normal current or a fault current, respectively. The proposed wavelet based diagnostic and protection technique for inverter fed interior permanent magnet (IPM) motors using the DSP board ds1102 is shown in Fig. 15. The proposed WPT and NN based faults diagnostic algorithm is written in the turbo C language. The algorithm uses a set of initialization and input/output (I/O) functions in order to initialize TMS320C31's on-chip timers and to access the ds1102's on board A/D and D/A converters. When a timer is started, the A/D converters of the DSP board continuously sample stator currents at the rate of 8 kHz. The samples of stator currents are sent to the memory of the DSP by the host PC, where they are squared and summed into one sample. This sample is placed into a circular buffer of size six. The six current data are processed using the filter coefficients of the mother wavelet '*db*3', and the biases and weights of the NN algorithm.

The procedure to implement the proposed WPT and NN based faults diagnostic algorithm using the DSP board ds1102 is shown in the flow chart of Figure 16. In the proposed technique, samples of three-phase stator currents are squared and summed into one sample at the beginning for minimizing the computational burden. The WPT is applied on squared samples of stator currents, and the feature vectors are calculated from the WPT coefficients. The feature vectors are given as inputs to the neural network. The NN algorithm determines the values of the network output using the trained weights and biases, and checks whether it is greater than the threshold or not in order to generate the trip signals for the circuit breakers.

5.4 Laboratory implementation of wavelet based faults diagnostic algorithm

The proposed wavelet based faults diagnostic technique is tested in real time using the experimental setup of Fig. 15. The experimental responses of the wavelet based faults diagnostic technique for supply and inverter fed IPM motors are shown in Figs. 17–19. Figures 17(*a*) and 17(*b*) show the test results for the case of single phasing of a supply fed IPM motor. Figures 17(*c*) and 17(*d*) contain the test results for the case of L-L fault of a supply fed IPM motor. The proposed faults diagnostic algorithm generated trip signal almost at the instant of fault occurrence without any delay. The experimental responses of the wavelet based faults diagnostic technique for inverter fed IPM motor are shown in Figs. 18(*a*)–18(*d*). It is clear from Figs. 18(*a*)–18(*d*) that for all fault cases disturbances are identified promptly and properly. However, the trip signal is initiated after three cycles of fault occurrence for the case of single phasing of inverter fed IPM motor of Figs. 18(*a*)–18(*b*). In addition, the algorithm generated trip signal after one cycle of fault occurrence for the case of line to line fault of Figs. 18(*c*)–18(*d*). These delays are due to the fact that the response time includes the executions of the proposed protection algorithm, the speed control algorithm, and the vector control algorithm for generation of logic signals of inverter switches. Figure 19(*a*) shows the phase-*a* current and experimental response of no trip signal of the hybrid wavelet packet transform (WPT) and neural network (NN) based faults diagnostic algorithm for step increase and step decrease of command speeds of the inverter fed IPM motor. Figure 19(*b*) shows the phase-*a* current and experimental response of no trip signal of the hybrid diagnostic algorithm for the sudden change of load torque of the inverter fed IPM motor. The hybrid algorithm identified these unfaulted conditions of Figs. 19(*a*)–19(*b*) as normal conditions and did not change the status of trip signal. Thus the proposed WPT and NN based hybrid algorithm correctly and promptly detected faulted and normal currents of both supply fed and inverter fed IPM motors.

The wavelet based faults diagnostic technique is also implemented on a three-phase, Y-connected, 1705 rpm, 1hp squirrel cage induction motor. The proposed algorithm is based on the identification of WPT coefficients of stator currents of different faulted and normal unfaulted conditions. The experimental responses of the faults diagnostic algorithm and three-phase stator currents are shown in Figs. 20(*a*)–20(*d*). Figures 20(*a*)–20(*b*) show the test results for single phasing, and Figures 20(*c*)–20(*d*) show the test results for line to ground fault of supply fed induction motors. The proposed WPT based faults diagnostic algorithm correctly and promptly classified faulted and normal currents of induction motor.

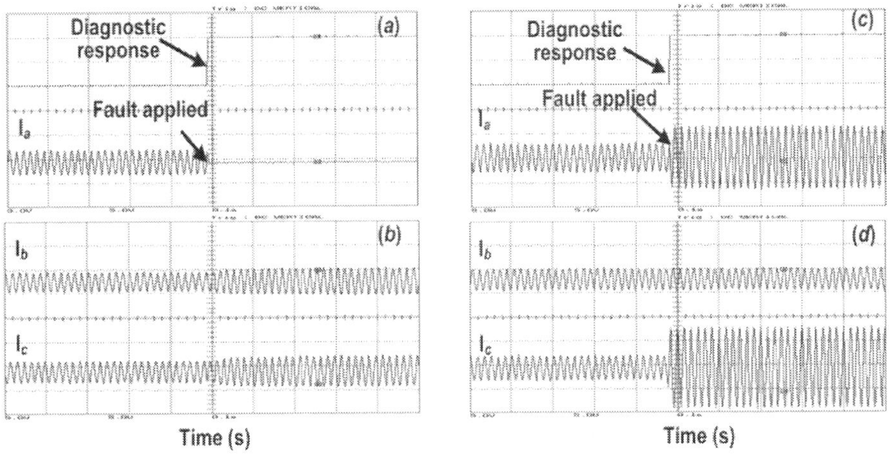

Fig. 17. Experimental responses of the WPT and NN based faults diagnostic technique for supply fed IPM motor: (*a*)-(*b*) single phasing condition and (*c*)-(*d*) line to line fault condition. (time: 0.1 s/div., trip signal: 5 V/div., I_a: 4.172 A/div., I_b: 4.66 A/div., and I_c: 4.82 A/div.)

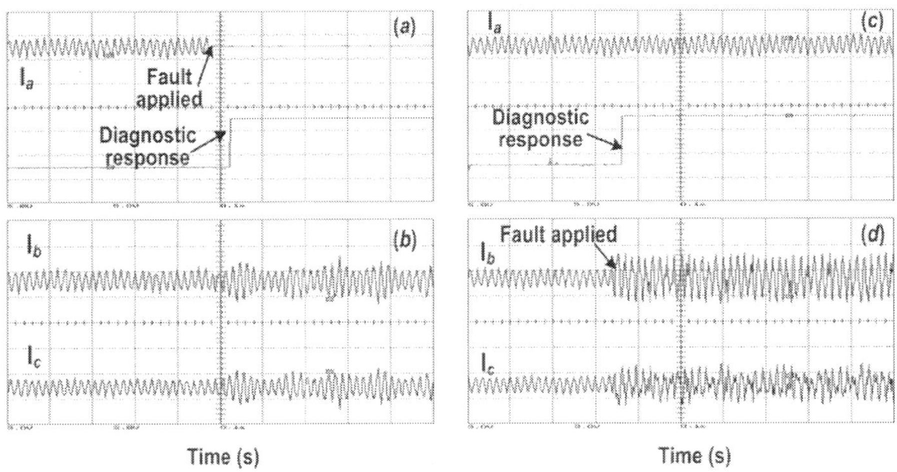

Fig. 18. Experimental responses of the WPT and NN based faults diagnostic technique for inverter fed IPM motor: (*a*)-(*b*) single phasing condition and (*c*)-(*d*) line to line fault condition. (time: 0.1 s/div., trip signal: 5 V/div., I_a: 4.172 A/div., I_b: 4.66 A/div., and I_c: 4.82 A/div.)

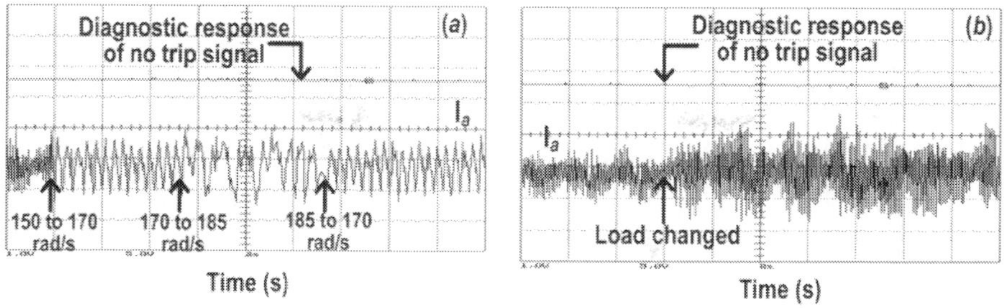

Fig. 19. Experimental responses of the WPT and NN based faults diagnostic technique for inverter fed IPM motor: (*a*) step changes of speed condition and (*b*) change of load torque condition. (time: 2 s/div., trip signal: 1 V/div., I_a: 4.172 A/div., I_b: 4.66 A/div., and I_c: 4.82 A/div.)

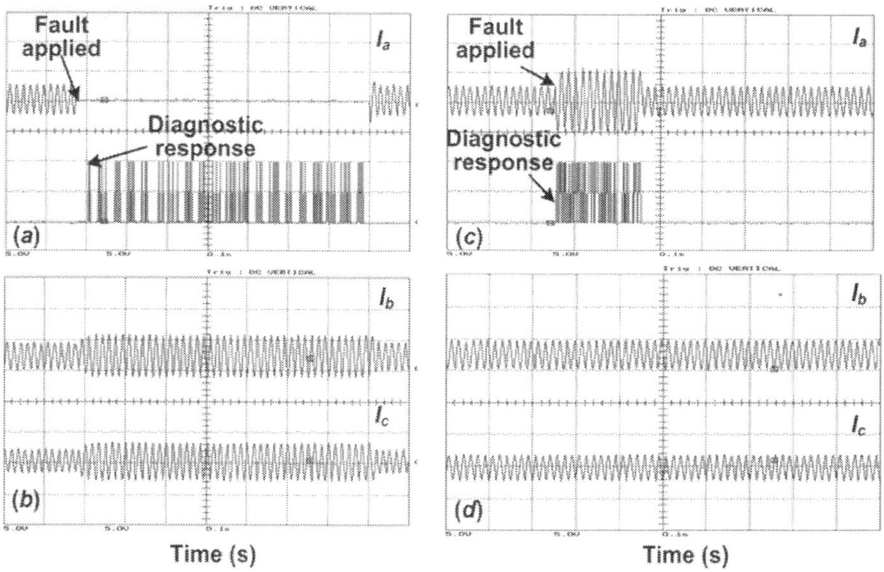

Fig. 20. Experimental responses of the WPT based faults diagnostic technique for supply fed induction motor: (*a*)-(*b*) single phasing condition and (*c*)-(*d*) line to ground fault condition. (time: 0.1 s/div., trip signal: 5 V/div., I_a: 4.172 A/div., I_b: 4.66 A/div., and I_c: 4.82 A/div.)

6. Conclusions and Remarks

In this chapter, a short review of conventional Fourier transforms and new wavelet based faults diagnostic and protection techniques for electric motors is presented. The new hybrid wavelet packet transform (WPT) and neural network (NN) based faults diagnostic algorithm is developed and implemented for electric motors. The proposed WPT and NN faults diagnostic algorithm based protection technique is implemented in real time using the DSP board ds1102 for both supply fed and inverter fed IPM motors. In the proposed faults diagnostic technique, the WPT feature coefficients of stator currents are used as inputs to a two-layer feed forward neural network. The WPT based faults diagnostic algorithm is developed and implemented for a squirrel cage induction motor. The performances of both hybrid and WPT based diagnostic algorithm are found satisfactory. The proposed techniques do not require any harmonic contents analysis, and these are independent of motor equivalent circuit model parameters. The wavelet based technique is quite fast and easy to implement. It also requires less computational memory for the on-line implementation.

7. References

Addison, P. S. (2002). *The Illustrated Wavelet Transform Handbook,* Institute of Physics Publishing, Bristol, UK.

Chow, T. W. S. & Hai, S. (2004). Induction machine fault diagnostic analysis with wavelet technique, *IEEE Transactions on Industrial Electronics,* Vol. 51, No. 3, pp. 558-565.

Chow, M.; Mangum, P. M. & Yee, S. O. (1991). A neural network approach to real-time condition monitoring of induction motors, *IEEE Transactions on Industrial Electronics,* Vol. 38, No. 6, pp. 448-453.

Douglas, H.; Pillay, P. & Ziarani, A. K. (2004). A new algorithm for transient motor current signature analysis using wavelets, *IEEE Transactions on Industry Applications,* Vol. 40, No. 5, pp. 1361-1368.

Dalpiaz, G. & Rivola, A. (1997). Condition monitoring and diagnostics in automatic machines: comparison of vibration analysis techniques, *Journal on Mechanical Systems and Signal Processing,* Vol. 11, No. 1, pp. 53-73.

Filippetti, F.; Franceschini, G.; Martelli, M. & Tassoni, C. (1988). An approach to a knowledge representation about induction machine diagnostics in expert systems, *Proceedings Conference Record of International Conference on Electric Machines (ICEM),* Vol. 3, pp. 289-296, Pisa, Italy, September 1988.

Hamid, E. Y. & Kawasaki, Z. I. (2002). Wavelet based data compression of power system disturbances using the minimum description length data, *IEEE Transactions on Power Delivery,* Vol. 17, No. 2, pp. 460-466.

Khan, M. A. S. K. & Rahman, M. A. (2009). Development and implementation of a novel fault diagnostic and protection technique for IPM motor drives, *IEEE Transactions on Industrial Electronics,* Vol. 56, No. 1, pp. 85-92.

Khan, M. A. S. K. & Rahman, M. A. (2008). Diagnostic and protection of inverter faults in IPM motor drives using wavelet transform, *Proceedings of IEEE International Conference on Electrical and Computer Engineering (ICECE),* pp. 176-180, Dhaka, Bangladesh, 20-22 December 2008. (a)

Khan, M. A. S. K. & Rahman, M. A. (2008). A new wavelet based diagnosis and protection of faults in induction motor drives, *Proceedings of IEEE International Power Electronics Specialist Conference (PESC),* pp. 1536-1541, Rhodes Island, Greece, 15-19 June 2008. (b)

Khan, M. A. S. K.; Ozgonenel, O. & Rahman, M. A. (2007). Wavelet transform based protection of stator faults in synchronous generators, *Journal of Electric Power Components and Systems*, Vol. 35, No. 6, pp. 625–637. (a)

Khan, M. A. S. K.; Radwan, T. S. & Rahman, M. A. (2007). Real-time implementation of wavelet packet transform based diagnosis and protection of three-phase induction motors, *IEEE Transactions on Energy Conversion*, Vol. 22, No. 3, pp. 647–655. (b)

Khan, M. A. S. K. & Rahman, M. A. (2007). Neuro-WPT based diagnosis and protection of three-phase IPM motor, *Proceedings of IEEE Power Engineering Society General Meeting*, pp. 1–8, Tampa, FL, 24-28 June 2007. (c)

Khan, M. A. S. K.; Ozgonenel, O. & Rahman, M. A. (2007) Diagnosis and protection of stator faults in synchronous generators using wavelet transform, *Proceedings of IEEE International Electric Machines and Drives Conference (IEMDC)*, pp. 184–189, Antalya, Turkey, 3–5 May 2007. (d)

Khan, M. A. S. K. (2006). *A wavelet packet transform based on-line technique for the protection of three-phase interior permanent magnet motors,* Masters' of Engineering Thesis, Memorial University of Newfoundland, St. John's, NL, Canada.

Kim, K. & Parlos, A. G. (2002). Induction motor fault diagnosis based on neuro-predictors and wavelet signal processing, *IEEE/ASME Transactions on Mechatronics*, Vol. 7, No. 2, pp. 201–219.

Luo, G. Y.; Osypiw, D. & Irle, M. (2003). On-line vibration analysis with fast continuous wavelet algorithm for condition monitoring of bearing, *Journal of Vibration and Control,* Vol. 9, No. 8, pp. 931–947.

Lasurt, I.; Stronach, A. F. & Penman, J. (2000). A fuzzy logic approach to the interpretation of higher order spectra applied to fault diagnosis in electrical machines, *Proceedings Conference Record of North American Fuzzy Information Processing Society (NAFIPS) Annual Meeting,* pp. 158–162, Atlanta, GA, 13-15 July 2000.

Leith, D.; Deans, N. D. & Stewart, L. I. D. (1988). Condition monitoring of electrical machine using real-time expert systems, *Proceedings Conference Record of International Conference on Electric Machines (ICEM),* vol. 3, pp. 297–302, Pisa, Italy, September 1988.

Mathworks. (2007). *Matlab: Neural Network and Signal Processing Tool Box,* Natick, MA.

Pöyhönen, S.; Negrea, M.; Arkkio, A.; Hyötyniemi, H. & Koivo, H. (2002). Fault diagnostics of an electrical machine with multiple support vector classifiers, *Proceedings of IEEE International Symposium on Intelligent Control (ISIC)*, vol. 1, pp. 373–378, Vancouver, BC, October, 2002.

Park, J.; Kim, D.; Kim, S.; Lee, D. & Chum, M. (2004). C-ANFIS based fault diagnosis for voltage-fed PWM motor drive systems, *Proceedings Conference Record of North American Fuzzy Information Processing Society (NAFIPS) Annual Meeting,* vol. 1, pp. 379–383, Banff, AB, 27–30 June 2004.

Rahman, M. A. (1980). Permanent magnet synchronous motors – a review of the state of design art, *Proceedings of International Conference on Electric Machines (ICEM)*, vol. 1, pp. 312–319, Athens, Greece, 15–17 September 1980.

Roux, W. L.; Harley, R. G. & Habetler, T. G. (2003). Detecting rotor faults in permanent magnet synchronous machines, *Proceedings Conference Record of IEEE Symposium for Diagnostics of Electrical Machines, Power Electronics and Drives (SDEMPED)*, pp. 198–203, Atlanta, GA, 24–26 August 2003.

Slemon, G. R. (1992). *Electric Machines and Drives*, Addison-Wesley, Boston, MA.

Schoen, R. R.; Habetler, T. G.; Kamran, F. & Bartheld, R. G. (1995). Motor bearing damage detection using stator current monitoring, *IEEE Transactions on Industry Applications*, Vol. 31, No. 6, pp. 1274-1279.

Tavner, P.; Ran, L.; Penman, J. & Sedding, H. (2008). *Condition Monitoring of Rotating Electrical Machines, Institute of Engineering and Technology*, ISBN 978-0-86341-739-9, London, UK.

Toliyat, H.; Abbaszadeh, K.; Rahimian, M. & Olson, L. (2003). Rail defect diagnosis using wavelet packet decomposition, *IEEE Transactions on Industry Applications*, Vol. 39, No. 5, pp. 1454–1461.

Vas, P. (1996). *Parameter Estimation, Condition Monitoring, and Diagnosis of Electrical Machines*, Calrendon, Oxford, UK.

Yang, D. M.; Stronach, A. F. & Macconnell, P. (2003). The application of advanced signal processing techniques to induction motor bearing condition diagnosis, *International Journal on Theoretical and Applied Mechanics*, Vol. 38, No. 2, pp. 297–308.

Yen, G. G. & Lin, K. -C. (2000). Wavelet packet feature extraction for vibration monitoring, *IEEE Transactions on Industrial Electronics*, Vol. 47, No. 3, pp. 650–667.

Yazici, B. & Kliman, G. (1999). An adaptive statistical time-frequency method for detection of broken bars and bearing faults in motors using stator current, *IEEE Transactions on Industry Applications*, Vol. 35, No. 2, pp. 442-452.

Zanardelli, W. G.; Strangas, E. G. & Aviyente, S. (2007). Identification of intermittent electrical and mechanical faults in permanent magnet AC drives based on time-frequency analysis, *IEEE Transactions on Industry Applications*, Vol. 43, No. 4, pp. 971–980.

Zanardelli, W. G.; Strangas, E. G. & Aviyente, S. (2005). Failure prognosis for permanent magnet ac drives based on wavelet analysis, *Proceedings of IEEE International Electric Machines and Drives Conference (IEMDC)*, pp. 64–70, San Antonio, TX, 15-18 May 2005.

Zanardelli, W. G.; Strangas, E. G.; Khalil, H. K. & Miller, J. M. (2002). Comparison of wavelet based methods for the prognosis of failures in electric motors, *Proceedings Conference Record of IEEE Workshop on Power Electronics in Transportation*, pp. 61–67, Auburn Hills, Michigan, 24–25 October 2002.

Zhengjia, H.; Jiyuan, Z.; Yibin, H. & Qingfeng, M. (1996). Wavelet transform and multi-resolution signal decomposition for machinery monitoring and diagnosis, *Proceedings Conference Record of IEEE Conference on Industrial Technology (ICIT)*, pp. 724–727, Shanghai, China, 2-6 December 1996.

13

Fault detection and diagnosis for in-vehicle networks

Jittiwut Suwatthikul, PhD.
National Electronics and Computer Technology Center (NECTEC)
Thailand

1. Introduction

Automotive vehicles are nowadays equipped with a significant number of networked electronic systems by which advanced vehicle control, elimination of bulky wiring, and sophisticated features can be achieved. Most of the features are enabled by the use of distributed electronic systems including sensors, switches, actuators and electronic control units (ECUs). In today's premium automobiles, there can be fifty or more individual ECUs communicating over multiplexed data networks such as Controller Area Network (CAN), Local Interconnect Network (LIN), FlexRay for X-by-wire applications (Kopetz & Bauer, 2003; Leen et al., 1999; Leen & Heffernan, 2002; Shrinath & Emadi, 2004).

However, as more features and ECUs are introduced, overall system complexity increases, in turn raising the likelihood for unpredictable or emergent behaviour that could not have been anticipated during ever shrinking vehicle development cycles. These cycles reduced from 48 months in 1985 to 24 months in 2005, and are expected to be 12 months in 2010 (Ortega et al., 2006). The consequences of the unpredictable behaviour or implementation errors would discourage brand loyalty and bring a manufacturer into disrepute. In addition, rising feature levels have resulted in the embedded software and electronic components becoming an increasing proportion of the total value of the vehicle. The average cost for in-vehicle electronic content increased from 2% of the total car price in 1974 to 23% in 2004, and is forecasted to reach 40% by 2010 (Ortega et al., 2006).

Against this background, vehicle manufacturers are striving to reduce costs and at the same time to improve levels of customer satisfaction. Work to improve test and validation of large distributed electronic systems has been ongoing for years (Athanasas & Dear, 2004; Ehret, 2003; Simonot-Lion, 2003). This has provided manufacturers with approaches to test and validation, with some degree of coverage. It is, however, still impossible to use an ideal test scheme that provides complete input test coverage to perform exhaustive testing and validation because of the large number of possible system states (Storey, 1997), possibly resulting in vehicles not working properly due to some design flaws and errors. This is compounded by general wear out of mechanical, electrical and electronic components. To date, on-board diagnostic systems (OBD) have come into play to cope with faults when vehicles are used by customers. OBD are integrated in ECUs to detect and diagnose vehicle faults such that diagnostic trouble codes (DTCs) relevant to the faults are set and logged in the ECUs' memory for later off-board, return-to-dealership-based fault analysis and rectification.

Despite the available OBD, diagnostic techniques have largely been focused on individual or defined areas of a vehicle, e.g. engine management, brakes and steering. In a vehicle where only a few ECUs and communication messages are deployed, traditional off-board diagnostics can be adequate. As system complexity continues to increase, off-board diagnostic approaches have become more costly and sometimes ineffective, resulting in high levels of "no-fault-found", incorrect component replacement and increased warranty costs. Recent years have seen research work on a paradigm shift from off-board dealership-based diagnosis and repair to on-board remotely-assisted diagnosis and in-vehicle repair (Amor-Segan et al., 2007). It is anticipated that the new on-board vehicle diagnosis scheme will improve customer expectations and satisfaction for vehicle reliability.

This chapter concerns fault detection and diagnosis (FDD) techniques applied to automotive electronic systems, especially focusing on faults in in-vehicle networks. FDD on a CAN network is demonstrated. Readers will be provided with knowledge on how vehicle faults are generally managed, and the trend of intelligent FDD in future vehicles. The rest of this chapter is divided into three main sections: (i) a brief introduction to in-vehicle electronic systems; (ii) FDD for in-vehicle electronic systems including those in component, feature and network levels; (iii) recent research on FDD techniques for in-vehicle networks.

2. In-vehicle electronic systems

Over the past two decades, the rapid growth in performance and reliability of electronic embedded systems has enabled vehicle manufacturers to implement complicated automotive control systems through the use of sophisticatedly integrated mechanical and electronic devices, so called mechatronics (Isermann, 2008). Vehicles are equipped with a variety of electronic devices performing different functions, and mostly transferring signals via electrical wiring.

2.1 In-vehicle electronics

State-of-the-art control algorithms and rapidly-improved semiconductor technology make it possible for modern vehicles to meet the demands for driver-assisted functions, safety, comfort and environment protection. A number of sensors are used to measure controlled variables as input signals for ECUs, e.g. engine speed, temperature. The input signals can be analogue, e.g. voltage signals from sensors, digital such as switch positions, or modulated e.g. Pulse Width Modulation (PWM) signals. With these input signals, ECUs calculate required parameters to adjust controlling devices such as actuators. Improved performance and additional functions are obtained by synchronising processes controlled by individual control units and by adapting their respective parameters to each other in real-time. An example of this type of function is a traction control system (TCS) which reduces the driving torque when the drive wheels spin (Robert Bosch GmbH, 2004a).

In 1902, a magneto-ignition system, the first on-board electrical system, was installed in a vehicle. It consisted of the magneto itself, an ignition distributor, ignition coils, spark plugs and cables. More than a decade later, Bosch had the first complete automotive electrical system ready for installation. The system comprised the magneto-ignition system with spark plugs, a starter, a generator, headlamps, a battery, and a regulator switch. This was the starting point of a progress towards a genuine on-board electrical and electronics system (Robert Bosch GmbH, 2004).

Today's in-vehicle electronic systems have become extensive and much more complicated. A large number of electronic devices and software contents are integrated in modern vehicles. Applications range from simple door or window control to remotely wireless data transfer between vehicles or between a vehicle and the infrastructure. Fig. 1 illustrates the exponentially soaring complexity of today's in-vehicle electronics.

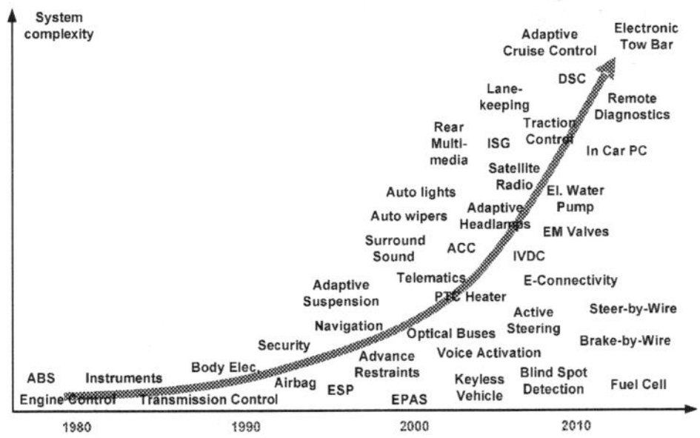

Fig. 1. Rapid increase in in-vehicle electronic systems (adapted from McMurran et al., 2006).

In-vehicle electronics can be broadly classified into four main categories corresponding to different functionalities, constraints and models: "Powertrain", "Chassis", "Body", and "Infotainment" including telematics, multimedia, various types of information and entertainment (Zurawski, 2006).

"Powertrain" refers to components that generate power and deliver it to the road surface. This includes the engine, transmission, clutches, driveshaft, etc. "Powertrain" sometimes simply refers to the engine and transmission, and the other components in the transmission. "Chassis" involves the systems that control the interaction of the vehicle with the road and the chassis. It is related to steering, braking and ride quality. This category includes systems such as ABS (Anti-lock Braking System), ESP (Electronic Stability Program) and the new forthcoming technology so-called X-by-Wire. "Body" refers to the electronic systems not directly involved in the movement of the vehicle. These are, for instance, the systems that control doors, windows, seats, boots, etc. "Infotainment" provides drivers with information and entertainment through HMI (Human Machine Interface). This enables drivers to see or exchange not only in-vehicle information and entertainment such as vehicle conditions and movies from an in-car video player, but also information from remote sources via telematics (Zurawski, 2006).

2.2 In-vehicle networks

Historically, the communication between simple electronic devices was mostly achieved by using point-to-point links, as shown in Fig. 2 a). Signals in the vehicle were transmitted and received among ECUs over non-multiplexed and hard-wired cables. This resulted in bulky, expensive and complicated wiring when dealing with the increasing use of ECUs, because the number of required communication channels grows exponentially with the number of ECUs (Navet et al., 2005). The attempt to eliminate wiring difficulties and to improve automotive distributed control systems became a challenge for automotive manufacturers. A wiring harness of a middle-class vehicle was roughly 1 mile long and included approximately 300 connectors with 2,000 pins (Robert Bosch GmbH, 2004a). The early 1980s saw the emerging solutions for vehicle networking, many of which were based on simple data transfer e.g. point-to-point or master-slave UART (Universal Asynchronous Receiver/Transmitter) (Leen et al., 1999).

As requirements in vehicle control grew quickly, and were mostly based on real-time control strategies, such simple network configurations and protocols became unwieldy. In the mid 1980s, Robert Bosch GmbH invented a robust automotive control network known as CAN. CAN is based on a bus configuration, as shown in Fig. 2 b), which allows ECUs connected on the bus to receive in-vehicle signals digitally encoded in CAN messages at almost the same time. This significantly enhances real-time applications in the vehicle.

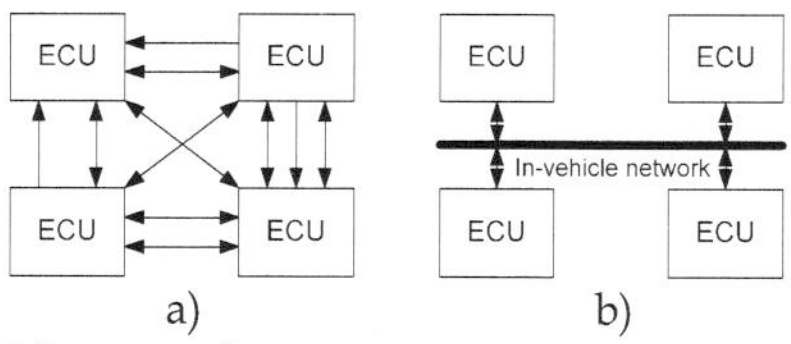

Fig. 2. Evolution of in-vehicle networks.

Whilst CAN is concerned in this chapter, more recently, there have also been several other communication protocols and network topologies used in different categories, depending on requirements and cost constraints of applications. These are, for example, Local Interconnect Network (LIN) for Body, Time-Triggered CAN (TTCAN) for Powertrain, FlexRay for X-by-wire applications, MOST (Media Oriented System Transport) for infotainment, and the short-range wireless communication Bluetooth. Fig. 3 shows a cost and speed comparison of different networks.

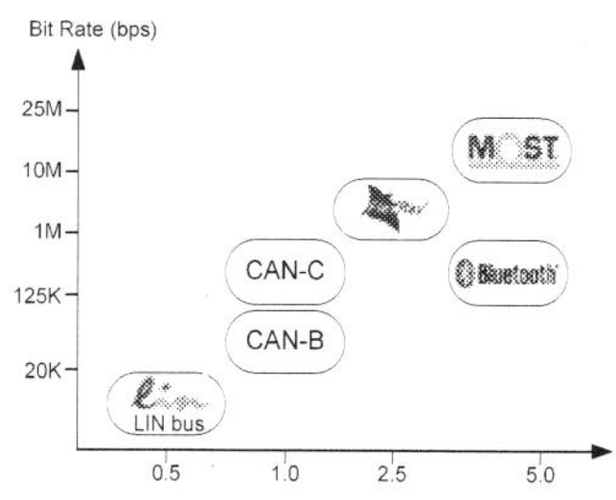

Fig. 3. Speed and cost comparison of the well-known in-vehicle networks.

CAN is currently one of the serial communication protocols widely used in automotive and industrial automation applications. Fig. 4 illustrates a CAN bus topology in a typical automotive application where a number of ECUs can be connected on a high-speed CAN (CAN-HS) and a low-speed CAN (CAN-LS) buses which are connected via a gateway.

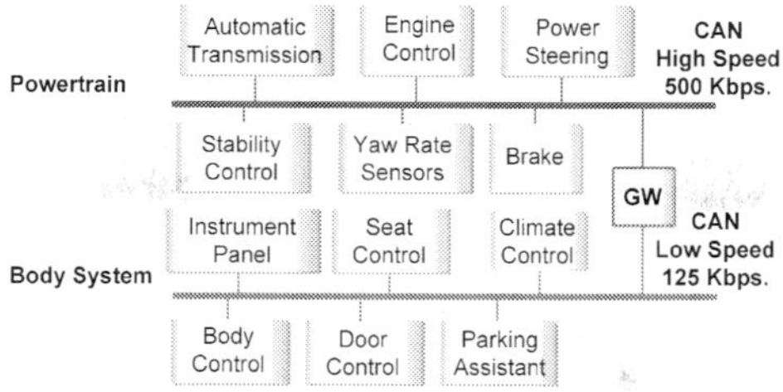

Fig. 4. CAN in automotive application.

A CAN communication controller is used in each ECU to control message transmission between ECUs connected on the same bus. CAN uses an arbitration feature, illustrated in Fig. 5, to control bus access in order to avoid transmission collision which causes communication errors. Messages which are assigned lower message identifications (IDs) have higher priorities to access the bus. For instance, in Fig. 5, node B loses the arbitration to node A because the third bit of node B (logic 1, so called "recessive") is replaced by the third bit of node A (logic 0, so called "dominant").

If errors occur during communication, an error management feature in the communication controller detects, handles and confines such errors. An error frame is transmitted to notify ECUs that errors have been detected so that the ECUs ignore the message recently present on the bus. The errors involve Bit Error, Bit Stuffing Error, CRC Error, Form Error and Acknowledgement Error. Effect of the errors is also limited by CAN controllers to prevent further communication failure. More details on CAN and these errors can be found in Lawrenz, 1997.

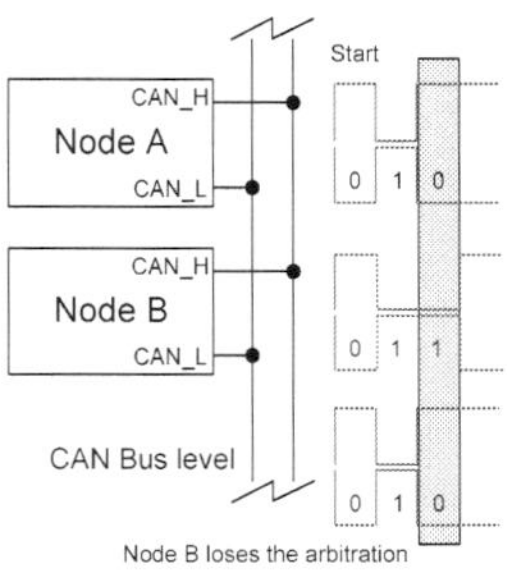

Fig. 5. CAN arbitration.

In case of communication loss, network-relevant DTCs are set and logged in the ECUs that expect the lost messages. Applications of CAN embrace not only vehicle control but also diagnostic services. Despite the common diagnosis standards such as ISO 9141 K-line, ISO 14230 KWP2000 (Keyword Protocol 2000), ISO 15765, etc., the trend of diagnostic services is towards diagnostics over CAN such as the standard defined in ISO 14229.

3. Fault detection and diagnosis for in-vehicle electronics

As mentioned earlier, the rapid growth in hardware and software content in today's vehicles results in increased overall system complexity. Fig. 6 illustrates the evolution of vehicle diagnostics together with the system complexity trend. Vehicle diagnostics with only conventional instruments, which has been used for simple measurement of normal or faulty electrical signals, is no longer effective. Enhanced in-vehicle diagnostic methods have been introduced in order that root causes of faults occurring during vehicle operation are efficiently detected and identified. For instance, OBD, which was initially intended for emission monitoring, is now able to provide logged DTCs and signal measurement for off-board diagnostics of other non-emission related functions.

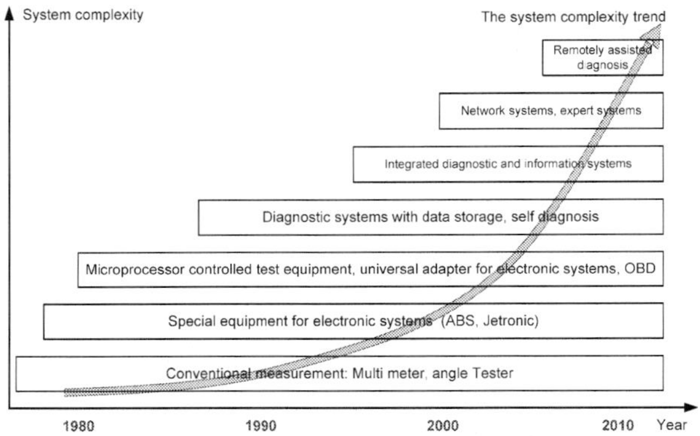

Fig. 6. Evolution of diagnostic test equipment (adapted from Robert Bosch GmbH, 2004b).

In-vehicle electronic and software control systems can operate on a number of different levels: (i) component level—individual ECUs, sensors, actuators and components; (ii) functional or feature level—such as braking, cruise control, stability control; (iii) network level—such as system-wide coordination and configuration, network wake-up and sleep, inter-network gateways and the network itself. As illustrated in Fig. 7, a diagnostic application in each level monitors different parameters. For instance, the diagnostic application in the component level monitors input and output signals and the battery voltage. ECU's internal operation and communication are monitored by the ones in the feature and the network levels respectively. Once faults are detected, tested and confirmed, some default or alternative signal values are used by the ECU in a "limp-home" mode where the vehicle can be driven to the nearest service centre. The diagnostic processor then provides a warning to the driver as well as logging DTCs and additional environment parameters, e.g. speed, temperature and timestamp. It should be noted that there are currently no actual processors in deployment. The diagnostic processor illustrated in Fig. 7 is a sub-function in each ECU.

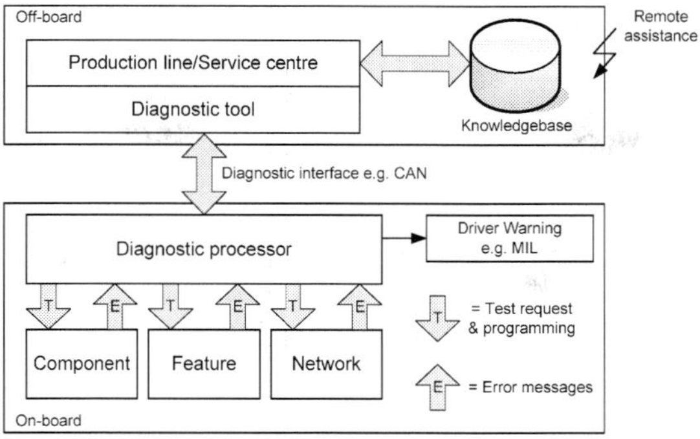

Fig. 7. Diagnostic scheme for in-vehicle electronic systems.

A flow chart of vehicle diagnostics is shown in Fig. 8. The logged DTCs are accessed off-board by a diagnostic tool at a service centre, and can be appropriately deleted once the faults are found and rectified. Fig. 9 illustrates an example of an off-board diagnostic tool used to read DTCs. If any further information about the vehicle repair is needed, the tool can be connected to another service centre or the manufacturer's central diagnostic knowledge base in order to request remote assistance.

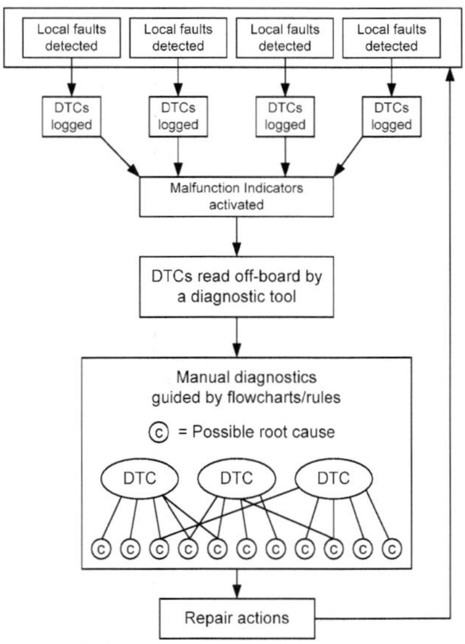

Fig. 8. Flow chart of conventional diagnostics.

Fig. 9. Diagnostic tool for reading DTCs.

Since this chapter is intended to focus on FDD for in-vehicle networks, this section next discusses FDD for the component and feature levels briefly, and then puts emphasis on the network level.

3.1 Component and feature levels

In the component and feature levels, FDD applications employ different methods for detecting functional software and physical hardware faults. Detection approaches range from simple threshold and out-of-range comparisons to learning algorithms for non-linear signal processing.

Input and output signals from sensors, connecting lines (signal paths), and actuators are checked. Short-circuits to steady-state voltage e.g. battery voltage and the device ground, and line interruptions are monitored. Measurements from sensors are examined whether they are in permissible ranges. If additional information is available, a plausibility check is performed by cross-checking two signals e.g. comparison of crankshaft and camshaft speeds. Internal ECU's operations are checked to ensure that the ECU works as specified. These checks are performed immediately after the vehicle being switched on, when the ECU operating or controlling the vehicle, and after the ECU finishing its operation. After faults are detected, some alternative values are used so that the ECU can continue its operation. Additionally, limp-home measures, e.g. limitation of engine speed, may be initiated to maintain driving safety, prevent consequence damage, and minimise emissions (Robert Bosch GmbH, 2004c).

Recent research work has proposed new methods for diagnosing root causes of faults in distributed electronic systems in vehicles. For example, firstly, Murphey et al., 2004 proposed the distributed diagnostic agent system (DDAS), a novel diagnostic architecture developed for automotive fault diagnosis. The DDAS, implemented for fault detection of ECU signals, consists of a vehicle diagnostic agent (VDA) and a number of signal diagnostic agents (SDAs), each of which is responsible for diagnosing one particular automotive signal using either a single or multiple signals. A novel fuzzy learning algorithm was implemented to learn to classify good and bad signal segments by supervised learning from good examples only in the training data set. This work used a case-based reasoning (CBR) approach in the VDA agent to find the cause of vehicle faults, by which effective diagnostic results have been obtained. Secondly, an intelligent diagnostic system (IDS) was developed for diagnosis of component faults in a multi-ECU environment by using model-based reasoning (MBR), and information taken from failure modes and effects analysis worksheet

(FMEA) (Foran & Jackman, 2005). This work tried to isolate a core faulty component instead of simply returning a series of faults. The system was tested by a particular list of test cases and was capable of identifying all scenarios in the list with 100% success rate.

3.2 Network level

A number of research experiments on network design, verification and test to prevent network level faults and to improve reliability have been reported (Tindell & Hansson, 1995; Temple, 1998; Navet & Song, 1999; Navet & Song, 2001; Gaujal & Navet, 2005; Buja et al., 2007). Hardware-in-the-loop facilities are playing an important role in the testing of networked electronic systems (Isermann et al., 1999; Kendall & Jones, 1999; Short & Pont, 2008). Methods more specific to network level testing have been discussed (Armengaud et al., 2004; Armengaud et al., 2005a; Armengaud et al., 2005b). This has provided device suppliers and manufacturers with testing approaches with some degree of test coverage and system validation. Despite this, it is inevitable that unexpected faults can still occur due to untested conditions or worn components.

3.2.1 Existing Network Management

It is well understood that generally automotive manufacturers do not develop ECUs themselves; rather they mostly outsource this task to different suppliers to do so in accordance with the manufacturers' specifications and existing standards. ECUs from different suppliers are then integrated and networked into vehicles by the manufacturers. One such well-known standard related to in-vehicle networks is Open System and Corresponding Interfaces for Automative Electronics Network Management (OSEK NM).
Due to the increasing number of ECUs deployed in vehicles, the Open Systems and the Corresponding Interfaces for Automotive Electronics (OSEX/VDX) working group has defined a standard for communication of ECUs in automotive applications, called Network Management or NM (OSEK/VDX, 2004), illustrated in Fig. 10. NM resides in the environment where an Operating System (OS) controls system interactions among different software layers. NM consists of algorithms defined by OSEK and protocol specific algorithms. NM can communicate with the application software in the application layer, and vice versa, by using an Application Program Interface (API) in order to enable (or disable) OSEK algorithms.

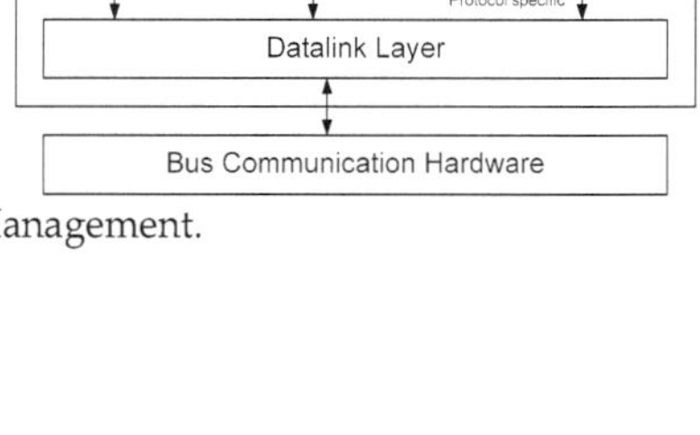

Fig. 10. OSEK Network Management.

The protocol specific algorithms are responsible for handling data transfer on the network by communicating with the datalink layer. OSEK NM is now a published ISO standard: ISO 17356-5: 2006. It provides standardised features which ensure the functionality of inter-networking by using standardised interfaces. NM implementations are incorporated in all networked nodes, e.g. ECUs. This means that a solution for NM can be implemented throughout the varieties of available ECU hardware. The status of the network is recorded and evaluated in all ECUs and thus each node features a determined behaviour regarding the network and the application concerned. NM supports diagnostic applications by providing acquired network status.

NM features two mechanisms for network status and fault monitoring—direct network management (DNM) by using dedicated messages with a token principle and indirect network management (IDNM) by monitoring application messages. The use of these mechanisms is dependent on a particular system. Processing of information collected by these mechanisms must be in accordance with requirements of the entire networked system.

Dedicated NM messages are used in DNM: Ring, Alive and Limp Home messages. Each node is monitored by other nodes on the network. Nodes transmit and receive NM messages via a logical ring in which a communication sequence for synchronisation is defined. Each node on the network has a unique ID normally available from 1 to 255. To set up a logical ring, a Ring message is sequentially passed from the lowest to the highest ID node which then passes the Ring back to the lowest ID node, as illustrated in Fig. 11. DNM also requires a broadcast type of network implementation so that every node can hear the messages that are being sent (Lemieux, 2001).

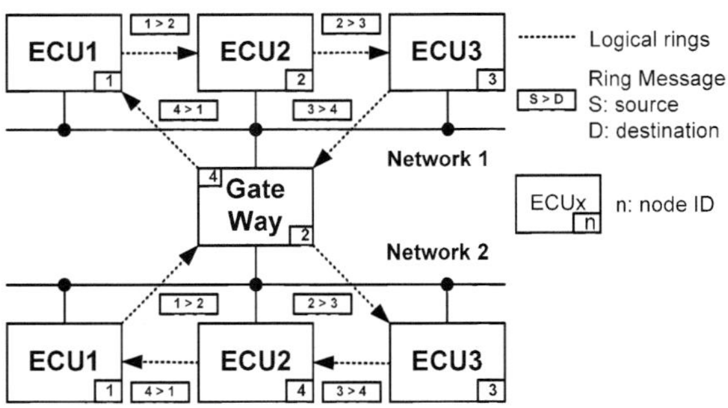

Fig. 11. DNM logical ring.

An Alive message identifies a new node present on the network and puts the system into a transient state. When the new node has identified all nodes on the network, the system changes into a stable state where the NM component is fully aware of the status of all nodes on the network. A faulty node transmits a Limp Home message cyclically until: (i) it is able to receive NM messages from other nodes correctly; or (ii) NM component stops; or (iii) the bus goes to sleep mode. Then the node enters a reset state and performs an NM initialisation. Other nodes that have received a Limp Home message update their configurations to identify the malfunction node being absent from the bus.

Although DNM provides useful information on network and node status, it requires a somewhat amount of computational resources. Thus, DNM may not be suited for some systems in which simple software algorithms and computational resource consumption are of concern. IDNM is therefore defined as an alternative mechanism for network management.

For IDNM, instead of dedicated NM messages circulating in a logical ring, application messages are monitored by nodes that receive the messages to determine the status of the transmitting node. In addition, two message monitoring methods are defined: a single timeout for observation (TOB) for all messages and an individual timeout per message. Single TOB is used by all nodes for monitoring monitored application messages to identify the node states. For instance, if a node is able to successfully transmit its monitored application message within TOB, the node is considered as "Not Mute". If monitored messages are successfully received before TOB elapses, the monitored node (the node that sent the messages) is "Present", shown in Fig. 12.

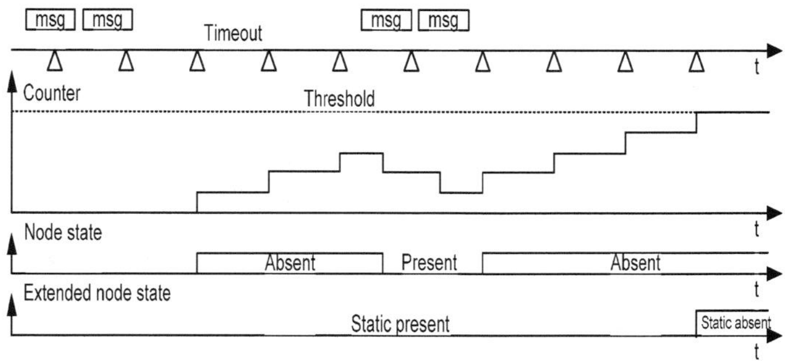

Fig. 12. Application message monitoring for IDNM.

As for an individual timeout per message, IDNM uses COM Deadline Monitoring, a feature of OSEK Communication Component, to monitor individual message timeouts. If a monitored message has not arrived within predefined time, COM component will signal NM to update its configuration to indicate that the monitored node is "Absent". Similarly, the COM component also signals NM if the local node has failed to transmit a message within a predefined time. As a result, that local node is classified as "Mute".

To identify the static states (extended states), a specific counter and a threshold level are used. A counter is incremented and decremented in accordance with a node state. For instance, if the node is "Absent", the counter will be incremented; if the node resumes being "Present" again, the counter will be decremented. When the counter reaches the threshold, NM identifies this state as "Node Absent Statically". NM modes, state transitions and timer definitions are also defined in the OSEK NM standard. Further details of OSEK NM can be found in OSEK/VDX, 2004 and Lemieux, 2001.

3.2.2 Message rate fault detection

Examples of CAN application message faults which are most likely to occur on in-vehicle networks are shown in Fig. 13 and are defined in detail in Table 1. In general, details of periodic messages are specified in a message database during a design process. For instance, a message 'M' is transmitted by an ECU 'E' at every 10 ms. This transmission period can be guaranteed by the use of special network software tools. However, message faults can still happen when the bus traffic quality is poor due to a large number of error frames, or in a system where no such special software tools are implemented in ECUs. It is critical when faults occur on messages which are used in real-time control loops. Therefore, network level fault detection becomes necessary to monitor the number of messages and their timings, and the number of error frames actually present on the network.

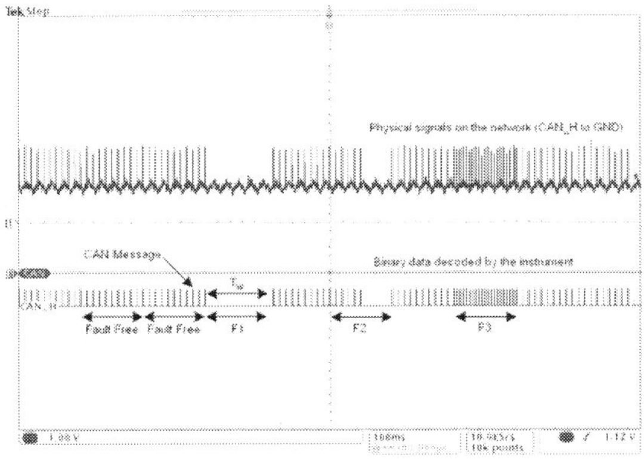

Fig. 13. CAN message faults.

An application message on the network is periodically monitored by the fault detection in a time window T_w (e.g. 100 ms as shown in Fig. 13). The intermediate message rate N_{mi} (in msg/s) is compared with the rate specified in a message database N_{si} and a threshold value N_{ti}, where i is a message identification number. T_w is chosen to satisfy the inequality:

$$T_W \geq 1/\mathrm{MIN}[N_{s0}, N_{s1}, \dots N_{sn}]$$

where n = 0 to the last message ID, e.g. 255.

Faults	Details
Fault free	Message is sent at the specified rate ($N_{si} - N_{ti} \leq N_{mi} \leq N_{si} + N_{ti}$)
F1: Timeout	Message is not sent within predefined time (Nm = 0)
F2: Missing-Message	Message is missing in a particular period (Nm < Ns - Nt)
F3: Too-Many-Message	Too many messages are transmitted (Nm > Ns + Nt)

Table 1. Details of application message faults.

The measurement of the message rates can be seen in the statistics window of CANoe, a CAN bus simulation and network design tool. The horizontal axis of the window displays hexadecimal message IDs and the vertical axis represents the message rates in msg/s. Fig. 14 shows the normal condition of the simulated network, where application messages with different message IDs are transmitted within the normal operating bands defined by:

$$N_{si} - N_{ti} \leq N_{mi} \leq N_{si} + N_{ti}$$

The fault detection is responsible for detecting any message rates that are out of the bands such as F1, F2 and F3 shown in Fig. 15. After faults are detected, DTCs and relevant snapshot data are logged in accordance with the manufacturer's definitions.

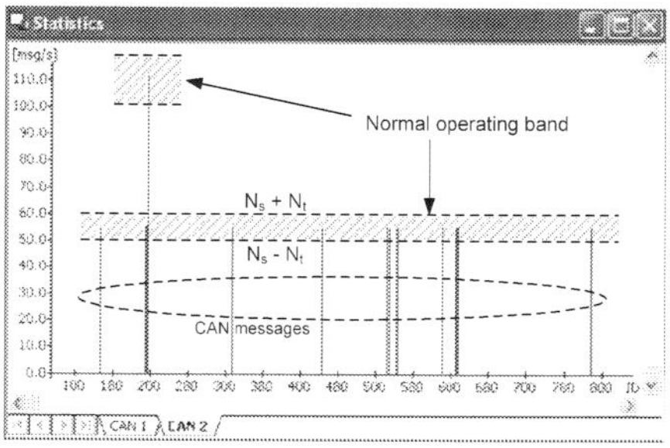

Fig. 14. Normal condition.

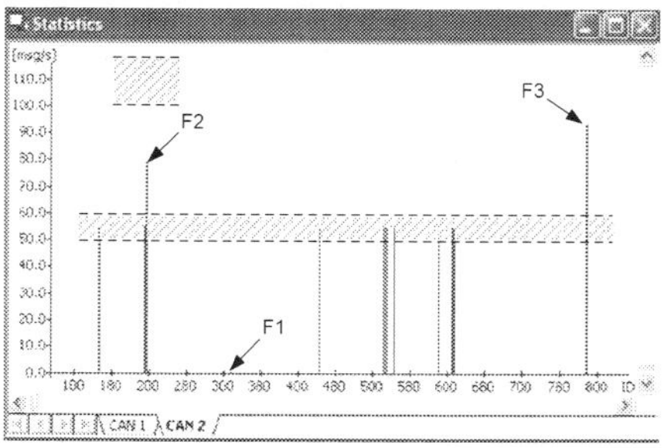

Fig. 15. Faulty condition.

3.2.3 Fault isolation with multiple discrete events

It is possible that any of the faulty electronic components on the communication link contributes to F1 and F2 presented above. This subsection demonstrates a method for isolating possible faulty components by inferring from multiple discrete events in the networked system.

In engineering systems where some basic knowledge of relationships between faults and symptoms is available, the knowledge can be expressed in rules based on fault-tree analysis (FTA) or event-tree analysis (ETA). A simple example a rule of '*FAULT1*' caused by two symptoms—'*SYMPTOM1*' and '*SYMPTOM2*' is as follows:

IF (*SYMPTOM1* AND *SYMPTOM2*) THEN *ERROR1*

IF (*ERROR1* OR *ERROR2*) THEN *FAULT1*

FTA is a graphical method that hierarchically represents a binary relationship between a fault ('top event') and symptom(s). Intermediate events relevant to the 'top event' are combined using logical operations. FTA is commonly used in hazard analysis in safety-critical systems as it can represent dependencies in the systems (Storey, 1997). Fig. 16 illustrates a simple fault tree of a headlamp failure in a vehicle, which involves several root causes that can be traced back. FTA is also widely used in fault diagnosis with inference methods. Fault diagnosis can be preformed by inferring Boolean values of symptoms from sensor measurements such as pressure and position switches, and fault-symptom relationships.

For in-vehicle network diagnosis, this subsection addresses how to utilise local message monitoring data in ECUs for isolating faulty components. Fault isolation with multiple discrete events by interpreting symptoms or intermediate events of the presence and correctness of message communication is discussed.

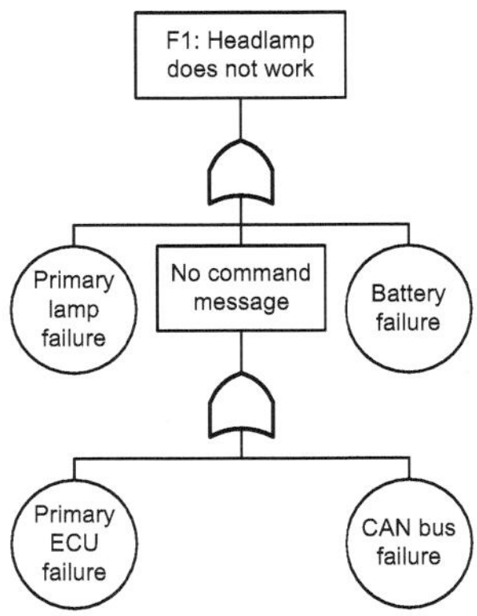

Fig. 16. Simple fault tree of a headlamp failure in a vehicle.

Conventionally, each ECU monitors particular messages it expects to receive. Faulty event data such as "timeout" events of the expected messages is locally stored in the ECU. When faults occur in a particular component which is not in a communication path between a monitoring ECU and a monitored message, the monitoring ECU will not recognise the faults and will not store any fault data. For instance, in the ECU diagram shown in Fig. 17, suppose that E5 expects M1 from E1 and the other ECUs do not monitor any messages from E5. If a "timeout" fault of M1 is detected and stored in E5, the root cause can be several components in the communication path e.g. E1, W1, C1, W6, W7, C5, W5 and E5. This list of possible faulty component(s) can be narrowed down if every ECU (i) locally monitors particular messages from the other ECUs even if the messages are not used in its control functions, and (ii) can provide local fault events to be analysed by diagnostic applications in a dedicated ECU, provided that there is a redundant communication channel available if the main channel is permanently unavailable.

A communication path is shown in Fig. 18 where a monitoring ECU receives a monitored message from a monitored ECU through several peripheral components. The peripheral components can be classified into three major groups: (i) components at the monitoring ECU, (ii) those at the monitored ECU and (iii) those shared among ECUs.
Consider the networked system in Fig. 17 where M5 is monitored by all ECUs, fault symptoms in a situation where M5 is not received by E1 - E4 and M1 - M4 are not received by E5 can be summarised in Table 2. Possible faulty components can be identified in relation to an individual symptom and the peripherals involved in the communication path.

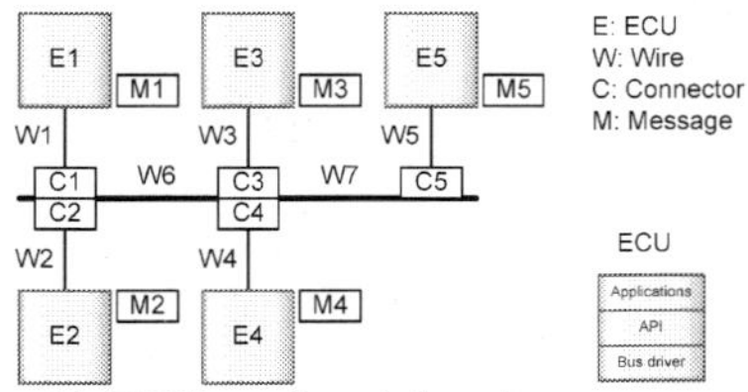

Fig. 17. In-vehicle network with ECUs and peripheral components.

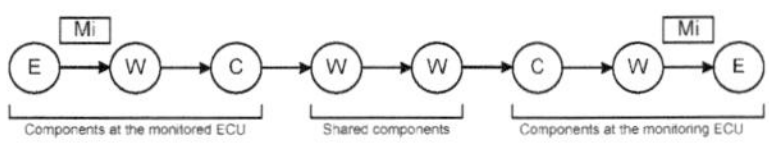

Fig. 18. Communication path.

		M1	M2	M3	M4	M5
S1	E1	1	1	1	1	0
S2	E2	1	1	1	1	0
S3	E3	1	1	1	1	0
S4	E4	1	1	1	1	0
S5	E5	0	0	0	0	1

Table 2. Fault symptom table: M5 not received by all ECUs except E5 itself.

where 1: M*i* is received by E*i*; 0: M*i* is not received by E*i*; S*i*: Symptom i.
As shown in Table 3, for instance, S1 can be caused by E1, W1, E1, E5, W5, C5, W6 and W7. After considering all of the discrete fault symptoms on the entire network (S1 – S5), W7 is involved in every symptom; therefore it is isolated as a faulty component.
Another example of fault isolation for multiple faults is provided in Table 4. In this example, M5 is not received by all ECUs except E5 itself; M1 - M2 are not received by E3 - E5; M3 - M5 are not received by E1 - E2; M3 - M4 are not received by E5. Similarly, after considering multiple discrete fault symptoms on the entire network, W6 and W7 can be isolated as faulty components.

	E1	W1	C1	E2	W2	C2	E3	W3	C3	E4	W4	C4	E5	W5	C5	W6	W7
S1	1	1	1	0	0	0	0	0	0	0	0	0	1	1	1	1	1
S2	0	0	0	1	1	1	0	0	0	0	0	0	1	1	1	1	1
S3	0	0	0	0	0	0	1	1	1	0	0	0	1	1	1	0	1
S4	0	0	0	0	0	0	0	0	0	1	1	1	1	1	1	0	1
S5	1	1	1	1	1	1	1	1	1	1	1	1	0	0	0	1	1
P	-	-	-	-	-	-	-	-	-	-	-	-	-	-	-	-	x

Table 3. Fault isolation table for the symptoms in Table 2.

where 1: component is in the communication path and can cause S*i*;
0: component is not in the communication path;
P: possible faulty components.

		M1	M2	M3	M4	M5
S1	E1	1	1	0	0	0
S2	E2	1	1	0	0	0
S3	E3	0	0	1	1	0
S4	E4	0	0	1	1	0
S5	E5	0	0	0	0	1

Table 4. Fault symptom table: M5 not received by all ECUs except E5 itself; M1 - M2 not received by E3 - E5; M3 - M5 not received by E1 - E2; M3 - M4 not received by E5.

	E1	W1	C1	E2	W2	C2	E3	W3	C3	E4	W4	C4	E5	W5	C5	W6	W7
S1	1	1	1	0	0	0	1	1	1	1	1	1	1	1	1	1	1
S2	0	0	0	1	1	1	1	1	1	1	1	1	1	1	1	1	1
S3	1	1	1	1	1	1	0	0	0	0	0	0	1	1	1	1	1
S4	1	1	1	1	1	1	0	0	0	0	0	0	1	1	1	1	1
S5	1	1	1	1	1	1	1	1	1	1	1	1	0	0	0	1	1
P	-	-	-	-	-	-	-	-	-	-	-	-	-	-	-	x	x

Table 5. Fault isolation table for the symptoms in Table 4.

In practice, however, the presented fault isolation technique can be implemented only if a redundant channel is available. Multiple fault events from all ECUs of the entire network are gathered by a dedicated ECU through either the existing network itself or the redundant channel when the main network is permanently unavailable, e.g. permanently broken wires. The redundant channel can be a wired or wireless link among the ECUs, e.g. FlexRay dual-channel topology or Bluetooth. Intermittent fault data could be used to represent a

deterioration statistics by using a counter to count the number of detected faults in the component.
This section has discussed in-vehicle network fault detection which ranges from simple application message monitoring to more robust and complicated network and ECU status monitoring e.g. OSEK DNM. After faults are detected, on-board diagnostic applications in ECUs can obtain current network or status by requesting NM, if implemented, to provide the status such as an ECU in "Limp-Home", "Present", or "Mute" mode. After gathering the status and information, and performing fault isolation, the diagnostic applications log network-relevant DTCs according to a description specified by the manufacturer. Network-relevant DTCs are commonly standardised to have a prefix 'U'. Some examples are shown below.

U0003: "High speed CAN communication bus (+) Open"
U0100: "Lost communication with ECU A"
U0301: "Software incompatibility with ECU B"
U0401: "Invalid data received from ECU C"

These DTCs are read by a specific tool to be guidance to engineers to find out root causes. In general, there can be several DTCs that relate to a single fault logged by all diagnostic levels. To correctly diagnose the real root causes, interpretation from experienced engineers or assistance from an advanced guiding method such as the work of Huang et al., 2008 may be required.

4. Recent research on FDD for in-vehicle networks

The previous section discussed how faults in automotive electronic systems are conventionally handled. However, network level faults can result from a number of causes. As illustrated in Fig. 19, failures within an in-vehicle electronic system can manifest themselves in many different ways and on a number of different levels. The causes can be directly from the network level itself or indirectly from the other two levels. For instance, if a message is not transmitted from an ECU as specified, this problem can be caused by an internal function in the ECU itself (feature level), the CAN controller or the electrical wiring (component level). It is noted that physical hardware faults of in-vehicle networks such as CAN cables short-circuit-to-ground or the battery are considered as component level faults.

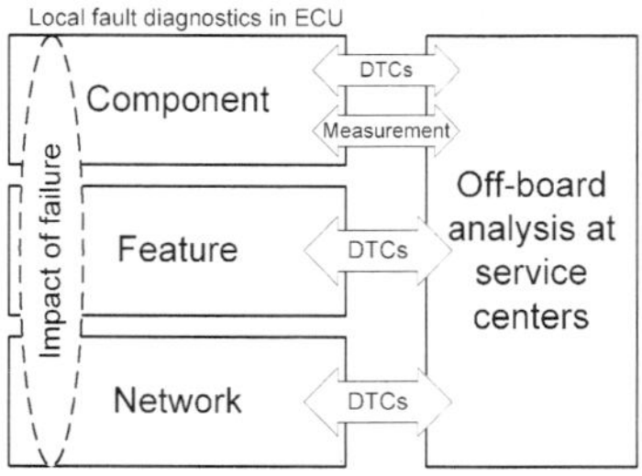

Fig. 19. Impact of failure across different diagnostic levels.

Vehicle diagnosis using DTCs intended for later off-board root cause analysis and repair actions can be unwieldy when it comes to dealing with a system comprising a large number of ECUs and messages. For instance, several components in the system or communication path can cause a DTC U0401 to be set. The data might have been sent correctly by the "ECU C" but it was probably corrupted on the bus by intermittent faults in the cable wires. When performing off-board diagnostics, it is almost impossible to simulate the same operating condition as when the faults were present. This subsection discusses recent research on fully on-board FDD techniques that are being implemented to manage faults in real-time and in a system-wide and network-wide perspective.

Recent years have seen a number of research projects on various areas related to intelligent vehicles such as vehicles with collision avoidance and self diagnosis capabilities. One of such projects is known as the Self-Healing Vehicle concept which outlines a required paradigm shift of fault diagnostics and management of in-vehicle electronic systems (Amor-Segan et al., 2007). The Self-Healing Vehicle concept, illustrated in Fig. 20, is intended to mitigate failures within the vehicle's embedded software and electronic control systems in order to "safely keep the vehicle on the road". The concept envisages a vehicle equipped with a standardised, general purpose, networked computing architecture that facilitates software and application mobility. This mechanism will be managed by an intelligent fault management system with access to remote support services via a telematics link. The Intelligent Black Box (iB^2) will efficiently use available information in a vehicle that is currently under-utilised. This is the large volume of real-time component-, feature- or network-related information flowing through the vehicle's networks. Intelligent use of this information will be a key factor in the accurate operation of whole-vehicle diagnostics and prognostics, and the overall effectiveness of the fault management system.

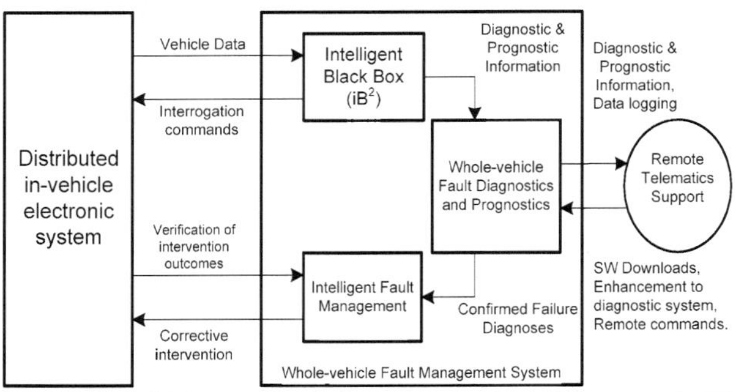

Fig. 20. The Self-Healing Vehicle concept (adapted from Amor-Segan et al., 2007).

The iB^2 is a system capable of performing the following important functions:

- Monitoring of a variety of data sources such as network conditions, system and component resources, internal ECU data, etc.;
- Intelligent data mining, data compression, filtering and analysis;
- Adapting its behaviour under the instruction from the whole-vehicle diagnostic and prognostic system;
- Proactive system observations for fault diagnosis and prognosis.

These functions will support the on-board whole-vehicle fault management in obtaining accurate diagnoses. Clearly, a new framework for fully on-board monitoring and utilising the system information—component-, feature-, and network-related information—becomes necessary.

As for the network level, network information can be analysed by an intelligent network diagnostic and prognostic process as illustrated in Fig. 21. As a result, accurate diagnoses can be generated, and an alarm is triggered when predicted healthiness of the network falls below a threshold value (T). Certainly, to achieve this, robustness, accuracy, and continuity of network level fault detection will need to be improved.

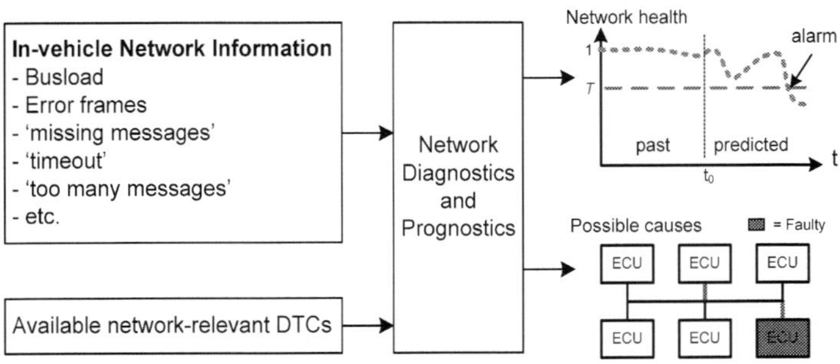

Fig. 21. In-vehicle network diagnostics and prognostics.

4.1 Evaluating network health

The network health could be indicated by the number of error frames. An error frame is transmitted when any of the communication error types is detected by an error management unit fitted in a CAN controller. Error detection embraces Cyclic Redundancy Check (CRC), bit monitoring, bit stuffing and frame checking. Errors can be caused by a number of reasons one of which is the presence of physical disturbances. When CAN buses are installed in a vehicle, there can be a situation where the physical wiring or connectors are damaged or gradually deteriorated by vibration or corrosion. This would cause the electrical signal to vary from specifications, thereby resulting in communication failures and error frames being transmitted.

Fig. 22 illustrates a captured CAN message from an oscilloscope displaying a signal distortion effected by analogue disturbances such as low resistance between CAN cables. The data can still be decoded by the oscilloscope and there are no error frames transmitted. Fig. 23 depicts a zoomed-in capture of the distorted signal. Fig. 24 shows a situation where there are more disturbances injected such that error frames are transmitted and the data cannot be decoded by the oscilloscope.

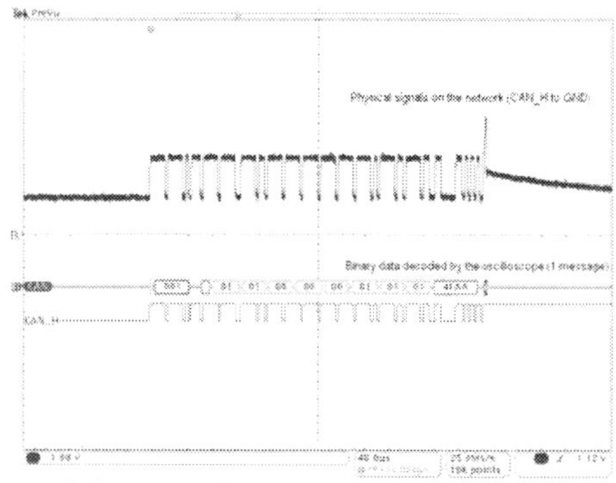

Fig. 22. Physical CAN signal disturbed by an analogue disturbance.

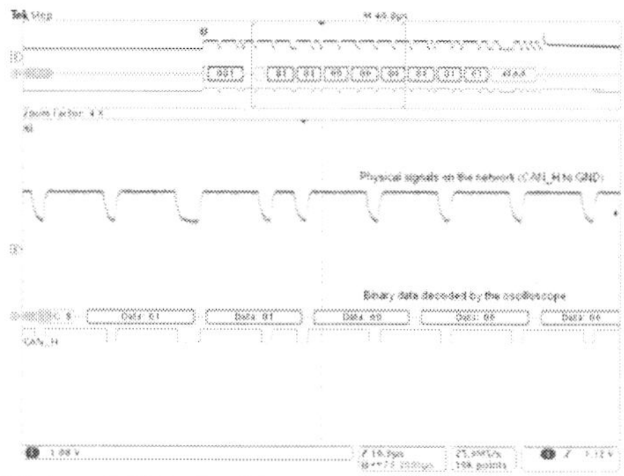

Fig. 23. Zoomed-in CAN signal disturbed by an analogue disturbance.

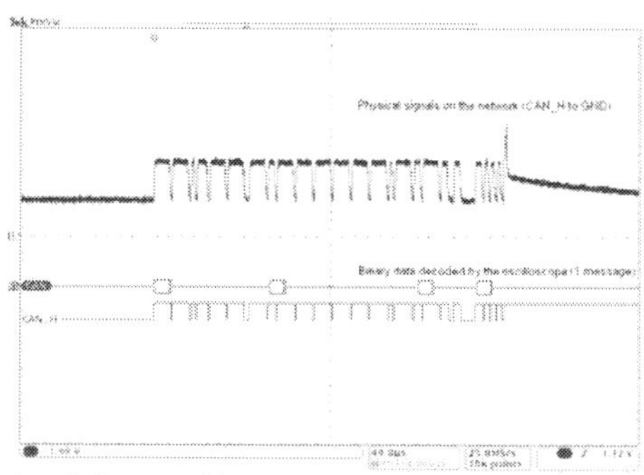

Fig. 24. Physical CAN signal distorted by more analogue disturbances.

The relationship between the number of error frames and the fault types remains under investigation at this stage. However, inference systems may be used to analyse the collected data from preliminary experiments. For instance, as shown in Fig. 25, an adaptive-network-based fuzzy inference system (ANFIS) can classify the number of error frames into such network health conditions as "unhealthy", "fair" and "healthy".

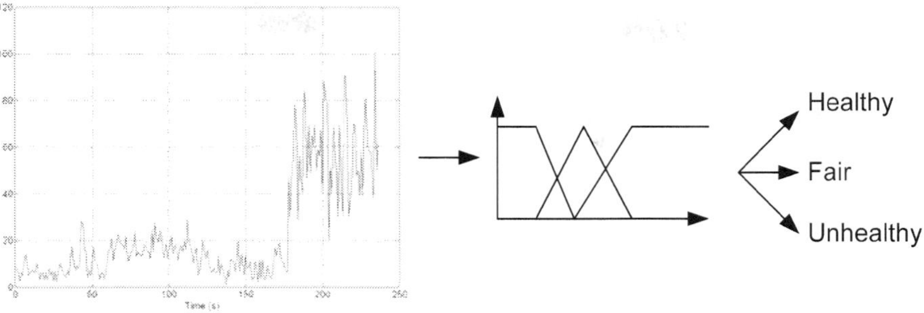

Fig. 25. Network health classification using ANFIS.

4.2 Adaptive network management

As discussed in the previous section, DNM provides network status to an application which is running in a networked ECU environment. With DNM, there is less overhead to the application and a high level of portability across multiple applications can be achieved. IDNM, on the other hand, is much simpler to implement. Despite being less robust and limited availability of node and network status, IDNM is an alternative mechanism for systems where rapid response from NM is not necessary, and simple software algorithms and computational resource consumption are of concern.

Traditionally, network diagnostic applications in ECUs only use either DNM or IDNM to provide them with node and network status. If DNM is implemented, some application message faults such as intermittent application message missing cannot be detectable because application messages are not monitored. Moreover, if NM messages are not available from a particular node as a result of its embedded software faults while application messages can still be transmitted, diagnostic applications in other nodes will no longer be able to detect faults from that faulty node. Conversely, if IDNM is implemented, as mentioned earlier, availability of node and network status is limited. This therefore would result in less fault coverage in a networked ECU environment. To enhance coverage, robustness and continuity of network level fault detection the Adaptive OSEK NM technique has been proposed (Suwatthikul, 2007). This technique is divided into two approaches: (i) switching DNM to IDNM, called Switched NM; and (ii) combining DNM and IDNM, called Combined NM.

The concept of combining both approaches is based on the fact that if the use of NM by applications is not fixed to DNM or IDNM, then the applications will benefit from using the advantages of both types of traditional OSEK NM as fault monitoring continues. As illustrated in Fig. 26, both approaches begin with variable and timer initialisation. For Switched NM, DNM is initially used as the main NM of the system. If DNM is not available for some reason such as NM API failure, the IDNM mode will be entered, i.e. application messages are monitored by a diagnostic application rather than using information from dedicated NM messages. For Combined NM, an application uses DNM and IDNM together such that application messages and NM messages are monitored at all time.

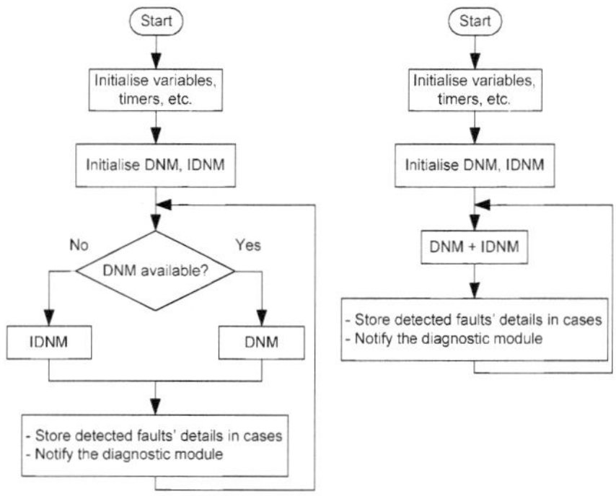

a) b)

Fig. 26. Adaptive OSEK NM: a) Switched NM; b) Combined NM.

Although Combined NM will inherently consume more computational resources than when traditionally using separate NM, the simulation results show that Combined NM can cover more fault scenarios, and provide more accurate network fault detection and improved robustness (Suwatthikul, 2007).

In conclusion, this chapter has provided readers with an idea of how FDD can be applied in practice, especially current and future applications in the automotive industry. Modern in-vehicle electronic systems have become much more complicated due to additional sophisticated features. This certainly requires more intelligent and robust FDD. Despite available FDD approaches from a number of non-automotive domains, the approaches cannot be simply applied to the automotive domain due to obvious constraints such as costs and different end users. The aerospace industry has employed advanced FDD for decades, involving model-based FDD, bespoke components and redundant systems, in particular for safety-critical applications. Inevitably, such approaches are extremely costly and therefore not well suited for applications in the automotive industry.

The direction of future research on FDD in automobiles tends to focus on cost-effective and intelligent approaches to whole-vehicle fault management, and new component and system architectures. For instance, the use of generic processing components to implement different fault management strategies in a vehicle that will achieve the required levels of resilience or fault tolerance, yet in a way that is cost-effective and realisable for the automotive domain.

5. Acknowledgements

The author would like to thank the International Automotive Research Centre, University of Warwick, for the support of equipment and tools used in the experiments.

6. References

Amor-Segan, M. L.; McMurran, R.; Dhadyalla, G. & Jones, R. P. (2007). Towards the Self-Healing Vehicle, *Proceeding of the 3rd IET Automotive Electronics Conference*, Coventry, June 2007.

Armengaud, E.; Steininger, A.; Horauer, M.; Pallierer, R. & Friedl, H. (2004). A monitoring concept for an automotive distributed network: the FlexRay example, *Proceeding of the 7th IEEE Workshop on Design and Diagnostics of Electronic Circuit and Systems*, Stara Lesna, April 2004. pp. 173-178.

Armengaud, E.; Steininger, A. & Horauer, M. (2005). Efficient stimulus generation for testing embedded distributed systems: the FlexRay example, *Proceeding of the 10th IEEE Int. Conference on Emerging Technology and Factory Automation*, Catania, September 2005, pp. 763-770.

Armengaud, E.; Rothensteiner, F.; Steininger, A.; Pallierer, R.; Horauer, M. & Zauner, M. (2005). A structured approach for the systematic test of embedded automotive communication systems, *Proceeding of International Test Conference 2005*, Texas, Nov. 2005, pp. 1-8.

Athanasas, K. & Dear, I. (2004). Validation of complex vehicle systems of prototype vehicles. *IEEE Transactions on Vehicular Technology*, Vol. 53, No. 6, pp.1835-1846.

Buja, G.; Pimentel, J. R. & Zuccollo, A. (2007). Overcoming babbling-idiot failures in CAN networks: A simple and effective bus guardian solution for the FlexCAN architecture. *IEEE Transactions on Industrial Informatics*, Vol. 3, No. 3, pp. 225-233.

Ehret, J. (2003). Validation of safety-critical distributed real-time systems. Dr.-Ing. thesis, Technical University of Munich, Munich.

Foran, T. & Jackman, B. (2005). An intelligent diagnostic system for distributed multi-ECU automotive control systems, SAE paper: 2005-01-1444, *SAE World Congress*, Michigan, April 2005.

Gaujal, B. & Navet, N. (2005). Fault confinement mechanism on CAN: Analysis and Improvements. *IEEE Transactions on Vehicular Technology*, Vol. 54, No. 3, pp.1103-1113.

Isermann, R.; Schaffnit, J. & Sinsel, S. (1999). Hardware-in-the-loop simulation for the design and testing of engine-control systems. *Control Engineering Practice*, Vol. 7, No. 5, pp. 643-653.

Isermann, R. (2008). Mechatronic systems—Innovative products with embedded control. *Control Engineering Practice*, Vol. 16, No. 1, pp. 14-29.

Kendall I. R. & Jones, R. P. (1999). An investigation into the use of hardware-in-the-loop simulation testing for automotive electronic control systems. *Control Engineering Practice*, Vol. 7, No. 11, pp. 1343-1356.

Kopetz, H. & Bauer, G. (2003). The time-triggered architecture. *Proceedings of IEEE*, Vol. 91, No. 1, pp.112-126.

Lawrenz, W. (1997). *CAN System Engineering: From Theory to Practical Applications*, Springer-Verlag, ISBN 0387949399, New York.

Leen, G.; Heffernan, D. & Dunne, A. (1999). Digital networks in automotive vehicle. *Computing and Control Engineering Journal*, Vol. 10, No. 6, pp. 257-266.

Leen, G. & Heffernan, D. (2002). TTCAN: A new time-triggered controller area network. *Microprocessors and Microsystems*, Vol. 26, No. 2, pp. 77-94.

Lemieux, J. (2001). *Programming in the OSEK/VDX Environment*, CMP Books, ISBN 1578200814, Kansas.

McMurran, R.; McKinney, F.; Tudor, N. J. & Milam, M. (2006). Dependable Systems of Systems, SAE paper: 2006-01-0597, *SAE World Congress*, Michigan, April 2006.

Murphey, Y. L.; Crossman, J. A.; Chen, Z. & Cardillo, J. (2003). Automotive fault diagnosis—Part II: A distributed agent diagnostic system. *IEEE Transactions on Vehicular Technology*, Vol. 52, No. 4, pp.1076-1098.

Navet, N. & Song, Y.-Q. (1999). Reliability improvement of the dual-priority protocol under unreliable transmission. *Control Engineering Practice*, Vol. 7, No. 8, pp. 975-981.

Navet, N. & Song, Y.-Q. (2001). Validation of real-time in-vehicle applications. *Computers in Industry*, Vol. 46, No. 2, pp. 107-122.

Navet, N.; Song, Y. Q.; Simonot-Lion, F. & Wilwert, C. (2005). Trend in automotive communication systems. *Proceedings of IEEE Special Issue Industrial Communication Systems*, Vol. 93, pp.1024-1223.

Ortega, E.; Heurung, T. & Swanson. R. (2006). System design from wires to warranty. *Automotive Electronics Magazine*, February, pp. 14-18.

OSEK/VDX (2004). Network Management: Concept and Application Programming Interface Version 2.5.3. [Online]. Available: http://www.osek-vdx.org.

Robert Bosch GmbH. (2004). *Automotive Electrics Automotive Electronics*, Professional Engineering Publishing Limited, ISBN 0470519371, Suffolk.

Robert Bosch GmbH. (2004). *Gasoline-Engine Management*, Robert Bosch GmbH, ISBN 0837610524, Bury St. Edmunds.

Robert Bosch GmbH. (2004). *Automotive Handbook*, Robert Bosch GmbH, ISBN 0768015138, Plochingen.

Short M. & Pont, M. J. (2008). Assessment of high-integrity embedded automotive control systems using hardware in the loop simulation. *Journal of Systems and Software*, Vol. 81, No. 7, pp. 1163-1183.

Shrinath, A. & Emadi, A. (2004). Electronic control units for automotive electrical power systems: Communication and networks. *Proceedings of IMechE Part D: Journal of Automobile Engineering*, Vol. 218, pp. 1217-1230.

Simonot-Lion, F. (2003). In-car embedded electronic architectures: how to ensure their safety, *Proceeding of the 5th IFAC International Conference on Fieldbus Systems and Their Applications*, Aveiro, July 2003.

Storey, N. (1997). *Safety-Critical Computer Systems*, Addison Wesley Longman, ISBN 0201427877, New York.

Suwatthikul, J; McMurran, R. & Jones, R. P. (2007). Adaptive OSEK Network Management for In-vehicle Network Fault Detection, *Proceeding of the 2007 IEEE International Conference on Vehicular Electronics and Safety*, Beijing, Dec. 2007.

Temple, C. (1998). Avoiding the babbling-idiot failure in a time-triggered communication system, *Proceeding of the 28th Annual International Symposium on Fault-Tolerant Computing*, pp. 218–225.

Tindell, K. & Hansson, H. (1995). Babbling idiots, the dual priority protocol, and smart CAN controllers, *Proceeding of the 2nd International CAN Conference*, pp. 7.22–7.28.

Zurawski, R. (2006). *Embedded Systems Handbook*, CRC Press, ISBN 0849328241, Florida.

14

Fault Detection of Actuator with Digital Positioner Based on Trend Analysis Method

Xiaobin Huang, Jizhen Liu and Yuguang Niu
North China Electric Power University
P.R. China

1. Introduction

The demanding on high safety, performance and reliability in controlled processes has becoming increasingly stringent in recent years. Control valves (or actuators) are widely used in industrial processes. As the final control elements, they are often installed in the technology nodes working in the harsh environment: high temperature, high pressure, humidity, pollution, chemical solvents, etc. Their malfunctions usually lead to poor control performance or process disturbance, even result in unqualified product. Therefore, the on-line detection and diagnosis of control valve should be applied to preserve the high-reliability of control valves due to the severity of its possible effects of failure on the processes.

The malfunctions of actuator mainly include fully failure, offset and bias, change of gain, serious hysteresis, and stick-slip fault. In the past two decades, there has existed a number of fault detection and diagnosis methods for actuators in process control systems. Some efforts involved model-based approaches: state estimation (Hoefling et al.,1995; Park T.G. et al., 2000; Edwin Engin Yaz & Asad Azemi,1998); parity equation (Massoumia et al.,1998; Mediavilla et al.,1997). These methods require relatively accurate mathematic models about the processes. However, it is very difficult to obtain accurate mathematic models in most industrial processes. Other studies focused on using neural networks (Patan,2001; Patan & Parisini,2003; Pawel et al.,2003), fuzzy logic, and signal analysis (Deibert,1994). An important issue that should be highlighted is that, there is no a method that can detect and diagnose all kinds of faults because various fault types may occur in control valves. most existing method requires process knowledge or user-interaction (Forsman & Stattin 1999; Hagglund,1995; Wallen,1997). Only few approaches do not need prior knowledge about the process (Horch, 1999).

In process control systems, actuators with digital positioner are widely used. Generally, few related signals can be sampled to process monitor systems. These signals are: the input and output signals related to the component itself, and the flow signal that controlled by the industrial actuator. In fact, these signals provide useful information about the operation of the actuator.

In this chapter, a series of methods based on trend analysis are proposed to detect typical faults of industrial actuators with digital positioners by using these three signals. Because there is no need to have prior knowledge about the control processes, these methods provided can be easily applied to the real processes.

2. Description of Actuator with Digital Positioners and Its Typical Faults

2.1 Description of actuators with digital positioners in process control systems

As a typical example, Fig.1 shows the air-operated actuator with digital positioner in most process control systems. It consists of three primary parts: digital positoner, air-driven part(or executive body), and valve body. The digital positoner receives valve travel setting signal (control signal U_d) from the controller in a control system, and set the actual valve travel (x) according to the valve travel setting signal by supplying air to the driven part. The valve steam is then driven by air pressure and moves to a specified position to achieve flow control.

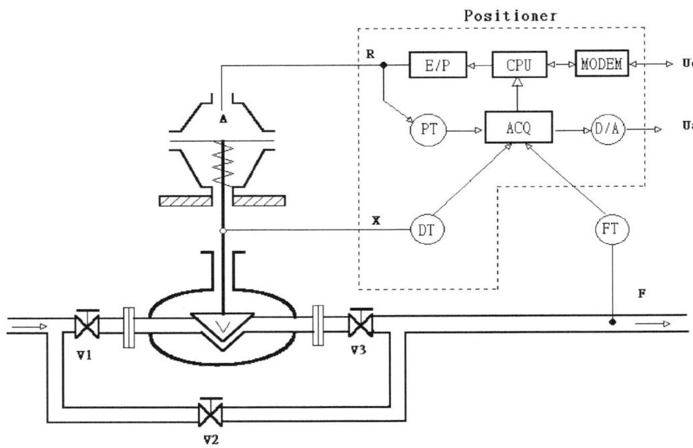

Fig. 1. Diagram of the air-operated actuator with digital positioner

The valve body is a liquid-limit component which includes valve seat and valve spool. When the valve stem moves to a specific position, the valve spool will also reach a corresponding opening point. The model of valve body can be described as follows:

$$Q/Q_{max} = f(x/L) \tag{1}$$

where Q_{max} is the maximum flow rate and L is the spool displacement when the valve spool is fully open, Q is the real flow rate, , x is the real spool displacement, x/L represents the spool opening, and Q/Q_{max} represents relative flow rate.

2.2 Generalized model of actuator with digital positioner

In actuator with digital positioner, important signals for fault detection are: valve travel setting signal, the actual valve travel, air pressure of driven part, inlet pressure of valve body, outlet pressure of valve body, and the flow rate through valve body. However, in most cases, only few signals can be sampled by DCS (distributed control system) or SCADA system. They are:

a) valve travel setting signal: CV
b) actual valve travel:X
c) flow rate through valve body: Q

For other kinds of actuator, such as electric valve actuators and hydraulic actuator, we get the same result: the three signals can be easily obtained. Because the three signals can represent most important running state of actuator, a generalized model of actuator with digital positoner is given in Fig.2.

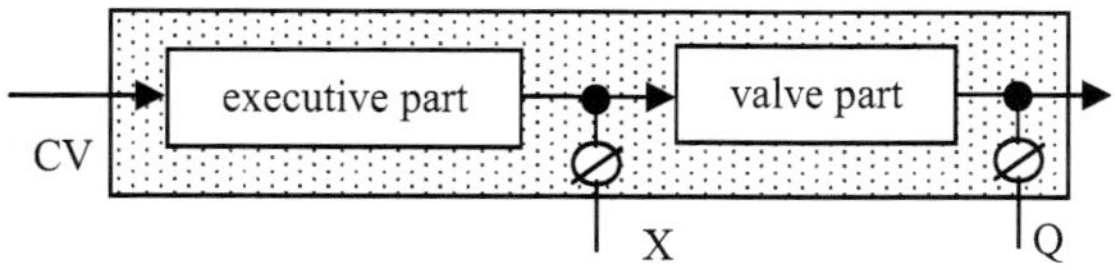

Fig. 2. Generalized model of actuator with digital positioner

When the actuator works well, the X will trace CV rapidly and smoothly, and the flow rate will keep a corresponding value to X. Once there are some malfunctions, the relationship among CV, X, and Q will change. Fault detection can be made by detecting these changes with the three signals.

2.3 Typical malfunctions of digital positioners

Malfunctions may occur in every part of the actuator. As an example, 19 typical fault types of the membrane air-operated actuator are given in table 1.

Valve body	Air-driven part	Digital positioner	Other fault
F1: valve body blocked	F8: valve stem bending	F12: electro- peumatic convertor failure	F16: air feed pressure declining
F2: Sediment in valve seat or valve spool	F9: overshadowed tight membrane	F13: position feedback sensor failure	F17: abnormal differential pressure of valve body
F3: valve body eroded	F10: membrane damage	F14: pressure sensor failure	F18: bypass valve open
F4: friction force increase in valve or bush	F11: spring failure	F15: positioner spring failure	F19:flow sensor failure
F5:external leakage (bush or valve cover			
F6: internal leakage(tight valve)			
F7: fluid evaporation			

Table 1. Typical malfunctions of the membrane air-operated actuator

These fault types F1 to F19 will make the actuator work in faulty state. Meanwhile, the relationship between the three signal CV, X, and Q will change in an abnormal way. These different faulty state of actuators can be classified generally into the following types: "Stick-slip" fault, "constant bias" fault, "change of gain" fault, "serious hysteresis" fault,"stuck" fault. Fig.3. to Fig.8 show the above typical fault types, in which the three signals behave in different ways.

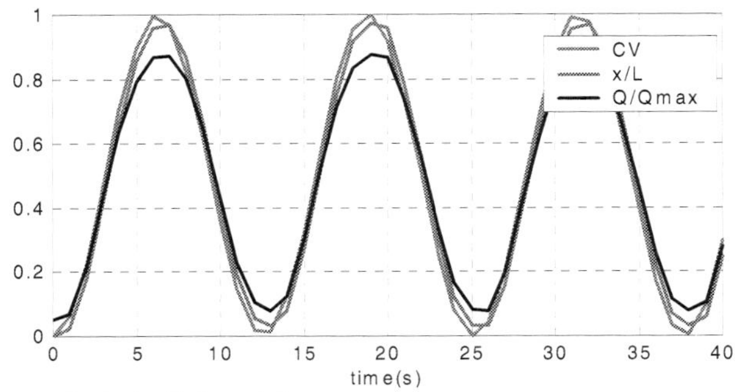

Fig. 3. Actuator works in fault-free state

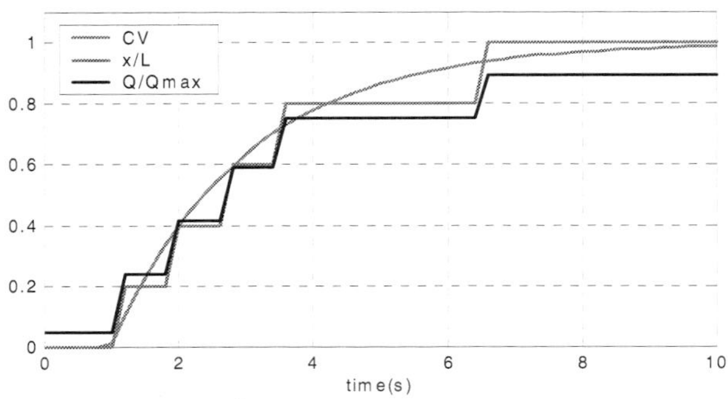

Fig. 4. Actuator works in "stick-slip" state

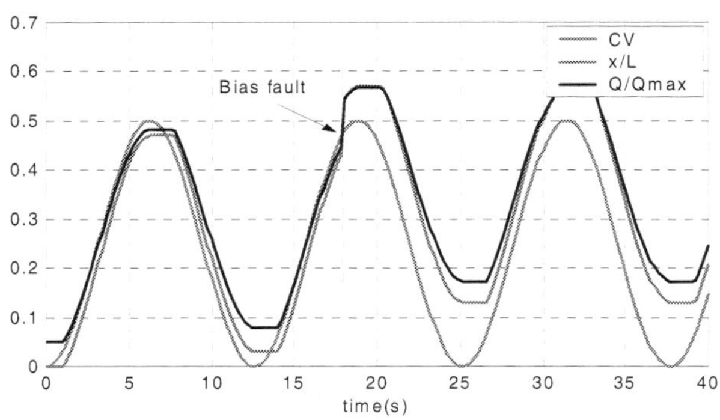

Fig. 5. Actuator works in "constant bias" state

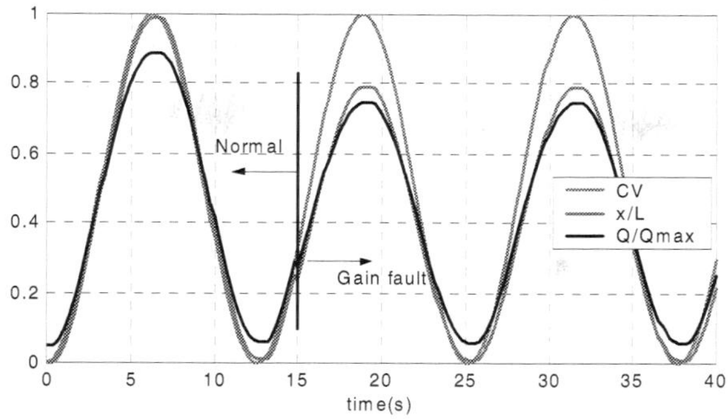

Fig. 6. Actuator works in "change of gain" state

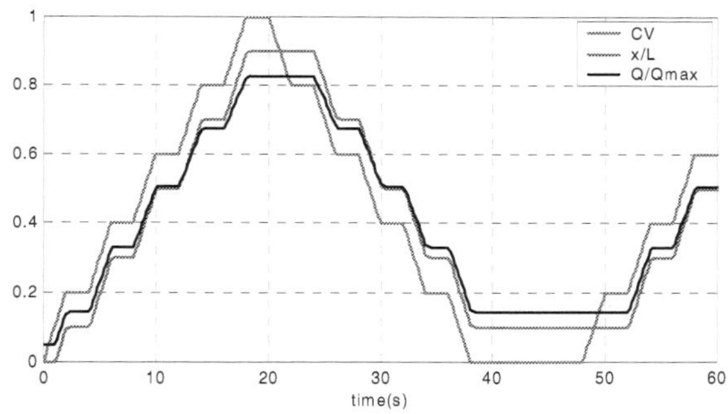

Fig. 7. Actuator works in "serious hysteresis" state

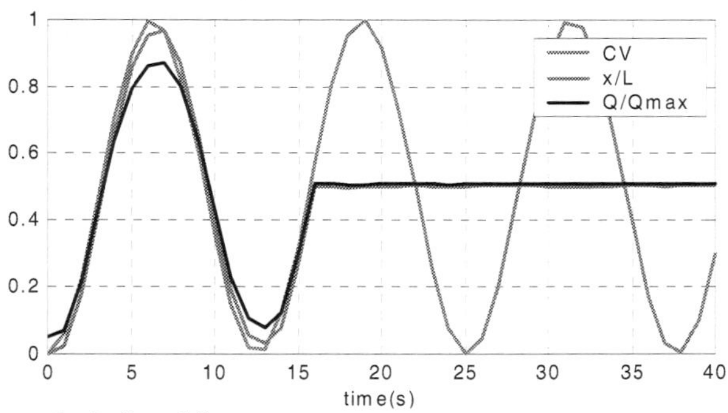

Fig. 8. Actuator works in "stuck" state

3. Online Fault Detection Methods Based on Trend Analysis

The online fault detection based on trend analysis methods adopts the three signals in the generalized model of actuator with digital positioner. In the real-world data is always noisy. It is necessary to filter the sampled data. Here, a first-order filter is used for data preprocessing. Sliding window is use to handle data of the three signals to accomplish online fault detection tasks.

3.1 "stick-slip" fault detection by trend analysis method

Under normal conditions, the valve stem moves smoothly; when "stick-slip" fault occurs, the valve travel varies in steps. Fig.9 shows simulation data of the valve travel variations during stick-slippage against time. The enlarged part in it is the behavior of the valve stem at the transition from sticking to slipping. Stick-slip fault is often caused by several factors (seal degradation, lubricant depletion, inclusion of foreign matter, and activation at metal sliding surfaces at high temperatures) that lead to change of friction characteristics of the valve stem when moving. Its seriousness is determined by the friction characteristics. Detailed analysis of this phenomenon can be seen in the past studies (Li C.B., 1982; Kagawa et al. 1993).

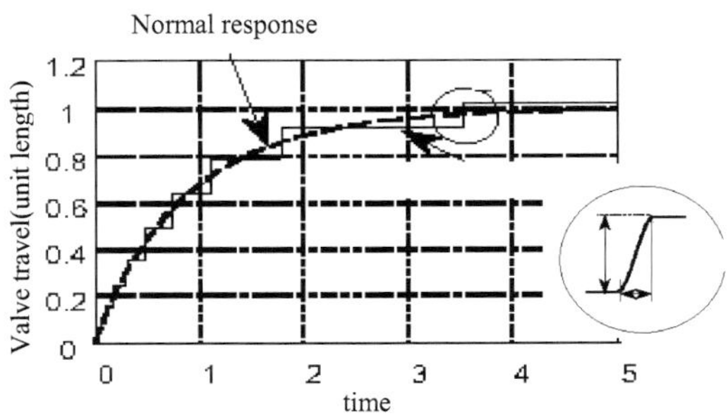

Fig. 9. Behavior of the valve travel during stick-slippage against time

It is very disadvantageous for the control valve when in the stick-slippage condition. In addition to its influence on control performance, such as oscillation in control loops, this 'moving-stopping-moving' state will damage the actuator and reduce its lifetime.

3.1.1 Principle of fault detection of stick-slippage

In order to detect the stick-slip fault, two signals are needed: actual valve travel signal and valve travel setting signal. There is an assumption that the above two measurement signals are fault -free. There is no need to have prior knowledge about the control processes.

The detection method is derived from the previously-described studies around stick-slip phenomena. Under normal operational state, since the movement period is extremely short compared with the response rate of the control valve, the valve stem moves smoothly, and its speed of movement is generally distributed as shown in Fig.10 corresponding to the normal response in Fig.9. When the valve malfunctions due to stick-slippage, the movement of the valve steam is just like "moving-stopping-moving", Fig.4 gives the speed distribution of the valve stem, where the valve stem speed is divided into two moving and stationary states, reflecting repetition of slipping and sticking conditions. From Fig.10 and Fig.11 we can see that at different speed (from small to large), the speed occurrence frequency shows different characteristics. This difference can be represented by the relationship between the mean and root mean square of the stem valve speed magnitudes. In the case of no stick-slippage, the mean is very close to the root mean square of the speed, but they distance from each other when stick-slip fault exists. It is therefore possible to detect valve malfunctions caused by stick-slippage, using the relationship between the differences in shape of these occurrence-frequency distributions.

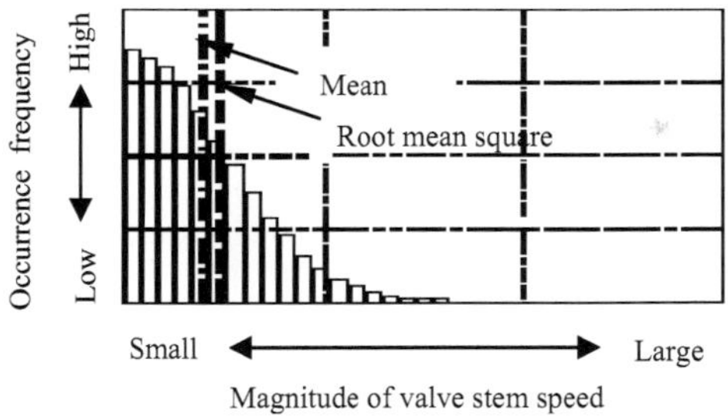

Fig. 10. Speed distribution of the valve stem when no stick-slip fault

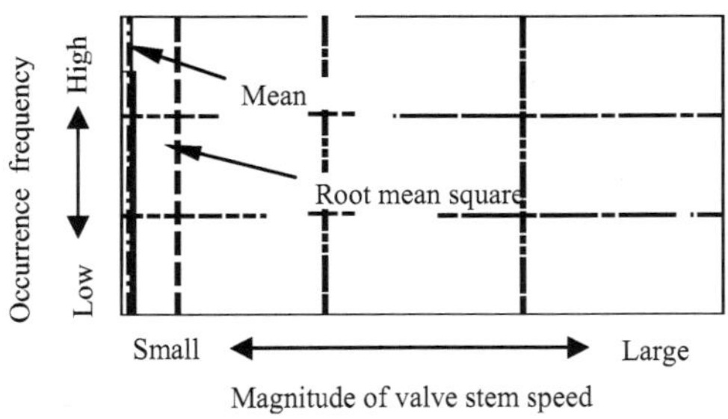

Fig. 11. Speed distribution of the valve stem when stick-slip fault occurs

After the data is filtered, assume s_i and c_i are respectively the values at the ith sampling time of the valve travel and the valve travel setting signals. For continuous and on-line monitoring the state of control valves, a fixed-length sliding window with N sample data is used to see whether the stick-slippage has occurred during the time from the current sample time to the past N sample time.

The ratio of the mean to RMS of the stem speed magnitudes can be then used as an indicator of stick-slippage because it theoretically related to the shape of valve stem speed occurrence frequency distribution. The mean of the stem movement speed $\bar{v}_p$ is calculated as follows:

$$\bar{v}_p = \frac{1}{N}\sum_{i=1}^{N} | v_i | \tag{2}$$

$$v_i = \frac{s_i - s_{i-1}}{dt} \tag{3}$$

where N is the length of observing data, v_i is the speed of ith sample time, dt is the sample time, s_i is the filtered value of the measurement of stem position at ith sample time.

The RMS of the stem speed is:

$$RMS_p = \sqrt{\sum_{i=1}^{N} |v_i|^2 /(N-1)} \tag{4}$$

and the ratio is:

$$r_p = \frac{RMS_p}{\bar{v}_p} \tag{5}$$

In fact, the input command signal to the actuator should be considered because the valve travel setting signal may behave like stick-slippage, even the speed distribution of stem in a healthy actuator will be similar with the case during stick-slippage. Similarly, the mean and RMS of the speed of command signal can be calculated like Equ.(2) to Equ.(4). The r_c is then obtained as:

$$r_c = \frac{RMS_c}{\bar{v}_c} \tag{6}$$

Defining the decision variables D_c and D_p as:

$$D_c = \begin{cases} 1 & r_c > \varepsilon \\ 0 & \text{else} \end{cases} \tag{7}$$

$$D_p = \begin{cases} 1 & r_p > \varepsilon \\ 0 & \text{else} \end{cases} \tag{8}$$

where ε is the tolerance value (or the threshold) to represent the seriousness of stick-slip fault, and the value 1 denotes stick-slip phenomenon, otherwise, the value is 0. The r_p (or r_c) will be close to 1 when no stick-slip fault occurs. When this kind of fault existing, they will increase according to the severity of stick-slip phenomenon. Larger value of r_p (or r_c) means more serious degree of the fault. A tolerance value should be set in order to avoiding frequent alarm in the case of slight stick-slippage, which is not fault. ε can typically be chosen between 2.5~6, and this is also dependent of specific application. The following boolean equation can be used to give a final decision to indicate whether the control valve suffers from stick-slippage fault.

$$\text{Stiction} = D_p \text{ AND } (\text{Not } D_c) \tag{9}$$

3.1.2 Experimental result

In this section, the proposed method will be evaluated on real world data sets, collected from different air baffles and control valves from the DCS in a power plant. The illustrative application of the proposed method is made on the steam superheater control system. There are two control valves installed in this control system as the digital positioners to control flex of water to reduce the temperature of the boiler steam. The first control valves is the valve-A which is installed on the left side in the superheated steam system, and the second on the right side is valve-B. Data are sampled during the operation of the process with the sampling time of 1 second, and calculated on-line. Two cases are illustrated with data related to valve A and B, where the length of sliding window is 100 and ε is chosen as 2.5. Stick-slippage occurs in valve-A while the input signal to the control valve is not like steps. Fig.12 shows 400 sets of sampling data of valve- A. The results of r_p and r_c is shown in Fig.13.

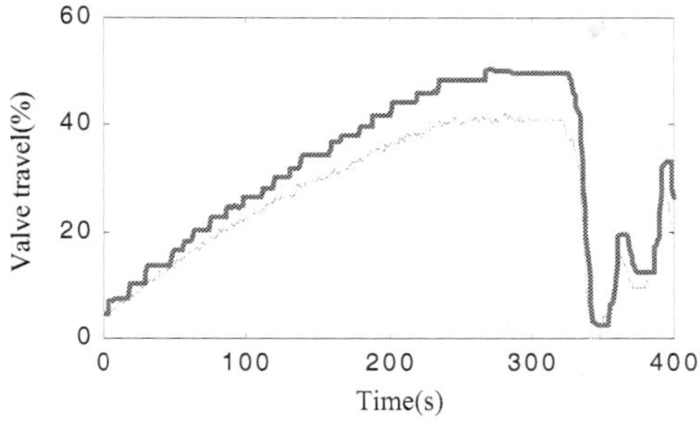

Fig. 12. Valve signals in control valve A, the valve travel setting signal (thin line), the actual valve travel signal (bold line), *dt*=1s

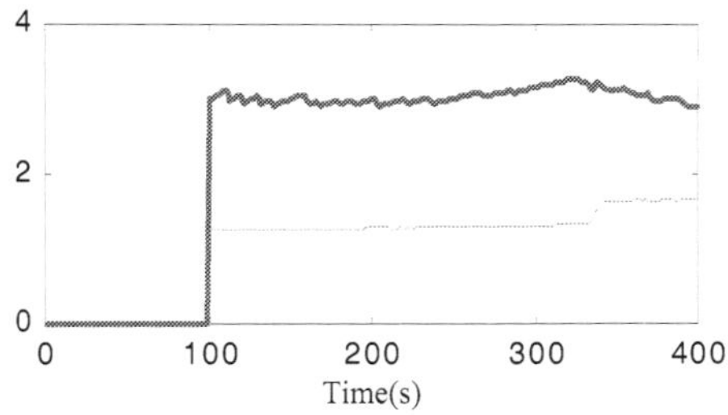

Fig. 13.Results of the ratio of rc (thin line), rp (bold line)

3.2 "serious hysteresis" fault detection by trend analysis method

Serious hysteresis fault comes from too large dead zone with the actuator. Generally, a healthy actuator have little dead zone. This feature shown in Fig.14 is useful for noise suppression, and avoiding frequent movement because of all kinds of signal noise. From Fig.7, obviously, only when the bias between the valve travel setting signal (CV) and the actual valve travel (X) exceeds dead zone (D_z), the valve stem will move. When too large dead zone exists, the actual valve travel cannot trace the valve travel setting signal. It also may result in bad control performance and even oscillation in control loops.

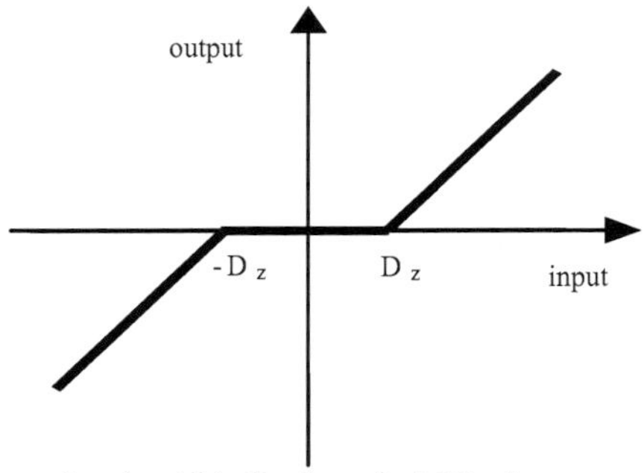

Fig. 14.Results of the ratio of rc (thin line), rp (bold line)

3.2.1 Principle of fault detection of serious hysteresis

From the point of signal trend analysis, when the input signal to actuator (CV) is not step-like signal, it is very difficult to evaluate the size of dead zone only from the two signals (CV and X). However, the size of dead zone can be easily detect from the X-Y graph between CV and X. When the gain of actuator is linear, when the input signal change from 0% to 100%, and then change from 100% to 0%, we can get a X-Y graph, the relationship between the CV and X is like a parallelogram. The dead zone D_z is just one half of width of the parallelogram.

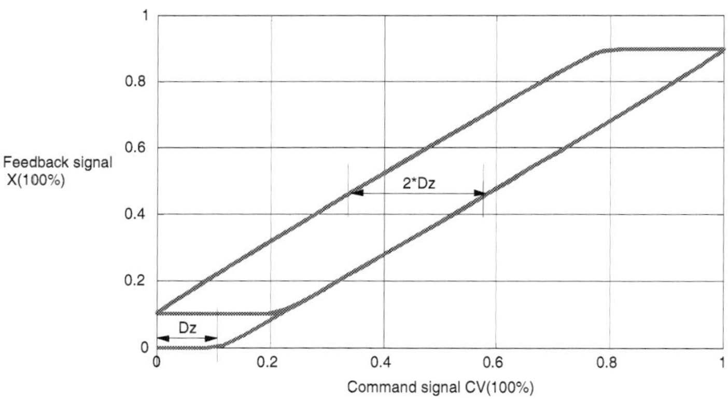

Fig. 15. X-Y graph with dead zone is D_z

Since there must be some non-linearity with practical actuators, the parallelogram representing dead zone will not be a standard graph. In this case, an average value or maximum value can be chosen as the dead zone.

Online fault detection procedure for this kind of fault is done as follows: detecting the CV which changes from a range of increasing and a range of decreasing, online recording the recent data about the actual valve travel signal in this stage. In these data, For each valve of actual valve travel, finding two CV values v_1 and v_2, one (v_1) is in the stage of increasing and the other is in the stage of descreasing. Assume there are N groups of v_1 and v_2 ea. The online estimation value of dead zone is:

$$D_z = \frac{1}{2N}\sum_{i=1}^{N}|v_{i1} - v_{i2}| \quad (10)$$

3.2.2 Experimental result

The normal dead zone of the testing actuator is 4%. Simulation result of fault detection method is shown in Fig. 16. In Fig.16, the serious hysteresis occurs at the time of 40 seconds, and the dead zone changes to 12%. After nearly 30 seconds, the estimation value of dead zone is stable at 12%.

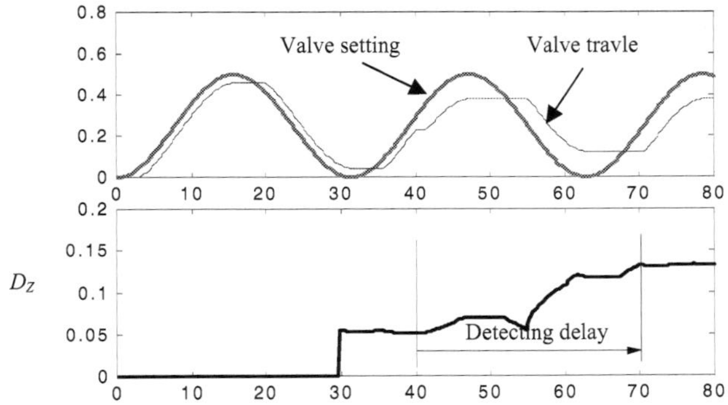

Fig. 16. Estimated dead zone D_z during the fault detection procedure

3.3 "stuck" fault detection by trend analysis method

When the "stuck" fault appears in actuators, the three signals related will change in the following two ways:

a) During a period of time, the valve travel setting changes largely, however, the actual valve travel does not change with the valve travel, and also the flow rate signal does not change.

b) During a period of time, the valve travel setting does not change, and the actual valve travel and flow rate signals keep no change, however, the bias between the valve travel setting and the actual valve become very large.

Online fault detection can be done according to the above two rules.

3.3.1 Principle of fault detection of "stuck" fault

The key point of "stuck" fault detection is how to judge whether one signal is stable or not. Some methods for judging stable state need statistical distribution characteristics of the signal itself. They are not practical in most cases because statistical characteristics can not be easily obtained. Here, we adopt a method of calculating stable factor of signals for this fault detection task.

For a given process variable Z, the specific steps of calculating stable factor is:

(1) In order to eliminate measurement noise, we can use sliding average filter to filter the process data, then get the filtered data Z_{fi};

(2) Get the maximum value Z_1 and the minimum value Z_2 from Z_{fi}:

$$Z_1 = \max_i(Z_{fi}), Z_2 = \min_i(Z_{fi}) \tag{11}$$

(3) Calculate the average value Z_m of data Z_{fi};

(4) Stable factor (SF) is defined as:

$$SF = 100(Z_1 - Z_2)/Z_m \tag{12}$$

Here we use the form of percent to describe the stable degree of signal. For objective comparison between two signals, data size should be the same.

In order to illustrate the effectiveness of SF, some testing experiments on 5 different signals are made, where s_1 is a stable signal with average value 10.0 and standard deviation 0.2; s2 is a signal with average value 10.0 and standard deviation 1.0; s3 a step signal; s4 is a ramp signal; and s5 is a fluctuant signal. These signal are shown respectively in Fig.17. The SF values of s1 to s5 are listed in table2. It can be seen that the stable factor is suitable for judging stable degree of process variables.

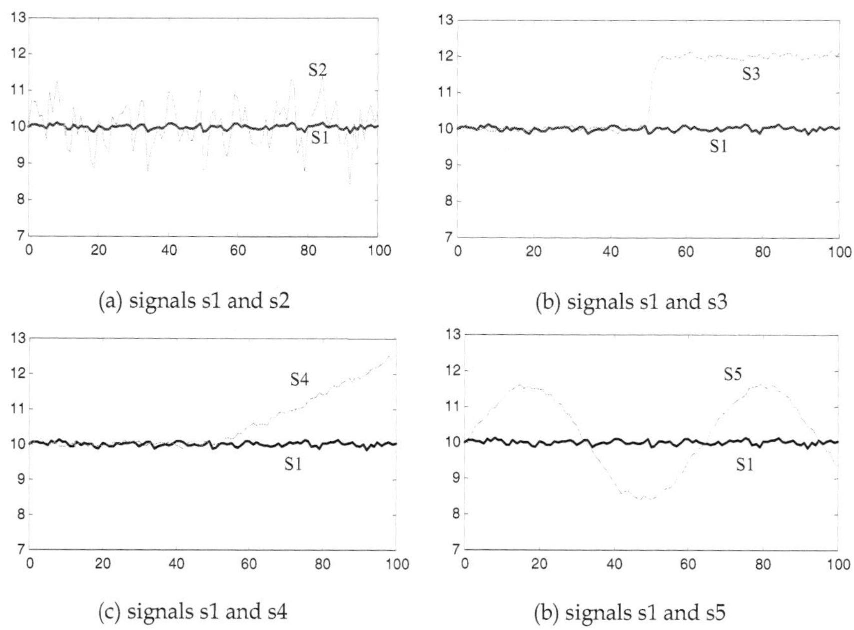

Fig. 17. The five typical signals for testing stable factor

Signal	s_1	s2	s3	s4	s5
SF value	3.02%	29.87%	20.76%	25.08%	31.80%

Table 2. Stable Factor values of s1 to s5

For one sliding window, online fault detection logic is:

(a) If SF of valve travel setting signal is distinctively larger than SF of valve travel, then the "stuck" fault exists;

(b) If the SF values of travel setting signal, valve travel, flow rate are small, but the average bias between valve setting and valve travel is distinctively large, then the "stuck" fault exists.

3.3.2 Experimental result

Fig.18 shows the three signals when stuck fault occurs. Sliding window is use to handle data of the three signals within a period of time, its length is 60 seconds. The fault detection result can be seen in Fig.19.

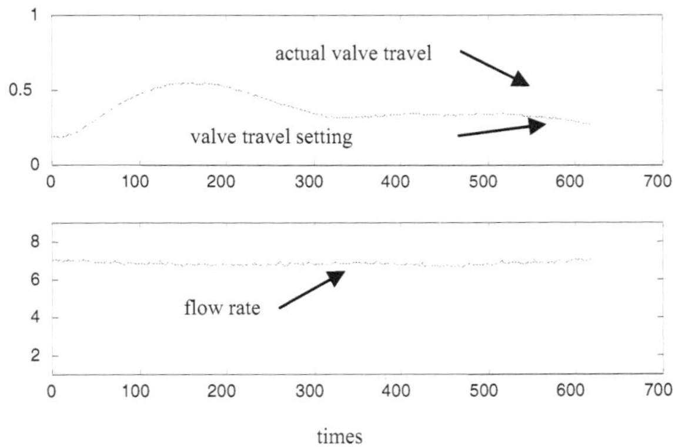

Fig. 18. Signals when the stuck fault occurs

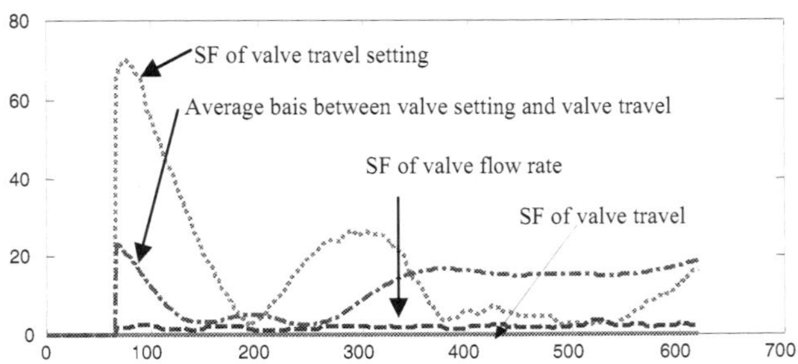

Fig. 19. Estimated stable factors during the fault detection procedure

3.4 "change of gain" fault detection by trend analysis method

The gain of an actuator describes the ratio relationship between its input and valve travel , and can be defined as follows:

$$a = \frac{\Delta x}{\Delta u} \tag{13}$$

where a is the gain factor, Δx is the change of valve travel, and Δu is the change of valve travel setting signal. Here we assume the actuator have only little nonlinearity, that is to say, the gain is nearly linear at different points within the range of the valve travel.

In normal state , the value of gain factor of actuator is close to 1.0. When the gain of actuator changes, the actual valve travel will not trace valve travel setting very well, it may also lead to poor control performance in control loops.

3.4.1 Principle of fault detection of "change of gain"

The process of fault detection is online calculating the gain factor in a sliding window. The estimated value of a can be defined as follows:

$$\hat{a} = \frac{1}{N-1} \sum_{i=1}^{N-1} \frac{x_{i+1} - x_i}{u_{i+1} - u_i} \tag{14}$$

where N is the length of sliding window, x_i is the i-th value of valve travel, and u_i is i-th value of valve travel setting.

3.4.2 Experimental result

Picture (a) in Fig.20 shows the signals of valve travel setting and valve travel when "change of gain" fault occurs. The gain of actuator changed from 1.0 to 0.7 at the time of 100 seconds. Sliding window is use to handle data of the three signals within a period of time, its length is 60 seconds. The online estimated gain factor result is given in Picture (b) of Fig.20.

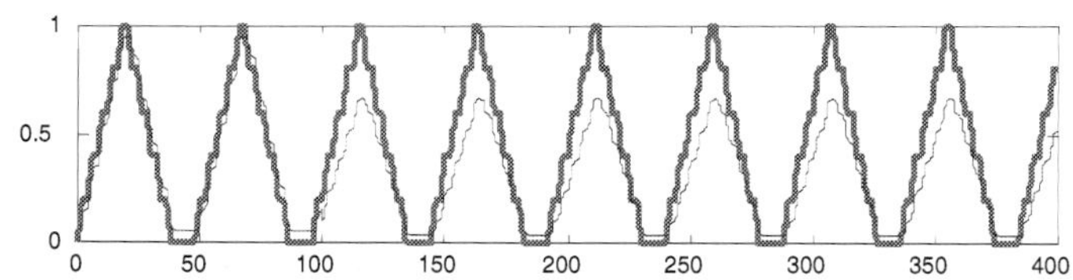

(a) signals of valve travel setting(bold line) and valve travel

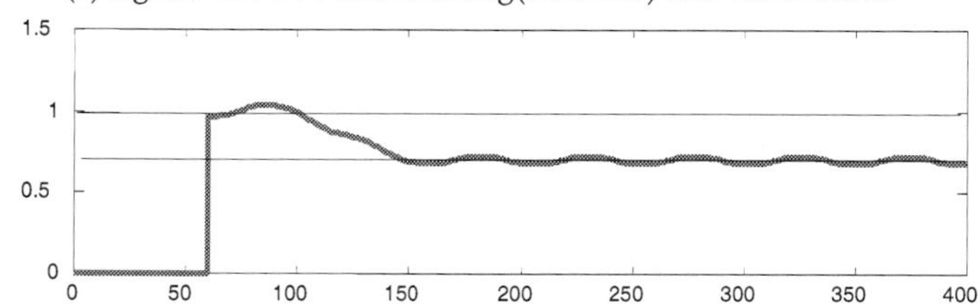

(b) estimated gain factor value of the actuator

Fig. 20. Estimated gain factors during the fault detection procedure

This fault is detected at the time of 160 seconds.

3.5 "constant bias" fault detection by trend analysis method

3.5.1 Principle of fault detection of "constant bias"

The process of fault detection is online calculating the
When "constant bias" fault occurs in an actuator, the bias between the valve travel setting signal and the valve travel will outrange the dead zone and keep a fixed value. Fig.5 shows this abnormal working state of an actuator.
The online fault detection is done by calculating between the valve travel setting signal and the valve travel in a sliding window. The estimated value of constant bias C_b can be defined as follows:

$$\hat{C}_b = \frac{1}{N}\sum_{i=1}^{N}\left|u_i - x_i\right| \tag{15}$$

where N is the length of sliding window, x_i is the i-th value of valve travel, and u_i is i-th value of valve travel setting.

3.5.2 Experimental result

Picture (a) in Fig.21 shows the signals of valve travel setting and valve travel when "const bias" fault occurs. When the actuator works healthy, C_b is small, 4% in this example. A const bias fault appeared at the time of 100 seconds, and C_b changes to 20%. The Sliding window is use to handle data of the three signals within a period of time, its length is 60 seconds. The online estimated const bias C_b is given in Picture (b) of Fig.21., and this fault is detected at the time of 160 seconds.

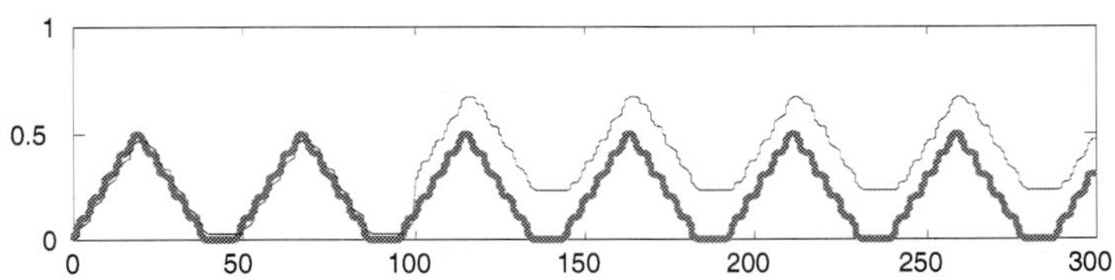

(a) signals of valve travel setting(bold line) and valve travel

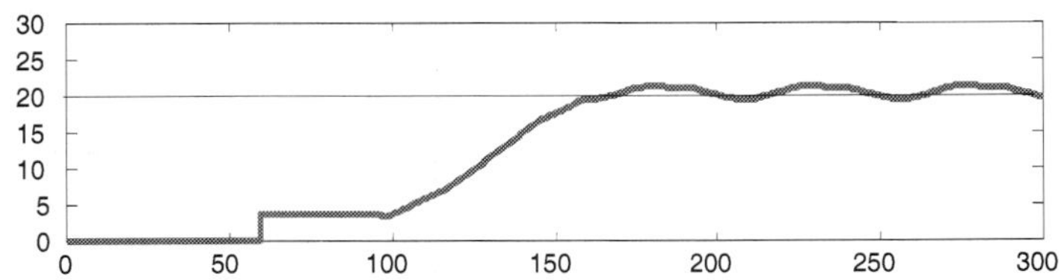

(b) estimated const bias value of the actuator

Fig. 21. Estimated const bias during the fault detection procedure

4. Discussions about trend analysis methods

4.1 Selection of the length of sliding window

The length of sliding window has an effect on the delay time of fault detection. The delay time will increase when the length of sliding window increases. However, using short sliding window will lead to mistaken fault detection results. The selection of the length of sliding window is a trade-off. Especially, we should chose enough length for the "stuck" fault detection to cover the up and down trip of the valve.

4.2 Selection of fault decision thresholds

How to choose suitable fault decision thresholds is another key point for fault detection. According to many experimental results by using real data sample from control processes, we give some guidance when using these trend analysis methods.

For "stick-slip" fault, ε is between 2.5 to 3.0 means that the actuator suffer minor failure, ε is greater than 4.0 means that the actuator suffer severe failure. For "stuck" fault, the SF is less than 5% means the signal is stable, the threshold of average bias between valve setting and valve travel should be greater than the normal value of dead zone. For "serious hysteresis" fault, the threshold should be 1.5 to 2.0 times of the normal value of dead zone. For "change of gain" fault, the variation should be 15% to 20% of the normal gain value. For "const bias" fault, the threshold should be chosen as 1.5 to 2.0 times of the normal value of dead zone.

4.3 Notes when using different trend analysis methods at the same time

In this chapter, a series of fault detection methods are proposed to deal with different typical fault of the actuator with digital positioner. Since each method is used to extract one kind of fault feature for corresponding fault type, they can be used at the same time. When there are more than one faults existing, they will give respectively fault decision results. In order to make it clear, much experimental work is done and the results are listed in Table3. Table3 shows the relationship between these five trend analysis methods for one same fault.

	stick-slip fault	serious hysteresis fault	const bias fault	change of gain fault	Stuck fault
stick-slip fault	----	no effect	no effect	no effect	minor effect
serious hysteresis fault	have effect when fault is severe	----	no effect	no effect	no effect
const bias fault	no effect	minor effect	----	have effect when fault is severe	minor effect
change of gain fault	minor effect	no effect	no effect	----	no effect
Stuck fault	no effect	no effect	no effect	no effect	----

Table 1. Relationship between these five trend analysis methods for one same fault

5. References

Hoefling, Th.; Pfeufer,Th. & Isermann, R. (1995). Observer and signal-processing approach to FDI for the industrial actuator benchmark test. *Control Engineering Practice*, vol. 3, pp. 1741-1746.

Park T.G.; Ryu J.S. & Lee K. S.(2000). Actuator fault estimation with disturbance decoupling. *IEE Proc. Of Control Theory Appl.*, vol. 147(5), pp.501-508.

Edwin Engin Yaz & Asad Azemi (1998). Actuator fault detection and isolation in nonlinear systems using LMI's and LME's. *Proceedings of American Control Conference* pp.1590-1594, Philadelphia, Pennsylvania, August 1998.

Massoumia, M.A. and W.E. Van der Velde (1998). Generating parity relation for detecting and identifying control systems component failures. *Journal of Guidance, Control and Dynamics*, vol. 11, pp. 60-65.

Mediavilla M., L.J. de Miguel, and P. Vega (1997). Isolation of multiplicative faults in the industrial actuator benchmark. *IFAC Symposium SAFEPROCESS '97,* Kingston Upon Hull, UK, pp. 855-860.

Patan K. (2001). Fault detection of the digital positioners using neural networks. *Proceedings of the 7th IEEE International Conference on Methods and Models in Automation and Robotics*, Porland, pp.1085-1090.

Patan, K., Parisini, T., Dynamic neural networks for actuator fault diagnosis: application to the damadics benchmark problem, *proceedings of IFAC Symposium SAFEPROCESS,* 2003, Washington, USA, pp.1077-1082.

Pawel, R., Syfert, M., Jegorov,S.(2003).On-line actuator diagnosis based on neural models and fuzzy reasoning: the damadics benchmark study, P*roceedings of IFAC Symposium SAFEPROCESS*, 2003, Washington, USA, pp.1083-1088.

Deibert, R. (1994). Model based fault detection of valves in flow control loops. *IFAC Symposium SAFEPROCESS '94,* Espoo, Finland, pp. 445-450.

Forsman, K and Stattin, A. (1999). A new method of detecting oscillation in control loops. *Proc. Of European Control Conference*, Karlsruhe.

Hagglund, T. (1995). A control-loop performance monitor. *Control Eng. Practice*, vol.3(11), 1543–1551.

Wallen, A. (1997). Valve diagnosis and automatic tuning. *Proceedings of the American Control Conference*, Albuquerque, New Mexico. pp. 2930-2934.

Horch, A.(1999). A simple method for oscillation diagnosis in process control loops. *Proceedings of the 1999 IEEE International Conference on Control Applications*, Hawai'i, USA, pp.1284-1289.

Li C.B., Pavelescu, D. (1982). The friction-speed relation and its influence on the critical velocity of stick-slip motion, Wear, vol. 82(3).

Kagawa et al. (1993). Analysis of electro-pneumatic positioner valve system considering non-linear characteristics of pilot valve. *Transition of the Soceity of Instrument and Control Engineerings*, vol. 29(11).

15

Rotor Fault Detection in Line-fed Induction Machines Using Complex Wavelet Analysis of Startup Transients

Fernando Briz, Michael W. Degner†, Pablo García,
David Díaz-Reigosa and Alberto Díez
University of Oviedo, †Ford Motor Company
Spain, †USA

1. Introduction

Three-phase inductions machines are the primary consumers of electric power in industrialized countries, they can typically consume between 40 to 50% of all the generated power (Thomsom et al., 2001). Diagnostic testing of induction machines is therefore of tremendous importance in many applications, and has been the focus of intense research for many years (Aller et al., 2002), (Antonino-Daviu et al., 2006), (Briz et al., 2008), (Benbouzid, 2000), (Douglas et al. 2003), (Douglas et al. 2004), (Douglas et al. 2005), (Douglas et al. 2005-b), (Faliang Niu & Jin Huang, 2005), (Nandi & Toliyat, 2005), (Thomson & Fenger, 2001), (Supangat et al., 2006).

Based on the results from failure surveys, the failure mechanism for induction machines can be divided into the following four general categories, with the percentage of total failure shown (Thomsom et al., 2001).

- Stator related: 40%
- Rotor related: 10%
- Bearing related: 40%
- Other: 10%

A number of methods have been developed for the purpose of detecting faults in induction machines, the ultimate goal being to prevent unexpected equipment downtime or severe equipment damage. Classification of the existing methods can be done according to several criteria, including:

- The type of fault that can be detected (stator, rotor, bearing, ...).
- The physical quantities that are measured and further processed for the detection (phase currents, voltages, temperature, infrared, vibration, axial flux, acoustic noise, ...).

This work was supported in part by the Research, Technological Development and Innovation Program of the Spanish Ministry of Science and Education-ERDF under grant MEC-ENE2007-67842-C03-01.

- If they work on-line or off-line. On-line methods do not stop the normal operation of the machine, while off-line methods require discontinuing the normal operation of the machine and often require moving it to a test rig.
- If they can work with constant and/or variable fundamental excitation, i.e., if they are valid for line-connected machines and/or inverter fed machines.

On-line methods are preferred in general for obvious reasons. Of special interest are methods that use terminal electrical quantities (currents/voltages), as they are easily measured. Methods of this type include negative sequence currents/impedances and motor current signature analysis (MCSA) (Kholer et al., 2002), (Lee et al., 2003), (Nandi & Toliyat, 2005), (Thomson & Fenger, 2001). These methods exploit the fact that an imbalanced machine, i.e., faulted, when fed with a balanced, three-phase voltage, produces specific components in the stator currents whose magnitudes and frequencies depend on the level of asymmetry and the cause of the fault condition.

Independent of their attractive properties, these methods have drawbacks that can limit their performance (Kholer et al., 2002), (Lee et al., 2003), (Nandi & Toliyat, 2005), (Thomson & Fenger, 2001). First, the line voltages feeding the machine often contain negative sequence components, which produce spurious current components caused not by imbalances in the machine but by the voltage feeding it. Although it is possible to measure the level of imbalance in the line voltages, it is not easy to accurately estimate their contribution to the overall current, since it depends on the machine's parameters and operating point. A second limitation of these techniques is the dependence of the fault related components on the operating point of the machine, and specifically, on the load level or slip. Several improvements have been proposed to mitigate these drawbacks (Lee et al., 2003), (Nandi & Toliyat, 2005).

1.1 Rotor fault detection

Rotor faults account for a relatively small portion (10%) of the overall failures occurring in induction machines. However, this portion is not uniform for all machines designs and applications. Broken rotor bars primarily occur in medium voltage motors with copper bar rotors. Direct line starting with heavy loads causes high thermal and mechanical stresses; pulsating mechanical loads such as reciprocating compressors or coal crushers (etc.) can subject the rotor cage to high mechanical stresses; an increased risk of rotor failure can exist in these cases (Loránd et al., 2004).

MCSA methods have successfully been applied to rotor fault detection. Damaged rotors produce spectral components in the stator current at frequencies that are function of the slip (and therefore of the rotor speed). The magnitude of these components is a function of the level of asymmetry of the rotor, from which the rotor condition can be evaluated. Regardless of its simple physical principles, use of MCSA for rotor fault detection presents limitations. When the machine operates with light load (small slips), the magnitude of the fault related components decrease, and their frequencies gets close to large fundamental excitation frequency related components, which result in them being difficult to separate. Also, oscillating loads or imbalances in coupled gears and/or mechanical transmissions can produce components in the currents at frequencies similar to those caused by faults, making them difficult to distinguish from a real fault.

The startup transient offers opportunities for performing rotor diagnostics of line-connected machines. When a machine is connected to the line, the startup transient is characterized by large stator (and rotor) currents, as well as by large slips (i.e., rotor speed significantly smaller than the excitation frequency). Damaged rotors create large rotor speed dependent components in the stator current during the startup transient. Analysis of the stator current of line-connected machines during startup provides, therefore, an excellent opportunity for performing rotor diagnostics.

Detection of damaged rotor bars using startup transients requires the use of methods capable of processing the transient signals. Wavelet functions are well suited for this purpose and wavelet based analysis has been proposed for machine diagnostics using startup transient currents (Aller et al., 2002), (Antonino-Daviu et al., 2006), (Briz et al., 2008), (Douglas et al. 2003), (Douglas et al. 2004), (Douglas et al. 2005), (Douglas et al. 2005-b), (Faliang Niu & Jin Huang, 2005), (Nandi & Toliyat, 2005), (Thomson & Fenger, 2001), (Supangat et al., 2006). In most of these works, *standard* wavelets transforms, i.e., wavelet families proven useful in other applications, are used. Discrete wavelet transform based filtering was used in (Aller et al., 2002), Daubechies-8 wavelet in (Douglas et al. 2003), (Douglas et al. 2004), (Douglas et al. 2005), (Douglas et al. 2005-b), (Supangat et al., 2006), Daubechies-40 wavelet in (Antonino-Daviu et al., 2006). In (Faliang Niu & Jin Huang, 2005), a continuous Morlet wavelet is proposed and the analysis was based on an estimate of the torque. This requires measurement of the stator voltage, as well as an estimate of the stator resistance, which significantly complicates the implementation of the method. A significant inconvenience of *standard* wavelets is that no clear criteria for selecting the wavelet function exist, with selection being made *ad hoc*. Also, although the wavelet based analysis can reveal differences between a healthy and a damaged machine, there is not a clear relationship between the physical characteristics of the signal and the results of wavelet based analysis.

Alternatively to the use of standard wavelets, it is possible to design wavelet transforms based on the physical properties exhibited by the currents when a damaged rotor exist (Briz et al., 2008). Since these wavelet respond to a known pattern, the design, as well as the selection of the parameters, are deterministic, with the interpretation of the results of the wavelet transform being straightforward.

A method for rotor fault detection using wavelet based analysis of the stator current complex vector during startup is presented in this work. The chapter is organized in 8 sections. Section 2 introduces the stator current complex vector of three-phase machines, with its nature during start-up transients for both healthy and faulty machines being discussed in section 3. Section 4 introduces complex vector wavelets designed to detect rotor fault related patterns of the stator current complex vector during the startup transient. Discussion on the design of the wavelet function, as well as experimental results demonstrating the viability and performance of the method are presented in Section 5. Section 6 analyses implementation aspects, the conclusions being presented in Section 7. Section 8 provides the reference list.

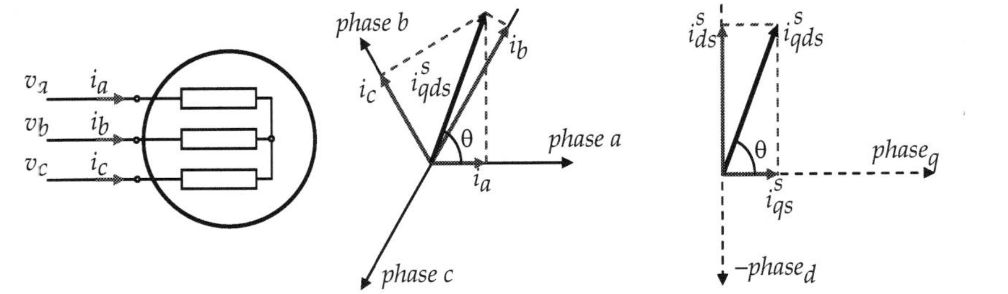

a) Three phase machine b) Three phase system c) Equivalent two-phase system

Fig. 1. Three phase machine and reference coordinates systems.

Table 1. Induction motor parameters (fe = 50 Hz)

Power Rating	0.9 kW, 4 poles
V rated/ I rated (wye)	400 V / 2.3 A (rms)
Rated speed	1425 rpm
# Stator Slots / #Rotor Slots	24 / 30

2. The stator current complex vector

The stator current complex vector is the quantity measured and used in this work for detecting rotor faults in induction machines. This section provides a basic background on the definition and utilization of complex vector variables for the analysis of ac electric machines.

Complex vector quantities (or alternatively qd-axis models) are widely used for the analysis, design, control and diagnosis of three-phase, ac machines and three-phase power systems (Novotny & Lipo, 1996). Given a set of three-phase currents i_a, i_b and i_c (Fig. 1a), the stator current complex vector is defined as:

$$i^s_{qds} = i^s_{qs} - j\, i^s_{qd} = \frac{2}{3}\left(i_a + i_b\, e^{j2\pi/3} + i_c\, e^{j4\pi/3} \right) \tag{1}$$

The transformation in (1) can be visualized as the currents i_a, i_b and i_c aligned with three non-orthogonal axes with a phase shift of 120º from each other (Fig. 1b), being transformed to an orthogonal *q-d* reference frame (Fig. 1c). The stator current complex vector (1) can be represented as a complex quantity or in cartesian form by using its real and imaginary components, i^s_{qs} and i^s_{ds} respectively.

Fig. 2a shows the three phase stator currents of the test machine operating in the steady-state. The parameters of the machine are shown in Table I. Fig. 2b shows the resulting q- and d-axis components obtained using the transformation defined by (1), while Fig. 2c shows the trajectory of the resulting complex vector. For the case of balanced, sinusoidal three-phase currents, the stator current complex vector consists of a *single rotating component*, with its magnitude being equal to the peak value of the phase currents, its frequency of rotation coinciding with the frequency of the phase currents.

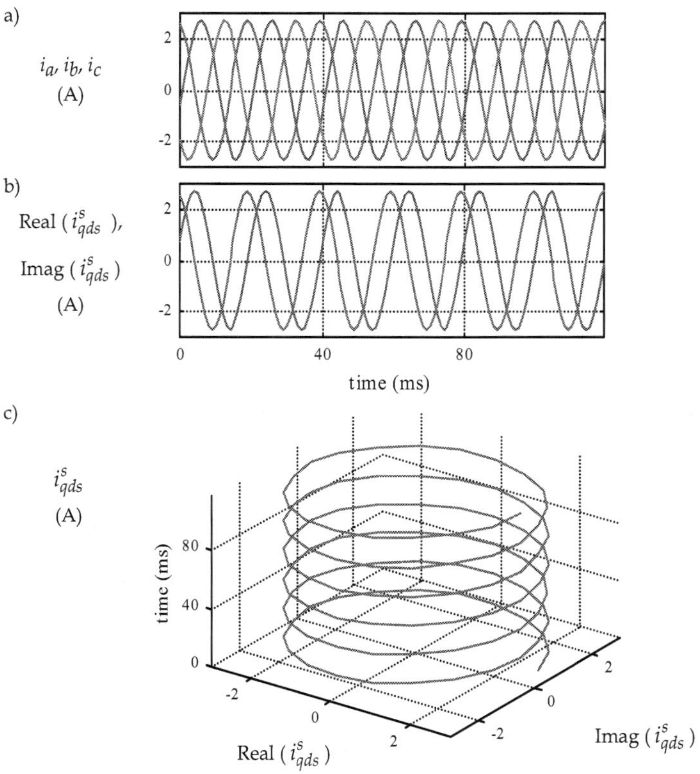

Fig. 2. a) phase current i_u, i_v, i_w, b) q and d-axis components of the stator current complex vector, and c) trajectory of the stator current complex vector with the machine operated in steady-state.

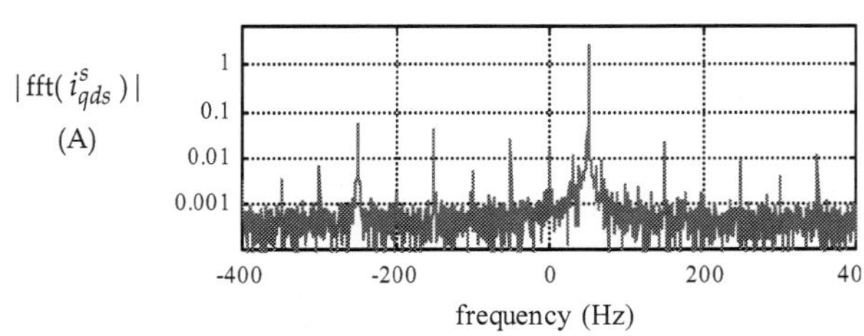

Fig. 3. Frequency spectrum (only magnitude) of the stator current complex vector.

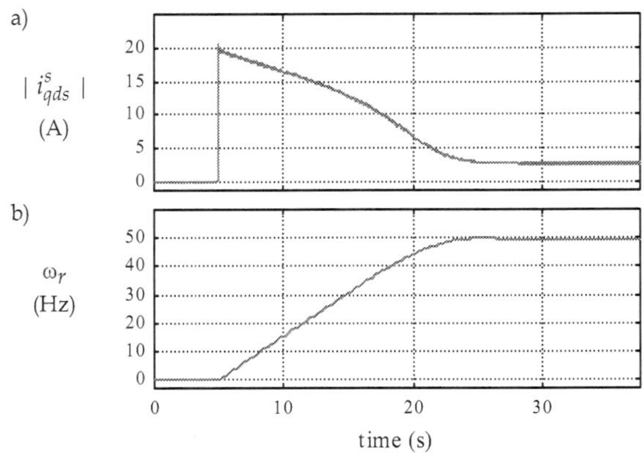

Fig. 4. a) Magnitude of the stator current complex vector and b) rotor speed during startup (400 V rms, 50 Hz), with the motor driving a large inertia load.

Fig. 3 shows the frequency spectrum of the stator current complex vector. Since it is a complex vector quantity, its frequency spectrum contains both positive (forward rotating) and negative (backward rotating) frequency components. It can be observed from the figure that the largest component is at a frequency of ω_e=50 Hz, which corresponds to the fundamental (line) frequency, with a number of additional harmonics of relatively small magnitude being also observable both at positive and negative frequencies.

3. Analysis of the stator current complex vector during startup

Fig. 4 shows the stator current complex vector magnitude and the rotor speed of the test machine after it was connected to the line. Since the transient stator current complex vector after connection is a non-stationary signal, conventional FFT based analysis of the type shown in Fig. 3 is not adequate. Instead, the Short Time Fourier Transform (STFT) can be used for these purposes (Benbouzid, 2000), (Briz et al., 2008). Fig. 5a shows the spectrogram of the stator current complex vector (1), obtained using the STFT, during the transient shown in Fig. 4, for the case of a healthy machine. A sampling frequency of 5 kHz, with a window width of 2048 samples was used.

The most significant component in the spectrogram is at the line frequency, ω_e=50 Hz, and corresponds to the fundamental current. Additional components of reduced magnitude (note the logarithmic scale of the color bar) can also be observed. Horizontal lines correspond to components that do not vary with the rotor speed and have several physical origins, e.g., saturation; additional components in the line voltages; interactions between the fundamental excitation and asymmetries intrinsic to the design of the machine; imbalances in the measurement circuits (current sensors and further electronics), etc. All of these components generally have a small magnitude compared to the signals of interest but they are still visible when a logarithmic scale is used (Briz et al., 2008).

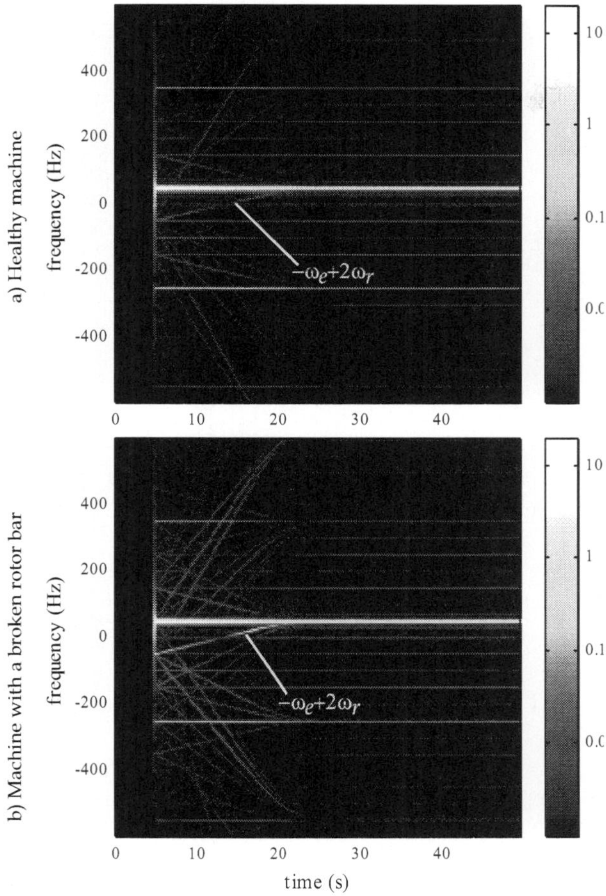

Fig. 5. Spectrogram of the stator current complex vector during startup with the motor driving a large inertia load, for the case of a) a healthy machine and b) machine with a broken rotor bar. Color bar units: Amps.

Rotor speed dependent components can also be observed in the spectrogram. Such components have previously been reported even for the case of healthy machines due to design and construction asymmetries (Benbouzid, 2000).

Fig. 5b shows the spectrogram of the stator current complex vector during a startup transient for the case of a machine with a broken rotor bar. The bar was *disconnected* by drilling the end ring, as shown in Fig. 6. The rotor was modified so that the continuity of the end-ring was maintained and the rotor laminations were not affected. Components of the stator current complex vector caused by a damaged rotor bar are given by (2) (Benbouzid, 2000), where ω_e is the electrical frequency, s is the slip, p the number of pole pairs,

Fig. 6. Rotor modified to break a rotor bar by drilling the end-ring.

and with $k/p = 1, -5, 7, -11, 13, \ldots$. The most important component, (3), denoted as ω_{brb-2}, is obtained from (2) with $k/p = 1$.

$$\omega_{brb} = \omega_e \left(k\frac{(1-s)}{p} \pm s \right) \tag{2}$$

$$\omega_{brb-2} = \omega_e\,(1-2s) = -\,\omega_e + 2\,\omega_r \tag{3}$$

Using (2) and (3), the stator current complex vector can be modeled as (4), where the first term on the right hand side represents the current of a healthy machine, while the rest of the terms represent the rotor fault-induced components.

$$i_{qds}^{s} = i_{qds\ healthy}^{s} + I_{brb-2}\, e^{j(-\theta_e+2\theta_r)} + \Sigma I_{brb-k}\, e^{j((-\theta_e+2\theta_r)k/p)} \tag{4}$$

It can be seen from Fig. 5b that rotor speed dependent components that are readily observable during the startup transient have a significantly reduced magnitude in the steady-state. In addition, all of the rotor speed dependent components converge to frequencies in the steady-state that are spectrally close to the fundamental excitation frequency or its harmonics. This makes their separation in the steady-state difficult, especially if the machine is not heavily loaded. It can be concluded then that significantly richer information for revealing a damaged rotor exists in the current vector during startup than exists during steady-state operation.

4. Detection of damaged rotor bards using complex vector wavelets

The separation of the information revealing a damaged rotor from the overall stator current requires adequate signal processing. The STFT, combined with a spectrogram representation of Fig. 5 is one option. However, while visually insightful, it is not easy to establish a metric that indicates the rotor condition. Wavelets are an efficient tool for pattern detection in transient signals. This can be accomplished by designing the wavelet to resemble the attributes of the signal to be detected.

4.1 Wavelet design

It has been shown that a damaged rotor induces, among others, a component, I_{brb-2} (4), in the stator current complex vector. This component is resembled during a startup transient by a wavelet function ψ, (5), consisting of a complex exponent multiplied by a windowing function, h.

$$\psi(t) = h(t)\ \mathrm{e}^{\,j((tqe(t)+2qr(t))} \tag{5}$$

The complex exponent is a function of the electrical angle, θ_e and the rotor angle, θ_r. These angles can be obtained by integration of the fundamental excitation frequency, ω_e, and the rotor speed, ω_r, respectively, (6). However, while ω_e is known in advance and constant, ω_r is practically never measured in line-connected machines. Because of this, an estimate of the rotor speed during the startup transient ω_{rw} needs to be used. For this work, the estimate $\omega_{rw}(t)$ was obtained using a simple polynomial curve fit method from the $-\omega_e+2\omega_r$ component in the spectrogram shown in Fig. 5a during a commissioning stage, the estimated rotor speed being shown in Fig. 4b. The angle θ_{rw} was then obtained by integrating $\omega_{rw}(t)$ (6).

$$\theta_e(t) = \int \omega_e\, dt\,,\ \theta_{rw}(t) = \int \omega_{rw}(t)\, dt \tag{6}$$

It is noted that the determination of the exact value of $\omega_{rw}(t)$ is not critical, as the wavelet (5) will later be scaled (i.e., stretched or shrunk) to effectively sweep the expected range of startup transient time lengths and shifted to find the exact time in which the startup transient occurred. This is done by defining θ_{rw} (7) and h as a function of *scale* (denoted as a) and *time translation* (denoted as b) (discussion about the design of h is presented in sub-section 4.3). The resulting wavelet function is given by (8).

$$\theta_{rw}(a,b) = \int \omega_{rw}((t-b)\,/\,a)\, dt \tag{7}$$

$$\psi(a,b) = h(a,b)\ \mathrm{e}^{\,j\,(tqe+2qrw(a,b))} \tag{8}$$

Since the wavelet definition (8) relies on an estimation of the *rotor speed shape*, the reliability of the method can be reduced if significant variations in the *rotor speed shape*, due to variations

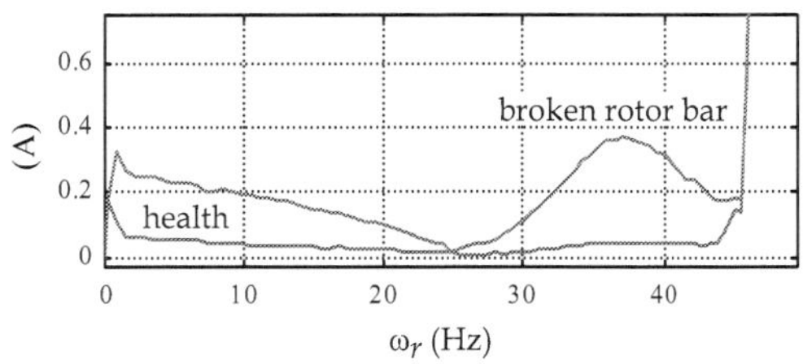

Fig. 7. Magnitude of the stator current complex vector component at $-\omega_e+2\omega_r$ (I_{brb-2} in (4)) vs. ω_r for the case of a healthy machine and a machine with a broken rotor bar, during the startup transients of Fig. 5.

in the load characteristics, occur. It is emphasized however that variations of the transient length do not prevent the method from providing reliable fault detection, as will be demonstrated in section 5.

4.2 Wavelet transform

Once ψ is defined, (8), the coefficient of the wavelet transform C is obtained using (9) (MathWorks Inc., 2007), with '*' standing for the complex conjugate. It is noted that, although the integral symbol is used in (9), it actually operates with sampled signals, with the integration being transformed into a summation in the practical implementation.

$$C(a,b) = \frac{1}{\sqrt{a}} \int \psi^*(a,b)\, i_{qds}^s \, dt \qquad (9)$$

This coefficient is a function of a, i.e., how much the base wavelet is stretched or shrunk, and b, i.e., how much the wavelet is shifted in time with respect to the signal being analyzed. Evaluation of (9) is made by changing a and b at short, regular steps, resulting therefore in a *continuous wavelet transform* (MathWorks Inc., 2007).

Selecting a and b in (9) is straightforward since they are directly related to characteristics of the signal being analyzed. The limits for a are related to the minimum and maximum startup transient time lengths, while the limits for b are related to how accurately the startup transient can be detected.

4.3 Selection of the windowing function

The windowing function h in (5) is required so that the wavelet, ψ, has a finite length and a smooth transition, since the complex exponential term in (5) has a constant magnitude equal to one. Fig. 7 shows the magnitude of the stator current complex vector $-\omega_e+2\omega_r$ component vs. ω_r for the case of a healthy machine and for the case of a machine with a broken rotor bar. They were obtained from signal processing of the spectrograms in Fig. 5a and 5b, respectively. It is interesting to notice that while the differences between the two signals are noticeable during the startup transient, both signals increase significantly and coincide for values of ω_r near ω_e (50 Hz). This is due to the fact that the frequency being tracked (3) approaches the fundamental frequency of the current, which has a much larger magnitude

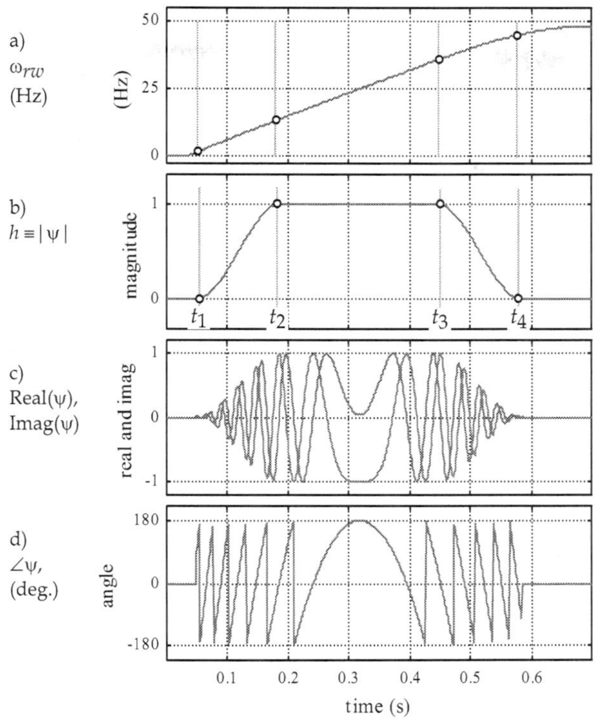

Fig. 8. Construction of the complex vector wavelet, ψ. a) estimated rotor speed, b) windowing function, h, c) real and imaginary components of ψ, and d) phase angle of ψ.

and is difficult to separate from the desired signal. A similar effect is also observed for values of ω_r near 0, caused by the spectral closeness of the $-\omega_e$ component.

To avoid *border effects*, h is chosen to be equal to zero for values of ω_{rw} near 0 and ω_e. This is schematically shown in Fig. 8. The window has two transition regions (t_1–t_2 and t_3–t_4) of variable magnitude and a mid-region (t2–t3) of constant magnitude. The transition regions are defined by (10)-(11), being similar to a *hanning* window. The real and imaginary components of the resulting wavelet function, (5), are shown in Fig. 8c, with its magnitude and phase in Fig. 8b and 8d, respectively.

$$h_{12} = \frac{1 - cos(\alpha_{12})}{2}, \text{ with } \alpha_{12} = \frac{(t - t_1)\pi}{(t_2 - t_1)} \text{ and } t_1 < t < t_2 \tag{10}$$

$$h_{34} = \frac{1+\cos(\alpha_{34})}{2}, \text{ with } \alpha_{34} = \frac{(t-t_3)\pi}{(t_4-t_3)} \text{ and } t_3 < t < t_4 \qquad (11)$$

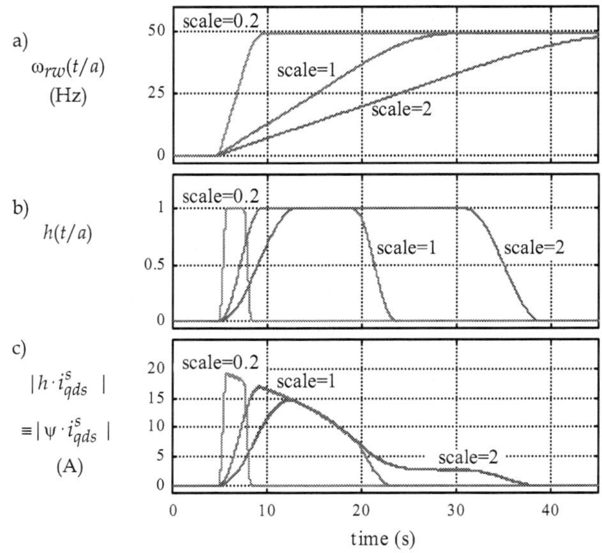

Fig. 9. a) Estimated rotor speed, b) windowing function and c) magnitude of windowed stator current complex vector resulting from applying the windowing functions to the stator current complex vector magnitude of Fig. 1. In all cases b=0.

Although this window will be used for all of the experimental results presented in this work, a variety of windowing functions can be used, each providing similar results (Briz et al., 2008).

5. Experimental verification of the method

The proposed method was tested using the machine in Table I. Fig. 9a and 9b shows the estimated rotor speed ω_{rw} and the windowing function, h, defined by (10)-(11), for different scales of a, with 0.2 and 2 representing the limits over which the wavelet transform (9) was evaluated, and 1 being a nominal case. Fig. 9c shows the magnitude of the windowed stator current complex vector for these values of a.

Fig. 10a and 10b show the magnitude of the C coefficient resulting from the wavelet transform for the case of a healthy machine and a machine with a broken rotor bar respectively. The estimated start of the transient was assigned a time t=0, the b (*time translation*) parameter of the wavelet transform, ψ, being evaluated in a range of –0.2 to 0.3 s around this value. The a parameter was varied within a range of 0.2 to 2 times the base function. The region of Fig. 10b that shows the most significant differences with respect to Fig. 10a is zoomed. Comparing Fig. 10a and 10b it can be observed that the coefficient of the wavelet transform for the case of a healthy machine (Fig. 10a) has a rather constant value (uniform color), independent of the values for *a* and *b*. This means that the wavelet transform

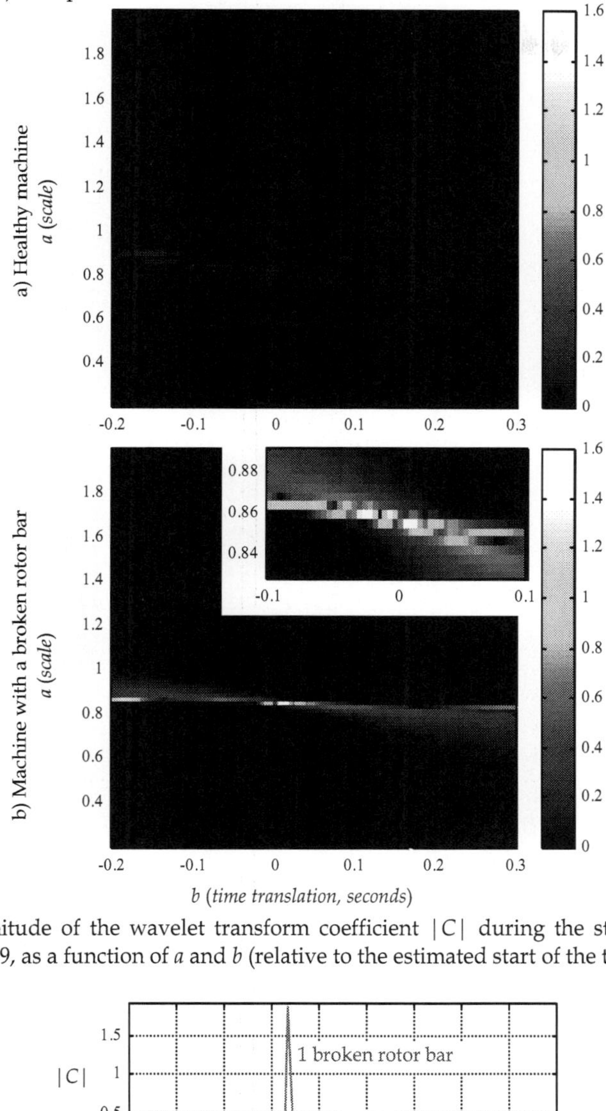

Fig. 10. Magnitude of the wavelet transform coefficient |C| during the startup transient shown in Fig. 9, as a function of *a* and *b* (relative to the estimated start of the transient).

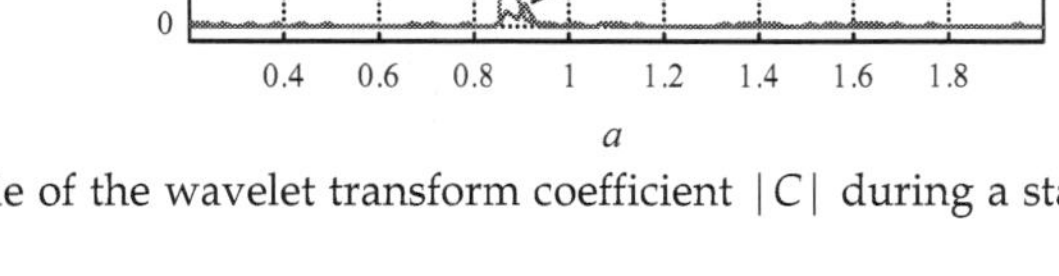

Fig. 11. Magnitude of the wavelet transform coefficient |C| during a startup transient, as a

function of *a* (*scale*), (*b*=–0.04 s), for the case of a healthy machine and a machine with a broken rotor bar.

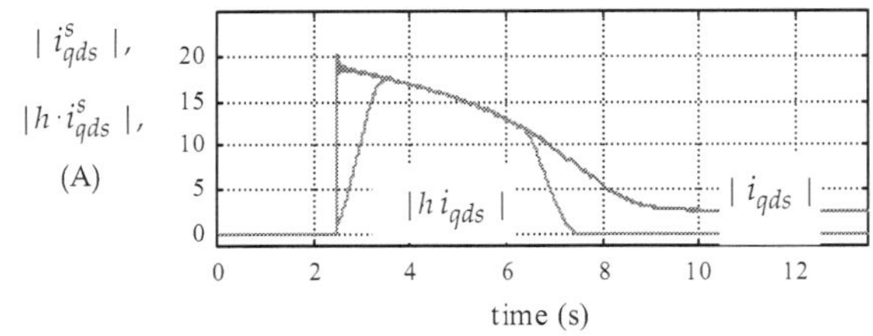

Fig. 12. Magnitude of the stator current complex vector and of the windowed stator current complex vector during startup transient.

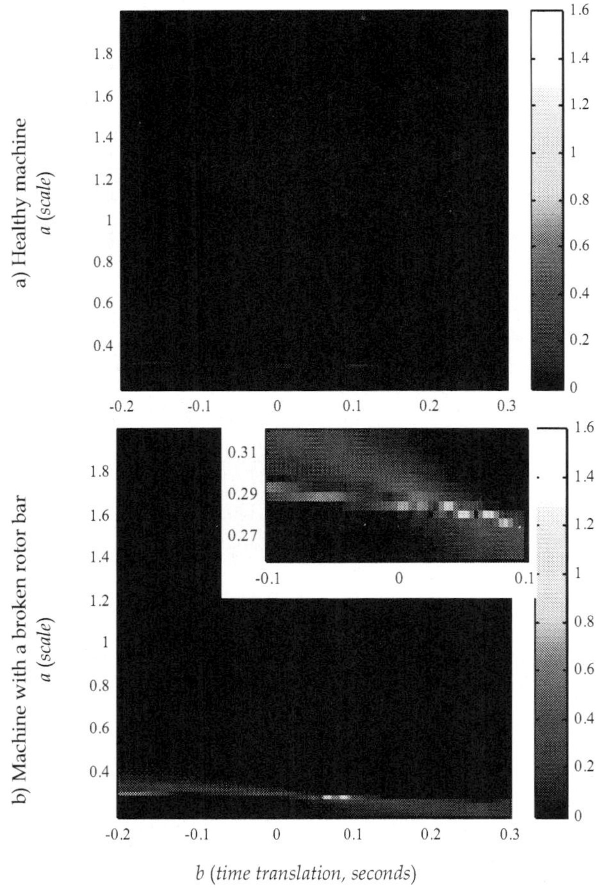

Fig. 13. Magnitude of the wavelet transform coefficient |C| during the startup transient shown in Fig. 12, as a function of *a* and *b* (relative to the estimated start of the transient).

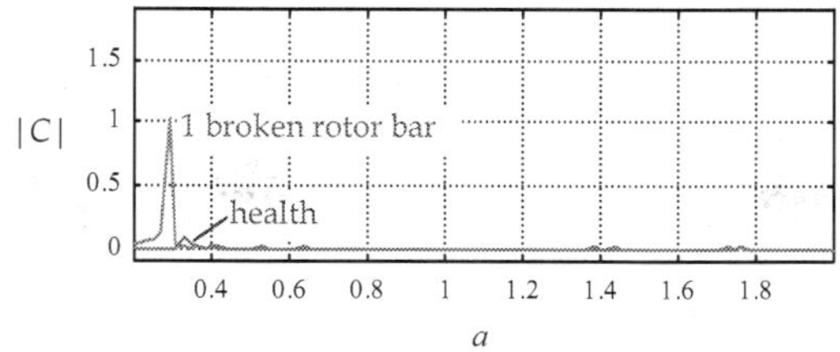

Fig. 14. Magnitude of the wavelet transform coefficient |C| during a startup transient, as a function of *a* (*scale*), (*b*=–0.04 s), for the case of a healthy machine and a machine with a broken rotor bar.

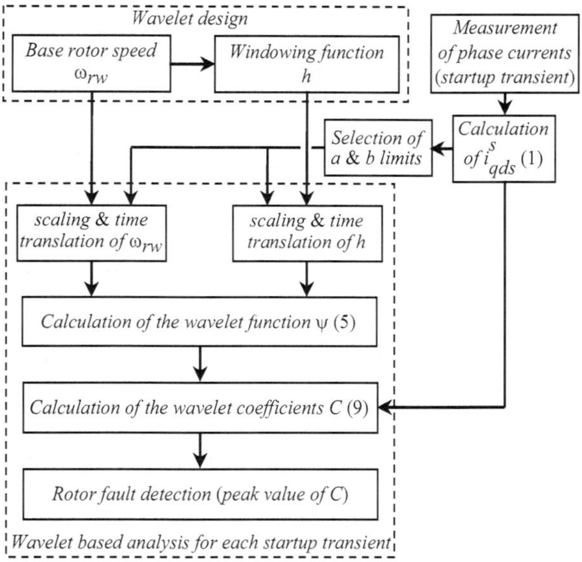

Fig. 15. Flowchart showing the implementation of the method.

effectively *eliminates* components of the stator current complex vector different from the one modeled by the wavelet function ψ. On the other hand, the coefficient C resulting from the wavelet transform for the case of a machine with a broken rotor bar has large values for well defined values of *a* and *b*, which reflects that good correlation exists between the wavelet, ψ, and the signal for these values of *a* and *b*.

Fig. 11 shows the value of C, as a function of *a*, from Fig. 10 for *b*=–0.04 s, for the cases of a healthy machine and a machine with a broken rotor bar. It can be observed that the maximum value of C is obtained for *a*=0.86, i.e., the startup transient was shorter than the wavelet base function.

It is concluded from these results that the proposed wavelet transform is highly sensitive to the properties exhibited by the current of a motor with a damaged rotor during startup. A metric as simple as the peak value of $|C|$ can be used to effectively indicate the condition of the machine.

5.1 Influence of the start up transient length

To assess the robustness of the method against variations in the startup transient characteristics, the analysis was repeated with the machine driving a significantly smaller inertia. Fig. 12 shows the magnitude of the stator current complex vector during the startup, as well as the windowed current. Fig. 13a and 13b shows the magnitude of the coefficient, C, for the case of a healthy machine and of a machine with a broken rotor bar, respectively. Regardless of the significant change in the startup transient's length, the wavelet transform in Fig. 13b clearly displays the presence of a fault. Fig. 14 shows the values of C, as a function of a, for b= 0.06 s, both for the healthy machine and for the machine with a broken rotor bar.

6. Implementation of the method

The proposed method requires the measurement of three current sensors for the implementation described in Section 3 (see Fig. 1a). The use of two current sensors is adequate for these purposes, however, and is a common practice, provided that no zero sequence current exists. It is equally valid for both delta- and wye-connected machines, and for machines with open as well as closed rotor slots. For the experiments presented in this work, conventional Hall effects current sensors and 12 bits A/D converters were used.

Fig. 15 shows the flowchart of the method. For the wavelet design, the rotor speed, ω_{rw}, during a startup transient is estimated and stored. The windowing function, h, is then designed as discussed in Section 5 and stored for later use. The stored ω_{rw} and h are then used for the wavelet based analysis each time that a startup occurs.

The peak value of $|C|$, is used to indicate the condition of the machine. This implies the establishment of a threshold that differentiates a healthy from a faulty machine. This cannot be easily done when no previous data from the machine of interest exists. It is recommended that C_{peak} be measured when a machine is first installed, or is known to be healthy, after that, increments or changes in the value of C_{peak} would indicate a deterioration of the rotor. Tracking the changes in C_{peak} could also be used for diagnostics in machines already installed and for which the condition of the rotor is unknown (with the assumption that the rotor starts out relatively healthy and any increase in C indicates a deterioration in the rotor bar health).

The experimental results shown in Fig. 10 and 13 used a wide range of values for a and b to better show the results of the method. However, to reduce the computational and time requirements of the method, the range of evaluation for a and b parameters can be significantly reduced by simple analysis of the startup transient current vector magnitude (Fig. 4 and 12). The start of the transient can typically be accurately established, which suggests that b could be restricted to a single value. However, it is convenient in practice to allow a narrow range of evaluation for b. This allows for compensation of effects like those caused by the non-ideal (non-repetitive) behavior of the breakers, the variation of the instant in time relative to the period of the phase voltages in which the breakers were ordered to close, as well as for inaccuracies in the rotor speed ω_{rw}. For the experiments presented in this paper, the time translation parameter b was changed in steps of 5 ms, with a range of variation in the order of ±0.1 s relative to the theoretical start of the transient.

As for the selection of the range of a, line-connected machines often show repetitive startup transients, allowing for narrow bounds to a, which would be specific for each application. For the case of machines showing significant variations in the startup transient length, simple threshold based analysis of the startup current can be used to dynamically adapt the limits for a for each startup transient. For the experiments presented in this paper, a was changed in steps of 0.02. The calculation of the C coefficient for each value of a and b, including scaling of h and ω_{rw}, took ≈0.06 s in a standard computer.

In the top-right corner of Fig. 10b and 13b the wavelet transform coefficient is shown with a and b restricted to a reduced range of values, which were automatically obtained by the analysis of startup transient current described above.

7. Conclusion

Broken rotor bar detection in a line-connected induction machine using complex wavelets to analyze the stator currents during startup transients was presented in this work. These wavelets allow for the accurate detection of fault related components in the stator current complex vector indicative of damaged rotors. In comparison with standard wavelet based methods, the wavelet design is made following a well defined procedure. Because of this, limits for the *scale* and *time translation* parameters of the transform are easily established, reducing the computational requirements of the transformation. In addition, interpretation of the results of the wavelet transform is physically insightful. Experimental results confirm the effectiveness of the method in detecting damaged rotor bars.

8. References

Aller, J. M., Habetler, T. G. , Harley, R. G. , Tallam, R. M. , and Lee, S. B. ,"Sensorless Speed Measurement of AC Machines using Analytic Wavelet Transform," *IEEE Transactions on Industry Applications*, vol. 38, pp. 1344–1350, Sept./Oct. 2002.

Antonino-Daviu, J.A., Riera-Guasp, M., Folch, J.R.; Palomares, M.P.M., "Validation of a new method for the diagnosis of rotor bar failures via wavelet transform in industrial induction machines", *IEEE Transactions on Industry Applications*, vol. 42, no. 4, pp:990–996, July-Aug. 2006.

Benbouzid, M.E.H., "A review of induction motors signature analysis as a medium for faults detection", *IEEE Transactions on Industrial Electronics.*, vol. 47, no 5, Oct. 2000, pp 984-993.

Briz, F., Degner, M. W. , García, P., Bragado, D., "Broken Rotor Bar Detection in Line-fed Induction Machines Using Complex Wavelet Analysis of Startup Transients", *IEEE Transactions on Industry Applications*, vol.: 44, nº 3, pp. 760-768, May. 2008.

Douglas, H.; Pillay, P.; Ziarani, A., "Detection of broken rotor bars in induction motors using wavelet analysis"; *IEEE International Electric Machines and Drives Conference IEMDC'03*, Vol. 2, pp: 923–928, June 2003.

Douglas, H.; Pillay, P.; Ziarani, A.K., "A new algorithm for transient motor current signature analysis using wavelets", *IEEE Transactions on Industry Applications*, vol 40, no. 5, Sept. 2004, pp 1361-1368.

Douglas, H.; Pillay, P., "The Impact of Wavelet Selection on Transient Motor Current Signature Analysis", *IEEE-International Conference on Electric Machines and Drives IEMDC*, 2005 pp 80-85.

Douglas, H., Pillay, P. and Ziarani, A. K. , "Broken Rotor Bar Detection in Induction Machines with Transient Operating Speeds," *IEEE Transactions on Energy Conversion*, vol. 20, no. 1, pp. 135-141, March 2005.

Faliang Niu; Jin Huang, "Rotor broken bars fault diagnosis for induction machines based on the wavelet ridge energy spectrum", *Proceedings of the Eighth International Conference on Electrical Machines and Systems ICEMS 2005*, Vol. 3, pp:2274 - 2277, Sept. 2005.

Kohler, J.L., Sottile, J., Trutt, F.C., "Condition Monitoring of Stator Windings in Induction Motors: Part I-Experimental Investigation of the Effective Negative-Sequence Impedance Detector", *IEEE Trans. on Ind. Appl.*, vol. 38, no. 5, pp 1447 - 1453, Sept. 2002.

Lee, S.B., Tallman, R.M., Habetler, T.G.,"A Robust, On-Line Turn-Fault Detection Technique for Induction Machines Based on Monitoring the Sequence Component Impedance Matrix", *IEEE Trans. on Power Electron*, vol. 18, n° 3, pp. 865-872, May 2003.

Loránd, S., Dobai, J.B., Biró, K.A., "Rotor faults detection in squirrel-cage induction motores by current signature analysis", *IEEE-TTTC-International Conference on Automation, Quality and Testing, Robotics*, May 13–15, 2004, Cluj-Napoca, Romania.

MathWorks Inc., *Wavelet Toolbox User's Guide*, Natick, MA, 2007.

Nandi, S.; Toliyat, H.A., "Condition monitoring and fault diagnosis of electrical machines-a review", *IEEE Trans. on Energy Conversion*, vol. 20, no 4, Dic. 2005, pp 719-729.

Novotny, D.W., Lipo, T.A., *Vector Control and Dynamics of AC Drives*, Oxford University Press, New York, 1996.

Supangat R., Ertugrul N., Soong W.L., Gray D.A., Hansen C. and Grieger J., "Detection of Broken Rotor Bars in Induction Motor using Starting-Current Analysis and Effects of Loading," *IEE Proceedings - Electric Power Appications*, Vol. 153, No. 6, November 2006.

Thomson, W.T., Fenger, M., "Current Signature Analysis to Detect Induction Motor Faults", *IEEE Industry Applications Magazine*, May/June 2001, pp. 26-34.

16

Recent Advances in Localization of Winding Deformation in a Transformer

Satish L
Indian Institute of Science
India

Ragavan K
Indian Institute of Technology Gandhinagar
India

1. Introduction

Power transformers are designed to withstand a variety of stresses and mechanical forces during their service life. Abnormal forces generated during short-circuits (occurring close to a transformer) is the main cause of deformation of winding and core. Rough transportation and unskilled handling is the other known cause. The cumulative effect of exposure to such abnormalities can lead to creation of a weak spot. Often, the damage leading to the creation of weak spot is not so severe to grossly impair its normal operation and hence immediately not perceivable. However, these weak spots provide favourable surroundings for the genesis of a fault, which grows and eventually leads to a catastrophic failure. In the present day scenario of deregulation and privatization, any unplanned outage is highly undesirable, and it is needless to elaborate its consequences. Therefore, a highly reliable and early detection of a winding deformation is paramount to enable timely initiation of preventive measures. Transformers are very bulky, so it is desirable to perform such status checks on-site and non-intrusively.

The normal operation of a transformer is the result of a synergic operation of several subsystems, such as, electrical, mechanical, thermal, insulation, etc. Over the years, monitoring and diagnostic tools have evolved to ascertain the status of these subsystems. So far as mechanical integrity of transformer windings and core is concerned, methods based on frequency response measurements, such as sweep frequency and transfer function, have a distinct advantage over the rest. Simply because, not only are the frequency-based methods highly sensitive in detecting even the slightest change in winding geometry (which constitutes a deformation), but can also be employed to estimate its location (as revealed by recent findings). This localization feature, although still in its infancy, makes this tool unique and worthy of further scrutiny. Driven by this motivation, recent progress by authors in localization of transformer winding deformation, using frequency-based technique, is presented.

2. Objective and Background

The frequency response of a transformer is highly sensitive to its winding geometry, distribution of winding capacitances and inductances, ground clearances, type of winding, and so on. Any physical change to the winding (either due to a deformation and/or a displacement) results in a change in some or all of these quantities, which in turn manifests as a change in the frequency response. More specifically, the peaks and troughs in the amplitude frequency response will exhibit a left or right shift. On some occasions, an additional peak may emerge or an existing one may disappear. Detection of any such deviation in frequency response, compared to the healthy frequency response (recorded earlier on the same transformer) implies that a winding damage might have occurred. Once such a condition is detected, the natural questions that arise are, (i) where along the winding is this damage located?, (ii) what is the extent of the damage and its seriousness? For answering these questions, obviously a non-invasive approach would be most attractive, as it would circumvent disassembly of the winding. Disassembling a winding is an expensive and time-consuming exercise, and this should evidently be the last resort. Therefore, the main objective is to demonstrate localization of winding deformation based on frequency response measurements.
A critical analysis of research performed so far, pertaining to this topic, reveals that all earlier efforts can essentially be grouped into the following categories, viz., detecting deformation in transformer windings (Lech & Tyminski, 1966; Dick & Erven, 1978; Rahimpour et al., 2003), assessing sensitivity and correlation between type of fault and measured quantity (Ryder, 2003; Florkowski & Furgal, 2003; Islam, 2000), and, developing circuit models and rational function approximations (Morched et al., 1993; Oguz Soysal, 1993; Gustavsen & Semlyen, 1999). As a matter of fact, there has been no previous attempt to localize deformation along a transformer winding using frequency response data, and hence it is worthy of consideration.

3. Approach

To achieve the goal of localization, it was considered worthwhile to construct a ladder network for representing the winding terminal behaviour, using the measured data. The reason for choosing a ladder network representation for this task is twofold-

1. The mutually coupled, lumped-parameter, ladder network is known to be best suited to represent all the intricacies of impulse behaviour of a transformer winding, and
2. It inherently captures the physical length of the winding, i.e. the physically continuous winding can be visualized as being mapped onto sections of the circuit, starting from line-end to the neutral-end.

In summary, the task essentially turns out to be a circuit synthesis exercise, subject to the constraint that the synthesized circuit must exhibit the same terminal characteristics as that measured. It is possible to synthesize an unique circuit by invoking the properties uniquely exhibited by driving-point functions and assuming a fixed circuit topology.
The approach essentially comprises of measuring the driving point impedance. From the amplitude frequency response plot of driving-point impedance, the frequencies corresponding to the peaks and troughs are identified and are referred to as open-circuit natural frequencies (ocnf) and short-circuit natural frequencies (scnf) respectively. For more details on extracting the ocnf and scnf values from the FRA plot, overcoming noise problems, etc, refer publication (Ragavan & Satish, 2007). This is nowadays a normal activity carried out by utilities and there exist commercial equipment to make these measurements. In addition to this, the effective

resistance, shunt capacitance and inductance are measured. Utilizing these measured data and some design information, the proposed method synthesizes a mutually-coupled, lumped parameter ladder network corresponding to each set of measurement. A comparison of these circuits with the reference reveals the location, as well as, the nature of damage.
When synthesizing a circuit, convergence is deemed to have occurred, when the differences between estimated and measured values of ocnf, scnf, effective capacitance and inductances are simultaneously less than a preset tolerance. Corresponding to every quantity, the error is calculated as-

$$\text{Percentage of Error} = 100 \times \frac{\text{Measured value - Estimated value}}{\text{Measured value}} \tag{1}$$

In addition to satisfying the tolerance, a manual inspection of the amplitude frequency response corresponding to the synthesized circuit and that measured is performed to ensure a one-to-one match throughout the frequency range of interest. Quite clearly, deformation localization in an actual multi-winding transformer is a very complex task, and poses several insurmountable difficulties. Here, a simplified version of the problem is discussed and solutions presented in three parts, with increasing order of complexity.

1. Initially, a model coil is considered. In this arrangement, discrete changes are introduced to a few elements at physically different positions along the model. Ability to accurately localize these changes is demonstrated.
2. Encouraged by the success and potential of the method, the approach is extended to an actual, single, continuous-disc interleaved winding of a transformer. Here, an iterative circuit synthesis approach is developed and localization of discrete changes, based on terminal measurements, is demonstrated.
3. The above iterative circuit synthesis approach is a brute-force method and takes tens of hours to synthesize even a 5-section ladder network. Obviously, localization accuracy achieved with such a small network is grossly inadequate and synthesizing large networks with this approach is simply ruled out. So, an entirely new, highly time-efficient algorithm (based on constrained optimization techniques) capable of synthesising large-sized ladder networks (within tens of seconds) is presented. Thus, in principle, it paves way for extending this method to multiple windings, which is the ultimate goal.

4. Specimens and Measurement Setup

Experiments were conducted on three specimens, viz. a model coil, a continuous-disc interleaved winding, and a continuous disc winding. The model coil consisted of a single-layer winding wound on an insulating air-core former of diameter 200 mm. The winding had 200 turns and taps were provided after every 20 turns. Using these taps, series and shunt capacitances could be externally connected to represent either a continuous disc, interleaved or partially interleaved winding. At the taps, specific changes at any location along the model coil could be introduced. The model coil along with the FRA measurement setup is shown in Fig. 1.
The other two specimens are actual transformer windings, specifically built for this purpose. They are wound on an insulating air-core former, and each had 16 discs with 10 turns per disc. The copper turn had a cross-section of 18 sq mm and the insulation thickness, duct spacing, etc. corresponded to 11-kV rating. It had a total height of 215 mm, inner and outer diameter of

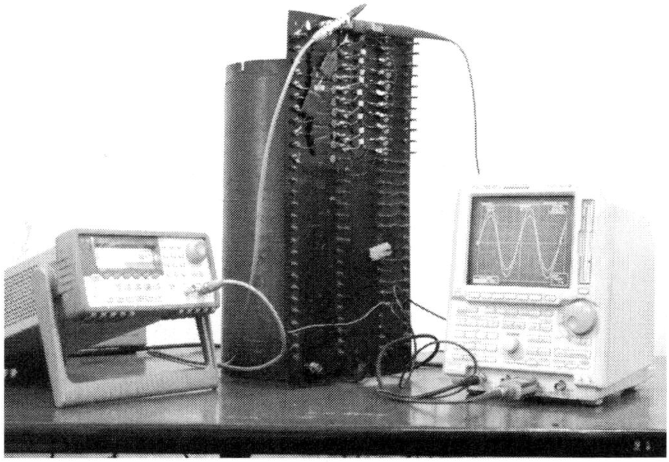

Fig. 1. Experimental arrangement for frequency response measurements on a model coil

260 and 350 mm respectively. Of the two actual windings, the first was fully interleaved, while the second was entirely of disc-type. These windings were placed on an aluminium sheet and another concentrically placed inside to represent the ground plane. Since, the magnetic flux at higher frequencies is predominantly confined to the outer surface of the core, it is justifiable to neglect the presence of the core in this work. In order to introduce discrete changes, the outer and top portions were kept uncovered. Experimental arrangement together with the interleaved winding is shown in Fig. 2.

The effective inductance and ground capacitance were measured by an LCR bridge at 1 kHz, while the dc resistance was measured by a multimeter. The driving point impedance was measured automatically (refer to Satish & Santosh C Vora, 2009 for details) by sweep frequency measurements (and cross verified by the manual sweep method) using (i) an arbitrary waveform generator producing 20 Vp-p sinusoid with frequency range of 0-80 MHz, (ii) a 150 MSa/s, 8-bit digital oscilloscope and (iii) a clamp-on current probe with sensitivity of 2 mA/mV and bandwidth of 450 Hz to 60 MHz. All signals were connected with a 50 Ω co-axial cable and the oscilloscope was terminated by 50 Ω. A sample of the amplitude frequency response for each specimen (reference case) is included while discussing the results.

5. Discrete Changes

Transformer winding deformation is classified into radial, axial, presence of eccentricity, loss of clamping pressure, and so on. The severity and extent of damage, among other factors, depends on the short circuit force magnitude. The damage suffered by the winding is, at times, confined to only a few discs or as most often observed, affects a majority of the discs in the entire winding. So far, in spite of several investigations, there seems to emerge no particular pattern that correlates the nature of winding damage to measurable quantities at

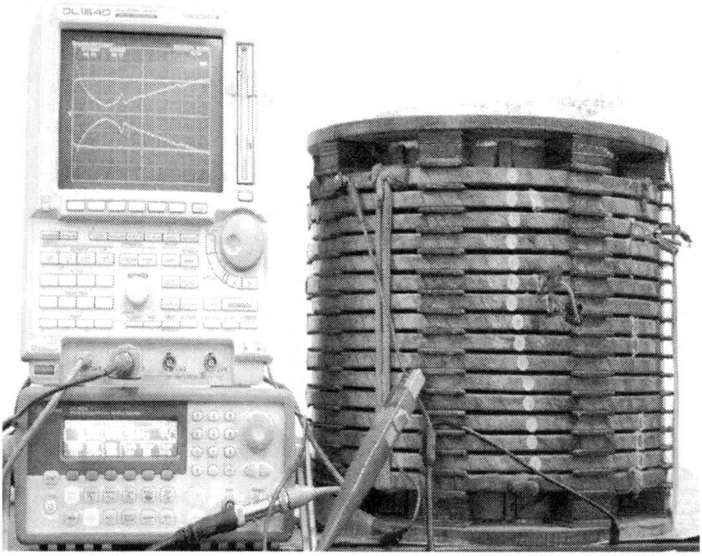

Fig. 2. Experimental arrangement for frequency response measurements on an interleaved winding

the terminals. Inability to generalize this wide variability makes any investigation, and in particular the task of localization, all the more complex. Nevertheless, to make a beginning in this regard, deformations considered in this work are discrete and confined to a small part of the winding. In other words, they are intended to be localized and specifically cause an increase or decrease of the capacitance or inductance in the winding. Since their location and extent is precisely known a priori, it assists in developing and fine-tuning the proposed methodology. Once an acceptable method to address this task emerges, further studies by incorporating actual deformations in windings can be undertaken.

Specifically, deformations in this work are represented by a discrete change of the following two types- (i) addition of some tens of pico-farads of capacitance at a particular disc to ground, resulting in a predominant increase in shunt capacitance, and/or (ii) short-circuiting of a few turns within a disc, introducing a predominantly inductive change. These changes introduced could be more than one in number and be physically occurring at different positions along the winding. Also, any combination of these changes could be introduced in the winding at a time. Thus, it will be interesting to investigate how simultaneous changes can be correctly localized. Also, there may or may not be any net change in the effective capacitance, after these changes are made. After discrete changes are incorporated, a new set of ocnf and scnf are measured along with other parameters. Lastly, the deformations considered in this work are not like the real ones (radial, axial etc.) arising in practice. Despite this, the effort is still considered noteworthy, since deformation localization based on terminal measurements on a transformer winding has, so far, not been attempted.

6. Experimental Results on a Model Coil

6.1 Underlying Principle

The first step in the proposed method involves representing the model coil by means of a ladder network, called the reference circuit. To achieve this, the following quantities of the model coil, resistance, R; equivalent air-core inductance, L_{eq}; effective shunt capacitance to ground, $C_{g,eff}$; and initial voltage distribution constant α are required. Among these, except α, all the remaining quantities are measurable at the input terminals of the model. The initial voltage distribution constant, α, can be got from design details. The steps involved in synthesizing a circuit to represent the model coil are as follows.

- Determine natural frequencies (ocnfs and scnfs) of the model coil by sweep frequency measurements.
- Determine the number of sections (N) of the ladder network to be synthesized.
- Estimate individual values of the elements of the ladder network. Let K_{ref} and L_{ref} be the nodal capacitance and inductance matrices respectively corresponding to the reference circuit.

Details of these steps along with the pseudo code for the algorithm are described in (Ragavan & Satish, 2007). For brevity these details have been omitted. Synthesis of the circuit after introduction of changes essentially consists of the following steps-

- The measurements are repeated after a few elements in the model coil are changed. From the frequency response, ocnfs and scnfs are identified. A comparison of these frequencies with those corresponding to the reference case is made. A deviation greater than 2% would indicate that a new circuit has to be synthesized.
- Using the new set of measured values another ladder network is synthesized. For this purpose, the matrices describing the reference circuit are used as the initial guess. Its elements are iteratively varied until the estimated and measured effective inductance, effective shunt capacitance, and the natural frequencies are well within 2% tolerance.
- Let K_{est} and L_{est} be the capacitance and inductance matrices respectively corresponding to the model coil after changes. A comparison of those matrices with K_{ref} and L_{ref} reveals the elements that have changed. The row and column of the changed element in the matrix reveals the position along the model coil where the changes have been made. Further, the extent of change and whether the change is confined to inductances and/or capacitances can also be ascertained.

6.2 Reference Circuit

The measured magnitude and phase plots of the driving-point impedance corresponding to a continuous-disc model coil (reference or healthy case) are presented in Fig. 3. From the plot, the ocnf and scnf can be extracted and are presented in Table 1. Then the following quantities, viz. $R = 8\ \Omega$, $L_{eq} = 6.98$ mH and $C_{g,eff} = 5.6$ nF were measured on the model coil.
Employing the method described in (Ragavan & Satish, 2007), a ladder network having the same set of natural frequencies as measured with the model coil is synthesized. Details of how the number of sections of the ladder network was decided are described in (Ragavan & Satish, 2007). The self and mutual inductances are estimated iteratively and presented in Table 2. Corresponding to the model coil shown in Fig. 4(a), a 6-section ladder network was synthesized and is shown in Fig. 4(b). The estimated driving-point impedance corresponding

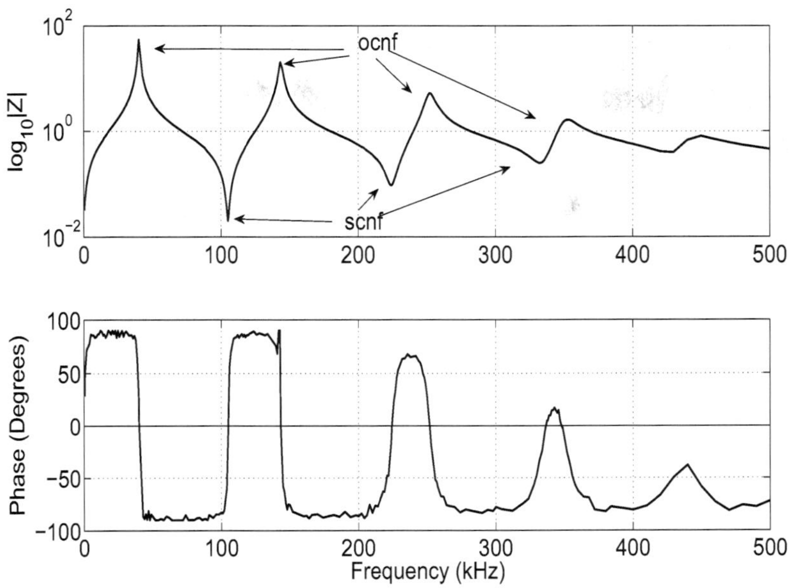

Fig. 3. Measured driving-point impedance for model coil (continuous disc representation) - Reference Case

ocnf	Measured	40.3	143	250.6	346
(kHz)	Estimated	40.48	142.58	249.42	346.41
scnf	Measured	105	223.8	336	-
(kHZ)	Estimated	105.19	224.79	334.44	-

Table 1. Measured and estimated natural frequencies for model coil (continuous disc representation) - Reference Case

L_s	$M_{i,i+1}$	$M_{i,i+2}$	$M_{i,i+3}$	$M_{i,i+4}$	$M_{i,i+5}$
0.4310	0.2392	0.1435	0.0947	0.0612	0.0496

Table 2. Estimated self and mutual inductances (in mH) for circuit in Fig. 4(b)

to the synthesized circuit is shown in Fig. 5. A comparison of the computed driving point impedance with the measured clearly shows the excellent match and accuracy achieved by the synthesized circuit. The next step is to introduce changes and repeat the exercise. The changes made in the model coil were confined only to capacitive changes, since, due to practical difficulties in the available model coil, desired inductive changes could not be made. However, case studies with inductive changes will be presented and discussed in Section 7.

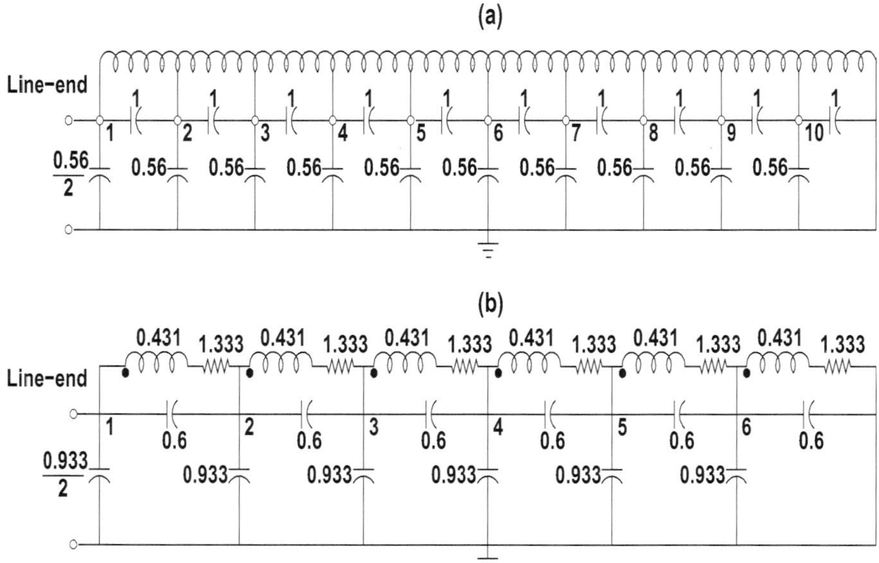

Fig. 4. Reference case (a) Model coil, (b) Synthesized circuit (Note: Elemental values are in Ω, nF, mH. Mutual inductances of synthesized circuit are given in Table 2)

6.3 Case-A, Capacitive Changes Occurring at Physically Connected Taps

In these experiments, capacitive changes are introduced at more than one tap in the model coil. For example, series capacitance between TAPS 3 and 4 was decreased from 1 to 0.56 nF, and shunt capacitance at TAP 4 was increased from 0.56 nF to 1 nF (Fig. 6(a)). Thus, there are two simultaneous changes occurring in the model coil which have to be localized. The new set of ocnf and scnf were measured and are indicated in Table 3. These changes introduced in the model coil manifest as a decrease and increase of the capacitances linked to the NODES 2 and 3 respectively, in the synthesized circuit shown in Fig. 6(b). Thus, it is evident that not only have all changes introduced been accurately localized, their nature has also been correctly identified. Some of the discrete changes introduced in the model coil were not correspondingly reflected by the same number of discrete changes in the synthesized circuit. This aspect should not be construed as an error. This arises because all taps in the model coil are not mapped on to discrete nodes in the synthesized circuit. This matter is explained while discussing mapping between model coil and synthesized circuit, in Section 6.6

6.4 Case-B, Capacitive Changes Occurring at Physically Separated Taps

In reality, mechanical deformations often result in damages that affect different parts of the winding. Hence, it is interesting to simulate such a situation to the extent possible. With this in mind, in this experiment, capacitances at TAPS 1 (line end) and 6 (middle) were simultaneously increased and decreased, respectively, by the same margin (Fig. 7(a)). As a result, the net change in effective capacitance was zero. The corresponding frequencies identified from

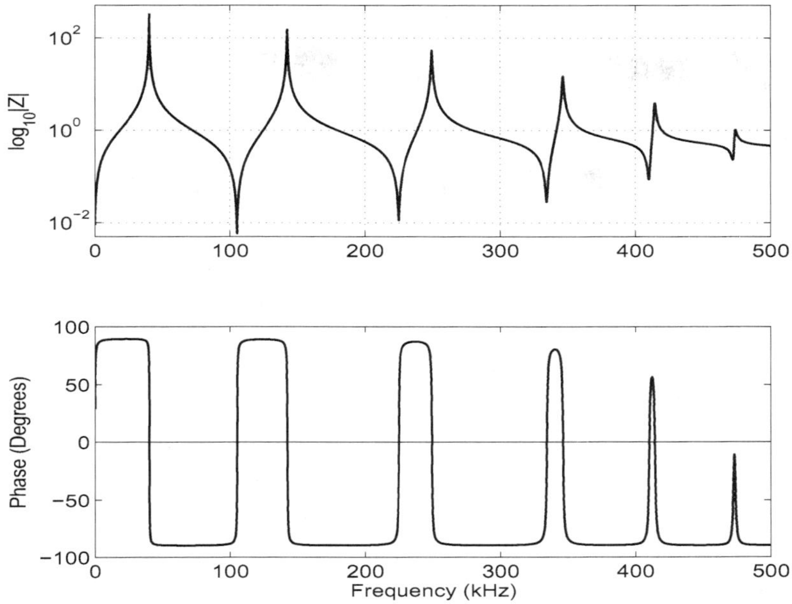

Fig. 5. Driving-point impedance estimated using synthesized circuit in Fig. 4(b)

ocnf	Measured	38.4	144.6	246.8	341
(kHz)	Estimated	39.15	143.68	244.74	346.82
scnf	Measured	101.3	216.7	331	-
(kHz)	Estimated	101.52	219.54	337.52	-

Table 3. Measured and estimated natural frequencies for model coil (continuous disc representation) - Fault Case A

FRA are mentioned in Table 4. The synthesized circuit for this case was obtained following

ocnf	Measured	38.8	141	245.3
(kHz)	Estimated	39.01	140.16	244.03
scnf	Measured	110	224.9	-
(kHz)	Estimated	109.51	224.81	-

Table 4. Measured and estimated natural frequencies for model coil (continuous disc representation) - Fault Case B

the same procedure and is shown in Fig. 7(b). From the results, it is found that the changes

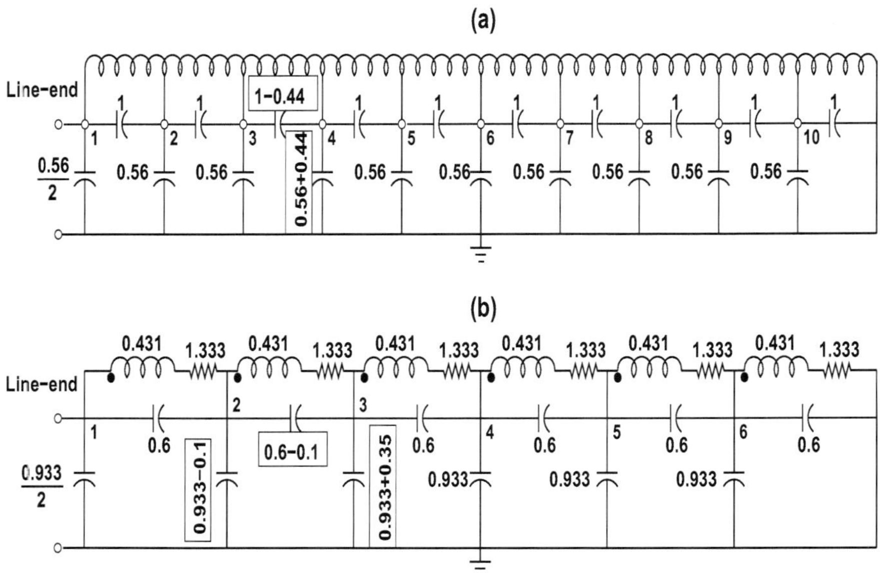

Fig. 6. Fault Case-A (a) Model coil, (b) Synthesized circuit (changes made and predicted are encircled)

were reflected at NODES 1 (line end) and 4 (middle) in the synthesized circuit. Despite no net change in the value of effective capacitance, the method was able to correctly identify the changes made in the model coil. This demonstrates the ability of the method to identify multiple changes occurring at physically different locations along the model coil.

6.5 Case-C, Experiments with Partially Interleaved Representation

The previous two case studies involved a uniform continuous disc representation. It would be interesting to examine the applicability of this method for a partially interleaved representation. In a partially interleaved winding, initial voltage distribution constant would be different for different parts of the winding. In this study, a partially interleaved representation (i.e. 20%) is considered as shown in Fig. 8(a), wherein the first two sections have double the series capacitance compared to the rest. Natural frequencies of this model coil are measured and presented in Table 5. Calculations showed that $N = 5$. Since the percentage of

ocnf	Measured	40.7	151.2
(kHz)	Estimated	40.9	149.27
scnf	Measured	107.62	-
(kHz)	Estimated	107.86	-

Table 5. Measured and estimated natural frequencies for model coil (partially interleaved representation) - Reference Case

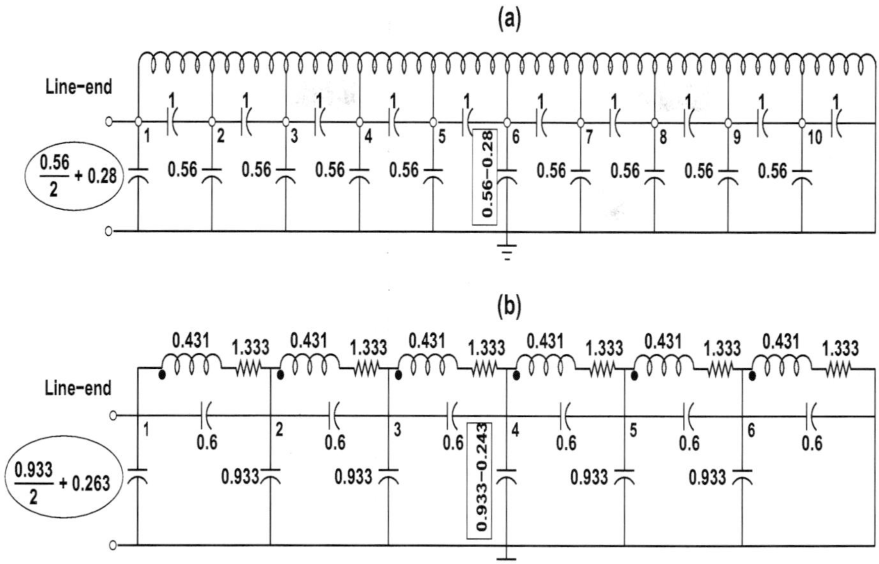

Fig. 7. Fault Case-B (a) Model coil, (b) Synthesized circuit (changes made and predicted are encircled)

interleaving is 20%, an attempt was made to synthesize a 5-section equivalent circuit in the ratio of 1:4 and is shown in Fig. 8(b).

With this model coil, case studies similar to those presented earlier were repeated. In all of them, it was observed that the changes estimated by the synthesized circuit corresponded to the changes introduced in the model coil. One sample result is presented here.

In Fig. 9(a), C_g at TAP 5 was decreased from 0.56 nF to 0.1 nF and C_s between TAPS 1 & 2 and 2 & 3 were increased from 1 nF to 2 nF. In Table 6, new set of measured natural frequencies for this case are given. The corresponding synthesized circuit is shown in Fig. 9(b). The net series capacitance change is only 0.5 nF, which is predicted as a 0.33 nF change in the synthesized circuit. The shunt capacitance change in the model coil is 0.46 nF and this is identified as a change of 0.4 nF in the synthesized circuit. Thus, the changes introduced in the model coil were correctly identified in the synthesized circuit shown in Fig. 9(b). The changed and predicted elements are encircled, in the figure. Thus, it has been shown that the method works equally well for both types of winding representations.

ocnf (kHz)	Measured	42.7	146.6
	Estimated	42.67	146.56
scnf (kHz)	Measured	112.63	-
	Estimated	112.71	-

Table 6. Measured and estimated natural frequencies for model coil (partially interleaved representation) - Fault Case C

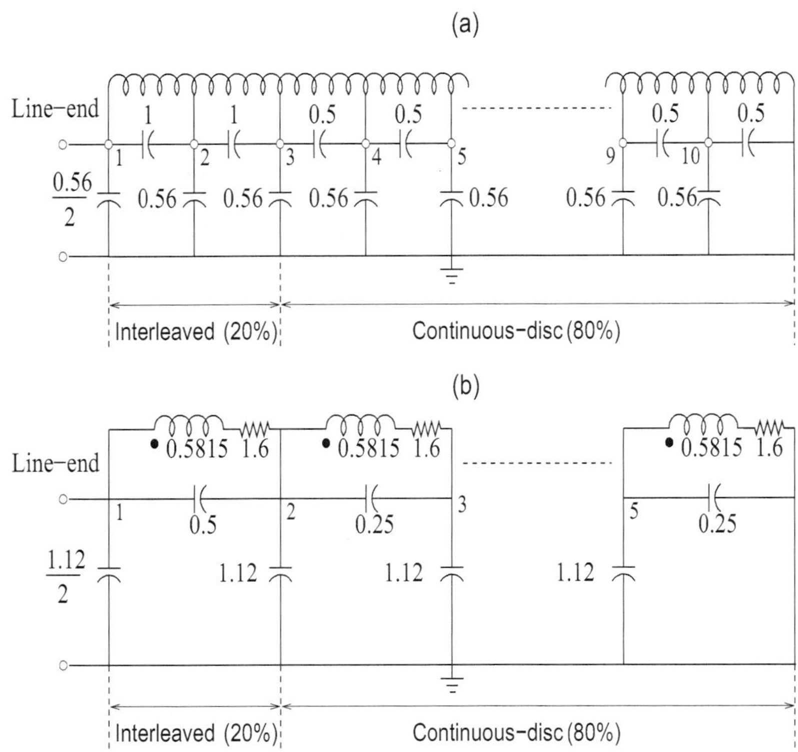

Fig. 8. Reference case (a) Model coil (partially interleaved), (b) Synthesized circuit

6.6 Mapping of Taps in Model Coil to Nodes in Synthesized Circuit

In the case studies corresponding to Fig. 6 and Fig. 7, the actual position of the changes introduced in the model coil were accurately localized, by observing the elements in the synthesized circuits that have changed. From these results, it emerges that a correlation exists between the model coil and the synthesized circuit. This is schematically shown in Fig. 10 along with the mapping results, which link changes introduced at taps in the model coil to the nodes in the synthesized circuit. The next step is to implement the proposed method on an actual transformer winding.

7. Experimental Results on an Actual Transformer Winding (Fully Interleaved)

When considering an actual winding, the following issues are to be addressed additionally as opposed to the model coil.

- In an actual winding, the capacitances (shunt and series) and inductance are all truly distributed, in contrast to discrete series and shunt capacitances used in a model coil and crucially depend on the physical disposition of discs/conductors, clearances to

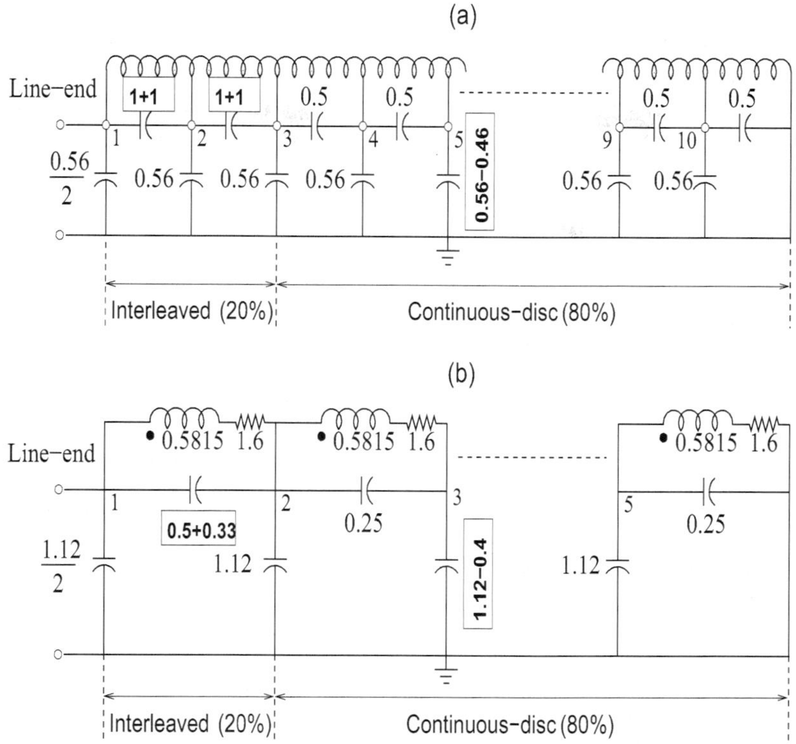

Fig. 9. Fault Case - C (a) Model coil (partially interleaved), (b) Synthesized circuit (changes made and predicted are encircled)

ground, thickness of insulation, etc. Hence, while estimating shunt capacitances, the condition of uniform distribution cannot be invoked here.

- The initial voltage distribution constant (α) in an actual winding, needs to be iteratively estimated, as opposed to a model coil (where series and shunt capacitances can be readily determined). For an interleaved winding (under consideration in the present case), lower and upper bounds for α can be chosen as two and four respectively.
- Capacitive changes (increase and/or decrease) in a model coil were manifested as capacitance changes at pertinent nodes (Ragavan & Satish, 2007). However, when inductance is changed (essentially short-circuiting a few turns), it does not manifest as a discrete change in the synthesized circuit, but affects the self and all mutual inductances (i.e. results in a widespread change to the entire inductance matrix). Therefore, suitable modifications have to be built into the circuit synthesis procedure to account for inductive and/or capacitive changes. Furthermore, due to loss of symmetry in the inductance matrix following an inductive change, each individual element of the in-

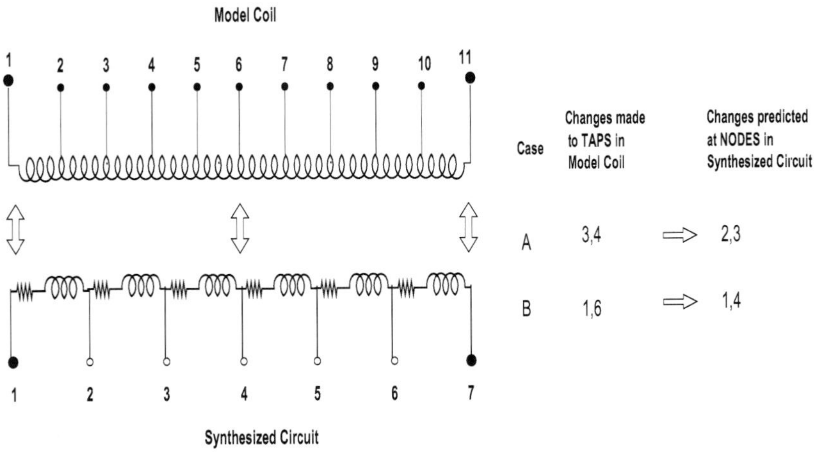

Fig. 10. Mapping of taps in model coil to nodes in synthesized circuit (series and shunt capacitances exist both in model coil and synthesized circuit, but those are not shown)

ductance matrix needs to be iteratively estimated, thereby significantly increasing the computational burden.

- Lastly, whenever an actual mechanical deformation occurs, in order to enable its correct representation, it is obvious that both inductive and capacitive changes have to be simultaneously considered. Keeping in mind this very essential requirement, while formulating a solution for an actual winding, simultaneous inductance and capacitance change feature was built in.
- Compared to the solution for model coil, the solution for an actual winding involves iterative estimation of several quantities, and each of them has to be chosen from a large search-space. To be precise, as opposed to estimating two parameters (viz. self and mutual inductances), a total of five parameters (viz. number of sections, initial voltage distribution constant, shunt capacitance, self and mutual inductances) have to be iteratively estimated in the present case. Thus, synthesizing a circuit for an actual winding is computationally intensive and requires a lot of computer time, even for a reasonably small-sized circuit to be synthesized.

For the above stated reasons, synthesizing a circuit for an actual transformer winding (from measured data) is a far more onerous task, compared to the model coil problem dealt in the previous section.

Lastly, introduction of mechanical deformation to the actual winding (done by other researchers such as Lech & Tyminski, 1966; Dick & Erven, 1978; Rahimpour et al., 2003; Ryder, 2003) was not possible in the present experimental setup. So, as a first approximation, only discrete changes (capacitive and/or inductive) were introduced at selected positions along the winding. Such a procedure achieves the desired goal of introducing a change in inductance and/or capacitance at specified positions on the winding. Furthermore, it assists in checking the correctness of the obtained results.

From measured values of ocnf, scnf, effective capacitance ($C_{g,eff}$), inductance (L_{eq}) and resistance (R), and a knowledge of the type of winding, a lumped-parameter ladder network is iteratively synthesized so as to closely match (within a specified tolerance) the measured driving-point impedance. The circuit thus obtained is called the reference circuit. After introducing a change, all the measurements are repeated. If any of the ocnf or scnf differs by more than 2% of the corresponding reference value, then a change is deemed to have occurred and a new circuit is synthesized. Details of the algorithm is described in (Satish & Subrat K Sahoo, 2009).

7.1 Reference Circuit

The interleaved winding in Fig. 2 is considered for this work. The driving-point impedance corresponding to the reference (or healthy) case was measured and is shown in Fig. 11[A]. From this, three ocnf peaks could be counted, and hence it was taken as the initial value for N. Since, it was an interleaved winding, the initial value of α was taken as 2, and allowed to vary up to 4, in small steps. The effective inductance, capacitance and resistance were measured and found to be 4.9 mH, 0.37 nF and 0.3 Ω respectively. Convergence is obtained when all the estimated and measured ocnf and scnf satisfy the 2% tolerance limit. The converged values of inductance and capacitance were 5 mH and 0.37 nF respectively. This is pictorially observable from the estimated impedance plot in Fig. 11[B]. The number of sections and the value of α were estimated as 5 and 2.8, respectively. Thus, out of a total of 16 discs, roughly 3 discs of the winding are mapped to each section of the synthesized circuit. The synthesized reference circuit is shown in Fig. 12, while the estimated self and mutual inductances are listed in Table 7.

L_S	$M_{i,i+1}$	$M_{i,i+2}$	$M_{i,i+3}$	$M_{i,i+4}$
0.440	0.220	0.090	0.085	0.080

Table 7. Estimated self and mutual inductances (in mH) for circuit in Fig. 12

7.2 Case-D, Inductive Change at Line-end

This example pertains to a condition wherein the inductance has changed predominantly. This is achieved by shorting the outermost turns of disc number 2 and 3. The effective inductance decreased by 200 μH while the effective capacitance remained unchanged. The driving-point impedance was measured and the identified ocnfs and scnfs are presented in Table 8. The

ocnf	Measured	201	399	523
(kHz)	Estimated	204.7	403.3	532.7
scnf	Measured	338	499	-
(kHz)	Estimated	334.8	507.9	-

Table 8. Measured and estimated natural frequencies for an actual transformer winding (interleaved) - Case D

deviation of natural frequencies is beyond 2% with respect to the reference and hence a new circuit (with topology fixed at 5 sections) needs to be synthesized. Following the procedure described, a new circuit was synthesized. The natural frequencies of the synthesized circuit

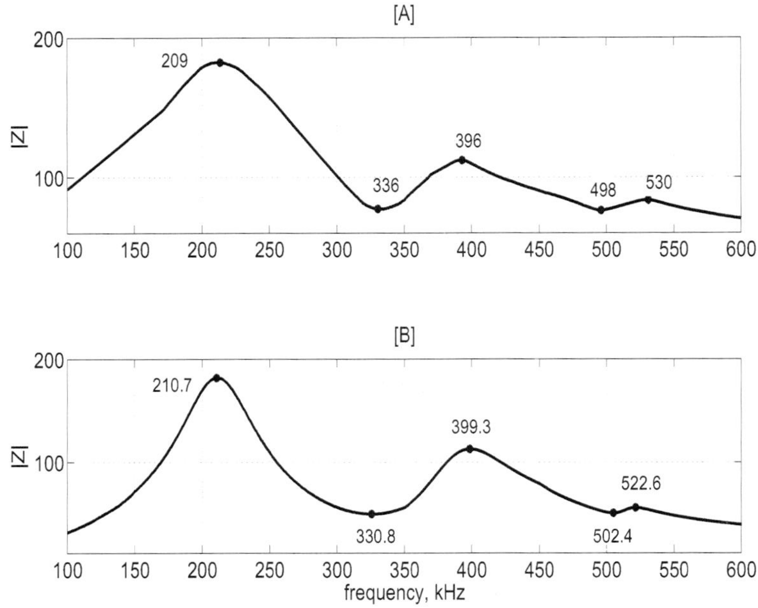

Fig. 11. Driving-point impedance for an actual transformer winding (Interleaved), Reference case, [A] Measured, [B] Estimated

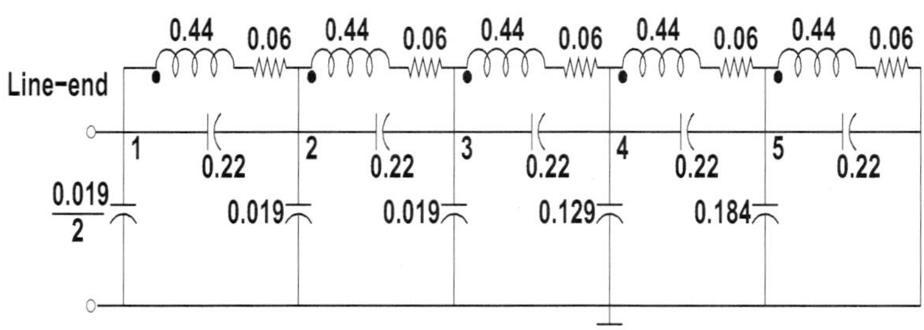

Fig. 12. Reference circuit pertaining to an actual transformer winding (interleaved) (Note: Elemental values are in Ω, nF, mH. Mutual inductances of synthesized circuit are given in Table 7)

are also presented in Table 8. The convergence criterion has been satisfied and the match between estimated and measured characteristics was observed to be good. The converged values of inductance matrix is given below.

$$L_{est,D} = \begin{bmatrix} 0.400 & 0.190 & 0.080 & 0.075 & 0.070 \\ 0.190 & 0.430 & 0.210 & 0.090 & 0.085 \\ 0.080 & 0.210 & 0.435 & 0.215 & 0.090 \\ 0.075 & 0.090 & 0.215 & 0.435 & 0.220 \\ 0.070 & 0.085 & 0.090 & 0.220 & 0.440 \end{bmatrix} \quad (2)$$

Node	Line		⇒	Neutral	
1	9	7	2	2	2
2	7	2	2	0	0
3	2	2	1	0	0
4	2	0	0	0	0
5	2	0	0	0	0

Table 9. Percentage of normalized deviation of inductance from that of reference: Case-D

From the synthesized circuit, it is seen that the capacitance matrix has remained unchanged (signifying fault is not capacitive), whereas, inductance matrix shows many changes. For easier interpretation, the normalized difference matrix (i.e. ratio of the difference between $[L_{est,D}]$ and $[L_{ref}]$ to L_{max}, where L_{max} is the diagonal element of $[L_{ref}]$) is calculated and shown in Table 9. From this it is observed that, maximum change in inductance has occurred at element L_{11}, which refers to Section 1 or line-end. So, the inductive change has been correctly identified, as well as, localized. As mentioned earlier, when there is a short-circuit (i.e. inductance changes), there is observed a widespread change to elements in $[L]$. This is evident in the difference matrix. But, the row and column index corresponding to maximum change (shown as shaded) points to the location of the change. Thus, localization and type of change have been accurately identified.

7.3 Case-E, Inductive Change in the Middle

This is an example wherein the type-of-change is same as in Case-D, but its position has been moved from the line-end (disc 2 and 3) to the middle (disc 8 and 9). All other aspects are the same. The overall change in effective inductance was 220 μH. Results for this case were obtained. Table 10 summarizes the measured and estimated ocnf and scnf.

ocnf (kHz)	Measured	216	399
	Estimated	219.1	407.6
scnf (kHz)	Measured	336	-
	Estimated	340.8	-

Table 10. Measured and estimated natural frequencies for an actual transformer winding (interleaved) - Case E

The match between them is evident. As anticipated, the capacitance matrix has remained unchanged while the inductance matrix shows changes. The normalized difference between

$[L_{ref}]$ and $[L_{est,E}]$ is calculated and is shown in Table 11. Compared to the results of the previous case, the maximum change has now shifted to exactly the middle of the winding, i.e. node 3. Thus, localization and type of change have been accurately identified.

Node	Line		⇒	Neutral	
1	1	2	0	0	0
2	2	2	2	0	0
3	0	2	16	14	3
4	0	0	14	9	7
5	0	0	3	7	1

Table 11. Percentage of normalized deviation of inductance from that of reference: Case-E

7.4 Case-F, Capacitive and Inductive Changes at Physically Different Locations

This case pertains to simultaneous inductive and capacitive changes, occurring at different locations. This was achieved by imposing a short-circuit between the outermost turns of disc 14 and 15, and by addition of a lumped capacitance of 50 pF between outermost turn of disc 6 and ground. With this condition imposed, swept frequency measurements were completed. The

ocnf	Measured	204	402	541
(kHz)	Estimated	200.6	403.8	537.3
scnf	Measured	354	508	-
(kHz)	Estimated	348.8	500.6	-

Table 12. Measured and estimated natural frequencies for an actual transformer winding (interleaved) - Case F

effective inductance showed a change of 200 μH, while effective shunt capacitance changed by 50 pF. Using the proposed approach, the following results were obtained. Table 12 is a comparison of measured and estimated natural frequencies pertaining to this case and from which it is evident that the agreement is reasonably good. The capacitance and inductance matrices of the synthesized circuit are given below from which location and extent of change can be easily determined.

Cg_1	Cg_2	Cg_3	Cg_4	Cg_5
0.019	0.059	0.029	0.129	0.184

Table 13. Estimated shunt capacitance (in nF) for an actual transformer winding (interleaved): Case-F

$$L_{est,F} = \begin{bmatrix} 0.440 & 0.215 & 0.090 & 0.080 & 0.065 \\ 0.215 & 0.435 & 0.215 & 0.085 & 0.070 \\ 0.090 & 0.215 & 0.435 & 0.210 & 0.080 \\ 0.080 & 0.085 & 0.210 & 0.390 & 0.165 \\ 0.065 & 0.070 & 0.080 & 0.165 & 0.350 \end{bmatrix} \quad (3)$$

Node	Line		⇒		Neutral
1	0	1	0	1	3
2	1	1	1	1	3
3	0	1	1	2	2
4	1	1	2	11	13
5	3	3	2	13	20

Table 14. Percentage of normalized deviation of inductance from that of reference: Case-F

From the normalized difference matrix presented in Table 14, it is seen that the last row and column has undergone the maximum change, signifying that the self-inductance in the last section (i.e. neutral-end) has changed. The estimated capacitance matrix (in Table 13) shows that shunt capacitance at node 2 and node 3 has increased by 40 pF and 10 pF respectively. Such a change is understandable, since disc number 6 maps to node 2 in the synthesized circuit. So, the overall estimation seems reasonably accurate, both in terms of localization, as well as, in discriminating the type of change. Also, two changes occurring at physically different locations could be localized.

7.5 Mapping of Winding Discs to Sections in Synthesized Circuit

Localization of inductive and/or capacitive changes introduced at various positions along a transformer winding was successfully demonstrated. The localization accuracy was uniformly good for all cases. As a consequence of these case studies, a mapping between winding discs and the sections of the synthesized circuit can be visualized, and this is shown in Fig. 13. From this, it is evident that approximately every three winding discs are mapped to one section in the synthesized circuit. Further, it is obvious that shorting of discs 6 and 7, 10 and 11, etc. would not map onto a single discrete node in the synthesized circuit. Hence, under such fault condition it is natural that the synthesized circuit would show some changes at the adjacent sections also.

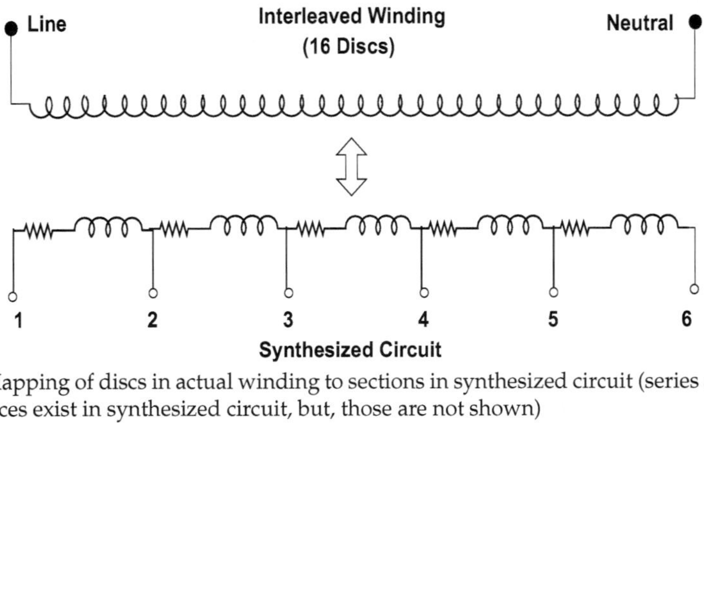

Fig. 13. Mapping of discs in actual winding to sections in synthesized circuit (series and shunt capacitances exist in synthesized circuit, but, those are not shown)

7.6 Scope for Improvement

The proposed method, involving circuit synthesis using measured data, required the selection of some parameters. In the absence of any other reliable method of predetermining them, an iterative solution was pursued. The upper and lower limits of each parameter and their increments were progressively fine-tuned by a trial and error method. Obviously, a compromise had to be made between time required for convergence and the degree of match achievable between measured and estimated characteristics. With a tolerance limit set to 2%, a reasonably good agreement was achievable. The method involved an iterative brute-force search strategy and was very time-consuming. As a matter of fact, the time taken to synthesize even a small 5-section ladder network was of the order of a few tens of hours. This aspect was indeed a shortcoming that prevented its further usage. To make this approach practically attractive, some means of reducing this processing time had to be explored. This matter is explained in the next section.

8. Time-Efficient Algorithm for Synthesizing Large-Sized Ladder Networks

A fully interleaved winding contains very few observable and measurable natural frequencies, typically 2 or 3. So, the circuit to be synthesized was small and could be handled by the iterative procedure. Whereas, in contrast, a disc winding contains a large number of regularly spaced natural frequencies, typically 8 to 10. The corresponding ladder network to be synthesized in this case is likely to have about 10 to 12 sections, i.e., a large-sized network. Hence, the FRA data pertaining to a disc winding was utilized here.
The objective of synthesizing a suitable ladder network corresponding to the measured terminal characteristics of a transformer winding can be construed as a constrained optimization problem. Through optimization techniques it is possible to overcome the shortcoming of the earlier described brute-force iterative search approach. The proposed solution approach, explained below, essentially involves formulating an objective function and minimizing it.

1. Formulation of the objective function
   As the goal is to synthesize a circuit having the same set of natural frequencies as that of the transformer winding, the objective function can be considered as the difference between the measured natural frequencies of the transformer winding ($f_{measured}$) and the estimated natural frequencies of the synthesized circuit ($f_{estimated}$). Hence, the objective function can be written as,

$$g = f_{estimated} - f_{measured} \tag{4}$$

   where,
   $f_{measured}$ - natural frequencies identified from measured FRA plot
   $f_{estimated}$ - natural frequencies estimated using synthesized circuit

2. Enforcing constraints
   As the problem is open-ended, obviously it will have multiple solutions. In order to limit the number of feasible solutions, the following constraints need to be enforced. Hence, the objective can be restated as, minimizing the objective function g, subject to the following constraints.

   - As the effective value of the shunt capacitance of the synthesized circuit should match with that of the measured value on the transformer winding, one of the equality constraints can be formulated as,

$$C_{g,eff}(estimated) = C_{g,eff}(measured) \tag{5}$$

- For the chosen configuration of the winding (as in Fig. 2), shunt capacitances can be chosen to be equal in magnitude up to 50% of the physical length of the winding, beyond which, it can be considered to monotonically increase in magnitude due to the closer vicinity of the ground plane to the discs closer to the neutral. This equality and inequality constraints on the disposition of shunt capacitances are mentioned below.

$$\begin{array}{ll} C_g(1) = C_g(2) = \cdots = C_g(\frac{N}{2}) & \\ C_g(k) < C_g(k+1), & k = \frac{N}{2}, \cdots, N-1 \end{array} \tag{6}$$

- Equality constraint on the effective value of the inductance of the winding is -

$$L_{eq}(estimated) = L_{eq}(measured) \tag{7}$$

- The mutual inductance between a given disc and other discs decreases as their separation distance increases. This fact can effectively be captured by describing an inequality constraint (connecting mutual inductances) as given below.

$$\begin{array}{l} m_1 < l_s \\ m_k < m_{k-1}, \quad k = 2, \cdots, N-1 \end{array} \tag{8}$$

The algorithm described above was coded and initially tested with FRA data presented in Section 7, on a fully interleaved winding. The reference circuit was obtained within a matter of few seconds and the estimated and measured characteristics matched very well. In order to test the capability of this alternative approach to synthesize large-sized ladder networks, the FRA data pertaining to a disc winding (with 16 discs) was considered next. It contained about 8-9 natural frequencies from which a 9-10 section ladder network could be anticipated to be synthesized. The important steps in realization of the circuit are explained in the following sections.

8.1 Circuit Synthesis for Disc Winding - Reference Case

The terminal characteristics of a disc winding pertaining to the un-faulted (or healthy) case is considered and the corresponding synthesized circuit is referred to as the reference circuit.

- Parameters pertaining to the disc winding are measured and mentioned below.
  R = 0.3 Ω, L_{eq} = 2.03 mH, $C_{g,eff}$ = 0.22 nF.
- With the neutral of the winding grounded, FRA is performed on the disc winding and is shown in Fig. 14(a). From the acquired FRA data, open-circuit and short-circuit natural frequencies are identified and tabulated in Table 15.

(NOTE: The measured natural frequencies of the disc winding show frequencies up to about 5.5 MHz. In practice one observes these natural frequencies well within 1.0 to 1.5 MHz. Since the transformer winding was small in size, it was natural to expect the natural frequencies to be scaled up by the same factor. Intentionally, no scaling was employed to the measured winding parameters. Therefore, using this raw measured data the resulting synthesized circuit had low values of capacitances and inductances. This was naturally anticipated. As the main aim of this section was to examine the

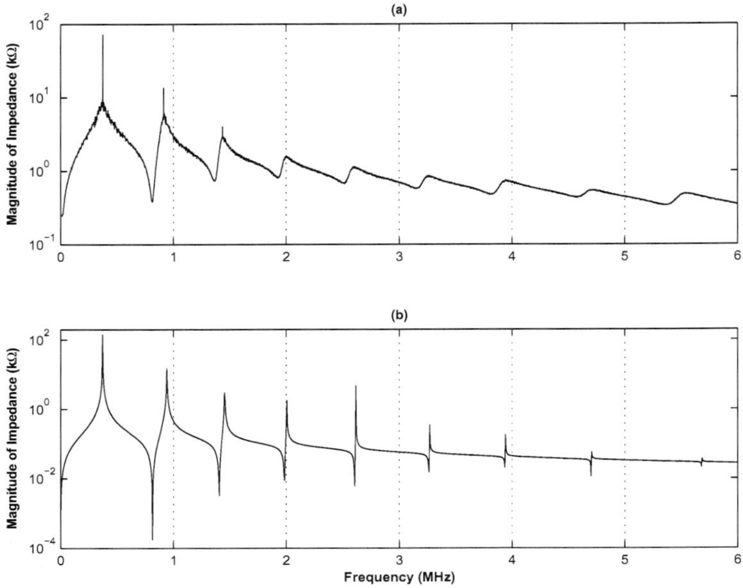

Fig. 14. Driving-point impedance for an actual disc winding, Reference Case, (a) Measured, (b) Estimated from synthesized circuit shown in Fig. 15

ocnf	Measured	370	927	1463	2033	2650	3314	4008	4765	5618
(kHz)	Estimated	371	942	1453	2005	2616	3270	3942	4703	5480
scnf	Measured	832	1391	1963	2552	3198	3872	4655	5332	-
(kHz)	Estimated	816	1405	1984	2605	3265	3939	4702	5480	-

Table 15. Measured and estimated natural frequencies for a disc winding - Reference case

capability to convert an FRA data into a synthesized circuit, no scaling of data was undertaken prior to using them.)

Since the FRA exhibits 9 open-circuit natural frequencies, number of sections in the circuit to be developed should be greater than or equal to 9. The proposed algorithm converges to the circuit described in Fig. 15. Values of mutual inductances between any two inductors of the synthesized circuit are given in Table 16. The circuit so synthesized has the natural frequencies (Fig. 14(b)) which are within 3% of tolerance level of those identified from FRA in Fig. 14(a).

L_s	$M_{i,i+1}$	$M_{i,i+2}$	$M_{i,i+3}$	$M_{i,i+4}$	$M_{i,i+5}$	$M_{i,i+6}$	$M_{i,i+7}$	$M_{i,i+8}$	$M_{i,i+9}$
0.0434	0.0325	0.0214	0.0142	0.0100	0.0067	0.0053	0.0037	0.0030	0.0023

Table 16. Self and mutual inductances (in mH) for circuit in Fig. 15

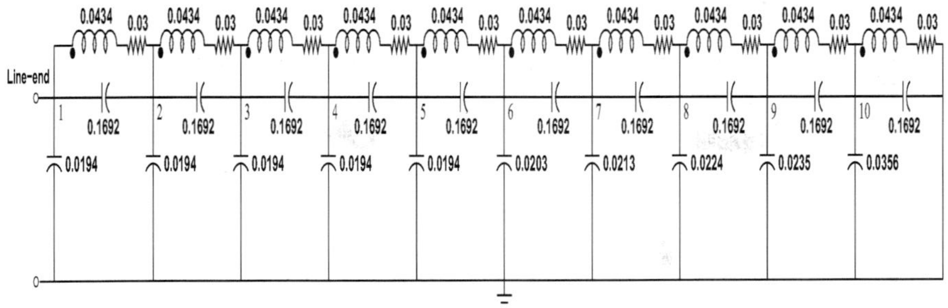

Fig. 15. Synthesized reference circuit pertaining to disc winding (Note: Elemental values are in Ω, nF, mH. Mutual inductances of synthesized circuit are given in Table 16)

8.2 Circuit Synthesis for Disc Winding After Changes - Fault Case

To simulate a fault case, an additional capacitance of 30 pF was added between the start of disc 8 and ground. How this change gets reflected in the synthesized circuit is explained below.

- Parameters pertaining to the changed system are determined.
  $R = 0.3\ \Omega$, $L_{eq} = 2.03$ mH, $C_{g,eff} = 0.249$ nF.
- For the changed system, FRA is performed and shown in Fig. 16(a). From the FRA data, the natural frequencies are identified and tabulated in Table 17.

ocnf (kHz)	Measured	351	870	1448	1981	2608	3245	3960	4691	5546
	Estimated	355	909	1453	1981	2616	3249	3942	4695	5452
scnf (kHz)	Measured	743	1368	1911	2512	3124	3820	4572	5351	-
	Estimated	756	1405	1963	2605	3241	3939	4694	5452	-

Table 17. Measured and estimated natural frequencies for a disc winding - Fault case

The algorithm converges to the synthesized circuit shown in Fig. 17 within 10-20 seconds. However, a few of the frequencies of the synthesized circuit (Fig. 16(b)) differ from those measured by 3-5%. To start with, this result can be taken as acceptable. This difference can be reduced further by fine-tuning the algorithm. However, it was not undertaken at the present moment. Perhaps, it may also be due to the fact that a winding with 16-discs is mapped on to a circuit with 10-sections. The accuracy achieved with the proposed optimization is reasonable. Further, it is worth mentioning that the algorithm converges within a very short time (in the order of a few tens of seconds). Hence, it is possible to address the following needs viz. finer representation of the winding, circuit realization of multi-winding transformer, etc. This was not achievable by the earlier approaches in spite of powerful computing machines. Research in this direction would continue and aid in localization of winding deformations and/or faults and in quantifying them in an actual multiwinding transformer.

9. Scope of Future Work

Some thoughts for future work revolve around the localization problem on multiple winding transformers and can be tackled as follows-

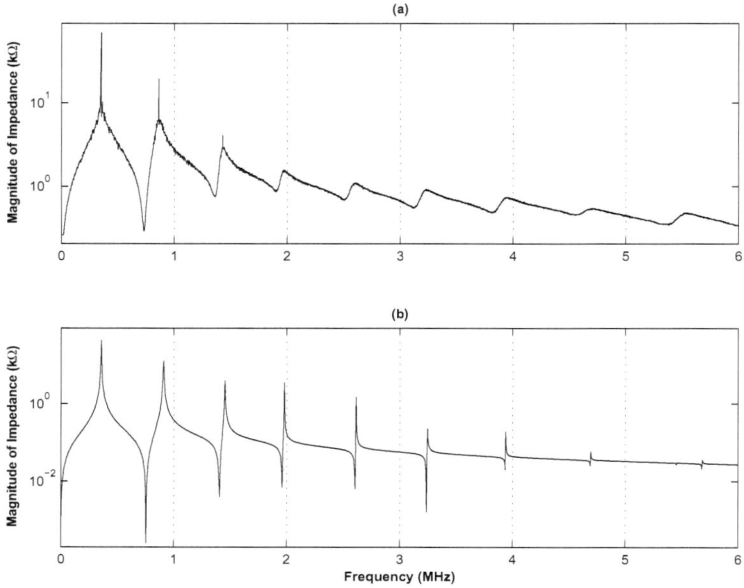

Fig. 16. Driving-point impedance for an actual disc winding, Fault Case, (a) Measured, (b) Estimated from synthesized circuit shown in Fig. 17

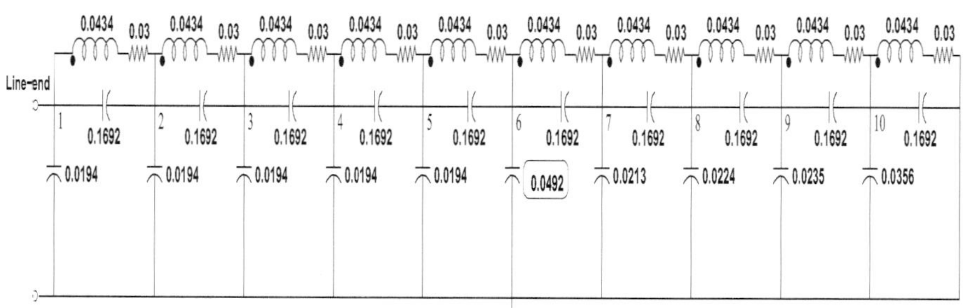

Fig. 17. Synthesized circuit pertaining to disc winding after changes (Note: Elemental values are in Ω, nF, mH. Mutual inductances of synthesized circuit are in Table 16)

1. A separate ladder network is required to represent each winding of a transformer and each of these networks must be linked to the immediate neighbour by the inter-winding capacitances.
2. To synthesize a ladder network corresponding to a winding, FRA must be conducted on that particular winding, keeping all the rest of the windings short-circuited and

connected to ground. Imposing this terminal condition ensures minimum influence of neighbouring windings on the tested winding during measurement.

3. A comparison of the subsequently synthesized ladder networks with the reference case can assist in localization of deformation.
4. A sacrificial transformer unit must be identified for this purpose, so that actual deformations (like radial, axial, etc.) can be incorporated to examine the true potential of the proposed method.

Future research efforts will pursue these ideas.

10. Conclusions

This chapter presents a mathematical approach to tackle the problem of localization of deformations in transformer windings. The approach essentially comprises of converting the measured driving-point impedance (via circuit synthesis) into a physically realizable, mutually coupled, ladder network that reproduces the measured characteristics as closely as possible. From a comparison of such synthesized circuits with the reference circuit, it is possible to assess the extent of damage the winding has undergone, in terms of the observed changes in the values of pertinent circuit elements.
The task of localizing discrete changes was demonstrated using a model winding and an actual transformer winding. The localization accuracy achieved was reasonably good in all the experimental cases presented. Nevertheless, it is impossible to use this iterative technique to address the localization problem in practice. This is mainly because, the process is extremely slow when large-sized circuits have to be synthesized, as it employs a brute-force search strategy. This shortcoming was recognized and a far more superior and time-efficient alternative was suggested. One such idea, based on constrained-optimization technique was presented and discussed at the end of this chapter. The algorithm is still in the developmental stages and hence only preliminary results are reported. In spite of it, from the results presented it is clear that this method is highly time-efficient. It required only a few tens of seconds to synthesize a mutually coupled, 10-section ladder network. Therefore, it seems to be well suited to synthesize many such large-sized circuits rapidly, corresponding to each winding in a transformer. This feature makes it an ideal candidate for addressing the ultimate problem of localization of winding deformation in multi-winding transformers. In final summary, it is believed that with the suggested method, we are better equipped to address the problem of deformation localization in an actual multi-winding transformer.

11. Acknowledgements

We thank research students, Mr. Saurav Pramanik and Mr. Santosh C Vora, of High Voltage Laboratory, Department of Electrical Engineering, Indian Institute of Science, India, for their assistance in conducting experiments and acquiring frequency response data pertaining to an actual disc winding used in this work.

12. References

[1] Dick, E. P. & Erven, C. C. (1978). Transformer Diagnostic Testing by Frequency Response Analysis. *IEEE Trans. Power Apparatus and Systems*, vol. PAS-97, Nov./Dec. 1978, pp. 2144-2153.

[2] Florkowski, M. & Furgal, J. (2003). Detection of Transformer Winding Deformations Based on the Transfer Function Measurements and Simulations. *Journal Meas. Sci. Tech.*, 14, 2003, pp. 1986-1992.
[3] Gustavsen, Bjorn & Semlyen, Adam (1999). Rational Approximation of Frequency Domain Responses by Vector Fitting. *IEEE Trans. Power Delivery*, vol. 14, no. 3, Jul. 1999, pp. 1052-1061.
[4] Islam, S. M. (2000). Detection of Shorted Turns and Winding Movements in Large Power Transformers using Frequency Response Analysis. *IEEE Power Engineering Society Winter Meeting. Conference Proceedings*, vol.3, pt. 3, 2000, pp. 2233-2238.
[5] Lech, W. & Tyminski, L. (1966). Detecting transformer winding damage- the low voltage impulse method. *Electrical Review*, no. 18, 1966.
[6] Morched, A.; Martf, L. & Ottevangers, J. (1993). A High Frequency Transformer Model for the EMTP. *IEEE Trans. Power Delivery*, vol. 8, no. 3, Jul. 1993, pp.1615-1626.
[7] Oguz Soysal, A. (1993). A Method for Wide Frequency Range Modeling of Power Transformers and Rotating Machines. *IEEE Trans. Power Delivery*, vol. 8, no. 4, Oct. 1993, pp. 1802-1808.
[8] Ragavan, K. & Satish, L. (2007). Localization of Changes in a Model Winding based on Terminal Measurements: Experimental Study. *IEEE Trans. Power Delivery*, vol. 22, no. 3, July 2007, pp. 1557-1565.
[9] Rahimpour, E.; Christian, J.; Feser, K. & Moheseni, H. (2003). Transfer Function Method to Diagnose Axial Displacement and Radial Deformation of Transformer Windings. *IEEE Trans. Power Delivery*, vol. 18, Apr. 2003, pp. 493-505.
[10] Ryder, S. A. (2003). Diagnosing transformer faults using frequency response analysis. *IEEE EI Magazine*, vol. 19, no. 2, Mar.-Apr. 2003, pp.16-22.
[11] Satish, L. & Subrat K. Sahoo (2009). Locating Faults in a Transformer Winding: An Experimental Study. *Electric Power Systems Research*, vol. 79, 2009, pp. 89-97.
[12] Satish, L. & Santosh C. Vora (2009). Amplitude Frequency Response Measurement - A Simple Technique. *IEEE Trans. Education*, 2009, in press.

17

Sensor fault detection and isolation by robust principal component analysis

Yvon THARRAULT, Gilles MOUROT, José RAGOT
Centre de Recherche en Automatique de Nancy
(CRAN) UMR 7039, Nancy Université, CNRS 2,
Avenue de la forêt de Haye. F-54 516 Vandoeuvre-lès-Nancy
France

Mohamed-Faouzi HARKAT
Université Badji Mokhtar – ANNABA
Faculté des Sciences de l'Ingénieur, Département d'Electronique BP. 12,
Annaba. 23000
Algérie

1. Introduction

Sensors are essential components of modern control systems. Any faults in sensors will affect the overall performance of a system because their effects can easily propagate to manipulative variables through feedback control loops and also disturb other process variables. The task for sensor validation is to detect and isolate faulty sensors and estimate fault magnitudes afterwards to provide fault-free values. Model-based methods constitute an important approach to sensor fault detection and isolation (FDI).

A model-based approach consists in generating residuals as the difference between the measurements and the estimates provided by the relationships existing between the various variables of the process. The analysis of these residuals may lead to detect and isolate the faulty sensors. Almost all conventional model-based methods presume the knowledge of an accurate model of the system, e.g. transfer function or system matrices in the state space representation. Principal component Analysis (PCA) is a data-driven method which is particularly well adapted to reveal linear relationships among the plant variables without formulating them explicitly and has also been employed for system identification. PCA has some other nice features. It can handle high dimensional and correlated process variables, provides a natural solution to the errors-in-variables problem and includes disturbance decoupling (Li & Qin, 2001). Moreover in the FDI field, Gertler & McAvoy (1997) have shown a close link between PCA and parity space method. Principal component analysis (PCA) has then been applied successfully in the monitoring of complex systems (Chiang & Colegrove, 2007; Harkat et al., 2006; Kano & Nakagawa, 2008).

PCA is used to model the normal process behavior from an empirical data set which is representative of a normal process operation. In general, the majority of the training data set is associated with such normal operating conditions. The remaining data (faulty data, data obtained during shutdown or startup periods or data issued from different operating modes) are referred to as *outliers*. Often, these outlying observations are not incorrect but they were made under exceptional circumstances. Therefore, they may disturb the correlation structure of the "normal data" and the result will be a model that does not accurately represent the process. The fact that multiple outliers can contaminate the model derived from a classical PCA has motivated the development of robust methods that are less affected by outliers. Large residuals from that robust fit indicate the presence of outliers. Once a robust model is determined, the next step deals with multiple fault detection and isolation. Indeed, outliers corresponding to either multiple faulty sensors or a priori unknown operating conditions affect many process variables.

This chapter is devoted to the problem of multiple fault detection and isolation. Section 2 presents the classical PCA principle and summarizes the benefits of different indices generally used for fault detection. Section 3, after a definition of outliers, introduces the main robust methods generally used. Next, a new robust method called MMRPCA for MM-estimator Robust Principal Component Analysis is proposed. It extends to all kinds of outliers the robust subspace estimator of Maronna (2005). Section 4 deals with multiple fault isolation. After a brief state of the art on fault isolation, structured residuals are generated for multiple fault isolation. These structured residuals are based on the reconstruction principle of process variables (Dunia et al., 1996; Wang et al., 2004a;b). However, instead of considering all the subsets of faulty variables (one up to all sensors), we determine the isolable multiple fault by evaluating the existence condition of these structured residuals. The proposed scheme avoids the combinatorial explosion of faulty scenarios related to the multiple faults to consider. In the last section 5 this method is applied on a simulated example in order to illustrate the different steps of our method.

2. Background

PCA is a widely used method for dimensionality reduction. Indeed, PCA transforms the data to a smaller set of variables (scores) which are the linear combinations of the original variables while retaining as much information as possible. Data redundancy stems from linear relation between process variables. The PCA model appears in the form of loadings, scores and variances. The eigenvectors (loadings) associated to the eigenvalues of the data covariance matrix span the representation and residual subspaces. The representation subspace (respectively residual subspace) associated to the first (respectively last) principal components (scores) describes significant variations of the process (respectively the noise in the data). Different statistics are defined on these two subspaces in order to detect faults. Qin (2003) proposed a unified representation of these fault detection indices.

For dynamic systems, since the current and past values of variables are related, Ku et al. (1995) proposed to include time lagged variables into the data matrix and to perform PCA on this augmented matrix in order to reveal dynamic linear relationships among the process variables. However, this straightforward application of PCA, called Dynamic PCA (DPCA), has several limitations as a monitoring approach for dynamic systems. Firstly, Li & Qin (2001) have shown that the DPCA method is consistent only if all variables have identical noise variance. Secondly, it is not clear how many time-lagged variables should be included into the augmented data matrix. These dynamic modeling problems have been approached by Li &

Qin (2001), Wang & Qin (2002), Qin & Wang (2006) in a instrumental variable framework. Thirdly, Kruger et al. (2004) have shown that DPCA produces correlated scores which lead to an undesired impact upon the fault detection ability in the representation subspace (production of false alarms). Xie et al. (2006) have then proposed to filter the principal scores obtained by DPCA by a Kalman innovation filter to remove the correlation between them. These points being outside the scope of this chapter, for the sake of simplicity, in the following, only classic PCA is considered in the following.

2.1 PCA modelling of systems

Let us consider $\mathbf{x}(k) = [x_1(k)\ x_2(k)\ \ldots\ x_m(k)]^{\mathrm{T}}$ the vector formed with m observed plant variables at time instant k. Define the data matrix $\mathbf{X} = [\mathbf{x}(1)\ \mathbf{x}(2)\ \ldots\ \mathbf{x}(N)]^{\mathrm{T}} \in \Re^{N\times m}$ with N samples $\mathbf{x}(k)(k = 1,\ldots,N)$ which is representative of a normal process operation. PCA determines an optimal linear transformation of the data matrix $\mathbf{X}$ in terms of capturing the variation in the data:

$$\mathbf{T} = \mathbf{X}\,\mathbf{P} \qquad \text{and} \qquad \mathbf{X} = \mathbf{T}\,\mathbf{P}^{\mathrm{T}} \tag{1}$$

with $\mathbf{T} = [\mathbf{t}_1\ \mathbf{t}_2\ \ldots\ \mathbf{t}_m] \in \Re^{N\times m}$, where the vectors $\mathbf{t}_i$ are called scores or principal components and the matrix $\mathbf{P} = [\mathbf{p}_1\ \mathbf{p}_2\ \ldots\ \mathbf{p}_m] \in \Re^{m\times m}$, where the orthogonal vectors $\mathbf{p}_i$, called loading or principal vectors, are the eigenvectors associated to the eigenvalues λ_i of the covariance matrix (or correlation matrix) $\mathbf{\Sigma}$ of $\mathbf{X}$:

$$\mathbf{\Sigma} = \mathbf{P}\mathbf{\Lambda}\mathbf{P}^{\mathrm{T}} \quad \text{with} \quad \mathbf{P}\,\mathbf{P}^{\mathrm{T}} = \mathbf{P}^{\mathrm{T}}\mathbf{P} = \mathbf{I}_m \tag{2}$$

where $\mathbf{\Lambda} = diag(\lambda_1 \ldots \lambda_m)$ is a diagonal matrix with diagonal elements in decreasing magnitude order.

The relations (1) are useful when the dimension ℓ of the representation subspace is reduced ($\ell << m$). Once the component number ℓ to retain is determined, the data matrix $\mathbf{X}$ can be approximated. For that, the different matrices are partitioned into the form:

$$\mathbf{P} = \left[\ \hat{\mathbf{P}}\ \middle|\ \tilde{\mathbf{P}}\ \right], \quad \hat{\mathbf{P}} \in \Re^{m\times \ell} \quad \text{and} \quad \tilde{\mathbf{P}} \in \Re^{m\times(m-\ell)} \tag{3}$$

$$\mathbf{\Lambda} = \begin{bmatrix} \hat{\mathbf{\Lambda}} & 0 \\ 0 & \tilde{\mathbf{\Lambda}} \end{bmatrix}, \quad \hat{\mathbf{\Lambda}} \in \Re^{\ell\times \ell} \quad \text{and} \quad \tilde{\mathbf{\Lambda}} \in \Re^{(m-\ell)\times(m-\ell)} \tag{4}$$

$$\mathbf{T} = \left[\ \hat{\mathbf{T}}\ \middle|\ \tilde{\mathbf{T}}\ \right], \quad \hat{\mathbf{T}} \in \Re^{N\times \ell} \quad \text{and} \quad \tilde{\mathbf{T}} \in \Re^{N\times(m-\ell)} \tag{5}$$

Equation (1) can be rewritten as:

$$\mathbf{X} = \hat{\mathbf{T}}\hat{\mathbf{P}}^{\mathrm{T}} + \tilde{\mathbf{T}}\tilde{\mathbf{P}}^{\mathrm{T}} = \hat{\mathbf{X}} + \mathbf{E} \tag{6}$$

with

$$\hat{\mathbf{X}} = \mathbf{X}\,\hat{\mathbf{C}} \qquad \text{and} \qquad \mathbf{E} = \mathbf{X}\,\left(\mathbf{I}_m - \hat{\mathbf{C}}\right) \tag{7}$$

where the matrix $\hat{\mathbf{C}} = \hat{\mathbf{P}}\,\hat{\mathbf{P}}^{\mathrm{T}}$ constitutes the PCA model.

The matrices $\hat{\mathbf{X}}$ and $\mathbf{E}$ represent, respectively, the modeled variations and non modeled variations of $\mathbf{X}$ based on ℓ components ($\ell < m$). The first ℓ eigenvectors $\hat{\mathbf{P}}$ constitute the representation subspace which describes the significant data variations whereas the last $(m - \ell)$ eigenvectors $\tilde{\mathbf{P}}$ constitute the residual subspace which describes the noises affecting the data. The identification of the PCA model thus consists in estimating its parameters by an eigenvalue/eigenvector decomposition of the matrix $\mathbf{\Sigma}$ and determining the number of principal components ℓ to retain.

A key issue to develop a PCA model is to choose the adequate number of principal components. Indeed, in most practical cases (noisy measurements), the small eigenvalues indicate the existence of linear or quasilinear relations among the process variables. However, the distinction between significant or insignificant eigenvalues may not be obvious due to modelling errors (disturbances and nonlinearities) and noise. Most methods to determine the number of principal components are rather subjective in the general practice of PCA (Qin & Dunia, 2000). Other methods are based on criteria actually used in system identification (Aikaike information criterion, minimum description length, ...) to determine the system order and emphasize the approximation of the data matrix $\mathbf{X}$ (see Valle et al. (1999) for a survey of these methods). However, the number of principal components has a significant impact on each step of the sensor fault detection and isolation scheme. Tamura & Tsujita (2007) proposed to choose ℓ which maximizes the sensitivity of fault detection indices to each sensor fault. This method can lead to monitor in parallel until m PCA models with various numbers of principal components. Qin & Dunia (2000) proposed to determine ℓ by minimization of the variance of the reconstruction error in the residual subspace. This variable reconstruction consists in estimating a variable from other plant variables using the PCA model, i.e. using the redundancy relations between this variable and the others. The reconstruction accuracy is thus related to the capacity of the PCA model to reveal the redundancy relations among the variables, i.e. to the number of principal components. This method will be retained here to determine ℓ.

2.2 Residual properties

After the PCA model has been built, we now examine its use for sensor fault detection and isolation. Let us consider now the fault propagation on the two signals obtained by the projection of the measurement vector $\mathbf{x}(k)$ onto the representation and the residual subspaces. In the presence of f faulty variables whose indices belong to the subset F, the measurement vector $\mathbf{x}(k)$ can be expressed as:

$$\mathbf{x}(k) = \mathbf{x}^o(k) + \epsilon(k) + \Xi_F\, \mathbf{f}(k) \tag{8}$$

where $\mathbf{x}^o(k)$ is true value vector, $\epsilon(k)$ is the zero mean i.i.d. measurement noise vector, $\mathbf{f}(k)$ is the fault magnitude vector (unknown) and Ξ_F is the matrix of the fault directions. This orthonormal matrix with dimension ($m \times$ f) is built with 0 and 1, where 1 indicates the faulty variables from the other variables (with 0). For example, for the subset of faulty variables $F = \{2, 4\}$ among 5 variables, matrix Ξ_F is formed as follows:

$$\Xi_F = \begin{bmatrix} 0 & 1 & 0 & 0 & 0 \\ 0 & 0 & 0 & 1 & 0 \end{bmatrix}^{\mathrm{T}}$$

Following decomposition (6), the measurement vector can also be represented as:

$$\mathbf{x}(k) = \hat{\mathbf{x}}(k) + \mathbf{e}(k) \tag{9}$$

where $\hat{\mathbf{x}}(k) = \hat{\mathbf{C}}\, \mathbf{x}(k)$ is the estimation vector, $\mathbf{e}(k) = \left(\mathbf{I_m} - \hat{\mathbf{C}}\right) \mathbf{x}(k)$ is the vector of estimation errors or residual vector.

The principal component vector is given by:

$$\mathbf{t}(k) = \mathbf{P}^{\mathrm{T}}\, \mathbf{x}(k) = \left[\hat{\mathbf{t}}(k)\ \ \tilde{\mathbf{t}}(k)\right] \tag{10}$$

where:

$$\hat{\mathbf{t}}(k) = \hat{\mathbf{P}}^{\mathrm{T}}\, \mathbf{x}(k), \tag{11}$$

$$\tilde{\mathbf{t}}(k) = \tilde{\mathbf{P}}^{\mathrm{T}}\, \mathbf{x}(k) \tag{12}$$

2.2.1 In the residual subspace

There is an equivalence between the residual vector and the last principal component vector $\tilde{\mathbf{t}}$:

$$\mathbf{e}(k) = \tilde{\mathbf{P}}\,\tilde{\mathbf{t}}(k) \quad \in \Re^{m\times 1} \tag{13}$$

So, it becomes simpler to work with this new residual vector $\tilde{\mathbf{t}}$ with dimension $(m-\ell)$. From (8) and (12), the residual vector is given by:

$$\tilde{\mathbf{t}}(k) \quad = \quad \tilde{\mathbf{P}}^{\mathrm{T}}\,\mathbf{x}^{\mathrm{o}}(k) + \tilde{\mathbf{P}}^{\mathrm{T}}\,\boldsymbol{\epsilon}(k) + \tilde{\mathbf{P}}^{\mathrm{T}}\,\Xi_F\,\mathbf{f}(k) \tag{14}$$

Since $\tilde{\mathbf{P}}^{\mathrm{T}}\,\mathbf{x}^{\mathrm{o}}(k) = 0$ then (14) becomes:

$$\tilde{\mathbf{t}}(k) \quad = \quad \tilde{\mathbf{P}}^{\mathrm{T}}\,\boldsymbol{\epsilon}(k) + \tilde{\mathbf{P}}^{\mathrm{T}}\,\Xi_F\,\mathbf{f}(k) \tag{15}$$

In the fault-free case, the expectation of the residual vector is zero. In the presence of faults, the expectation of the residual vector is no longer zero and the fault affects all the components of the residual vector. However if a variable is not correlated with others, its projection onto the residual subspace will be very small and then it will be very difficult to detect if this variable is faulty in the residual subspace. To conclude with this residual, it is possible to detect a fault (provided that residual is sufficiently sensitive to the fault) but it is difficult to isolate the faulty sensors.

Typically two main fault detection indices are used to monitor these residuals. The SPE (squared prediction error) is a statistic which measures the lack of fit of the PCA model to the data. At time k, the detection index SPE is given by:

$$SPE(k) = \tilde{\mathbf{t}}(k)^{\mathrm{T}}\tilde{\mathbf{t}}(k) \tag{16}$$

This quantity suggests the existence of an abnormal situation in the data when:

$$SPE(k) > \delta_\alpha^2 \tag{17}$$

where δ_α^2 is a control limit for SPE with a significance level α (Nomikos & MacGregor, 1995) or estimated using the historical data.

The SPE is formed by summing the square of residuals obtained from PCA model. However, modelling errors could be projected onto the residual subspace which results in residuals with a higher variance than the others. Then the SPE will be heavily biased in favour of those residuals with the largest residual variance whereas the residuals with the smallest residual variances are most useful for sensor fault diagnosis because they are associated with linear relationships. In this case using the indicator SWE (squared weighted error) can be more judicious because it takes into account the variances of the residual (Westerhuis et al., 2000), the problem inversion of near-zero singular eigenvalues seldom occuring in practice. At time k, the detection index SWE is given by:

$$SWE(k) = \tilde{\mathbf{t}}(k)^{\mathrm{T}}\tilde{\boldsymbol{\Lambda}}^{-1}\,\tilde{\mathbf{t}}(k) \tag{18}$$

This quantity suggests the existence of an abnormal situation in the data when:

$$SWE(k) > \chi^2_{m-\ell,\alpha} \tag{19}$$

where $\chi^2_{m-\ell,\alpha}$ is a Chi-2 upper control limit with $m-\ell$ degrees of freedom for a significance level α.

2.2.2 In the representation subspace

From (8) and (12), the principal score vector is given by:

$$\hat{\mathbf{t}}(k) \quad = \quad \hat{\mathbf{P}}^{\mathrm{T}}\,\mathbf{x}^{\mathrm{o}}(k) + \hat{\mathbf{P}}^{\mathrm{T}}\,\boldsymbol{\epsilon}(k) + \hat{\mathbf{P}}^{\mathrm{T}}\,\Xi_F\,\mathbf{f}(k) \tag{20}$$

The faults affect all the components of $\hat{\mathbf{t}}(k)$. Moreover, the term $\hat{\mathbf{P}}^{\mathrm{T}}\,\mathbf{x}^{\mathrm{o}}(k)$ describing the significant variations of the data, can mask the fault influence on the principal score vector except if the fault amplitude is high relative to this term which depends on the operating point $\mathbf{x}^{\mathrm{o}}(k)$. The T^2 statistic is the main fault detection index to monitor the principal scores. At time k, the detection index T^2 is given by:

$$T^2(k) = \hat{\mathbf{t}}(k)^{\mathrm{T}}\hat{\boldsymbol{\Lambda}}^{-1}\,\hat{\mathbf{t}}(k) \tag{21}$$

This quantity suggests the existence of an abnormal situation in the data when:

$$T^2(k) > \chi^2_{\ell,\alpha} \tag{22}$$

where $\chi^2_{\ell,\alpha}$ is a Chi-2 upper control limit with ℓ degrees of freedom for a significance level α.

2.2.3 In the entire space

However, according to the magnitude and the kind of faults, faults can be only projected onto the representation or the residual subspace and then affect only one of the previous indices. Rather than monitoring different fault detection indices, combined indices are preferred. The Mahalanobis distance can be used as a measure of variability in the entire space. At time k, the Mahalanobis distance D^2 is given by:

$$D^2(k) = \mathbf{x}(k)^{\mathrm{T}}\boldsymbol{\Sigma}^{-1}\,\mathbf{x}(k) \tag{23}$$

This quantity suggests the existence of an abnormal situation in the data when:

$$D^2(k) > \chi^2_{m,\alpha} \tag{24}$$

where $\chi^2_{m,\alpha}$ is a Chi-2 upper control limit with m degrees of freedom for a significance level α. Previously we have seen that faults affect all the components of a residual or principal score vector. Therefore it is rather difficult to isolate the faulty sensors by using these vectors. The problem of isolation enhancement will be addressed in Section 4.

3. Robust PCA

In the classical approach, the first principal component corresponds to the direction in which the projected observations have the largest variance. The second component is then orthogonal to the first one and again maximizes the variance of the data points projected on it. Continuing in this way produces all the principal components which correspond to the eigenvectors of the empirical covariance matrix. From a regression point of view, PCA also constructs the optimal orthogonal linear projections (in terms of mean squared error) from the eigenvectors of the data covariance matrix. The performance of PCA model is then based on the accurate estimation of the covariance matrix from the data which is very sensitive to abnormal observations.

In practice one often tries to detect outliers by using diagnostic tools starting from a classical fitting method. However, classical methods can be affected by outliers so strongly that the resulting fitted model does not allow to detect the deviating observations. This is called the

masking effect. Additionally, some good data points might even appear to be outliers, which is known as swamping. To avoid these effects, the goal of robust PCA methods is to obtain principal components that are not influenced much by outliers. Large residuals from that robust fit indicate the presence of outliers.
After having described outliers using PCA in the following section, the robust PCA methods are introduced.

3.1 Characteristics of outliers using PCA

The different types of outliers are classified according to their effects on the construction of the PCA model. They are explained in figure 1 considering a system with 3 variables and 2 principal components. Thus, 4 types of Observations can be distinguished (Hubert et al., 2005):

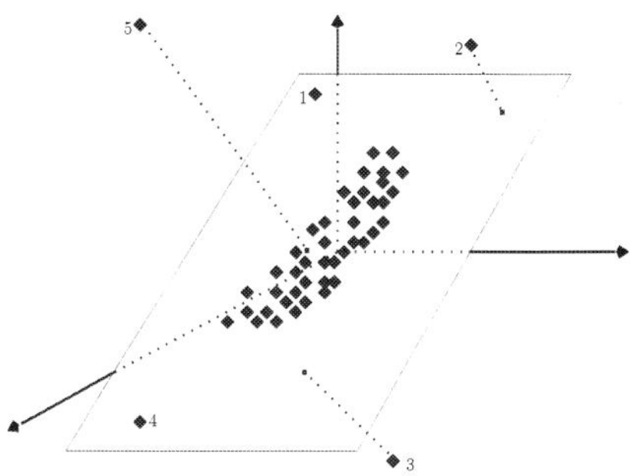

Fig. 1. Illustration of different types of outliers for an example with 3 variables and 2 principal components

- *Regular observations:* observation which belong to a homogeneous group that is close to the representation subspace.
- *Good leverage points:* Outliers close to the representation subspace (small projection onto the residual subspace) but far from the regular observations such as the observations 1 and 4 of figure 1.
- *Orthogonal outliers:* Outliers whose orthogonal distance to representation subspace is large (large projection onto the residual subspace) but not visible with its projection onto the representation subspace, like the observation 5 of figure 1.
- *Bad leverage points:* Outliers that have a large orthogonal distance (large projection onto the residual subspace) and whose projection onto the representation subspace is remote from the typical projections, such as the observations 2 and 3 of figure 1.

To construct a robust model with respect to outliers, the influences of outliers in the construction of the PCA model have to be minimised.

3.2 Classical robust PCA methods

Several ways of robustifying principal components have been proposed (Daszykowski et al., 2007; Filzmoser et al., 2008; Rousseeuw et al., 2006). To enable the comparison of different robust methods, measures of performance are necessary. One such performance measure for

robust methods is the *breakdown point* (Donoho & Huber, 1983) which can be defined as the maximal fraction of outlying objects in the data that the estimator can handle in order to yield acceptable estimates. For instance, the breakdown point of the mean estimator with 0% being the smallest possible. For such breakdown point a single outlier can completely corrupt the estimate. Conceptually, it is impossible to distinguish between the good and the bad parts in the data if the fraction of outliers becomes larger than 50%. In the following, we focus on robust methods with the highest breakdown point, i.e. close to 50%.
To robustify PCA with respect to outliers, M-estimators (Huber, 1964) can be used. These estimators minimise a more general objective function than the classical criterion on Mean Square Error. However, the breakdown point of M-estimators is limited by $\frac{1}{m+1}$, with m the number of variables. Then this method is not very robust when considering large systems (large m). The robust methods can be grouped as follows :

- A first group of robust PCA methods is obtained by replacing the classical covariance matrix by a robust covariance estimator, such as the minimum covariance determinant (MCD) estimator (Rousseeuw, 1987). The MCD looks for those h observations in the data set whose classical covariance matrix has the lowest possible determinant. The user-defined parameter h is the number of fault-free data among all the data and determines the robustness but also the efficiency of the resulting estimator. The computation of the MCD estimator is non-trivial and naively requires an exhaustive investigation of all h-subsets out of the N observations. This is no longer possible for large N or in high dimension. Rousseeuw & Van Driessen (1999) constructed a much faster algorithm called FAST-MCD which avoids such a complete enumeration. It is obtained by combining a basic subsampling and iterative scheme with MCD estimator. Another example is Caussinus et al. (2003) who define a "local" matrix of variance in the sense that the suggested form tends to emphasize the contribution of close observations in comparison with distant observations (outliers). As for the MCD method, this method has a parameter setting β, which depends on the number of outliers a priori unknown.
- A second approach to robust PCA uses Projection Pursuit (PP) techniques. These methods maximize a robust measure of data spread to obtain consecutive directions on which the data points are projected (Croux et al., 2007; Croux & Ruiz-Gazen, 2005; Huber, 1964; Li & Chen, 1985). The main step of these algorithms is then to search for the direction in which the projected observations have the largest robust spread to obtain the first component. The second component is then orthogonal to the first and has the largest robust spread of the data points projected on it. Continuing in this way produces all the robust principal components. To make these algorithms computationally feasible, the collection of directions to be investigated are restricted to all directions that pass through the robust center of the data and a data point or through two data points. However the robust directions obtained are approximations of the true ones. To improve the speed of algorithms, a PCA compression to the rank of the data is performed as a first step. According to the authors, these algorithms can deal with both low and high dimensional data.
- Another robust PCA method is to combine the two previous robust approaches like the method proposed in Hubert et al. (2005) and called ROBPCA. This method combined ideas of both projection pursuit and robust covariance estimation based on FAST-MCD algorithm. It first applied projection pursuit techniques in the original data space. These results are then used to project the observations onto a subspace of small to moderate

dimension. Within this subspace robust covariance estimation is applied. According to the authors, this algorithm is a powerful tool for high dimensional data when the number of variables is greater than the number of observations. The authors also used a diagnostic plot to visualize and classify the outliers. It plots the squared Mahalanobis distance versus the orthogonal distance of each observation to the representation subspace.

- The last proposals for robust PCA include the robust LTS-subspace estimator and its generalizations (Maronna et al., 2006). The idea behind these approaches consists in minimizing a robust scale of the orthogonal distances of each observation to the representation subspace, similar to the LTS estimator, S-estimators and many others in regression. These methods are based on iterative procedures for which there remains the problem of starting values. For example, for the LTS-subspace estimator, the classical PCA is performed on the h observations with the smallest orthogonal distance to the PCA subspace. Its drawbacks are the same as the MCD-estimator: a high computational cost, the choice of the user-defined parameter h and the starting values. Like MCD-estimator, a FAST-LTS algorithm has been proposed.

All these methods have a tuning parameter which changes with the fraction of outliers in the data. However, this information is unknown. To fill this gap, we propose a new robust method called MMRPCA for MM-estimator Robust Principal Component Analysis. Thus, a MM-estimator is used to determine a robust model. This estimator is a combination of two M-estimators: a M-estimator to estimate the model and a second M-estimator to estimate the robust scale of the residual. This estimator, calculated with an iterative algorithm is initialized with a robust estimator of the covariance matrix (Caussinus et al., 2003). Then fault detection tools are used, from the robust model for finding outliers in the data. The influences of outliers can then be eliminated and the resulting PCA model becomes unbiased. Moreover, a robust procedure is proposed to determine the number of principal components.

3.3 MMRPCA Method (MM-estimator Robust Principal Component Analysis)

Our approach consists in carrying out PCA directly on the data possibly contaminated by outliers. For that, a simple robust estimator, called MM-estimator, is used. However, this estimator is computed by an iterative procedure. Thus a good initialization parameter procedure is needed to avoid local minima. To initialize this MM-estimator a robust covariance matrix is first calculated.

3.3.1 Robust covariance matrix

Caussinus et al. (2003) define a "local" matrix of covariance in the sense that the suggested form tends to emphasize the contribution of close observations in comparison with distant observations (outliers). The matrix is defined in the following way:

$$\boldsymbol{Q} = \frac{\sum\limits_{i=1}^{N-1} \sum\limits_{j=i+1}^{N} \Omega(i,j)(\boldsymbol{x}(i) - \boldsymbol{x}(j))(\boldsymbol{x}(i) - \boldsymbol{x}(j))^{\mathrm{T}}}{\sum\limits_{i=1}^{N-1} \sum\limits_{j=i+1}^{N} w(i,j)} \tag{25}$$

where the weights $\mathbf{\Omega}(i,j)$ themselves are defined by:

$$\mathbf{\Omega}(i,j) = \exp\left(-\frac{\beta}{2}(\boldsymbol{x}(i) - \boldsymbol{x}(j))^{\mathrm{T}}\mathbf{\Sigma}^{-1}(\boldsymbol{x}(i) - \boldsymbol{x}(j))\right) \tag{26}$$

β being a tuning parameter to reduce the influence of the observations faraway, as recommended by the authors, β value is equal to 2. For $\beta = 0$, the robust covariance matrix $\boldsymbol{Q}$ is equal to $2\,\mathbf{\Sigma}$. And for a high value of β, only the closest observations are taken into account in the robust covariance matrix $\boldsymbol{Q}$.

3.3.2 MM-estimator

Two M-estimators are used, one to estimate the minimum of the objective function and another one for the estimation of the robust residual scale. The general MM-estimator minimizes the following objective function with the constraint $\tilde{\boldsymbol{P}}^{\mathrm{T}}\tilde{\boldsymbol{P}} = \boldsymbol{I}_{m-\ell}$ (Maronna, 2005):

$$\frac{1}{N}\sum_{k=1}^{N}\rho\left(\frac{\boldsymbol{r}(k)}{\hat{\sigma}}\right) \tag{27}$$

with $\boldsymbol{r}(k) = ||\tilde{\boldsymbol{P}}\,\boldsymbol{x}(k) - \tilde{\boldsymbol{P}}\,\boldsymbol{\mu}||^2$ is the residual, $\boldsymbol{\mu} = \frac{1}{N}\sum_{k=1}^{N}\boldsymbol{x}(k)$ is the mean, $\hat{\sigma}$ is the robust scale of the residual $\boldsymbol{r}(k)$ and the function $\rho{:}\Re^{+} \rightarrow [0,1]$ is nondecreasing, with $\rho(0) = 0$, $\rho(\infty) = 1$, and differentiable. $\tilde{\boldsymbol{P}}$ is the eigenvector matrix of the robust covariance matrix $\boldsymbol{S}$ (29) corresponding to its $m-\ell$ smallest eigenvalues. Then the weighted mean $\boldsymbol{\mu}$ and the covariance $\boldsymbol{S}$ are defined as follows:

$$\boldsymbol{\mu} = \frac{\sum_{k=1}^{N}\boldsymbol{w}(k)\boldsymbol{x}(k)}{\sum_{k=1}^{N}\boldsymbol{w}(k)} \qquad \text{with } \boldsymbol{w}(k) = \dot{\rho}\left(\frac{\boldsymbol{r}(k)}{\hat{\sigma}}\right) \tag{28}$$

$$\boldsymbol{S} = \sum_{k=1}^{N}\boldsymbol{w}(k)(\boldsymbol{x}(k) - \boldsymbol{\mu})(\boldsymbol{x}(k) - \boldsymbol{\mu})^{\mathrm{T}} \tag{29}$$

where $\dot{\rho}(x) = \frac{\partial\rho(x)}{\partial x}$.
Then the scale factor $\hat{\sigma}$ is defined as the solution to:

$$\frac{1}{N}\sum_{k=1}^{N}\rho\left(\frac{\boldsymbol{r}(k)}{\hat{\sigma}}\right) = \delta \tag{30}$$

with $\delta \in [0,1]$. This parameter δ is directly related to the fraction of outliers in the data *i.e.* the breakdown point. To maximize the breakdown point, Maronna et al. (2006) define δ as follows:

$$\delta = \frac{N - m + \ell - 1}{2N} \qquad \text{if } N >> m \text{ then } \delta \approx 0.5 \tag{31}$$

Usually, when the number of observations (N) is larger than the number of variables (m) then parameter δ is chosen equal to 0.5. By defining a weight function $\boldsymbol{w}_{\sigma}$ in the following way:

$$\boldsymbol{w}_{\sigma}(\boldsymbol{r}) = \begin{cases} \rho(\boldsymbol{r})/\boldsymbol{r} & \text{if } \boldsymbol{r} \neq 0 \\ \ddot{\rho}(0) & \text{if } \boldsymbol{r} = 0 \end{cases} \tag{32}$$

where $\ddot{\rho}(x) = \frac{\partial^2 \rho(x)}{\partial x^2}$.
The solution of equation (30) is expressed by the following equation:

$$\hat{\boldsymbol{\sigma}} = \frac{1}{N\delta} \sum_{k=1}^{N} w_{\sigma} \left(\frac{\boldsymbol{r}(k)}{\hat{\boldsymbol{\sigma}}} \right) \boldsymbol{r}(k) \tag{33}$$

One notices that an iterative resolution of equation (33) is needed to determine the scale factor $\hat{\boldsymbol{\sigma}}$.
We chose the function ρ (27) as the Bisquare function ($\boldsymbol{r}$ represents the square of residuals) because it allows the cancellation of the influence of outliers.

$$\rho(\boldsymbol{r}) = \min\{1, 1 - (1 - \boldsymbol{r})^3\} \tag{34}$$

and

$$w_{\sigma}(\boldsymbol{r}) = \min\{3 - 3\boldsymbol{r} + \boldsymbol{r}^2, 1/\boldsymbol{r}\} \tag{35}$$

However, this method is only robust to fault with a projection onto the residual subspace. Then to be robust to all kinds of faults, a similar approach in the representation subspace is used. In that case the MM-estimator maximizes the following objective function with the constraint $\hat{\boldsymbol{P}}^{\mathrm{T}} \hat{\boldsymbol{P}} = \boldsymbol{I}_{\ell}$:

$$\frac{1}{N} \sum_{k=1}^{N} \rho \left(\frac{||\hat{\boldsymbol{P}}^{\mathrm{T}} \boldsymbol{x}(k) - \hat{\boldsymbol{P}}^{\mathrm{T}} \boldsymbol{\mu}||^2}{\hat{\sigma}} \right) \tag{36}$$

To ensure the elimination of any fault disturbing the residual subspace when the MM-estimator in the representation subspace is used, the minimum between the weight determined with the first MM-estimator (robust to fault with a projection onto the residual subspace) and the weight obtained with the second MM-estimator (robust to fault with a projection onto the representation subspace) is used (line 26 of table 1). The algorithm of the MMRPCA method is described in table 1.

Finally, to improve the estimation of the covariance matrix, and thus the PCA model, a last weighting step is implemented by using the Mahalanobis distance to eliminate outliers.

$$\begin{cases} w(k) = 1 & \text{if} \quad D^2(k) \leq \chi^2_{m,\alpha} \\ \quad\;\;\; = 0 & \text{else} \end{cases} \tag{37}$$

where $D^2(k)$ is the Mahalanobis distance
The robust mean $\boldsymbol{\mu}$ and the variance matrix $\boldsymbol{S}$ are then defined as follows:

$$\boldsymbol{\mu} = \frac{\sum_{k=1}^{N} w(k)\boldsymbol{x}(k)}{\sum_{k=1}^{N} w(k)} \tag{38}$$

$$\boldsymbol{S} = \left(\sum_{k=1}^{N} w(k)(\boldsymbol{x}(k) - \boldsymbol{\mu})(\boldsymbol{x}(k) - \boldsymbol{\mu})^{\mathrm{T}} \right) \Big/ \left(\sum_{k=1}^{N} w(k) - 1 \right) \tag{39}$$

However, this algorithm requires the number of principal components to be known. Hence, a robust method to find the number of principal components is introduced.

1.	$it = 1$ and $\sigma_0 = \infty$;
2.	Compute $\tilde{\boldsymbol{P}}$ the eigenvector matrix of the robust covariance matrix $\boldsymbol{Q}$ corresponding to its $m - \ell$ smallest eigenvalues;
3.	Compute $\boldsymbol{a} = median(\boldsymbol{X}\,\tilde{\boldsymbol{P}})$;
4.	Do until convergence of Δ;
5.	Compute $\boldsymbol{r}(k) = \|\|\tilde{\boldsymbol{P}}\boldsymbol{x}(k) - \boldsymbol{a}\|\|^2$ for $k = 1...N$;
6.	Compute $\hat{\sigma}$ from (33);
7.	If $it > 1$, set $\Delta = 1 - \hat{\sigma}/\sigma_0$;
8.	Set $\sigma_0 = \hat{\sigma}$;
9.	Compute $w(k) = \dot{\rho}\left(\boldsymbol{r}(k)/\hat{\sigma}\right)$ for $k = 1...N$;
10.	Compute $\boldsymbol{\mu}$ from (28);
11.	Compute $\boldsymbol{S}$ from (29);
12.	Compute $\tilde{\boldsymbol{P}}$ the eigenvector matrix of the covariance matrix $\boldsymbol{S}$;
13.	corresponding to its $m - \ell$ smallest eigenvalues;
14.	Compute $\boldsymbol{a} = \tilde{\boldsymbol{P}}^{\mathrm{T}}\boldsymbol{\mu}$;
15.	Set $it = it + 1$;
16.	End do;
17.	Set $\boldsymbol{w}_{res} = \boldsymbol{w}$, $it = 1$ and $\sigma_0 = \infty$;
18.	Compute $\hat{\boldsymbol{P}}$ the eigenvector matrix of the robust covariance matrix $\boldsymbol{S}$ corresponding to its ℓ largest eigenvalues;
19.	Compute $\boldsymbol{a} = median(\boldsymbol{X}\,\hat{\boldsymbol{P}})$;
20.	Do until convergence of Δ;
21.	Compute $\boldsymbol{r}(k) = \|\|\hat{\boldsymbol{P}}\boldsymbol{x}(k) - \boldsymbol{a}\|\|^2$ for $k = 1...N$;
22.	Compute $\hat{\sigma}$ from (33);
23.	If $it > 1$, set $\Delta = 1 - \hat{\sigma}/\sigma_0$;
24.	Set $\sigma_0 = \hat{\sigma}$;
25.	Compute the $\boldsymbol{w}(k) = \dot{\rho}\left(\boldsymbol{r}(k)/\hat{\sigma}\right)$ for $k = 1...N$;
26.	Set $\boldsymbol{w} = min(\boldsymbol{w}, \boldsymbol{w}_{res})$;
27.	Compute $\boldsymbol{\mu}$ from (28);
28.	Compute $\boldsymbol{S}$ from (29);
29.	Compute $\hat{\boldsymbol{P}}$ the eigenvector matrix of the covariance matrix $\boldsymbol{S}$;
30.	corresponding to its ℓ largest eigenvalues;
31.	Compute $\boldsymbol{a} = \hat{\boldsymbol{P}}^{\mathrm{T}}\boldsymbol{\mu}$;
32.	Set $it = it + 1$;
33.	End do.

Table 1. Algortithm of the MMRPCA method

3.4 Robust determination of the principal component number

The number ℓ of principal components to choose is obtained by minimizing the normalized VRE (variance of reconstruction error) with respect to the number ℓ (Qin & Dunia, 2000), the criterion is then :

$$J(\ell) = \sum_{j=1}^{m} \frac{\boldsymbol{\xi}_j^{\mathrm{T}}\,(\boldsymbol{I}_m - \hat{\boldsymbol{C}})\,\boldsymbol{S}\,(\boldsymbol{I}_m - \hat{\boldsymbol{C}})\,\boldsymbol{\xi}_j}{\left(\boldsymbol{\xi}_j^{\mathrm{T}}(\boldsymbol{I}_m - \hat{\boldsymbol{C}})\boldsymbol{\xi}_j\right)^2} \tag{40}$$

with $\ell = 1, \ldots, m-1$, $\hat{\boldsymbol{C}}$ the robust PCA model, $\boldsymbol{S}$ the robust covariance matrix and $\boldsymbol{\xi}_j$ the reconstruction direction ($\boldsymbol{\xi}_j = [0\ ...\ 1\ ...\ 0]^{\mathrm{T}}$ where value 1 is at the jth position). Qin & Dunia (2000) show that this criterion may present a minimum in the interval $[1, m]$. The algorithm to determine both the number of principal components ℓ and the robust model is summarized in table 2.

1.	**Compute the robust covariance matrix Q (25);**
2.	**Set $\ell_{ini} = 1$, ℓ_{ini} is the number of principal component used to construct the robust model with the MMRPCA method;**
3.	**Do until $\ell_{ini} < m$;**
4.	**Compute the robust covariance with the MM-estimator S (39);**
5.	**Find the minimum of the normalized VRE (ℓ varying from 1 to $m - \ell$);**
6.	**Set ℓ_{ini}=ℓ_{ini}+1;**
7.	**End do;**
8.	**The smallest value obtain by minimizing the normalized VRE is associated with the number of principal component to use.**

Table 2. Algorithm to determine the number of principal components

4. Fault isolation

After the presence of faults has been detected, it is important to identify these faults and to apply the necessary corrective actions to eliminate the abnormal data. In the PCA framework, the well known isolation approaches are residual enhancement, contribution plot and variable reconstruction methods.

In structured design, each residual responds to a specific subset of faults, and thus a specific subset of the residuals responds to each fault, resulting in a unique fault code. Structured residuals may be obtained from a full PC model by algebraic transformation. But they can also be generated by first specifying residual structures and then obtaining subsystem models (partial PCA models), each corresponding to a residual (Huang & Gertler, 1999; Qin & Weihua, 1999). However, for a high dimensionality process, it is not always possible to find the residual structuration that enables to obtain the desired isolation properties because these properties are only defined according to the occurrence of the faults in the residuals without taking into account the sensitivities of the residuals to the faults.

Contribution plots are well known diagnostic tools for fault isolation. The most common indices used for fault diagnosis with contribution plots are SPE and T^2. Contribution plots on SPE indicate the significance of the effect of each variable on this index at different sampling times. If a sample vector $\mathbf{x}$ has an abnormal SPE, the components of this vector that appear to have a significant contribution are investigated. A contribution plot on principal scores indicates the significance of the effect of each variable on the T^2 index. The variables with the largest contribution are considered as major contributors to the fault. The contribution plots are very easy to calculate, with no prior knowledge required to generate the plots. Prior knowledge, however, is often used and required to interpret the plots. As explained by MacGregor & Kourti (1995), the contribution plots may not explicitly identify the cause of an abnormal condition. The reason is that when there is no fault, the contributions are uneven across variables. Therefore, a fault in a normally small contribution variable may not have the largest contribution unless the fault magnitude is very large. This can be a source of misdiagnosis.

An alternative approach for fault isolation is the variable reconstruction method proposed by Dunia et al. (1996). The reconstruction based approach eliminates the fault when the actual fault direction is used for reconstruction. In the case of arbitrary process fault directions the reconstruction-based approach brings a fault indicator within the normal control limit (completely removes the effect of the fault when the faulty variables are reconstructed).

4.1 Reconstruction approach

In this section the variable reconstruction approach is presented. The reconstruction $\mathbf{x}_R(k)$ of an observation $\mathbf{x}(k)$ is defined as follows:

$$\mathbf{x}_R(k) = \mathbf{x}(k) - \Xi_R \mathbf{f}_R \tag{41}$$

with $\mathbf{f}_R$ the fault magnitude (unknown) and the matrix Ξ_R indicating the reconstruction directions. This matrix is orthonormal with dimension ($m \times$ r), with r the number of component to reconstruct, and is built with 0 and 1, where 1 indicates the reconstructed variables from the other variables (with 0). For example, to reconstruct the set of variables $R = \{2,4\}$ among 5 variables, matrix Ξ_R is formed as follows:

$$\Xi_R = \begin{bmatrix} 0 & 1 & 0 & 0 & 0 \\ 0 & 0 & 0 & 1 & 0 \end{bmatrix}^T$$

The reconstruction can be formulated by an optimization problem of the detection indicator with respect to fault $\mathbf{f}_R$ according to the direction Ξ_R. So, according to the indicator used for detection SPE, SWE, T^2 or D^2, several types of reconstruction can take place.

The estimation of the fault magnitude $\mathbf{f}_R$ is obtained by solving the following optimization problem, according to the detection indicator:

$$\hat{\mathbf{f}}_R = \arg\min_{\mathbf{f}_R} \left\{ \mathbf{x}_R^T(k) \mathbf{\Phi} \mathbf{x}_R(k) \right\} \tag{42}$$

where

- In reconstruction according to SPE indicator, we have: $\mathbf{\Phi} = \tilde{\mathbf{P}}\tilde{\mathbf{P}}^T$,
- In reconstruction according to SWE indicator, we have: $\mathbf{\Phi} = \tilde{\mathbf{P}}\tilde{\mathbf{\Lambda}}^{-1}\tilde{\mathbf{P}}^T$,
- In reconstruction according to T^2 indicator, we have: $\mathbf{\Phi} = \hat{\mathbf{P}}\hat{\mathbf{\Lambda}}^{-1}\hat{\mathbf{P}}^T$
- In reconstruction according to D^2 indicator, we have: $\mathbf{\Phi} = \mathbf{P}\mathbf{\Lambda}^{-1}\mathbf{P}^T$

Given a subset R, the expression for the reconstruction $\mathbf{x}_R(k)$ of the vector $\mathbf{x}(k)$ is then expressed by:

$$\mathbf{x}_R(k) = \mathbf{G}_R\, \mathbf{x}(k) \tag{43}$$

with $\mathbf{G}_R = \left(\mathbf{I} - \Xi_R(\Xi_R^T \mathbf{\Phi} \Xi_R)^{-1} \Xi_R^T \mathbf{\Phi}\right)$

Faults can only be projected onto the representation or the residual subspace. To isolate all kinds of faults, reconstruction according to D^2 (Mahalanobis distance) indicator is considered ($\mathbf{\Phi} = \mathbf{P}\mathbf{\Lambda}^{-1}\mathbf{P}^T$). We define $D_R^2(k)$ as the Mahalanobis distance, calculated after the reconstruction of the subset R of variables which is given by:

$$D_R^2(k) = \mathbf{x}_R^T(k) \mathbf{P}\mathbf{\Lambda}^{-1}\mathbf{P}^T \mathbf{x}_R(k) \tag{44}$$

The system is considered normal if:

$$D_R^2(k) \leq \chi^2_{m-r,\alpha} \tag{45}$$

where $D^2_R(k)$ is the fault detection indicator (Mahalanobis distance), calculated after the reconstruction of a subset R of variables, and $\chi^2_{m-r,\alpha}$ is a chi-2 control limit with $m-r$ degrees of freedom for a significance level α.

If we write $\mathbf{x}_R$ (43) in the case where the matrix of the reconstruction directions is reorganized as follows (using column permutations):

$$\Xi_R = \left[\underset{(r\times r)}{\mathbf{I}_1} \;\middle|\; \underset{((m-r)\times r)}{0} \right]^{\mathrm{T}} \in \Re^{m\times r} \tag{46}$$

with $\mathbf{I}_1 \in \Re^{r\times r}$ an identity matrix. Then $\mathbf{P}$ and $\mathbf{\Lambda}$ are split in four parts:

$$\mathbf{P}^{\mathrm{T}} = \left[\begin{array}{c|c} \underset{(r\times r)}{\mathbf{P}^{\mathrm{T}}_{11}} & \underset{(r\times(m-r))}{\mathbf{P}^{\mathrm{T}}_{12}} \\ \hline \underset{((m-r)\times r)}{\mathbf{P}^{\mathrm{T}}_{21}} & \underset{((m-r)\times(m-r))}{\mathbf{P}^{\mathrm{T}}_{22}} \end{array} \right] \in \Re^{m\times m} \tag{47}$$

$$\mathbf{\Lambda} = \left[\begin{array}{c|c} \underset{(r\times r)}{\mathbf{\Lambda}_1} & \underset{(r\times(m-r))}{0} \\ \hline \underset{((m-r)\times r)}{0} & \underset{((m-r)\times(m-r))}{\mathbf{\Lambda}_2} \end{array} \right] \in \Re^{m\times m} \tag{48}$$

Using definition (43), the reconstruction $\mathbf{x}_R$ of the vector $\mathbf{x}$ is written as follows:

$$\mathbf{x}_R = \left[\begin{array}{c|c} 0 & \left(\Xi_R^{\mathrm{T}}\Xi_R\right)^{-1} \left(\mathbf{P}_{11}\mathbf{\Lambda}_1^{-1}\mathbf{P}^{\mathrm{T}}_{12} + \mathbf{P}_{21}\mathbf{\Lambda}_2^{-1}\mathbf{P}^{\mathrm{T}}_{22}\right) \\ \hline 0 & \mathbf{I}_2 \end{array} \right] \mathbf{x} \tag{49}$$

with $\mathbf{I}_2 \in \Re^{m-r\times m-r}$ an identity matrix.

This form highlights two characteristics. First, the reconstructed vector $\mathbf{x}_R$ is constituted by the r reconstructed variables and a copy of the $m-r$ remaining variables. Secondly, the reconstructed variables are estimated without using their own measurement.

4.1.1 Reconstruction condition

To reconstruct a fault, it must be at least projected into the representation subspace ($r \leq \ell$) or onto the residual subspace ($r \leq m-\ell$). This implies that the number of reconstructed variables r must respect the following inequality:

$$r \leq \max(m-\ell, \ell) \tag{50}$$

4.1.2 Structured residual generation

In a diagnosis objective, residuals are generated for fault detection and isolation. Considering equations (8) and (43), then $\mathbf{x}_R$ can be expressed as:

$$\mathbf{x}_R(k) = \left(\mathbf{I} - \Xi_R(\Xi_R^T\mathbf{\Phi}\Xi_R)^{-1}\Xi_R^T\mathbf{\Phi}\right) \left(\mathbf{x}^{o}(k) + \boldsymbol{\epsilon}(k) + \Xi_F\mathbf{f}\right) \tag{51}$$

As the fault influence is generally unknown, we have to consider all possible reconstruction directions Ξ_R

- If the reconstruction directions Ξ_R (8) are the same as the fault directions, i.e. if $R = F$, then $D_R^2(k)$ is under the detection threshold ($D_R^2 < \chi^2_{m-r,\alpha}$), indeed, the fault influence is removed:

$$\left(\mathbf{I} - \Xi_R(\Xi_R^T \boldsymbol{\Phi} \Xi_R)^{-1} \Xi_R^T \boldsymbol{\Phi}\right) \Xi_R = \mathbf{0} \tag{52}$$

and the reconstruction is expressed as:

$$\mathbf{x}_R(k) = \left(\mathbf{I} - \Xi_R(\Xi_R^T \boldsymbol{\Phi} \Xi_R)^{-1} \Xi_R^T \boldsymbol{\Phi}\right) (\mathbf{x}^o(k) + \boldsymbol{\epsilon}(k)) \tag{53}$$

- If the reconstruction directions Ξ_R are different from the fault directions, then $D_R^2(k)$ is higher than the detection threshold $\chi^2_{m-r,\alpha}$ if the projection of the reconstruction directions are not collinear to the fault projection onto the residual subspace and onto the representation subspace.

For the faulty observation k, the faulty variables subset $\hat{R}$ is determined as follows:

$$\hat{R} = \arg_{R \in \Im} D_R(k) < \chi^2_{m-r,\alpha} \tag{54}$$

with $\Im$ the set of possible reconstruction directions.

4.2 Fault isolation

The proposed fault isolation procedure consists in two steps: offline and online steps. The offline step consists of a priori analysis of all reconstruction directions which allows to determine the isolable faults (useful reconstructions). The second online step consists in using the useful reconstruction directions (corresponding to isolable faults) for multiple fault isolation.

4.2.1 Offline step

All the directions of reconstruction Ξ_R have to be explored for fault isolation. The maximum reconstruction number can be calculated as follows:

$$\sum_{r=1}^{\max(m-\ell,\ell)-1} C_m^r \tag{55}$$

where C_m^r denotes the combination of r from m.

This number takes only into account the number of reconstructions in the different subspaces. However, collinear projections have the same fault signature. Then we will analyze the angles between the different projections of reconstruction directions. The largest primary angle θ between two subspaces of the same size is linked to the concept of distance between these two subspaces (Golub & Van Loan, 1996).

This distance is defined in the representation subspace $d(R_i, R_j)$ and in the residual subspace $\tilde{d}(R_i, R_j)$ as follows:

$$d(R_i, R_j) = ||\hat{\Xi}_{R_i}(\hat{\Xi}_{R_i}^T \hat{\Xi}_{R_i})^{-1} \hat{\Xi}_{R_i}^T - \hat{\Xi}_{R_j}(\hat{\Xi}_{R_j}^T \hat{\Xi}_{R_j})^{-1} \hat{\Xi}_{R_j}^T||_2 \tag{56}$$

$$\tilde{d}(R_i, R_j) = ||\tilde{\Xi}_{R_i}(\tilde{\Xi}_{R_i}^T \tilde{\Xi}_{R_i})^{-1} \tilde{\Xi}_{R_i}^T - \tilde{\Xi}_{R_j}(\tilde{\Xi}_{R_j}^T \tilde{\Xi}_{R_j})^{-1} \tilde{\Xi}_{R_j}^T||_2 \tag{57}$$

with $\hat{\Xi}_{R_i} = \hat{\Lambda}^{-1/2} \hat{\mathbf{P}}^T \Xi_{R_i}$, $\tilde{\Xi}_{R_i} = \tilde{\Lambda}^{-1/2} \tilde{\mathbf{P}}^T \Xi_{R_i}$ and R_i and R_j correspond to two sets of variable reconstructions.

Analyzing these distances, then the isolable fault can be determined a priori. Hence, a global indicator $\mathcal{K}$ is built.

$$\mathcal{K}(R_i, R_j) = \max\{(d(R_i, R_j), \tilde{d}(R_i, R_j)\} \tag{58}$$

Thus, if $\mathcal{K}(R_i, R_j)$ is close to zero, it means that the projections of the set of reconstructed variables R_i and R_j are collinear in the residual subspace and in the representation subspace. It means that a fault for the sets of reconstructed variables R_i or R_j are not isolable. The process to detect useful directions of reconstruction can be summarized as follows:

1. $r = 1$
2. Calculate for all available directions ($R_i \in \Im$ and $R_j \in \Im$) the indicator $\mathcal{K}(R_i, R_j)$ (58). The smaller the value of this indicator, the higher the magnitude of the fault has to be to ensure fault isolation. And if this indicator is equal to zero, then only a set of potentially faulty variables may be determined, i.e. the faulty variables are associated to the indices R_i or R_j or R_i and R_j. Thus, it is only required to study one single subset of directions, for example R_i.
3. $r = r + 1$
4. While $r \leq \max(\ell, m - \ell) - 1$ do to the step 2

This analysis of the structure of the model allows to determine a priori the isolable faults. The number of useful reconstructions can then be greatly reduced.

4.2.2 Online step

If a fault is detected on the $D^2(k)$ indicator:

1. For $r = 1$, (number of reconstructed variables)
2. Compute $D_R^2(k)$, where $R \in \Im$ is a subset (of r variables) of useful reconstruction selected in the offline step:
   - if for a particular reconstruction direction Ξ_R, $D_R^2(k) \leq \chi^2_{m-r,\alpha}$, variables in this subset R are the faulty variables. Isolation procedure is stopped.
   - Otherwise, if there are more than the r faulty variables, go to step 3
3. r=r+1
4. While $r \leq max(m - \ell, \ell) - 1$, go to step 2
5. The fault is not isolable (more than $r = max(m - \ell, \ell) - 1$ variables can be faulty)

5. Numerical example

5.1 Data generation

We consider here the situation in which several faults affect different variables at the same time. The matrix $\mathbf{X}$ includes $N = 450$ observations of a vector x with $m = 9$ components generated in the following way:

$$\begin{aligned}
&x_{i,1} = 1 + v_i^2 + \sin(i/3), \quad x_{i,2} = 2\sin(i/6)\cos(i/4)\exp(-i/N), \quad v_i \sim \mathcal{N}(0,1) \\
&x_{i,3} = \log(x_{i,2}^2), \quad x_{i,4} = x_{i,1} + x_{i,2}, \quad x_{i,5} = x_{i,1} - x_{i,2} \\
&x_{i,6} = 2x_{i,1} + x_{i,2}, \quad x_{i,7} = x_{i,1} + x_{i,3}, \quad x_{i,8} \sim \mathcal{N}(0,1), \quad x_{i,9} \sim \mathcal{N}(0,1)
\end{aligned} \tag{59}$$

On the data thus generated were added realizations of random variables with centred normal distribution and standard deviations equal to 0.02 as well as faults δx_1 represented by a bias of amplitude equal to 20% of the amplitude of the variable, δx_2, δx_3 represented by a bias of amplitude equal to 10% of the amplitudes of the variables, δx_8 represented by a bias of amplitude equal to 150% of the amplitude of the variable. Faults are defined on specific time intervals: observations from 50 to 100 (interval I_1) for the variable x_1, observations from 150 to 200 (interval I_2) for the variables x_2 and x_3, observations from 250 to 300 (interval I_3) for the variables x_8.

5.2 Robust PCA

To determine the number of principal components, the robust approach using the VRE, proposed in the section 3.4, is used. Five principal components are selected ($\ell = 5$). The robust model is then built. Figure 2 shows the measure of the first variable x_1 with its estimation obtained with the classic PCA model and the robust PCA model and the associated residual (measure - estimate).

This figure shows that in the fault-free case (for example interval from 300 to 450) the robust residuals are centred on zero while, with the classical PCA, the residuals are not centred on zero. It means that a smaller fault magnitude can be detected much better using the robust method than using the classical method. This shows the advantage of using a robust approach.

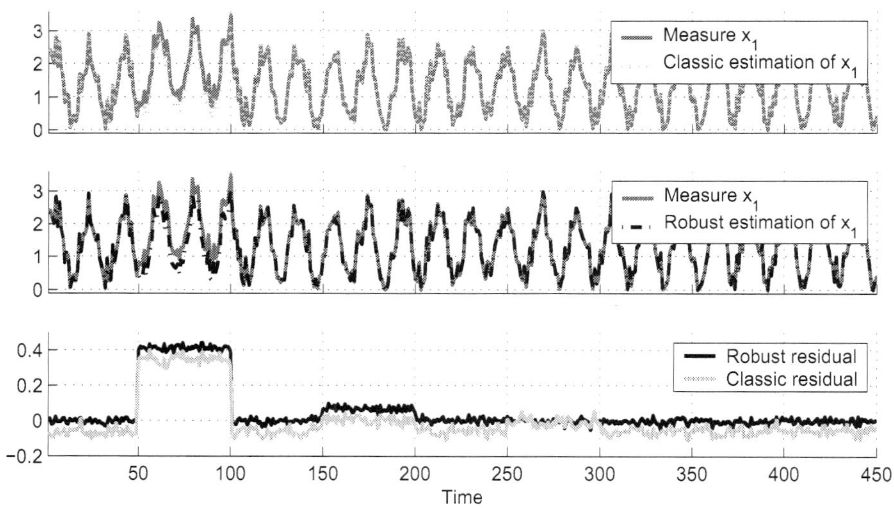

Fig. 2. Measure and estimation of x_1

5.3 Fault detection and isolation

5.3.1 Offline step

The Mahalanobis distance is used to detect and to isolate the faults. Considering the dimension of the residual subspace and of the representation subspace, we cannot reconstruct simultaneously more than five variables ($\max(m-\ell, \ell)$). The maximum number of reconstructions is then equal to 255 (55). Table 3 shows the values of the global indicator $\mathcal{K}$ (58) with $r = 1$,

i.e. only one variable is reconstructed. The two sets R_1 and R_2 contain the indices of the reconstructed variables. The smaller the value of this indicator $\mathcal{K}$, the higher the magnitude of the fault has to be to ensure fault isolation. All the values of $\mathcal{K}$ are not null, so all the faults on one variable are isolable.
For all the directions of reconstruction ($r = 2, 3, 4$) this indicator is calculated. A case where $\mathcal{K}$ is close to zero is detected between $D^2_{1,3}$ and $D^2_{1,7}$. Then the fault signatures of these two directions are identical ($D^2_{1,3} = D^2_{1,7}$). Therefore, only one indicator is useful to detect this fault, for example $D^2_{1,3}$. Moreover, we concluded that the signatures of reconstruction directions taking into account these sets are identical ($D^2_{1,3,4} = D^2_{1,4,7}$, $D^2_{1,3,6} = D^2_{1,6,7}$, ...). The number of useful reconstructions can be reduced to 168.

K		R_1								
		1	2	3	4	5	6	7	8	9
	1	0	1.00	0.99	0.99	0.89	0.94	0.99	1.00	1.00
	2		0	1.00	0.95	0.72	0.97	1.00	1.00	0.98
	3			0	1.00	0.98	0.95	0.23	1.00	1.00
R_2	4				0	0.98	0.75	1.00	1.00	0.99
	5					0	0.99	0.99	1.00	0.99
	6						0	0.96	1.00	1.00
	7							0	1.00	1.00
	8								0	1.00

Table 3. Indicator $\mathcal{K}$ for $r = 1$

5.3.2 Online step

Figure 3 shows the Mahalanobis distance divided by its detection threshold obtained for a significance level of $\alpha = 99\%$, i.e. a fault is detected if the normalised Mahalanobis distance is greater than one. Then, the faults on intervals I_1, I_2 and I_3 are detected only by using the robust Mahalanobis distance (constructed with the robust model). Let us remark that the classical Mahalanobis distance is not able to detect the faults.

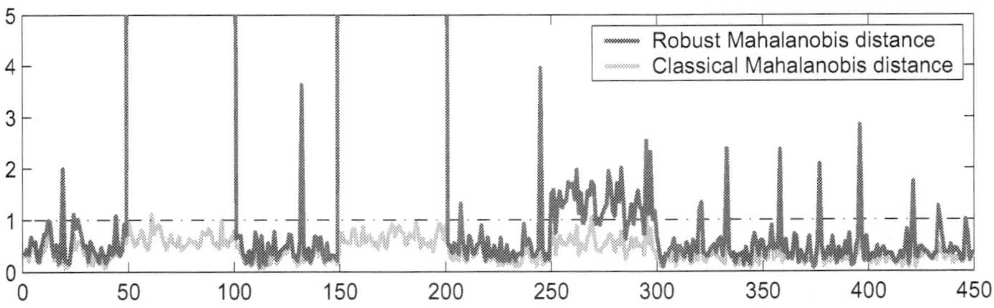

Fig. 3. Fault detection with Mahalanobis distance

Once the faults are detected, we will try to isolate them. First, all useful reconstruction directions for the reconstruction of one variable (r=1) are calculated. The first graph of the figure 4 shows the global indicator D^2_1 (44). For the observations of the interval I_1 this distance is close to the value 0 and thus shows the absence of outliers in the variables used for the reconstruction, i.e. all the variables except x_1. Let us note that the two other groups of observations (I_2,

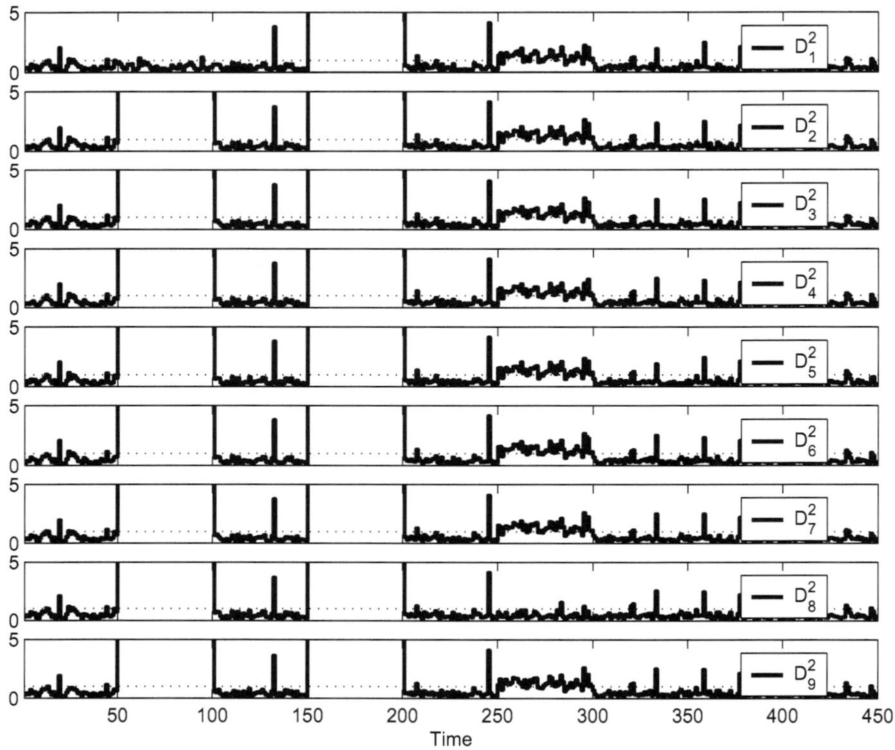

Fig. 4. Fault isolation with Mahalanobis distance

I_3) are affected by faults, but we don't know exactly which components of the measurement vector are faulty. Finally, by taking into account the fault presence in the three intervals, the examination of the first graph of the figure 4 helps us to conclude that:

- in each interval I_2, I_3 other variables than x_1 are faulty.

Other reconstructions are built and are interpreted in a similar way. Figure 4 shows all values of the indicator D_R^2 for useful reconstruction directions always in the case $r = 1$. From the reconstruction of one variable, the diagnosis is as follows:

- in the interval I_1, x_1 is faulty,
- in the interval I_2, more than one variable is faulty,
- in the interval I_3, x_8 is faulty.

Since the fault on interval I_2 is not isolate, then the useful reconstructions with two variables are performed. Figure 5 shows some values of the indicator D_R^2 for useful reconstruction directions with $r = 2$. From the reconstruction of two variables, the diagnosis is the following:

- in the interval I_2, x_2 and x_3 are faulty,

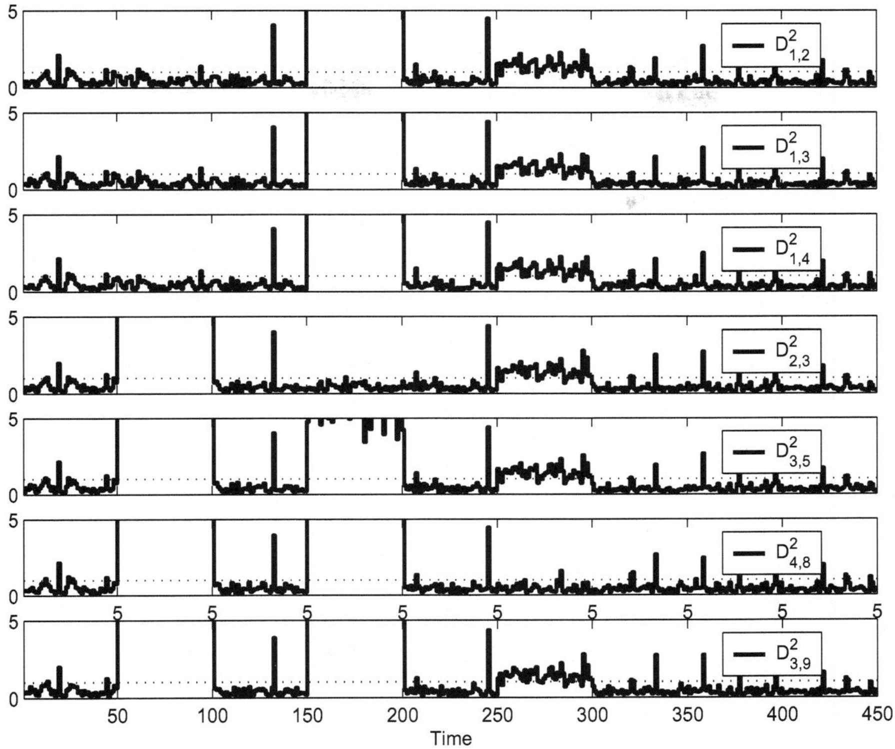

Fig. 5. Fault isolation with Mahalanobis distance

- in the intervals I_1 and I_3 the previous conclusions are confirmed.

If all the faults are isolated, then it is not useful to reconstruct more variables. Table 4 summarizes the conclusions resulting from the D_R^2 analysis (figures 4 and 5). Symbol 0 denotes the fault absence and symbol $\times$ denotes the fault presence in the considered interval.

	I_1	I_2	I_3
D_1^2	0	$\times$	$\times$
$D_{2,3}^2$	$\times$	0	$\times$
D_8^2	$\times$	$\times$	0

Table 4. Fault signatures

6. Conclusion

Principal components analysis reduces the data representation subspace and enables the determination of the redundancy relationships (linear relations among the variables). The re-

dundancy relations are then used to detect and isolate the faulty data. PCA is constructed with fault-free data from a decomposition in eigenvalues and eigenvectors of a covariance matrix.
However, real data sets are usually not fault-free then the covariance matrix is disturbed by outliers. In order to reduce the sensitivity of the model to all kinds of outliers (with a projection onto the representation or the residual subspace), a fast two-step algorithm is proposed. First, a MM-estimator is used to determine a robust model. This estimator is computed by using an iterative re-weighted least squares (IRWLS) procedure. This algorithm is initialized from a very simple estimate derived from a one-step weighted covariance estimate. Therefore, a model robust with respect to outliers is constructed. Secondly, structured residuals are generated for multiple fault detection and isolation. These structured residuals are based on the reconstruction-projection principle. For fault isolation, the proposed scheme avoids the combinatorial explosion of faulty scenarios related to multiple faults. Indeed, instead of considering all combinations of one up to all sensors, we limit the maximum number of faulty scenarios to consider by evaluating the existence condition of structured residuals. Therefore, the detectable faults and the isolable faults are determined as well as the different faulty scenarios for which it is not possible to distinguish the faulty variables. This procedure has been applied on one example, with single and multiple faults. The presence of approximately 30 percent of outliers authorizes a correct estimation of the principal components, then the estimation is not very sensitive to outliers. The method is efficient for fault detection and isolation.

7. References

Caussinus, H., Hakam, S. & Ruiz-Gazen, A. (2003). Projections révélatrices contrôlées : Groupements et structures diverses, *Revue de statistique appliquée* **51**(1): 37–58.

Chiang, L. H. & Colegrove, L. F. (2007). Industrial implementation of on-line multivariate quality control, *Chemometrics and Intelligent Laboratory Systems* **88**(2): 143–153.

Croux, C., Filzmoser, P. & Oliveira, M. (2007). Algorithms for projection-pursuit robust principal component analysis, *Chemometrics and Intelligent Laboratory Systems* **87**(2): 218–225.

Croux, C. & Ruiz-Gazen, A. (2005). High breakdown estimators for principal components: the projection-pursuit approach revisited, *Journal of Multivariate Analysis* **95**(1): 206–226.

Daszykowski, M., Kaczmarek, K., Vander Heyden, Y. & Walczak, B. (2007). Robust statistics in data analysis - a review: basic concepts, *Chemometrics and Intelligent Laboratory Systems* **85**(2): 203–219.

Donoho, D. L. & Huber, P. J. (1983). *A Festschrift for Eric Lehmann*, Wadsworth, Belmont, CA, chapter The notion of breakdown point.

Dunia, R., Qin, S. & Edgar, T. F. (1996). Identification of faulty sensors using principal component analysis, *American Institute Of Chemical Engineers Journal* **42**(10): 2797–2812.

Filzmoser, P., Maronna, R. & Werner, M. (2008). Outlier identification in high dimensions, *Computational Statistics & Data Analysis* **52**(3): 1694–1711.

Gertler, J. & McAvoy, T. (1997). Principal component analysis and parity relations - a strong duality, *IFAC Symposium on Fault Detection, Supervision and Safety of Technical Processes, SAFEPROCESS'97, Hull,UK.* 837-842.

Golub, G. H. & Van Loan, C. F. (1996). *Matrix computations*, The Johns Hopkins University Press, Baltimore and London.

Harkat, M.-F., Mourot, G. & Ragot, J. (2006). An improved PCA scheme for sensor FDI: Application to an air quality monitoring network, *Journal of Process Control* **16**(6): 625–634.

Huang, Y. & Gertler, J. (1999). Fault isolation by partial PCA and partial NLPCA, *14th triennial world congress, IFAC'99, Beijing, P. R. China.*

Huber, P. (1964). Robust estimation of a location parameter, *Annals of Mathematical Statistics* **35**: 73–101.

Hubert, M., Rousseeuw, P. & Van den Branden, K. (2005). ROBPCA: a new approach to robust principal component analysis, *Technometrics* **47 (1)**: 64–79.

Kano, M. & Nakagawa, Y. (2008). Data-based process monitoring, process control, and quality improvement: Recent developments and applications in steel industry, *Computers & Chemical Engineering* **32**(1-2): 12–24.

Kruger, U., Zhou, Y. & Irwin, G. W. (2004). Improved principal component monitoring of large-scale processes, *Journal of Process Control* **14**: 879–888.

Ku, W., Storer, R. & Georgakis, G. (1995). Disturbance detection and isolation by dynamic principal component analysis, *Chemometrics and Intelligent Laboratory Systems* **30**: 179–196.

Li, G. & Chen, Z. (1985). Projection-pursuit approach to robust dispersion matrices and principal components: Primary theory and Monte Carlo, *Journal of the American Statistical Association* **80**(391): 759–766.

Li, W. & Qin, S. J. (2001). Consistent dynamic pca based on errors-in-variables subspace identification, *Journal of Process Control* **11**: 661–678.

MacGregor, J. F. & Kourti, T. (1995). Statistical process control of multivariate processes, *Control Engineering Practice* **3**(3): 403–414.

Maronna, R. A. (2005). Principal components and orthogonal regression based on robust scales, *Technometrics* **47**(3): 264–273.

Maronna, R. A., Martin, R. & Yohai, V. J. (2006). *Robust Statistics: Theory and Methods*, John Wiley & Sons, Ltd.

Nomikos, P. & MacGregor, J. F. (1995). Multivariate SPC charts for monitoring batch processes, *Technometrics* **37**: 41–59.

Qin, S. J. (2003). Statistical process monitoring: basics and beyond, *Journal of Chemometrics* **17**(8-9): 480–502.

Qin, S. J. & Dunia, R. (2000). Determining the number of principal components for best reconstruction, *Journal of Process Control* **10**: 245–250.

Qin, S. J. & Weihua, L. (1999). Detection, identification and reconstruction of faulty sensors with maximized sensitivity, *American Institute Of Chemical Engineers Journal* **45**(9): 1963–1976.

Qin, S. & Wang, J. (2006). Closed-loop subspace identification using the parity space, *Automatica* **42**: 315–320.

Rousseeuw, P. (1987). *Robust regression and outliers detection*, John Wiley & Sons Inc.

Rousseeuw, P., Debruyne, M., Engelen, S. & Hubert, M. (2006). Robustness and outlier detection in chemometrics, *Analytical Chemistry*, **36**: 221–242.

Rousseeuw, P. & Van Driessen, K. (1999). Fast algorithm for the minimum covariance determinant estimator, *Technometrics* **41**(3): 212–223.

Tamura, M. & Tsujita, S. (2007). A study on the number of principal components and sensitivity of fault detection using PCA, *Computers and Chemical Engineering* **31**: 1035–1046.

Valle, S., Li, W. & Qin, S. J. (1999). Selection of the number of principal components: the variance of the reconstruction error criterion with a comparison to other methods, *Industrial Engineering Chemical Research* **38**: 4389–4401.

Wang, H., Jiang, N. & Yang, D. (2004a). Process monitoring in principal component subspace: part 1 - fault reconstruction study, *IEEE international conference on systems, man and cybernetics, The Hague, Netherlands, 10-13 october.*

Wang, H., Jiang, N. & Yang, D. (2004b). Process monitoring in principal component subspace: part 2. fault identification and isolation study, *IEEE international conference on systems, man and cybernetics, The Hague, Netherlands, 10-13 october.*

Wang, J. & Qin, S. (2002). A new subspace identification approach based on principal component analysis, *Journal of Process Control* **12**: 841–855.

Westerhuis, J., Gurden, S. & Smilde, A. (2000). Standardized q-statistic for improved sensitivity in the monitoring of residuals in MSPC, *Journal of Chemometrics* **14**(4): 335–349.

Xie, L., Kruger, U., Lieftucht, D., Littler, T., Chen, Q. & Wang, S. (2006). Statistical monitoring of dynamic multivariate process - part 1. modeling autocorrelation and cross-correlation, *Industrial Engineering Chemical Research* **45**: 1659–1676.

18

Single Line-to-Ground Fault Detection in Face of Cable Proliferation in Compensated Systems

Wan-Ying Huang, Robert Kaczmarek and Jean-Claude Vannier
Supelec
France

1. Introduction

In case of low ohmic SLG faults in compensated distribution networks, the directional function can be assured by transient relays which compare polarities of zero-sequence charging components at main frequency, either of voltage versus currents on each feeder (Nikander et al., 1995) or of their products on different feeders (Coemans & Maun, 1995). Some difficulties can arise, however, in the presence of violent discharging transients. They grow with inception angle θ, which determines pre-fault values of the faulty line voltage. The discharging currents are absent at θ=0°, but then they reach the charging currents' level when the inception angle is several degrees (Fig. 1), and grow rapidly. The most favorable conditions for generation of important transients are with low resistance fault when it occurs at inception angle 90° in a capacitive network, be it a cabled system or a mixed one composed of cables and lines.

In such conditions, we can expect the extraction of main frequency component out of the charging currents to be more delicate an operation in cables than in lines. The charging components in overhead lines (Lehtonen, 1998) were reported to reach amplitudes 10-15 times the rated frequency amplitude whereas the discharging currents were estimated as several percent of charging components. However, when replacing lines with cables, all the other conditions unchanged, the discharging currents will be more important than the charging ones, with frequency span between them diminishing.

A relevant example comparing amplitudes and frequencies of the main discharging and charging currents (Fig. 2) presents the amplitudes ratio A_{dis}/A_{ch} rising from 0.25 (lines) to 1.5 (cables) and the frequency ratio f_{dis}/f_{ch} diminishing from 11 to 6. The reason of these tendencies is with specific values of zero-sequence capacitances in cables and lines.

Obviously, it is simpler a task to isolate a paramount and somewhat isolated component. The consequence for the transient relays, when applied in capacitive systems, can be an uncertain choice of window and trigger for acquisition and heavy standards on extraction of the relevant charging components.

The disturbing presence of the discharge components can turn into an opportunity to make a correct directional decision. This opportunity is offered by rigorous waveform disposition in initial propagation zone, usually unexplored in distribution systems. However, as the high frequency acquisition procedures become simpler and cheaper, we are tempted by the travelling wave regime to get directional function.

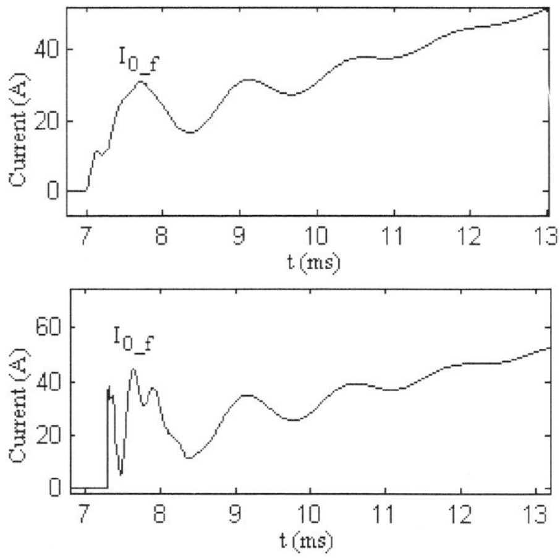

Fig. 1. Initial faulty feeder zero-sequence currents in cables. The discharging currents of higher frequency are superimposed on the charging ones of lower frequency. Upper figure: θ=0°, no discharging components. Bottom: θ=5°, the discharging current's amplitude reach the level of charging components.

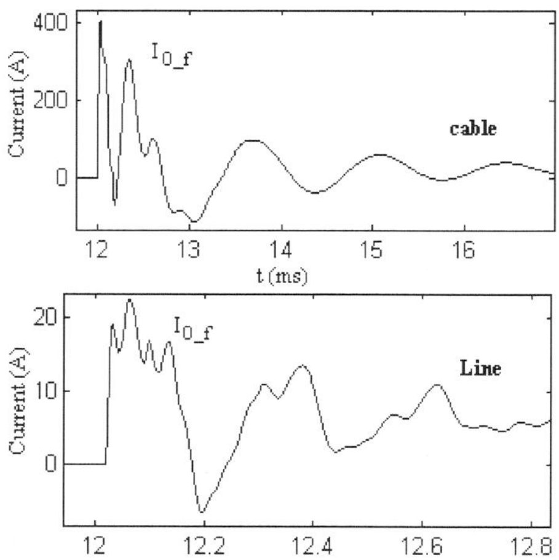

Fig. 2. Faulty feeder zero-sequence currents in cables and lines of the same length and power transfer, with 1Ω fault resistance and 90° inception angle. The discharging components in cables are of higher amplitudes than the charging components and the frequency ratio are different in cables and in lines.

Strong capacitive currents are troublesome also in case of resistive faults, where relays discriminate faulty and sound feeders upon presence of active current component in residual currents. This can be achieved either by comparing signs of projections of zero-sequence currents (Griffel et al., 1997; Welfonder et al., 2000) or looking for phase advance (Bastard et al., 1992; Segui et al., 2001) of the faulty zero-sequence current over the sound ones. However, in steady state the strong capacitive currents diminish this phase advance, with possible inhibition of the discrimination capacities of relays.
Then the way to reestablish these capacities is to exploit data recorded in transient regime, where the apparent phase difference is more important than in the steady state.
Similarly, some difficulties can arise with estimation of fault distance in strongly capacitive systems by exploiting the "main frequency" of charging components (Coemans et al., 1993), or by using the "resonance frequency" of the system (Welfonder et al., 2000), with aid of full zero-positive-negative sequence equation set.
The delicate problem of identification and extraction of the main frequency can be spared when the complete transient waveform is analyzed (Huang & Kaczmarek, 2008), rather than only one of its components. Then the system resistances' damping effect can be taken into account as a relevant parameter, possibly contributing to evaluation of the fault circuit parameters (e.g., with curve fitting).
The transients as they are, generated by faults, carry sufficient information not only for detection of the faulty feeder, but also for evaluation of the fault distance.
The procedures which follow were modeled and simulated in EMTP using frequency dependant parameters.

2. Directional function of discharging currents

We consider a radial network (Fig. 3) with no discontinuities in feeders' impedances. The network is supplied through a transformer with secondary winding grounded through Petersen coil. An SLG fault is modeled as a resistance R_f.
At fault occurrence an initial voltage wave annulling the phase voltage travels along the faulty phase of the faulty feeder with negative amplitude, accompanied by current wave also of negative amplitude. Their shape will be modeled by multiple refractions and reflections from busbar, fault and loads. The faulty phase current arriving on busbar reflects and the resulting current is equally distributed among faulty phases of all the sound feeders. These faulty phase currents impose their waveform upon residuals in each feeder. Consequently, initial currents on the sound feeders will be measured with the same polarity and opposed to the polarity on the faulty feeder.
On each sound feeder, the initial polarity regime will be over with first busbar reflection of wave getting back from loads. Its overall travel time is $2l/v$, with l (feeder's length) and v (wave velocity). This is the time interval where the current on the faulty phase of any sound feeder keeps its initial polarity unchanged. The shortest distance l_{ss} is with the shortest sound feeder and thus we get the duration of the initial polarity $\delta_{ip}=2l_{ss}/v$, where v is the maximal modal velocity.
The parameter $f_{ip}=1/\delta_{ip}$ determines the minimal sampling frequency necessary to get one point in the zone of a rigorous polarity disposition of residual currents.
For example, with the shortest length of a cable feeder being 2km, this frequency can be of 20 - 30 kHz. For practical reasons, the frequency of about 100kHz is to be reckoned with.

In cables the propagation phenomena take place both in cores and sheaths. The fault we refer to is a core-to-sheath (and, eventually, -to-ground) piercing, with sheaths grounded on supply side or on both sides. In each of these cases the traveling waves are very similar and the initial polarity zone clearly exposed (Fig. 4).

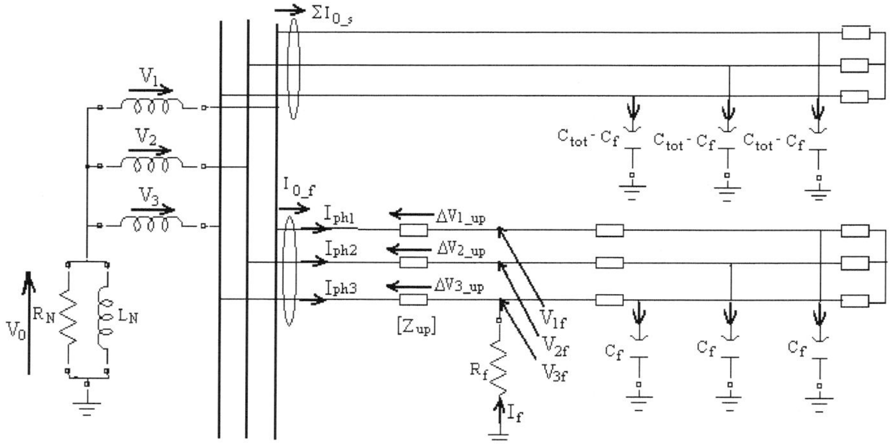

Fig. 3. Phase-to-ground fault in a radial network; all the sound feeders aggregated into one

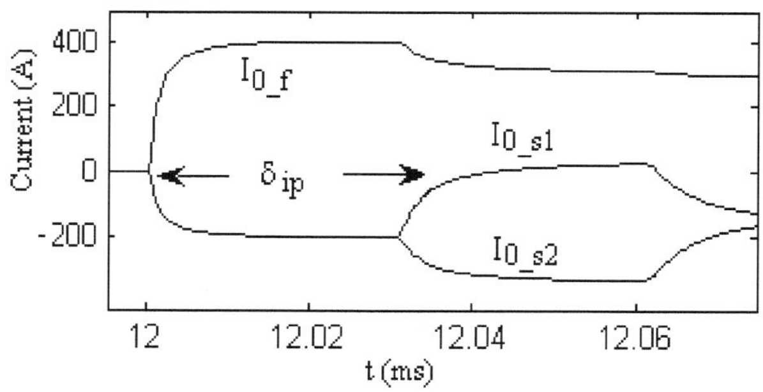

Fig. 4. Initial polarity zone in three feeders' network, zero-sequence currents. The index s is for "sound" and f- for "faulty" feeder

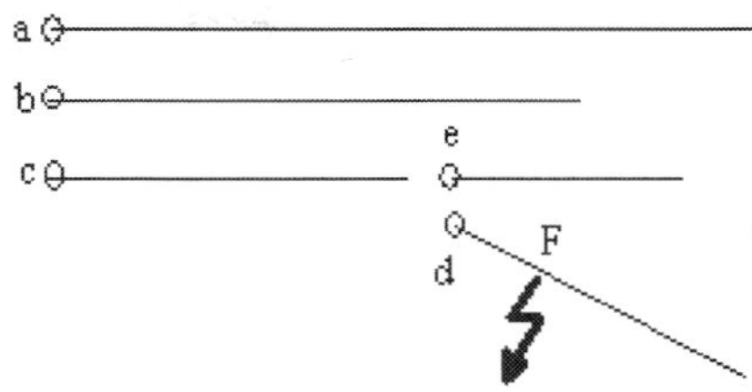

Fig. 5. Phase-to-ground fault F on the feeder "d" in a network with laterals

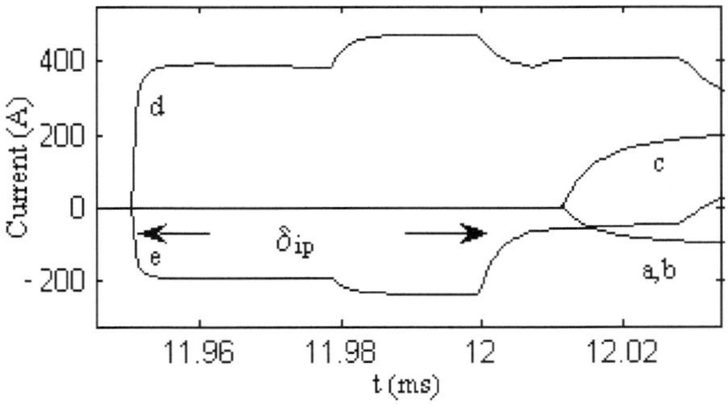

Fig. 6. Residual currents within initial polarity zone in the network of Fig.5

The initial polarity time interval does not depend neither on fault position, nor fault resistance value. If scrutinized on all feeders it can point out from busbar to fault in systems with laterals (Kaczmarek & Huang, 2005). In order to start tracing of the faulty feeder, the initial polarities of residual currents in all busbar connected feeders have to be compared. The beginning of the faulty chain is pointed with a unique sign, called "witness sign", being different from polarity of all the other feeders. Then we follow current sensors on feeders with the same "witness" polarity. The fault is at the end of the chain. This will be illustrated in the case of an SLG fault on the feeder d in a five feeders network (Fig. 5).

The rising current profiles have been recorded first on feeders d and e (Fig. 6). When the traveling waves arrive to busbar, the sign of the current measured on the feeder **c** is different than those measured on the feeders **a** and **b**. The feeder **c,** disclosing the "witness sign", points out to the feeder **d** as the fault location.

All methods based on analysis of traveling waves are highly sensitive to impedance mismatches, what results in severe conditions on their application in distribution networks. Feeders with single tee joints can be analyzed relatively easily, unless the initial polarity zone is too short to be detected. On the contrary, multiple tee joints and joints between cables and lines in mixed systems make the detection problem inextricable in terms of traveling waves' analysis.

3. Directional function of Phase difference

3.1 Principle

The steady state SLG fault regime can be analyzed on equivalent residual circuit (Fig. 7), where V'' is the voltage over a SLG fault emplacement in absence of the fault, I_{0_f} is the zero-sequence current on the faulty feeder, ΣI_{0_s} is the sum of zero-sequence currents on all the sound feeders, and I_N is the neutral point current composed of a resistive and an inductive components.

During permanent fault regime, an active current component is present in zero-sequence current I_{0_f} on the faulty feeder. The resulting phase difference between the faulty and the sound zero-sequence currents I_{0_s} (Fig. 8) is the basis of traditional wattmetric method of detection.

In low capacitive lines the phase advance can be almost 90°, because under the effect of compensation we have (Fig. 7):

$$I_{0\_f} = -I_N - \sum I_{0\_s} = -I_{c\_f} - I_{RN} \quad (1)$$

where the faulty residual I_{0_f} is dominated by its active component I_{RN}.

It is then easy to take direction decision. In cables, however, the faulty feeder capacitive current I_{c_f} dominates the composition of I_{0_f} and diminishes readability of the phase advance, particularly with fault on a long feeder or in case of system over tuning.

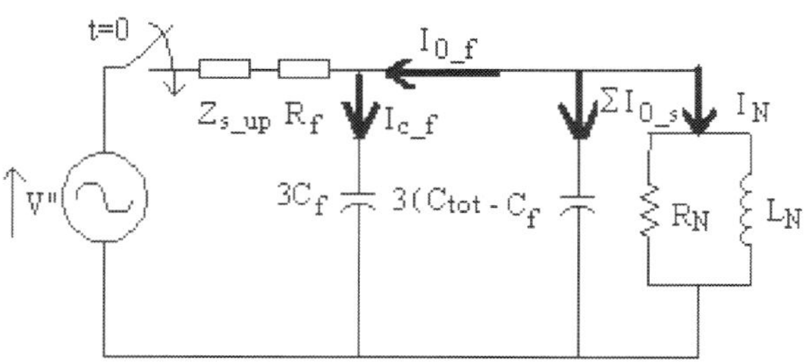

Fig. 7. Equivalent residual circuit of the faulty network on Fig. 3

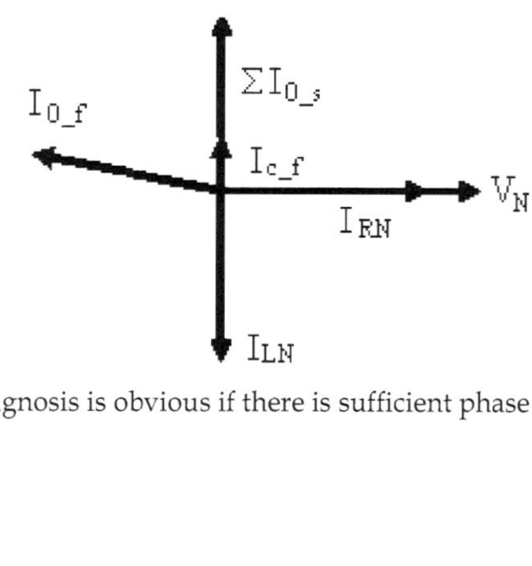

Fig.8. Faulty feeder diagnosis is obvious if there is sufficient phase advance of I_{0_f} over I_{0_s}.

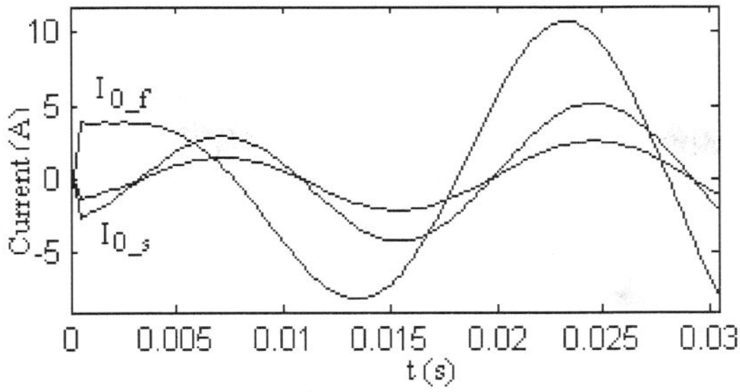

Fig. 9 Faulty and sound current residuals in case of resistive fault, non filtered curves (R_f =1kΩ, θ=90°)

We have simulated a three feeders' cable system of capacitive currents 15A+10A+5A, the fault installed on the 15-A feeder, with 100% tuning and the fault resistance values ranging from 1Ω to 1000Ω. In all cases, the actual phase advance in steady state was less than one sample step at 600Hz sampling frequency, this frequency being used in certain relays.

Fortunately, when the phase advance in steady state becomes undetectable, we can look upon an analogous parameter in transient regime, where it can be much larger.

This is a consequence of the way the transient regime develops, beginning just after the fault inception with phase opposition between faulty and sound current residuals and finishing in steady state with a slight phase advance of the faulty residual over all the sound ones. This development is correlated with evolution of the neutral point current I_N smoothly growing from zero to its permanent value. During the first millisecond after fault inception it can grow very slowly, particularly with resistive faults, because of high values of the neutral point elements L_N and R_N.

On the contrary, the feeders' line-to-ground capacitors C_f and C_s charge and discharge vigorously. During a short time interval, the neutral point current I_N is negligible comparing to capacitive zero-sequence currents.

The latter being under the same charging conditions, the faulty zero-sequence current is initially in phase opposition to the sound feeders zero-sequence currents; see (2):

$$I_{0\_f} = -\left(\sum I_{0\_s}\right) \tag{2}$$

and proceeds toward zero level (Fig. 9) with different polarities. The identification of faulty feeder operates then with aid of following algorithm.

3.2 Algorithm

We detect the slopes of filtered residuals at their first zero crossing after the fault inception. If all but one witness the same slope sign, then we can declare the latter as the faulty one without even controlling its zero crossing:

$$\textbf{IF } I_k(t_0) = 0 \textbf{ for } k = 1 \cdots n-1$$
$$\textbf{AND } \text{sgn}\left(\frac{dI_1}{dt}\right) = \text{sgn}\left(\frac{dI_2}{dt}\right) = \cdots \text{sgn}\left(\frac{dI_{n-1}}{dt}\right) \textbf{ at } t = t_0 \tag{3}$$
$$\textbf{THEN } \textit{the } n^{th} \textit{ feeder is the faulty one}$$

This is a one shoot procedure, without possibility of verification. On the other hand, it is a conclusive test, as the matter goes about unambiguous identification of slopes' signs.

4. Distance estimation by curve fitting

4.1 Extended Zero-Sequence Circuit (EC) for overhead line

We consider a compensated radial network (Fig. 3) supplied through a delta - star transformer grounded with Petersen coil. A SLG fault through represented by resistance R_f is installed on one of the feeders at the distance l_f from busbar on the phase 3.
The new equivalent circuit which we have developed for fault distance estimation in overhead lines (Huang & Kaczmarek, 2008) (Fig.7), is supplied with the inception voltage

$$V'' = -V_3 + E_1 + E_2 \tag{4}$$

and can easily be calculated from the pre-fault parameters. The correction component E_1 stands for the voltage drop related to the faulted phase load current over the up-stream impedance

$$E_1 = Z_{s\_up} I_{l3} \tag{5}$$

where the up-stream self impedance is

$$Z_{s\_up} = \frac{1}{3}\left(Z_{p\_up} + Z_{n\_up} + Z_{0\_up}\right) \tag{6}$$

with Z_{p_up}, Z_{n_up}, and Z_{0_up} as positive, negative and zero sequence impedances. The voltage E_2 stands for the voltage drop related to the sum of the faulted phase load currents of all the n feeders over the internal impedance of sources

$$E_2 = Z_{Th} \sum_n I_{l3} \tag{7}$$

Values of inception voltage in case of a 10kV network with total feeders' length 240km, compared to values issued from simulation, are presented in Table 1.

Fault position	0	0.5	1
Calculated inception voltage (16)	5507	5183	4874
Simulated inception voltage	5520	5173	4841
Error [%]	-0.2	+0.2	+0.7

Table 1. Equation-based inception voltage versus simulated one

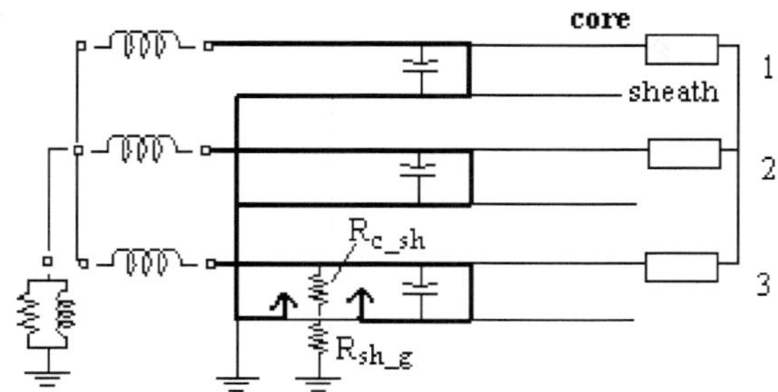

Fig. 10 The core-sheath-ground fault model split into two resistances

4.2 What cables change?

We consider a system with cables grounded on busbar side, the fault being installed (Fig. 10) between core, sheath and ground. The fault evaluating rapidly toward a solid one, the capacitive currents core-sheath choose the core-sheath fault rather than the sheath-ground-neutral point grounding impedance to get back to their original capacitances. The sheath-ground currents being weak, we can ignore the fault resistance sheath-ground.

The simulation confirms independence of currents from the sheath-ground fault resistance, the results being almost the same for one ohm core-sheath resistance R_{c_sh} and different values of the sheath-ground fault resistances R_{sh_g}. Consequently, in our development we ignore the sheath-ground fault resistance.

The corresponding equivalent circuit is apparent to that of overhead lines (Fig.7). The inception voltage V" (8) is composed of the Thevenin equivalent voltage V_{Th} :

$$V'' = -V_{Th} + E_1 + E_2 \tag{8}$$

the voltage drop E_1 related to the faulted phase load current over the up-stream impedance

$$E_1 = (Z_{c\_up} - Z_{c\_sh\_up})I_{l3} \tag{9}$$

with Z_{c_up} as up-stream core self impedance and $Z_{c\_sh\_up}$ as up-stream core-sheath mutual impedances. The E_2 related to the sum of the faulted phase load currents of all the n feeders over the internal impedance of sources

$$E_2 = Z_{Th} \sum_n I_{l3} \tag{10}$$

The corresponding up-stream self impedance Z_{s_up} in Fig.7, should be replaced by ($Z_{c\_up}+Z_{sh\_up}-2Z_{c\_sh\_up}$) in case of cables with Z_{sh_up} being up-stream sheath self impedance.

Fault position	0	0.5	1
Calculated inception voltage (22)	5591	5427	5263
Simulated inception voltage	5585	5433	5280
Error [%]	+0.1	-0.1	-0.3

Table 2. Equation-based inception voltage versus simulated values

Acc [%]	**EMTP** (reference value) R_f [Ω], l_f [0…1], θ [°]	**EC** R_f [Ω], l_f [0…1], θ [°]	l_f error [% l]
95	1, 0.2, 30	2.3, 0.20, 29	0
100	1, 0.5, 60	4.6, 0.47, 57	-3
105	1, 0.8, 90	12.5, 0.74, 85	-6
95	300, 0.2, 60	306, 0.17, 58	-3
100	300, 0.5, 90	269, 0.68, 87	+18
105	300, 0.8, 30	301, 0.84, 26	+4
95	600, 0.2, 90	584, 0.28, 88	+8
100	600, 0.5, 30	602, 0.52, 27	+2
105	600, 0.8, 60	607, 0.82, 56	+2
95	700, 0.2, 90	649, 0.46, 89	+26
100	700, 0.5, 30	704, 0.52, 27	+2
105	700, 0.8, 60	706, 0.82, 56	+2
95	1000, 0.2, 30	1004,0.21, 28	+1
100	1000, 0.5, 60	1002, 0.52, 57	+2
105	1000, 0.8, 90	986, 0.90, 86	+10
95	3000, 0.2, 60	2983, 0.27, 29	+7
100	3000, 0.5, 90	3074, 0.37, 87	-13
105	3000, 0.8, 30	2993, 0.87, 26	+7

Table 3. Fault on the 22.5km feeder

This way of calculating the inception voltage in cables is satisfactory as show Table 2 with results in a 10kV cable system of total feeders' length 14.5 km.

4.3 Fault distance evaluation by three-parameter fitting

This is a three parameters minimization problem, where the actual fault resistance R_f , fault position l_f, and inception angle θ are given by the equivalent circuit's best fitting curve, with EMTP currents taken for reference data. The EMTP currents are calculated with frequency dependant parameters.

The algorithm has been tested on overhead and cable line radial networks with a SLG fault. In an eight feeders line system of lengths (22.5+24+26+28+32+34+36+37.5)km, at 95…105% tuning, inception angles from 0° to 90°, 10MVA total load and the fault resistance up to 3kΩ we get the fault position with less than 10% mean error in relation to the fault position. Table 3 presents the cases with fault on the feeder of median length.

Acc [%]	**EMTP** (reference value) R_f [Ω], l_f [0…1], θ [°]	**EC** R_f [Ω], l_f [0…1], θ [°]	l_f error [%]
95	1, 0.2, 30	1.0, 0.22, 30	+2
100	1, 0.5, 60	1.2, 0.50, 60	0
105	1, 0.8, 90	2.8, 0.75, 90	-5
95	20, 0.2, 60	20, 0.20, 60	0
100	20, 0.5, 90	23, 0.42, 90	-8
105	20, 0.8, 30	20, 0.80, 30	0
95	50, 0.2, 90	52, 0.12, 90	-8
100	50, 0.2, 30	50, 0.51, 30	+1
105	50, 0.8, 60	50, 0.79, 60	-1
95	200, 0.2, 30	201, 0.17, 30	-3
100	200, 0.5, 60	203, 0.41, 90	-9
105	200, 0.8, 90	203, 0.67, 59	-13

Table 4. Fault on the 6.05km cable feeder

In cables high fault resistances are not relevant, the insulation breakdown being generally definitive, quickly developing to permanent solid fault. In tests on a 3 feeders cable network of feeders' length (3.63; 4.84; 6.05)km, 10MVA total load, at 95…105% tuning and inception angles from 0° to 90°, we get the fault position with average error less than 10% up to R_f=200Ω (Table 4).

5. Conclusion

We think that strong capacitive currents, generating unfavorable conditions for traditional protection relays in compensated systems, can be exploited as carriers of relevant information.

Whenever extremely rapid information is required, we can find it when treating the discharging currents as useful for treatment, at a price of higher sampling frequency. After fault inception we dispose of several tens of microseconds to get the data in wave propagation area. This can make sense in simple distribution systems, with single ramification per feeder and homogenous line impedance.

We can also make useful the presence of strong capacitive currents in permanent fault regime, where these currents occult detection of faulty feeder. In such cases, the diagnosis based on phase advance of the faulty residual current over the sound residuals is better assured when tracing the corresponding apparent phase advance in transient regime.

Both propositions need the current data be centralized; they work on few data points in a very short time span. These drawbacks are price for matching consequences of cable proliferation in resonant grounded distribution networks.

When simulating the fault currents, we usually exploit large possibilities of dedicated packages like EMTP. However, in components in compensated distribution system these currents can be usefully analyzed also with aid of an equivalent circuit, which permits an evaluation of SLG fault distance. We have calculated fault position with several percent average errors, for fault resistance up to 3kΩ in overhead lines or up to 200Ω in cables.

6. References

Bastard, P.; Bertrand, P.; Emura, T. & Meunier, M. (1992). The Technique of Finite-Impulse-Response Filtering Applied to Digital Protection and Control of Medium Voltage Power System. *IEEE Trans. Power Delivery,* Vol. 7, No. 2, (April 1992) pp. 620-626, ISSN 0885-8977

Coemans, J.; Philippot, L. & Maun, J.-C. (1993). Fault distance computation for isolated or compensated networks through electrical transient analysis, *Proceedings of Int. Conf. Power System Transients IPST,* pp. 250-254, 1993

Coemans, J. & Maun, J.-C. (1995). Using the EMTP and the Omicron to design a transients based digital ground-fault relay for isolated or compensated networks, *Proceedings of Int. Conf. Power System Transients IPST,* pp. 270-275, Lisbon, Sept. 1995

Griffel, D.; Harmand, Y.; Leitloff, V. & Bergeal, J. (1997). A new deal for safety and quality on MV networks. *IEEE Trans. Power Delivery,* Vol. 12, No. 4, (Oct. 1997) pp. 1428-1433, ISSN 0885-8977

Huang, W.-Y. & Kaczmarek, R. (2008). Equivalent Circuits for an SLG Fault Distance Evaluation by Curve Fitting in Compensated Distribution Systems. *IEEE Trans. Power Delivery,* Vol. 23, No. 2, (April 2008), pp. 601-608, ISSN 0885-8977

Kaczmarek, R. & Huang, W.-Y. (2005). Directional Function in Distribution Networks through Wave Propagation, *Proceedings of Int. Conf. PowerTech,* pp. , St Petersburg, June 2005

Lehtonen, M. (1998). Fault management in electrical distribution systems. *Final Report of the CIRED working group WG03,* Espoo, Finland, December 1998

Nikander, A.; Lakervi, E. & Suontausta, J. (1995). Applications of transient phenomena during earth faults in electricity distribution networks, *Proceedings of Int. Conf. Energy Management and Power Delivery,* pp. 234-239, Singapore, Nov 1995

Segui, T.; Bertrand, P.; Guillot, M.; Hanchin, P. & Bastard, P. (2001). Fundamental Basis for Distance Relaying with Parametrical Estimation, *IEEE Trans. Power Delivery,* Vol. 16, No. 1, (Jan. 2001) pp. 99-104, ISSN 0885-8977

Welfonder, T.; Leitloff, V.; Feuillet, R. & Vitet, S. (2000). Location strategies ans evaluation of detection algorithms for earth faults in compensated MV distribution systems, *IEEE Trans. Power Delivery,* Vol. 15, No. 4, (Oct. 2000), pp. 1121-1128, ISSN 0885-8977

19

Condition Monitoring of Power Plant Milling Process using Intelligent Optimisation and Model Based Techniques

Jihong Wang, Jianlin Wei and Shen Guo
University of Birmingham
UK

1. Introduction

Coal preparation is the first step in the whole process of coal-fired power generation. A typical milling process is illustrated in Figure 1. Also, coal-fired power stations nowadays are required to operate more flexibly with more varied coal specifications and regularly use coal with higher volatile contents and Biomass materials (Livingston 2004); this greatly increases the risks of explosions or fires in milling plants. The power stations are also obliged to vary their output in response to the changes of electricity demands, which results in more frequent mill start-ups and shut-downs. In many cases, coal mills are shutdown and then restarted before they have cooled adequately, which creates a potential fire hazard within the mill. Frequent start-ups and shut-downs of mills will also have an impact on power plant operation safety. Mill fires could occur if the coal stops flowing in the mill and the static deposit is heated for a period of time. Fires in out-of-service mills can cause explosions on mill starts. Fires in running mills can cause explosions on shut-downs. The result of a study indicated that as many as 300 "explosions" were occurring annually in the US pulverized coal industry (Scott, 1995). Especially, adding higher volatile Biomass materials greatly increases the chance of mill fires and explosions. The UK PF Safety Forum had recently reported an increase in the frequency of mill explosions in the UK. Operational safety and efficient combustion require better understanding to the milling process.
However, coal mills have been paid much less attention in research compared with boilers, generators, and other power generation system components. It is difficult to identify if there will be a fire in the mill. Outlet temperature and CO are established methods of detecting fires in mills, but at present they are not very effective for detecting small fires. The CO detection system becomes ineffective when the mill is in service due to dilution effects caused by primary air flow and associated oxygen content in the mill.
A wide range of literature survey shows that there are only a few reports on mathematical models of milling processes. A detailed milling process description can be found in Scott *et al.* 1995. An approximated linear transfer function model was obtained by Bollinger *et al*, in 1983. Mill modelling using system identification method was reported in 1984 (Corti *et al.* 1984). With specially designed input signals, a linear discrete time model was obtained by Cheetham, *et al* in 1990, in which system time-delay was considered. An approximated

linear time varying mill model was derived by Fan *et al.* in 1994. A polynomial matrix model was recently reported in 2000 (Hamiane *et al.* 2000). However, almost all the reported work describes the milling process by approximated linear mathematical models, which can not reflect the nonlinear features of coal mill systems. The complex nature of a milling process, together with the complex interactions between coal quality and mill conditions lead to immense difficulties for obtaining an effective mathematical model of a milling process.

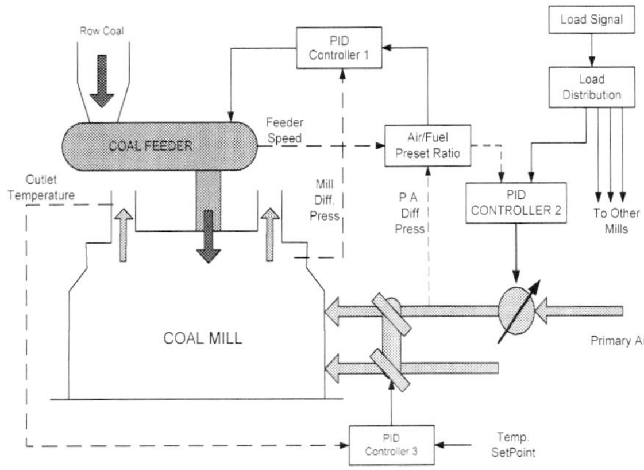

Fig. 1. A typical coal mill system

With the development of computer technology, advanced distribute control and monitoring systems enable the information data to be collected from the major plant components, e.g. mills, boilers and generators. The available measurements for coal mills include inlet/outlet temperature, PA (primary air) differential pressures, volume flow rate of coal into mills, primary airflow rate in mills *etc.* The availability of such a large volume of data offers an opportunity to identify the characteristics of mill processes and to develop a suitable coal mill model. A nonlinear coal mill model was derived based on the dynamic analysis of the coal milling process by Zhang *et al.* in 2002, which is a multi-input multi-output (MIMO) non-linear system. A machine learning method was adopted for identification of system parameters and the data employed in parameter identification covers a wide range of milling processes and also was measured on-site at power plants. Following the progress of early stage study, further studies have been carried out. It is noticed that there were no specific terms in the output temperature equation to represent the time delays caused by heat inertia in the system responses. Generally speaking, thermal processes always have big inertias, which are reflected in the system responses with obvious time delays. So the coal mill model was then improved by considering the temperature inertia terms (Wei *et al.* 2003). A multi-segment mill model was developed for the vertical spindle mills which was reported in Wei *et al.* 2007. The chapter summarises the research achievement at Birmingham in mill modelling, condition monitoring, on-line implementation, on-site test results and incident prediction.

2. Coal Mill Modelling for the Normal Grinding Process

The procedure adopted for coal mill modelling in this paper can be broken down into the following steps:

1). to derive the basic mill model dynamic equations through analysing the milling process, applying physics and engineering principles, and integrating the knowledge of experienced engineers;

2). to identify unknown parameters using evolutionary computation and system simulation techniques using the on-site measurement data;

3). to analyse the simulation results and interpret the parameters identified through the discussions between the researchers and experienced engineers;

4). to improve the mill model, that is, to go back to Step 2 if any modification is required, or to conduct further simulation to validate the model and go back to Step 3.

A nonlinear mathematical model for normal mill grinding process was developed, which were based on the following assumptions: a) The pulverizing mechanism in the mill is simplified and coal classification is not considered; b) Grinding and pneumatic transport in the milling process are separated into two stages; c) Coal size is grouped into only two categories named pulverized and un-pulverized coal. The mill model for the grading process can be described by the following equations (Zhang *et al.* 2002, Wei *et al.* 2007):

$$W_{air}(t) = 10 \cdot \sqrt{\Delta P_{pa}(t) \cdot \frac{273}{273 + T_{in}(t)} \times \frac{28.8}{22.4}} \quad (1)$$

$$W_c(t) = K_{fs}.F_s(t) \quad (2)$$

$$W_{pf}(t) = K_{16}\Delta P_{pa}(t) M_{pf}(t) \quad (3)$$

$$\dot{M}_c(t) = W_c(t) - K_{15}M_c(t) \quad (4)$$

$$\dot{M}_{pf}(t) = K_{15}M_c(t) - W_{pf}(t) \quad (5)$$

$$P(t) = K_6 M_{pf}(t) + K_7 M_c(t) + K_8 \quad (6)$$

$$\Delta P_{mill}(t) = K_9 \Delta P_{pa}(t) + \Delta P_{mpd}(t) \quad (7)$$

$$\Delta\dot{P}_{mpd}(t) = K_{11}M_{pf}(t) + K_{12}M_c(t) - K_{13}\Delta P_{mpd}(t) \quad (8)$$

$$\dot{T}_{out}(t) = [K_1 T_{in}(t) + K_2] \cdot W_{air}(t) - K_3 W_c(t) - [K_4 T_{out}(t) + K_5] \cdot [W_{air}(t) + W_c(t)] + K_{14}P(t) + K_t T_{out}(t) \quad (9)$$

The variables and parameters in the above equations are defined as:

ρ :	Primary air density ($kg \;/\; m^3$)
M_c :	Mass of coal in mill (*kg*)
M_{pf} :	Mass of pulverized coal in mill (*kg*)
T_{out} :	Outlet temperature of coal mill (*deg C*)
ΔP_{mill} :	Mill differential pressure (*mbar*)
ΔP_{mpd} :	Mill product differential pressure (*mbar*)
W_{pf} :	Mass flow rate of pulverized coal outlet from mill (*kg/s*)
P :	Mill current (*A*)

ΔP_{pa} : Primary air differential pressure (*mbar*)

W_c : Mass flow rate of coal into mill (*kg/s*)

F_s : Coal feeder speed (mm/s)

T_{in} : Inlet temperature of coal mill (*deg C*)

W_{air} : Primary air flow rate into coal mill (*kg/s*)

K_i : Model unknown parameters to be identified.

A brief explanation to the coal mill model (1) ~ (9) is given below.

- Equation 1 represents that the PA (primary air) flow rate is equal to the rate of air flow delivered by the primary fans, which is proportional to the square root of the PA differential pressure and the density of air under the mill inlet temperature.
- Equation 2 represents that the flow rate of coal into coal mill is equal to the rate at which coal delivered by feeder times the feeder speed coefficient K_{fs}, which depends on the sizes of different feeders (for a small feeder, K_{fs} = 0.16 kg/mm and for a large feeder, K_{fs} = 0.24 kg/mm).
- Equation 3 represents that the flow rate of PF (pulverized fuel) out of mill is equal to the rate that PF was carried out of mill by the primary air flow, which is proportional to the mass of the pulverized coal in mill and the differential pressure produced by the primary air fan.
- Equation 4 represents that the mass change rate of the coal in mill is proportional to the difference between the coal flow into the mill the fraction of coal that is converted into pulverized fuel $K_{15} M_c(t)$.
- Equation 5 represents that the change of mass of pulverized fuel in mill is proportional to the difference between the fraction of coal converted into pulverized and the pulverized coal flow outlet from the mill.
- Equation 6 represents that the total amount of mill current consumed to run the mill motor is equal to the sum of mill current required to grind over surface area, mill current to pulverized coal, and mill current to run empty mill.
- Equation 7 represents that the mill differential pressure is resulted by the differential pressure produced by the primary air fan and the mill product differential pressure.
- Equation 8 represents that the changes in the mill product differential pressure is proportional to the pressure due to pulverised fuel in mill, proportional to the pressure due to coal in mill, and the pressure of the previous time step.
- Equation 9 represents the changes in mill outlet temperature is the results of heat transferring balance. It increases with the heat contributed by hot primary air entering mill and the heat generated by grinding and also decreases with the heat lost to coal and moisture entering mill and the heat lost to hot primary air and pulverized fuel leaving mill. Basically this equation represents the heat balance model of the coal mill system, which is demonstrated in Fig. 2.

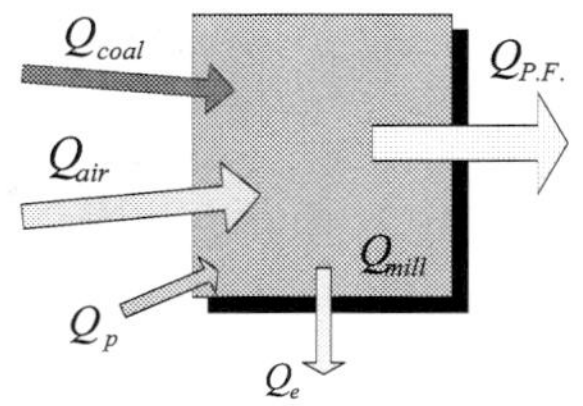

Fig. 2. Heat balance model of coal mill system

The heat of coal from inlet, Q_{coal}, is represented by the term K_3W_c, the heat of primary air from inlet, Q_{air}, is represented by the term $K_1T_{in}W_{air} + K_2W_{air}$, the heat generated by the coal grinding process is represented by the term $K_{14}P$, and the heat of the pulverized coal outlet from the mill is represented by term: $K_4T_{out}(t)\cdot[W_{air}(t)+W_c(t)]+K_5\cdot[W_{air}(t)+W_c(t)]$. The heat emitted from the mill body to the environment Q_e is neglected in the model. The term K_tT_{out} represents the approximation to the time delay inherent in the thermodynamics processes.

For the purpose of identifying the system parameters, the measured variables are organised into two groups - system inputs and outputs. The input variables of the model include the coal flow into the mill, primary air differential pressure and primary air inlet temperature. The output variables include mill differential pressure, outlet temperature and mil current. In order to identify the sixteen unknown coefficients of the coal mill mathematical model, the Genetic Algorithms (GAs) is adopted (Wang *et al* 2004). It has been proved that GAs is a robust optimisation method for this particular identification problem. The single-population real-value genetic algorithm was chosen and the fitness function shown in Formula (13) is employed for this purpose. The fitness function compromises the errors between the normalized coal mill measured outputs and the normalized model simulated outputs. Following the scheme of the coefficients identification shown in Fig. 3, the sixteen unknown coefficients are identified which are summarised in Table 1. More detailed information about the model coefficients identifications can be found in Wei *et al.* 2007. Define:

$$e_1(t) = {}^nT_{out}(t) - {}^n\hat{T}_{out}(t) = \frac{T_{out}(t)}{{}^{TopR}T_{out}} - \frac{\hat{T}_{out}(t)}{{}^{TopR}T_{out}} \tag{10}$$

$$e_3(t) = {}^n\Delta P_{mill}(t) - {}^n\Delta\hat{P}_{mill}(t) = \frac{\Delta P_{mill}(t)}{{}^{TopR}\Delta P_{mill}} - \frac{\Delta\hat{P}_{mill}(t)}{{}^{TopR}\Delta P_{mill}} \tag{11}$$

$$e_2(t) = {}^nP(t) - {}^n\hat{P}(t) = \frac{P(t)}{{}^{TopR}P} - \frac{\hat{P}(t)}{{}^{TopR}P} \tag{12}$$

Then the fitness function is described as follows:

$$fitness = \frac{1}{N}\sum_{t=0}^{N}\{\ W_1\times|e_1(t)| + W_2\times|e_2(t)| + W_3\times|e_3(t)|\ \} \tag{13}$$

where:

W_1, W_2, W_3: Weighting coefficients.

${}^{n}\hat{T}_{out}(t)$, ${}^{n}\hat{P}(t)$, ${}^{n}\Delta\hat{P}_{mill}(t)$	Normalized simulated outputs of the mill model at time t.
${}^{n}T_{out}(t)$, ${}^{n}P(t)$ ${}^{n}\Delta P_{mill}(t)$	Normalized measured outputs of the mill model at time t.
$T_{out}(t)$, $P(t)$ $\Delta P_{mill}(t)$	Measured outputs of the mill model at time t.
$\hat{T}_{out}(t)$, $\hat{P}(t)$ $\Delta\hat{P}_{mill}(t)$	Simulated outputs of the mill at mill model at time t.
${}^{TopR}T_{out}$, ${}^{TopR}P$ ${}^{TopR}\Delta P_{mill}$	Top ranges of the variables.
W_1, W_2, W_3:	Weighting coefficients.
N:	Number of measured data points.

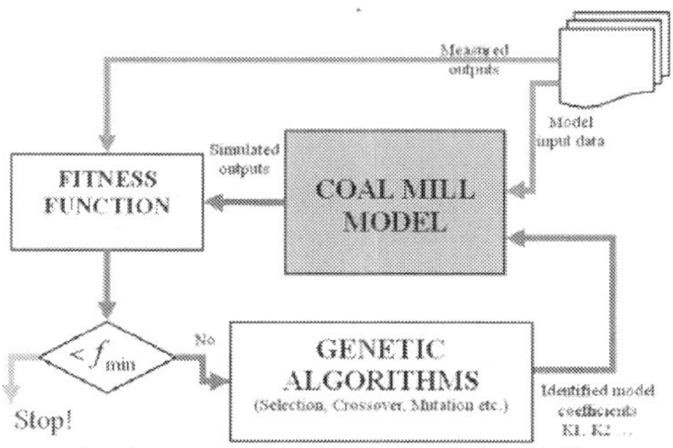

Fig. 3. Schematic diagram for the procedure of the model's coefficients identification

System parameter using the modified mill model	
$K_1 = 0.00061927802462$	$K_2 = 0.08961428118773$
$K_3 = 0.00383781469345$	$K_4 = 0.00155335789736$
$K_5 = -0.08634812577403$	$K_6 = 0.01712514044766$
$K_7 = 0.00293718071483$	$K_8 = 30.17329476338669$
$K_9 = 5.54900011072344$	$K_{11} = 0.00170000000000$
$K_{12} = 0.00056524986674$	$K_{13} = 0.08677360557200$
$K_{14} = 0.03250567038921$	$K_{15} = 0.00513313270266$
$K_{16} = 0.00261803749309$	$K_t = -0.05724453482491$

Table 1. Identified coefficients for the normal running coal mill model

The simulated results using data collected from power plant are shown in Fig. 4 ~ Fig. 5, where the solid line represents the coal mill measured outputs and the dash-dot line represents the model simulated outputs.

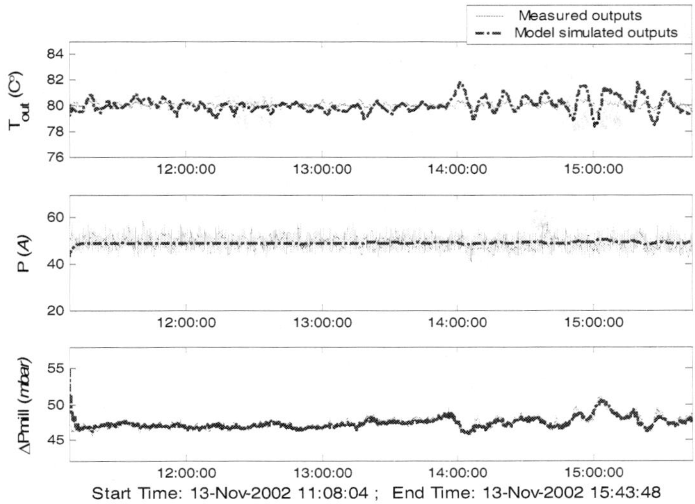

Fig. 4. Coal mill model simulation results using data set 1 (normal grinding stage)

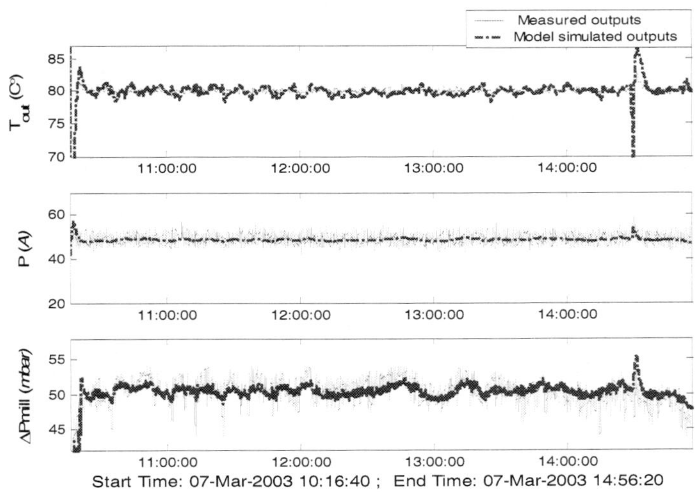

Fig. 5. Coal mill model simulation results using the data 2 (normal grinding stage)

From the simulation studies shown above, it can be seen that the model simulated outputs follow the coal mill measured outputs very well during the coal mill normal grinding period (see Figs. 4 and 5).

3. Multi-Segment Coal Mill Model

According to different operating stages of a coal mill, a multi-segment coal mill model is developed, which covers six different segments. The structure of the six-segment model is illustrated in Fig.6, where the whole milling process is divided into six-sessions. All the different working stages of the coal mill system (e.g. the start-up, the steady-state, the shut-down and the idle stages) are considered in this model. The coal mill idle stage is modelled by the model segment 0 in the multi-segment model; coal mill start-up stage is modelled by the model segment I and II; coal mill steady state stage is modelled by the model segment III; coal mill shut-down stage is modelled by the model segment IV and V.

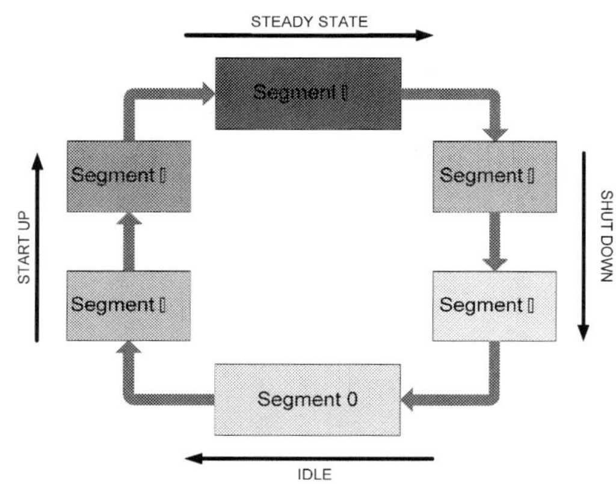

Fig. 6. Six-segment coal mill model

All the models for different segments are connected/switched by a number of operation triggers. The triggers are inherently determined by the mill operation procedures introduced in the last section of this paper. Detailed descriptions for each segment model are given in the following sections.

3.1 Coal Mill Model – Segment I

The coal mill model segment I represents the 'warm-up' process in the mill start-up procedure. During this period, the mill temperature is set to the point of the warming-up value, the PA Fan Damper is opened to the warming-up position, PA different pressure is set to a point to be higher than 4 mbar, and the mill lube oil pump starts running. The switch trigger for this segment is set to be the moment that the Primary Air Damper, the Temperature Damper are just open, which could be represented modularly by the beginning of the Step 3 as shown in Figure 7. The coal mill model segment I is described by Equation (14) ~ Equation (22).

$$W_{air}(t) = 10 \cdot \sqrt{\Delta P_{pa}(t) \cdot \frac{273}{273 + T_{in}(t)} \times \frac{28.8}{22.4}} \tag{14}$$

$$W_c(t) = K_{fs}.F_s(t) \tag{15}$$

$$W_{pf}(t) = K_{I\_16}\Delta P_{pa}(t)M_{pf}(t) \tag{16}$$

$$M_c(t) = M_c(t_{1\_0}) \tag{17}$$

$$\dot{M}_{pf}(t) = -W_{pf}(t) \tag{18}$$

$$P(t) = 0 \tag{19}$$

$$\Delta P_{mill}(t) = K_{I\_9}\Delta P_{pa}(t) + \Delta P_{mpd}(t) \tag{20}$$

$$\Delta\dot{P}_{mpd}(t) = K_{I\_11}M_{pf}(t) + K_{I\_12}M_c(t) - K_{I\_13}\Delta P_{mpd}(t) \tag{21}$$

$$\dot{T}_{out}(t) = [K_{I\_1}T_{in}(t) + K_{I\_2}]W_{air}(t) - [K_{I\_4}T_{out}(t) + K_{I\_5}]\cdot W_{air}(t) + K_{I\_t}T_{out} \tag{22}$$

where:

$M_c(t_{1\_0})$: The initial value of mass of coal in mill at the beginning of segment 1 (kg)

K_{I_i}: Model coefficients to be identified respect to model segment I

The other notations are same as explained in Section 2.

Comparing with the normal grinding mill model shown in Section 2 of this paper, there are several significant modifications made, which are: a). The raw coal in mill M_c is set to be constant value $M_c(t_{1\_0})$ through this segment (see Equation 17) since no grinding happens during this segment; b). The changing rate of pulverized coal in mill $\dot{M}_{pf}$ is set to be negatively proportional to the mass flow rate of pulverized coal outlet from mill W_{pf} only (see Equation 18) since there is not any M_{pf} to be generated by grinding during this segment; c). The coal mill current P is set to be zero since the coal mill motor is still off during this segment, where there is no current consumed to run the mill motor; d). The mill temperate is represented by Equation 22, which is similar to Equation 9 shown in Section 2.

3.2 Coal Mill Model – Segment II

The coal mill model segment II represents the 'pre-grinding' process in the mill start-up procedure. During this period, the coal mill grinding motor is started to pre-grind the coal left in the mill, where the coal feeder is still off. The switch trigger of the segment II is set to be the moment that the mill grinding motor is commanded to start which could be represented by the modular - beginning of the Step 7 as shown in Figure 7. The segment IIcoal mill model can be described by Equation (23) ~ Equation (31)

$$W_{air}(t) = 10\cdot\sqrt{\Delta P_{pa}(t)\cdot\frac{273}{273+T_{in}(t)}\times\frac{28.8}{22.4}} \tag{23}$$

$$W_c(t) = K_{fs}.F_s(t) \tag{24}$$

$$W_{pf}(t) = K_{II\_16}\Delta P_{pa}(t)M_{pf}(t) \tag{25}$$

$$\dot{M}_c(t) = -K_{II\_15}M_c(t) \tag{26}$$

$$\dot{M}_{pf}(t) = K_{II\_15}M_c(t) - W_{pf}(t) \tag{27}$$

$$P(t) = K_{II\_6} M_{pf}(t) + K_{II\_7} M_c(t) + K_{II\_8} \tag{28}$$

$$\Delta P_{mill}(t) = K_{II\_9} \Delta P_{pa}(t) + \Delta P_{mpd}(t) \tag{29}$$

$$\Delta \dot{P}_{mpd}(t) = K_{II\_11} M_{pf}(t) + K_{II\_12} M_c(t) - K_{II\_13} \Delta P_{mpd}(t) \tag{30}$$

$$\dot{T}_{out}(t) = [K_{II\_1} T_{in}(t) + K_{II\_2}] W_{air}(t) - [K_{II\_4} T_{out}(t) + K_{II\_5}] W_{air}(t) + K_{II\_14} P(t) + K_{II\_t} T_{out} \tag{31}$$

where:

K_{II_i}: Model coefficients to be identified respect to model segment II

The other notations have the same meanings as described in Section 2.

Comparing with the standard coal milling model shown in Section 2 of this paper, there are some modifications in this segment, which are shown as follows: a). The raw coal in mill M_c is self-reducing due to the pre-grinding of the coal mill system, which is represented by Equation 26; b). The mill temperate is represented by Equation 31, which is derived based on the heat balance principle of the coal mill system as described in Section 2.

3.3 Coal Mill Model – Segment III

The coal mill model segment III represents the steady state milling stage. During this period, the primary air fan, coal mill grinding motor and the coal feeder *etc.* have come into the steady state milling stage. The switch trigger for model segment III is set to be the moment that the mill start-up sequence is completed, which could be represented by the modular at the beginning of the Step 11 as shown in Figure 7. The segment III coal mill model is described by Equation (32) ~ Equation (40).

$$W_{air}(t) = 10 \cdot \sqrt{\Delta P_{pa}(t) \cdot \frac{273}{273 + T_{in}(t)} \times \frac{28.8}{22.4}} \tag{32}$$

$$W_c(t) = K_{fs} . F_s(t) \tag{33}$$

$$W_{pf}(t) = K_{III\_16} \Delta P_{pa}(t) M_{pf}(t) \tag{34}$$

$$\dot{M}_c(t) = W_c(t) - K_{III\_15} M_c(t) \tag{35}$$

$$\dot{M}_{pf}(t) = K_{III\_15} M_c(t) - W_{pf}(t) \tag{36}$$

$$P(t) = K_{III\_6} M_{pf}(t) + K_{III\_7} M_c(t) + K_{III\_8} \tag{37}$$

$$\Delta P_{mill}(t) = K_{III\_9} \Delta P_{pa}(t) + \Delta P_{mpd}(t) \tag{38}$$

$$\Delta \dot{P}_{mpd}(t) = K_{III\_11} M_{pf}(t) + K_{III\_12} M_c(t) - K_{III\_13} \Delta P_{mpd}(t) \tag{39}$$

$$\begin{aligned} \dot{T}_{out}(t) = {} & [K_{III\_1} T_{in}(t) + K_{III\_2}] W_{air}(t) - K_{III\_3} W_c(t) - [K_{III\_4} T_{out}(t) + K_{III\_5}] \\ & \cdot [W_{air}(t) + W_c(t)] + K_{III\_14} P(t) + K_{III\_t} T_{out} \end{aligned} \tag{40}$$

where:

K_{III_i}: Model coefficients to be identified respect to model segment III

The other notations have the same meanings as described in Section 2 .

The equations are same as the normal grinding mill model shown in Section 2, which represents the same stage of the coal milling process - coal milling steady state stage.

3.4 Coal Mill Model – Segment IV

The coal mill model segment IV represents the 'grinding-delay' process in the mill shut-down procedure. During this period, the coal feeder is switched off but the grinding motor is still kept on grinding the coal remained in the mill. The switch trigger of model segment IV is set to be the moment that the coal feeder is completely switched off, which could be represented by the block at the beginning of Step 19 as shown in Figure 8. The segment IV coal mill model is described by Equation (41) ~ Equation (49).

$$W_{air}(t) = 10 \cdot \sqrt{\Delta P_{pa}(t) \cdot \frac{273}{273 + T_{in}(t)} \times \frac{28.8}{22.4}} \tag{41}$$

$$W_c(t) = K_{fs}.F_s(t) \tag{42}$$

$$W_{pf}(t) = K_{IV\_16}\Delta P_{pa}(t)M_{pf}(t) \tag{43}$$

$$\dot{M}_c(t) = -K_{IV\_15}M_c(t) \tag{44}$$

$$\dot{M}_{pf}(t) = K_{IV\_15}M_c(t) - W_{pf}(t) \tag{45}$$

$$P(t) = K_{IV\_6}M_{pf}(t) + K_{IV\_7}M_c(t) + K_{IV\_8} \tag{46}$$

$$\Delta P_{mill}(t) = K_{IV\_9}\Delta P_{pa}(t) + \Delta P_{mpd}(t) \tag{47}$$

$$\Delta\dot{P}_{mpd}(t) = K_{IV\_11}M_{pf}(t) + K_{IV\_12}M_c(t) - K_{IV\_13}\Delta P_{mpd}(t) \tag{48}$$

$$\dot{T}_{out}(t) = [K_{IV\_1}T_{in}(t) + K_{IV\_2}]W_{air}(t) - [K_{IV\_4}T_{out}(t) + K_{IV\_5}]W_{air}(t) + K_{IV\_14}P(t) + K_{IV\_t}T_{out} \tag{49}$$

where:

K_{IV_i}: Model coefficients to be identified respect to model segment IV

The other notations have the same meanings as described in Section 2.

Comparing with the standard coal milling model shown in Section 2 of this paper, there are several significant modifications made in this segment, which are: a). The raw coal in mill M_c is self-reducing due to the grinding-delay of the coal mill system, which is represented by Equation 44; b). The mill temperate is represented by Equation 49, which is similarly modelled by the heat balance model of the coal mill system same as Equation 9.

3.5 Coal Mill Model – Segment V

The coal mill model segment V represents the 'cool-down' process in the mill start-up period. During this period, the mill temperature is set to the point of the cool-down value, PA fan induction regulator is closed to 15%, and the PA fan damper is set to the mill warming position. The switch trigger for this segment is set to be the moment that the mill temperature is set point to the cool-down value, which is illustrated in the block at the beginning of the Step 21 in Figure 8. The segment V mill model is described by Equation (50) ~ Equation (59).

$$W_{air}(t) = 10 \cdot \sqrt{\Delta P_{pa}(t) \cdot \frac{273}{273 + T_{in}(t)} \times \frac{28.8}{22.4}} \quad (50)$$

$$W_c(t) = K_{fs}.F_s(t) \quad (51)$$

$$W_{pf}(t) = K_{V\_16}\Delta P_{pa}(t)M_{pf}(t) \quad (52)$$

$$\dot{M}_{cDelay}(t) = -K_{V\_15}M_{cDelay}(t) \quad (53)$$

$$M_c(t) = M_c(t_{5\_0}) - [M_{cDelay}(t_{5\_0}) - M_{cDelay}(t)] \quad (54)$$

$$\dot{M}_{pf}(t) = K_{V\_15}M_{cDelay}(t) - W_{pf}(t) \quad (55)$$

$$P(t) = 0 \quad (56)$$

$$\Delta P_{mill}(t) = K_{V\_9}\Delta P_{pa}(t) + \Delta P_{mpd}(t) \quad (57)$$

$$\Delta\dot{P}_{mpd}(t) = K_{V\_11}M_{pf}(t) + K_{V\_12}M_c(t) - K_{V\_13}\Delta P_{mpd}(t) \quad (58)$$

$$\dot{T}_{out}(t) = [K_{V\_1}T_{in}(t) + K_{V\_2}]W_{air}(t) - [K_{V\_4}T_{out}(t) + K_{V\_5}] \cdot W_{air}(t) + K_{V\_t}T_{out} \quad (59)$$

where

$M_c(t_{5\_0})$: The initial value of mass of coal in mill at the beginning of segment V.

M_{cDelay}: A term to indicate the grinding delay caused by the inertias of the grinding components

K_{V_i}: Model coefficients to be identified respect to model segment V

The other notations have the same meanings as described in Section 2.

Comparing with the standard coal milling model shown in Section 2 of this paper, some modifications have been made in the model segment V, which is shown as follows: a) At the beginning of the cool down process, the raw coal in mill M_c is self-reducing for a short period of time due to the grinding-delay caused by the grinding inertias. Afterward it stays at a constant value through the segment as the grinding stops completely. Over the segment V, the raw coal in mill, M_c, is calculated by Equations 53 ~ 54, in which M_{cDelay} indicates the grinding delays caused by the inertias of the grinding components; b) The changing rate of pulverized coal in mill, $\dot{M}_{pf}$, is set to be negatively proportional to the mass flow rate of pulverized coal outlet from mill, W_{pf}, and proportional to the pulverized coal generated by the grinding delays due to the inertias of the grinding motor M_{cDelay}, which is modelled by Equation 55; c) The coal mill current P is set to be zero since the coal mill motor is switched off during this segment; d). The mill temperate is modelled by Equation 59, which is derived based on the heat balance principle.

3.6 Coal Mill Model – Segment 0

The coal mill model segment 0 represents the 'idle' stage of the milling process. During this period, the mill grinding motor, the coal feeder, the lube oil pump *etc.* are switched off. This segment model represents the natural cool down process of the coal mill system. The switch trigger for this segment is set to be the moment that the coal mill shut-down sequence completed, which is illustrated in Figure 7 (Step 0). The segment coal mill model is described by Equation (60) ~ Equation (67).

$$W_{air}(t) = 10 \cdot \sqrt{\Delta P_{pa}(t) \cdot \frac{273}{273 + T_{in}(t)} \times \frac{28.8}{22.4}} \tag{60}$$

$$W_c(t) = K_{fs}.F_s(t) \tag{61}$$

$$W_{pf}(t) = 0 \tag{62}$$

$$M_c(t) = M_c(t_{0\_0}) \tag{63}$$

$$M_{pf}(t) = M_{pf}(t_{0\_0}) \tag{64}$$

$$P(t) = 0 \tag{65}$$

$$\Delta \dot{P}_{mill}(t) = K_{0\_9} \Delta P_{pa}(t) + K_{0\_p} \Delta P_{mill}(t) \tag{66}$$

$$\dot{T}_{out}(t) = [K_{0\_1} T_{in}(t) + K_{0\_2}] W_{air}(t) - [K_{0\_4} T_{out}(t) + K_{0\_5}] \cdot W_{air}(t) - K_{0\_e}(T_{out} - T_e) + \frac{K_{0\_t}}{M_c(t) + M_{pf}(t)} T_{out} \tag{67}$$

where

$M_c(t_{0\_0})$: The initial value of mass of coal in mill at the beginning of segment 0 (*kg*)

$M_{pf}(t_{0\_0})$: The initial value of pulverized mass of coal in mill at the beginning of segment 0 (*kg*)

T_e: Temperature of the environment (°C)

K_{0_i}: Model coefficients to be identified respect to model segment 0

The other notations have the same meanings as described in Section 2.

In this segment, all the components of the coal mill system are off, there is no grinding and blowing happens, so that M_c, M_{pf} *etc.* should be constants through this segment. Further more, the coal mill temperature and the pressure are gradually dropping as described in Equations 66 and 67.

Comparing with the normal grinding coal milling model shown in Section 2, some modifications are made which are: a) In segment 0, all the components of the coal mill system is idle, so that M_c, M_{pf}, and P are set to be constants or zero, which are described by Equations 61 ~ 64; b) PA fan damper and coal mill valves are closed solidly during the idle stage, so coal mill pressure is naturally dropping, which is indicated in Equation 66. c) Equation 67 represents the naturally self-cooling process of the coal mill system through this segment. It is developed corresponding to the heat balance of the coal mill system as well, where the term $K_{0\_e}(T_{out} - T_e)$ represents the heat lost to the environment from the mill system via the mill body and valves *etc.*

4. Parameter identification and model validation

After the multi-segment coal mill model has been structured, the next task is to identify the unknown parameters associated with the six segment model. Then it is necessary to conduct the simulation to compare the performance of the new multi segment mill model. The validation of the new multi-segment model should be conducted as well. This section covers system parameter identification, simulation studies, and model validation.

4.1 Parameter identification and simulation study

A number of on-site data sets that cover the whole processes are chosen for parameter identification and simulations studies. An example using the data collected from a power plant is given as follows. The results obtained have been compared with the previous single grinding model, especially, the periods of start-up, normal grinding operation, shut-down and idle periods. The system parameters identified using this set of data are summarised in Table 3 ~ Table 8.

Identified coefficients of model Segment I		
$K_{I\_1} = 0.00072541773658$	$K_{I\_2} = 0.04000000000000$	$K_{I\_4} = 0.00057219041327$
$K_{I\_5} = -0.0167545 8545105$	$K_{I\_9} = 5.60489077201246$	$K_{I\_11} = 0.00012470913837$
$K_{I\_12} = 0.00002265487671$	$K_{I\_13} = 0.07943094801093$	$K_{I\_16} = 0.00010750292422$
$K_{I\_t} = -0.01329789102972$		

Table 3. Identified coefficient for model segment I

Identified coefficients of model Segment II		
$K_{II\_1} = 0.00080694873686$	$K_{II\_2} = 0.03085438464572$	$K_{II\_4} = 0.00056967623620$
$K_{II\_5} = -0.01389776245975$	$K_{II\_6} = 0.00658219296696$	$K_{II\_7} = 0.00500000000000$
$K_{II\_8} = 29.46408084086195$	$K_{II\_9} = 6.32278529817438$	$K_{II\_11} = 0.00008854456902$
$K_{II\_12} = 0.00036159145893$	$K_{II\_13} = 0.02000000000000$	$K_{II\_14} = 0.09036443542765$
$K_{II\_15} = 0.00068284301758$	$K_{II\_16} = 0.00144872777708$	$K_{II\_t} = -0.0566973 9327072$

Table 4. Identified coefficient for model segment II

Identified coefficients of model Segment III		
$K_{III\_1} = 0.00061927802462$	$K_{III\_2} = 0.08961428118773$	$K_{III\_3} = 0.00383781469345$
$K_{III\_4} = 0.00155335789736$	$K_{III\_5} = -0.08634812577403$	$K_{III\_6} = 0.01712514044766$
$K_{III\_7} = 0.00293718 07148$	$K_{III\_8} = 30.1732947 6338669$	$K_{III\_9} = 5.54900011072344$
$K_{III\_11} = 0.00170000000000$	$K_{III\_12} = 0.00056524986674$	$K_{III\_13} = 0.08677360557200$
$K_{III\_14} = 0.03250567038921$	$K_{III\_15} = 0.00513313270266$	$K_{III\_16} = 0.00261803749309$
$K_{III\_t} = -0.0572445 3482491$		

Table 5. Identified coefficient for model segment III

Identified coefficients of model Segment IV		
$K_{IV\_1} = 0.00100000000000$	$K_{IV\_2} = 0.03814538663991$	$K_{IV\_4} = 0.00110256287195$
$K_{IV\_5} = -0.0127384 9970596$	$K_{IV\_6} = 0.00502786796442$	$K_{IV\_7} = 0.01271951516693$
$K_{IV\_8} = 26.1933187675971$	$K_{IV\_9} = 6.02804165980328$	$K_{IV\_11} = 0.00001000000000$
$K_{IV\_12} = 0.00021827836644$	$K_{IV\_13} = 0.00594054009917$	$K_{IV\_14} = 0.06082368997338$
$K_{IV\_15} = 0.00500000000000$	$K_{IV\_16} = 0.00284411656238$	$K_{IV\_t} = -0.0495682 7263247$

Table 6. Identified coefficient for model segment IV

Identified coefficients of model Segment V		
K_{V_1} = 0.00093729 490970	K_{V_2} = 0.03715004 727992	K_{V_4} = 0.00109096 339731
K_{V_5} = -0.0174981 6468507	K_{V_9} = 5.97603398 05287	K_{V_11} = 0.00001000 000000
K_{V_12} = 0.00001000 000000	K_{V_13} = 0.12948317 846770	K_{V_15} = 0.15002431869507
K_{V_16} = 0.00031424 102569	K_{V_t} = -0.0084228 3447028	

Table 7. Identified coefficient for model segment V

Identified coefficients of model Segment 0		
K_{0_1} = 0.00153223 119637	K_{0_2} = 0.05315223 812253	K_{0_4} = 0.00239019 497476
K_{0_5} = -0.0135178 0190605	K_{0_9} = 0.35000000 00000	K_{0_p} = -0.0586331 9829717
K_{0_t} = -0.0974950 6071560	K_{0_e} = 0.00027743 032875	

Table 8. Identified coefficient for model segment 0

The simulation results using the data collected from a power plant are shown at Fig.7 ~ Fig.8, which covers all the start-up, normal running, shut-down, idle periods.

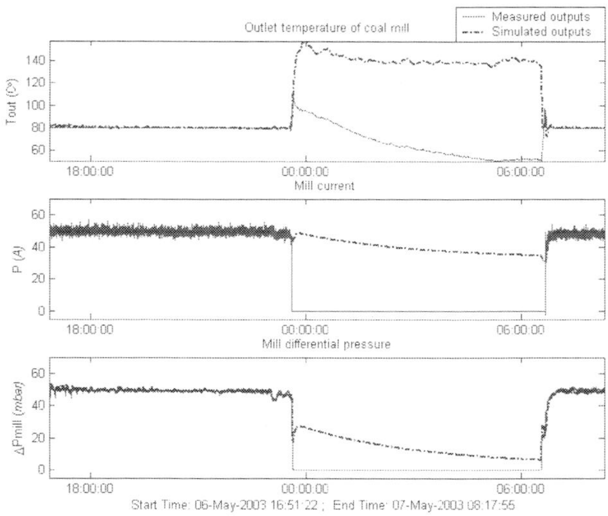

Fig. 7. Simulation results for the overall processes using coal mill single model

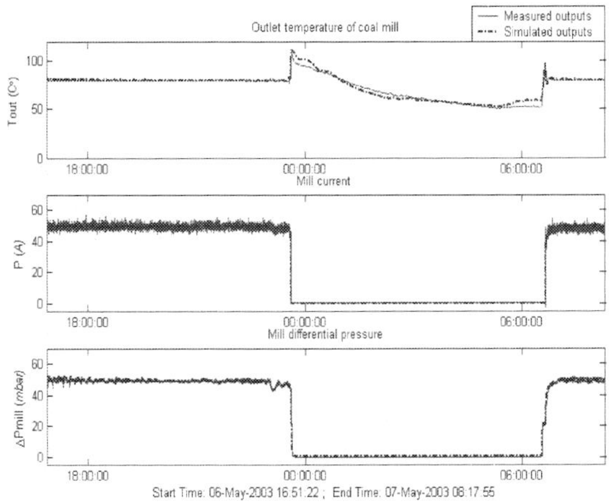

Fig. 8. Simulation results for the overall processes using multi-segment model

From the simulation results shown in Fig.7 and Fig.8, it is clearly that the six segment model gives the predicted system output more close to the measured system output comparing with the the single grinding mill model. Especially, during the periods of starting-up and shutting-down, the multi-segment model gives the accurate results while the single grinding model fails to predict the system outputs.

4.2 Model validation

The coal mill multi-segment model's validations have been carried out based on a number of on-site data sets that include start-up/shut-down dynamic processes. The validation results are shown in Fig. 9 and Fig. 10. The results are convincing.

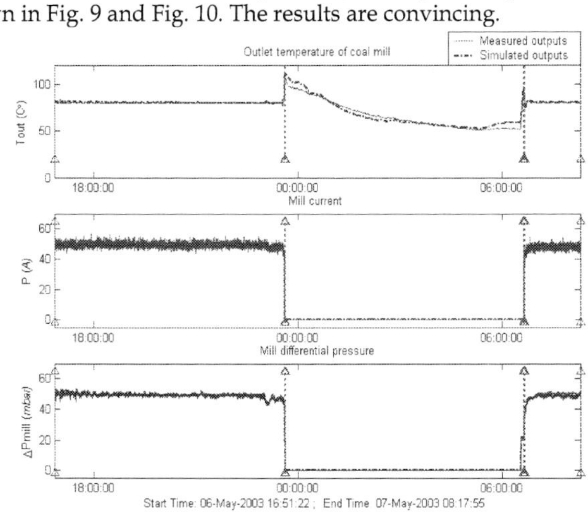

Fig. 9. Model validation uing a different data set

5. Mill diagnostic and fault detection

The model developed has been on-line implemented to give a real-time mill predicted outputs. The on-line model can be used for mill condition monitoring in the following ways: 1) to compare the mill measured and predicted outputs. If there are unusual differences between these two variables, the fault may occure and alarm should be raised. 2) to monitor the intermideat variable - coal in mills. If the coal inside the mill increases to an unusual high level, that indicates that too much coal inside the mill, which may cause problems and they should be monitoed closely. 3) to identify any unusual changes of model parameters. From the simulation study shown in the above section, it is noticed that the model parameter may need to be updated more frequently which is due to coal quality changes or biomass material mixture. This inspired a new idea for identification of the key model parameters related to the key variables in a short parameter updating period. This idea has been applied to Tube-ball mill based on a modified mathematical model.

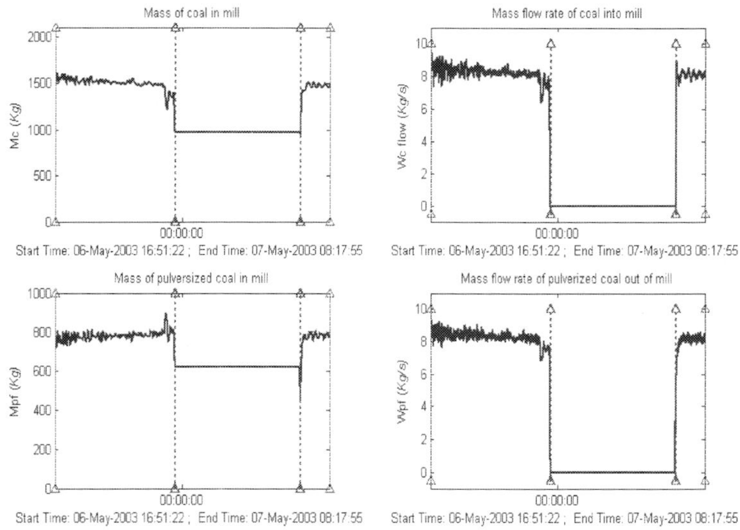

Fig. 10. Model validation using a different data set, intermidiate variables

Further study has been carried out to investigate if this parameter updating scheme can be adapted for condition monitoring. The parameter K_t in (9) is chosen to be updated in every 5 minutes as shown. The idea underline for this adaptation is to see any rough change or large deviation from the base value can indicate the mill condition changes. From the simulation, it is noticed for a particular data set, the parameter K_t has an extreme sharp change during a period of time. Also, the mill performance during this period of time varied violently. The measured and predicted responses are shown in Fig. 11. To identify what has happened at the power plant, we have had a discussion with the plant engineers. It is confirmed that the abnormal variation is caused by the big chunk of bio-mass fuel coming into the mill and melted later on. This is confirmed from other cases as well. So the potential uneven distribution of biomass mixing could be possibly identified and the potential fire incident can be picked up early to avoid mill broken down.

6. Conclusion

A complete multi-segment coal mill model was presented in this chapter. The model considered heat, mass, power balances involved in the milling process. The mill model can represent the whole milling process including the start-up and shut-down periods. Then the model is validated using on-line measurement. By adopting this multi-segment coal mill model in power plants, the power plant engineers will be offered a non-stop model for monitoring the coal mill system The model can estimate the inmeasurable intermediate variables which is very valuable for condition monitoring and fault detection. So the model will contribute for better mill control actions and prediction of system faults. The multi-segment model is implemented on-line usiing C++ language. Based on the mathematical model and its on-line implementation, a new diagnose method is introduced to diagnose the mill operation condition. From the current results, the model could predict the uneven mixture of biomass material with coal inside the mill from the variation trend of the key parameters, which can then predict potential fire incidents.

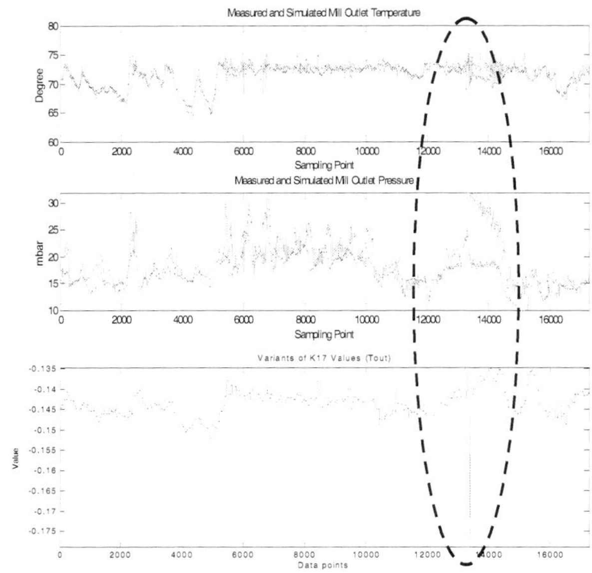

Fig. 11. Model simulated outputs, with K_t re-identified in very 5 minutes

7. References

Bollinger, K.E. & Snowden, H.R. (1983). The experimental determination of coal mill models, *IEEE Trans. Power App. Syst.*, Vol.PAS-102, June, 1983, pp1473-1477.

Cheetham, R.G.; Billings, S.A.; Bransby, M.L., & Owens, D.H. (1990). *Multivariable modelling of pulverised fuel mills from experimental PRBS data, Report Part I and II*, 1990, London, U.K.: Central Electrical Generating Board.

Corti, L.; De Marco, A. & Zizzo, A. (1984). Mathematical model of a coal mill, in *Proc. Int. Conf. Power Plant Simulation*, pp19-21, Nov. 1984, Cuernavaca, Mexico.

Fan, G.Q. and Rees, N. W. (1994). Control of vertical spindle mills in coal fired power plants, in *Proc. Ind. Commercial Power Syst. Tech. Conf.*, pp276-280, 1994, Beijing, China.

Hamiane, M. (2000). Identification of a pulverized fuel mill from normal operating records using a multivariable polynomial matrix model, *International Journal of Modelling and Simulation*, Vol. 20, No. 3, 1994, pp227-235.

Livingston, W.R. (2004). The current status of biomass co-firing at coal fired power stations in Britain, *the 2nd World Conference and Exhibition on Biomass for Energy Industry and Climate Protection*, May 2004, Rome, Italy.

Scott, D. H. (1995). *Coal Pulverises – Performance and Safety*, IEA Coal Research Publications, ISBN 92-9029-254-7, London.

Wang, J.; Wang, J.D.; Daw, N. & Wu, Q.H. (2004). Identification of pneumatic cylinder friction parameters using genetic algorithms, *IEEE/ASME Transactions* on *Mechatronics*, Vol 9. No.1, 2004, pp100-107.

Wei J.L.; Wang, J.H. & Wu, Q.H. (2003). Further Study of Coal Mill Modelling by Machine Learning Based on On-site Measurement, *ICSE conference 2003*, P746 ~ 741, Coventry, UK.

Wei, J.; Wang, J. & Wu, Q. H. (2006). Multi-segment coal mill modelling and on-line implementation, *International Journal of Modelling, Identification and Control*, , Vol.1, No.3, 2006, pp206-213.

Wei, J. L.; Wang, J. & Wu, Q. H. (2007). Development of a multi-segment coal mill model using an evolutionary computation technique, *IEEE Transactions on Energy Conversion*, Vol. 22. 2007, pp718-727.

Zhang, Y.G.; Wu, Q.H.; Wang, J.; Oluwande, G.; Matts, D. & Zhou, X.X. (2002). Coal mill modelling by machine learning based on on-site measurement, *IEEE Trans on Energy Conversion*, Vol.17, No.4, December, 2002, pp.549-555.

20

ROBUST AND REDUCED ORDER H-INFINITY FILTERING VIA LMI APPROACH AND ITS APPLICATION TO FAULT DETECTION

Young-Man Kim and John M. Watkins
Department of Electrical Engineering and Computer Science, Wichita State University United States of America

1. Introduction

Filters that estimate the state variables of a system are important tools for control and signal processing applications. Early work in the area assumed that the system dynamics were known and external disturbances were white noise with known statistical properties. In contrast to traditional Kalman filters, H_∞ filters do not require knowledge of the statistical properties of the noise. H_∞ filters are more robust to disturbances and modeling uncertainties than Kalman filters. Thus, in practical applications where disturbances may not be known exactly and system uncertainties may appear in modeling, the H_∞ technique is often used (Fu et al., 1994).

It is important to consider filter order when fast data processing is necessary. The reduced order filter is often desirable because it reduces the filter complexity and real time computational burdens in many applications. In (Grigoriadis & Watson, 1997) the reduced order H_∞ filtering problem was studied via an LMI (Linear Matrix Inequality) approach, but only for a specific linear time invariant plant model without model uncertainties. In (Bettayeb & Kavranoglu, 1994) the reduced order filter problem was studied in an H_∞ setting, but the H_∞ problem was formulated as distance problem.

Because a wide variety of problems arising in system and control theory can be reduced to an optimization problem involving LMIs and LMIs can be solved numerically very efficiently, LMIs have been used extensively in the controls field. In (Tuan et al., 2000) the robust reduced order filtering problem was studied in an H_2 setting via an LMI approach. In this study, the H_∞ approach is used because it is known that H_∞ approach is more robust to model uncertainties than H_2 (Kim & Watkins, 2006). Less conservative results can be achieved, and a computational example is given to show this.

This filtering technique can be used for fault detection filter design. As science and technology develops, the reliability and security of complex systems becomes more important. Thus, on-line monitoring of faults as they occur during operation of a dynamic system is necessary. In this study, estimator based fault detection methods will be the focus. The key to estimator based fault detection is to generate a fault indicating signal (residual) using input and output signals from the monitored system (Chen & Zhang, 1991). However, there is always a model-reality mismatch between plant dynamics and the model used for the residual generation (Chen & Patton, 1997). The robustness of residual depends on its fault sensitivity. The residual should be sensitive to faults but insensitive to modeling uncertainties and disturbances (Zhong et al., 2003). To produce the residual signal, an observer is usually used. In the fault detection literature this observer is often called a fault detection filter to emphasize the relationship with the filtering concept. In this study, robust fault detection filter (RFDF) design is formulated as a multi-objective H_∞ optimization for a polytopic uncertain system.

In (Casavola et al., 2005a), RFDF design was formulated as a multi-objective H_∞ optimization only for the full order case. In (Casavola et al., 2005b), RFDF was formulated as a quasi-LMI only for the full order case. In this study, the order of the RFDF is reduced using LMI techniques and the detection performance is compared with the full order filter (Kim & Watkins, 2007).

This paper is organized as follows: In Section 2, notations are introduced. In Section 3, the preliminary and main results for the H_∞ filter design are given. In Section 4, the preliminary and main results for the fault detection filter design are given. Numerical examples of H_∞ filter design and the fault detection filter design are shown in Section 5. Concluding remarks can be found in Section 6.

2. Notation

The notation that is used here is quite standard. R is the field of real numbers, R^n is a real vector with dimension n and $R^{m\times n}$ is a real matrix with dimensions $m \times n$. RH_∞ is the subspace of L_∞ with real and rational functions that are analytic and bounded in the open right-half plane, where L_∞ is the set of functions bounded on $j\omega$-axis including $\omega = \infty$. B^T is the transpose of matrix B. Symbol, *, stands for terms that are induced by symmetry, e.g.,

$$\begin{bmatrix} S+(*) & * \\ M & Q \end{bmatrix} K^T \equiv \begin{bmatrix} S+S^T & M^T \\ M & Q \end{bmatrix} K^T$$

BRL stands for Bounded Real Lemma, which is standard in robust control theory (Gahinet et al., 1996).

3. H_∞ Filter design

3.1 Preliminary result

The general filtering configuration can be depicted as in Figure 1 where G is the plant, F is the filter that will be designed, d is an uncertain disturbance that includes process and measurement noise, z is the signal to be estimated, z_F is the estimate of z, e is the estimation error, and y is the measured output.

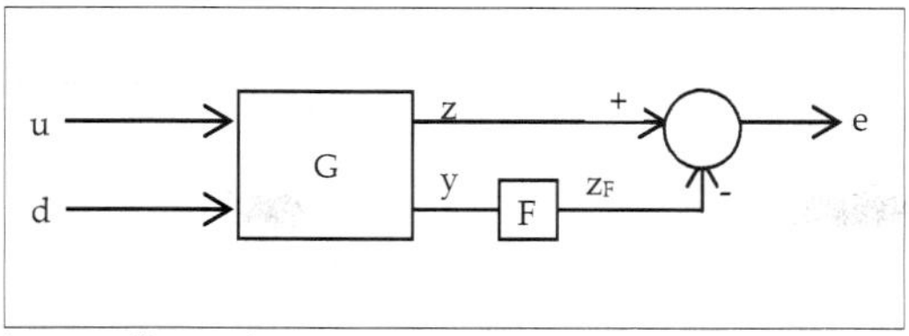

Fig. 1. General filtering configuration

A linear time invariant plant (G) described by

$$G : \begin{Bmatrix} \dot{x}(t) = Ax(t) + Bd(t) \\ y(t) = Cx(t) + Dd(t) \\ z(t) = Lx(t) \end{Bmatrix} \tag{1}$$

will be considered where $x(t) \in R^n$ is the state, $y(t) \in R^p$ is the measured output, $z(t) \in R^q$ is the estimated output, $d(t) \in R^m$ is the disturbance, and A, B, C, D and L are of appropriate dimensions. To include plant model uncertainties, we formulate them in the form of a polytopic model as follows.

$$\begin{bmatrix} A & B \\ C & D \\ L & 0 \end{bmatrix} \in \left\{ \begin{bmatrix} A(\alpha) & B(\alpha) \\ C(\alpha) & D(\alpha) \\ L(\alpha) & 0 \end{bmatrix} = \sum_{i=1}^{s} \alpha_i \begin{bmatrix} A_i & B_i \\ C_i & D_i \\ L_i & 0 \end{bmatrix}, \alpha \in \Gamma \right\} \tag{2}$$

where Γ is a unit simplex such that

$$\Gamma \equiv \left\{ (\alpha_1, \cdots, \alpha_s) : \sum_{i=1}^{s} \alpha_i = 1, \alpha_i \geq 0 \right\} \tag{3}$$

The formulation is a convex combination, so it is suitable for the LMI approach. This convex bounded polytopic mathematical description of model uncertainty is sufficiently general to include many uncertain systems with practical appeal. The filter F is attached to the system as follows:

$$F : \begin{Bmatrix} \dot{x}_F(t) = A_F x_F(t) + B_F y(t) \\ z_F(t) = L_F x_F(t) \end{Bmatrix} \tag{4}$$

where $x_F(t) \in R^k$ is the filter state $(k \leq n)$ and $A_F \in R^{k \times k}$, $B_F \in R^{k \times p}$ and $L_F \in R^{q \times k}$ are the filter matrices that are to be synthesized.

Definition 3.1:
The L_2 norm of a vector valued function $f(t)$ is defined:

$$\|f\|_{L_2} = \sqrt{\int_0^\infty f^T(t) f(t) dt} \tag{5}$$

The goal of the H_∞ optimal filtering problem is to find a filter F to minimize the worst case estimation error energy $\|e\|_{L_2}$ over all bounded energy disturbance d, where $e = z - z_F$, that is

$$\min_F \sup_{d \in L_2 - \{0\}} \frac{\|e\|_{L_2}}{\|d\|_{L_2}} \tag{6}$$

Using the induced L_2-gain property of the H_∞ norm, this problem is equivalent to the following H_∞ norm minimization problem

$$\min_F \|T_{ed}\|_\infty \tag{7}$$

where T_{ed} is the transfer function from disturbance d to the estimation error e and the H_∞ norm is defined as the largest gain over all frequencies such that

$$\|T_{ed}(s)\|_\infty \equiv \sup_\omega \sigma_{\max}(T_{ed}(j\omega)) \tag{8}$$

where $\sigma_{\max}$ denotes maximum singular value of the given function. The γ-suboptimal H_∞ filtering problem is to find a filter F such that

$$\|T_{ed}\|_\infty < \gamma \tag{9}$$

where γ is a given positive scalar.
To find the transfer function T_{ed}, (1) and (4) can be rewritten in augmented form as (10) and (11).

$$\dot{x}_{cl} = \begin{bmatrix} \dot{x} \\ \dot{x}_F \end{bmatrix} = \begin{bmatrix} Ax + Bd \\ A_F x_F + B_F y \end{bmatrix}$$

$$= A_{cl} x_{cl} + B_{cl} d \tag{10}$$

$$e = z_{cl} = z - z_F = L_{cl} x_{cl} \tag{11}$$

where

$$x_{cl} \equiv \begin{bmatrix} x \\ x_F \end{bmatrix},\ A_{cl} \equiv \begin{bmatrix} A & 0 \\ B_F C & A_F \end{bmatrix},\ B_{cl} \equiv \begin{bmatrix} B \\ B_F D \end{bmatrix},\ \text{and}\ L_{cl} \equiv \begin{bmatrix} L & -L_F \end{bmatrix} \tag{12}$$

The transfer function of closed system, T_{ed}, can be found as:

$$e \equiv z_{cl} = L_{cl} x_{cl}$$

$$= L_{cl}(sI - A_{cl})^{-1} B_{cl} d \equiv T_{ed} d \tag{13}$$

where

$$T_{ed} \equiv L_{cl}(sI - A_{cl})^{-1} B_{cl} \tag{14}$$

Using the well-known BRL (Gahinet & Apkarian, 1994), the condition in (9) with (14) can be described as:

$$\begin{bmatrix} PA_{cl} + A_{cl}^T P & PB_{cl} & L_{cl}^T \\ B_{cl}^T P & -\gamma I & 0 \\ L_{cl} & 0 & -\gamma I \end{bmatrix} < 0 \tag{15}$$

where A_{cl}, B_{cl} and L_{cl} are given in (12) and P is a symmetric positive definite matrix variable. Equation (15) satisfies the design requirement as follows:

a) internal stability: for d=0, the state vector x_{cl} of closed loop system tends to zero as time goes to infinity.

b) performance: The H_∞ norm, $\|T_{ed}\|_\infty$, is less than specified positive scalar γ.

Statement (a) implies quadratic stability and (b) implies quadratic performance. Remember that A_{cl}, B_{cl} and L_{cl} depend on parameter α. Thus, statement (a) should be replaced by quadratic stability in the sense of parameter dependency and statement (b) replaced by quadratic performance in the sense of parameter dependency.

Definition 3.2 (Gahinet et al., 1996):
The system of (10) and (11) is said to have AQS (affine quadratic stability) if there exist a positive symmetric affine-parameter dependent Lyapunov matrix

$$P(\alpha) = \sum_{i=1}^{s} \alpha_i P_i \tag{16}$$

where $P_i, (i = 1, 2, \cdots, s)$ are symmetric, such that

$$A_{cl}(\alpha)^T P(\alpha) + P(\alpha) A_{cl}(\alpha) < 0, \forall \alpha \in \Gamma \qquad \blacksquare \tag{17}$$

Definition 3.3 (Gahinet et al., 1996):
The system of (10) and (11) is said to have AQP (affine quadratic H_∞ performance) if there exists a positive symmetric affine-parameter dependent Lyapunov matrix (16) such that

$$\begin{bmatrix} A_{cl}(\alpha)^T P(\alpha) + P(\alpha) A_{cl}(\alpha) & P(\alpha) B_{cl}(\alpha) & L_{cl}(\alpha)^T \\ B_{cl}(\alpha)^T P(\alpha) & -\gamma I & 0 \\ L_{cl}(\alpha) & 0 & -\gamma I \end{bmatrix} < 0 \tag{18}$$

that holds for all admissible parameter $\alpha_i (i = 1, 2, \cdots, s)$ ■

3.2 Main result

The reduced-order H_2 filtering problem was studied in (Nagpal et al., 1987), and the reduced-order H_∞ filtering problem for discrete time systems was studied in (O'Brien & Kiriakidis, 2004), but in this section we will develop an H_∞ filtering approach in LMI framework that is applicable and easy to compute. Let's reconsider (18). The matrix inequality in (18) is nonlinear because the terms containing the P, A_{cl} and B_{cl} variables multiply the unknown variable P with the unknowns that are included in A_{cl} and B_{cl}. Finsler's Lemma is a standard tool to separate filter variables from the Lyapunov matrix P (Tuan et al., 2000). The LMI formulation in Theorem 3.4 is very useful for solving the robust filtering problem.

***Theorem* 3.4** (Tuan et al., 2000):
The LMI

$$P > 0, \begin{bmatrix} A^T P + PA & PB & C^T \\ B^T P & Q_{11} & Q_{12} \\ C & Q_{12}^T & Q_{22} \end{bmatrix} < 0 \tag{19}$$

has a feasible decision variable P if and only if, for any choice of $\mu > 0$, the following LMI is feasible in the decision variables V and P,

$$\begin{bmatrix} -(V+V)^T & V^T A + P & V^T B & 0 & V^T \\ A^T V + P & -\mu P & 0 & C^T & 0 \\ B^T V & 0 & Q_{11} & Q_{22} & 0 \\ 0 & C & Q_{12}^T & Q_{22} & 0 \\ V & 0 & 0 & 0 & -P/\mu \end{bmatrix} < 0 \tag{20}$$

Proof: refer to (Tuan et al., 2000) ∎

From ***Theorem* 3.4,** (18) can be written as

$$\begin{bmatrix} -(V+V^T) & V^T A_{cl} + P & V^T B_{cl} & 0 & V^T \\ A_{cl}^T V + P & -\mu P & 0 & L_{cl}^T & 0 \\ B_{cl}^T V & 0 & -\gamma I & 0 & 0 \\ 0 & L_{cl} & 0 & -\gamma I & 0 \\ V & 0 & 0 & 0 & -P/\mu \end{bmatrix} < 0 \tag{21}$$

The parameter α is omitted. The slack variable V has been introduced to separate the variable P from the filter design variables.
To proceed further, the V and P variables should be partitioned as

$$V=\begin{bmatrix} V_{11} & V_{12} \\ V_{21} & V_{22} \end{bmatrix} \tag{22}$$

$$P=\begin{bmatrix} P_1 & P_3^T \\ P_3 & P_2 \end{bmatrix} \tag{23}$$

Because the reduced order case is being considered, the filter order k, will be less than or equal to the plant order n. The partitioned sub-matrices in (22) will be dimensionalized as $V_{11}(n\times n)$, $V_{12}(n\times k)$, $V_{21}(k\times n)$, and $V_{22}(k\times k)$. Now, we need to enforce some special structure on V_{21} as

$$V_{21}=\begin{bmatrix} \tilde{V}_{21} & 0_{k\times(n-k)} \end{bmatrix} \tag{24}$$

where $\tilde{V}_{21}$ is a $k\times k$ matrix. We can replace V, P, A_{cl}, B_{cl} and L_{cl} in (21) with (12), (22), (23) and (24). After that, we perform a congruence transformation with the transformation matrix

$$diag\begin{bmatrix} I & V_{22}^{-1}\tilde{V}_{21} & I & V_{22}^{-1}\tilde{V}_{21} & I & I & I & V_{22}^{-1}\tilde{V}_{21} \end{bmatrix} \tag{25}$$

This yields (26)

$$\begin{bmatrix}
-(V_{11}+V_{11}^T) & -(S_2^T+\tilde{S}_1^T) & V_{11}^TA+\tilde{B}_FC+\hat{P}_1 & \tilde{A}_F+\hat{P}_3^T & V_{11}^TB+\tilde{B}_FD & 0 & V_{11}^T & \tilde{S}_1^T \\
* & -(S_1+S_1^T) & S_2A+\hat{B}_FC+\hat{P}_3 & \hat{A}_F+\hat{P}_2 & S_2B+\hat{B}_FD & 0 & S_2 & S_1^T \\
* & * & -\mu\hat{P}_1 & -\mu\hat{P}_3^T & 0 & L^T & 0 & 0 \\
* & * & * & -\mu\hat{P}_2 & 0 & -\hat{L}_F^T & 0 & 0 \\
* & * & * & * & -\gamma I & 0 & 0 & 0 \\
* & * & * & * & * & -\gamma I & 0 & 0 \\
* & * & * & * & * & * & -\frac{1}{\mu}\hat{P}_1 & -\frac{1}{\mu}\hat{P}_3^T \\
* & * & * & * & * & * & * & -\frac{1}{\mu}\hat{P}_2
\end{bmatrix}<0 \tag{26}$$

where

$$S_1=\tilde{V}_{21}^TV_{22}^{-T}\tilde{V}_{21} \tag{27}$$

$$\tilde{S}_1=\begin{bmatrix} S_1 & 0_{k\times(n-k)} \end{bmatrix} \tag{28}$$

$$S_2=\tilde{V}_{21}^TV_{22}^{-T}V_{12}^T \tag{29}$$

$$\hat{B}_F=\tilde{V}_{21}^TB_F \tag{30}$$

$$\tilde{B}_F = \begin{bmatrix} \hat{B}_F \\ 0_{(n-k)\times p} \end{bmatrix} \tag{31}$$

$$\hat{P}_1 = P_1 \tag{32}$$

$$\hat{P}_2 = \tilde{V}_{21}^T V_{22}^{-T} P_2 V_{22}^{-1} \tilde{V}_{21} \tag{33}$$

$$\hat{P}_3^T = P_3^T V_{22}^{-1} \tilde{V}_{21} \tag{34}$$

$$\hat{A}_F = \tilde{V}_{21}^T A_F V_{22}^{-1} \tilde{V}_{21} \tag{35}$$

$$\tilde{A}_F = \begin{bmatrix} \hat{A}_F \\ 0_{(n-k)\times k} \end{bmatrix} \tag{36}$$

$$\hat{L}_F = L_F V_{22}^{-1} \tilde{V}_{21} \tag{37}$$

$$\hat{P} = \begin{bmatrix} \hat{P}_1 & \hat{P}_3^T \\ \hat{P}_3 & \hat{P}_2 \end{bmatrix} \tag{38}$$

Remark 3.5:
The H_∞ filtering solvability condition in (18) is reformulated as the feasibility problem of (26) with respect to $\hat{A}_F$, $\hat{B}_F$, $\hat{L}_F$, S_1, S_2, V_{11}, $\hat{P}$ and γ where $\hat{P} > 0$. ■

Remark 3.6:
The filter matrices A_F, B_F and L_F can be derived by means of the following procedure.

(i) Compute V_{22} and $\tilde{V}_{21}$ by solving the factorization problem

$$S_1 = \tilde{V}_{21}^T V_{22}^{-1} \tilde{V}_{21} \tag{39}$$

(ii) Compute A_F, B_F and L_F

$$A_F = \tilde{V}_{21}^{-T} \hat{A}_F \tilde{V}_{21}^{-1} V_{22} \tag{40}$$

$$B_F = \tilde{V}_{21}^{-T} \hat{B}_F \tag{41}$$

$$L_F = \hat{L}_F \tilde{V}_{21}^{-1} V_{22} \quad ■ \tag{42}$$

Now, we need to remember that we had a polytopic uncertain system. As explained in (Tuan et al., 2000), a parameter dependent Lyapunov matrix $P(\alpha)$, such as

$$P(\alpha) = \sum_{i=1}^{s} \alpha_i P_i(\alpha), \alpha \in \Gamma \tag{43}$$

is symmetric positive definite for all admissible values of α, if and only, if this holds for each P_i. Therefore, we need to check the solvability of (18) only at the vertices, $i = 1, 2, \cdots, s$. This gives us the final form (44) of the LMI (26).

$$\begin{bmatrix} -(V_{11}+V_{11}^T) & -(S_2^T+\tilde{S}_1^T) & V_{11}^T A_i+\tilde{B}_F C_i+\hat{P}_{1,i} & \tilde{A}_F+\hat{P}_{3,i}^T & V_{11}^T B_i+\tilde{B}_F D_i & 0 & V_{11}^T & \tilde{S}_1^T \\ * & -(S_1+S_1^T) & S_2 A_i+\hat{B}_F C_i+\hat{P}_{3,i} & \hat{A}_F+\hat{P}_{2,i} & S_2 B_i+\hat{B}_F D_i & 0 & S_2 & S_1^T \\ * & * & -\mu\hat{P}_{1,i} & -\mu\hat{P}_{3,i}^T & 0 & L_i^T & 0 & 0 \\ * & * & * & -\mu\hat{P}_{2,i} & 0 & -\hat{L}_F^T & 0 & 0 \\ * & * & * & * & -\gamma I & 0 & 0 & 0 \\ * & * & * & * & * & -\gamma I & 0 & 0 \\ * & * & * & * & * & * & -\frac{1}{\mu}\hat{P}_{1,i} & -\frac{1}{\mu}\hat{P}_{3,i}^T \\ * & * & * & * & * & * & * & -\frac{1}{\mu}\hat{P}_{2,i} \end{bmatrix} < 0 \quad (44)$$

$(i = 1, 2, \cdots, s)$

Consequently, the minimum upper bound γ for the reduced order H_∞ filter can be found by solving the LMI optimization problem

$$\min_{V_{11}, S_1, S_2, \hat{A}_F, \hat{B}_F, \hat{L}_F, \gamma, \hat{P}_i} \{\gamma : (44)\} \quad (45)$$

In summary, the γ-suboptimal H_∞ reduced order filter for polytopic uncertain system can be solved if and only if (44) holds for all vertices, $i = 1, 2, \cdots, s$ and the minimum value can be found from (45). Also, the triple (A_F, B_F, L_F) defining the k_{th} order filter is obtained from (i) and (ii) in ***Remark 3.6.***

4. Fault detection filter design

4.1 Preliminary result

Let's consider the following uncertain continuous time linear system described by

$$\Sigma : \begin{Bmatrix} \dot{x}(t) = Ax(t) + B_u u(t) + B_f f(t) + B_d d(t) \\ y(t) = Cx(t) + D_u u(t) + D_f f(t) + D_d d(t) \end{Bmatrix} \quad (46)$$

where $x(t) \in R^n$ is the state, $y(t) \in R^m$ is the measured output, $f(t) \in R^{n_f}$ is the fault, $d(t) \in R^{n_d}$ is the bounded disturbance, and $u(t) \in R^{n_u}$ is the control signal. Actuator and component faults are modeled by $B_f f(t)$, and sensor faults are modeled by $D_f f(t)$. Plant model uncertainties are modeled in the form of a polytopic model as follows

$$\begin{bmatrix} A & B_u & B_f & B_d \\ C & D_u & D_f & D_d \end{bmatrix} \in \left\{ \begin{aligned} &\begin{bmatrix} A(\alpha) & B_u(\alpha) & B_f(\alpha) & B_d(\alpha) \\ C(\alpha) & D_u(\alpha) & D_f(\alpha) & D_d(\alpha) \end{bmatrix} \\ &= \sum_{i=1}^{s} \alpha_i \begin{bmatrix} A_i & B_{ui} & B_{fi} & B_{di} \\ C_i & D_{ui} & D_{fi} & D_{di} \end{bmatrix}, \alpha \in \Gamma \end{aligned} \right\} \quad (47)$$

where Γ is a unit simplex such that

$$\Gamma \equiv \left\{ (\alpha_1, \cdots, \alpha_s) : \sum_{i=1}^{s} \alpha_i = 1, \alpha_i \geq 0 \right\} \tag{48}$$

Because this formulation is a convex combination, it is suitable for an LMI approach.
Here we assume that the above polytopic system possesses the affine quadratic stability that was introduced earlier. Other assumptions that are made for our purpose are that (C, A) is detectable and $\begin{bmatrix} A - j\omega I & B_d \\ C & D_d \end{bmatrix}$ has full rank for all ω. The assumption that (C, A) is detectable is standard. The assumption that $\begin{bmatrix} A - j\omega I & B_d \\ C & D_d \end{bmatrix}$ has full row rank for all ω ensures that $G_{yd} = \begin{bmatrix} A & B_d \\ C & D_d \end{bmatrix}$ has no zeros on the $j\omega$ -axis.
The proposed fault detection filter (FDF) will have the form

$$F : \left\{ \begin{aligned} \dot{x}_F(t) &= A_F x_F(t) + B_F \begin{bmatrix} y(t) \\ u(t) \end{bmatrix} \\ z(t) &= L_F x_F(t) + H_F \begin{bmatrix} y(t) \\ u(t) \end{bmatrix} \end{aligned} \right\} \tag{49}$$

The residual $r(t)$ is defined as $r(t) \equiv z(t) - y(t)$. Thus, the RFDF design problem can be described as designing the filter (A_F, B_F, L_F, H_F) such that the residual $r(t)$ is as sensitive as possible to the fault $f(t)$ and as robust as possible to the unknown input $d(t)$, control input $u(t)$ and polytopic model uncertainties.
There are a number of schemes to approach the RFDF problem. In (Ding & Frank, 2003) it is formulated as an optimization problem. Unfortunately, this approach is difficult to extend to systems with modeling uncertainty (Zhong et al., 2003). Another approach is to formulate the RFDF problem as an H_∞ model-matching problem (Casavola et al., 2003). However, this approach may produce a false alarm in a no fault situation (Ding et al., 2000). Thus, we use multi-objective H_∞ optimization (Casavola et al., 2005a, Casavola et al., 2005b) based on the model-matching formulation.
Multi-objective H_∞ optimization based on the model-matching formulation can be described as follows. As previously explained, the generated residual should be sensitive to the faults, but insensitive to the uncertainties and noises. To indicate the residual's sensitivity to the faults, a new H_- index was proposed by (Patton & Hou, 1999). The H_- index is defined as the minimum nonzero singular value of a transfer matrix. However, this is not a norm (Chen & Patton, 1999). Therefore, the model-matching problem needs to be introduced to relate the minimum gain from fault to residual with the residual error (Niemann & Stoustrup, 2000). If we use the residual error instead of the residual, it will be more convenient for optimizing the detection filter F. The residual error is defined as

$$\begin{aligned}\tilde{r} &= W(s)f - r \\ &= W(s)f - \{T_{rf}(s)f + T_{ru}(s)u + T_{rd}(s)d\} \\ &= \begin{bmatrix} W(s) - T_{rf}(s) & \vdots & -T_{ru}(s) & \vdots & -T_{rd}(s) \end{bmatrix} \begin{bmatrix} f \\ u \\ d \end{bmatrix}\end{aligned} \tag{50}$$

where $W(s)$ is called reference model (Frisk, 2001) and the transfer matrices, T_{rf}, T_{ru} and T_{rd}, are defined as

$$T_{rf} = F\begin{bmatrix} G_{yf} \\ 0 \end{bmatrix} - G_{yf}\,,\; T_{ru} = F\begin{bmatrix} G_{yu} \\ 0 \end{bmatrix} - G_{yu}\,,\; T_{rd} = F\begin{bmatrix} G_{yd} \\ 0 \end{bmatrix} - G_{yd}$$

The reference model, $W(s)$, is an RH_∞ transfer matrix (Chen & Patton, 1999). The idea of a reference model has successfully been used to describe signal behavior in other fields like controller design and adaptive control. As discussed in (Frisk, 2001), the main function of the reference model is to describe the desired behavior of the residual vector r with respect to the faults f. For example, if we want to detect faults in the frequency range between 0 and 2 radians/s with a -20dB/decade roll-off at the higher frequencies, the reference model, $W(s)$ is given as $\frac{2}{s+2}$. Using the approach in (Casavola et al., 2005a), $W(s)$ in (50) is placed between the plant and the filter so that the filter can track faults with its specific feature. This can increase the robustness of the filter to the specific faults, if the designer can choose $W(s)$ suitably. Thus, the block diagram for the residual error, $\tilde{r}(t)$, with tracking filter $W(s)$ is shown in Figure 2.

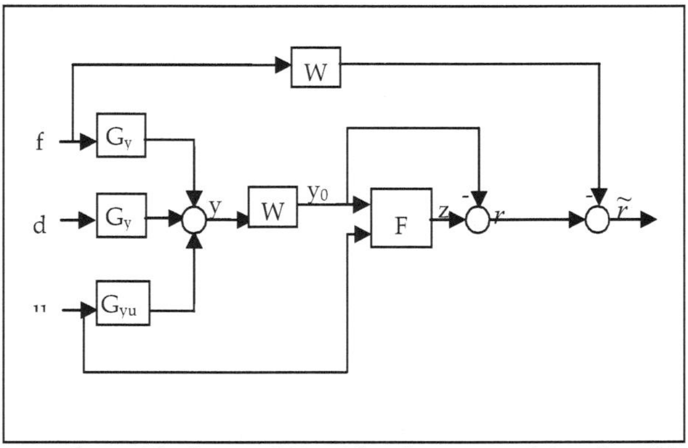

Fig. 2. Block diagram for residual error $\tilde{r}(t)$ with reference model and tracking filter

With this background, we can describe the RFDF design problem using multi-objective H_∞ optimization as follows.

Definition 4.1:
Given positive scalars α_d, α_f and α_u, the RFDF design problem using multi-objective H_∞ optimization is defined as finding $F(s)$ that satisfies $\min_{F(s)}(\alpha_d\gamma_d + \alpha_f\gamma_f + \alpha_u\gamma_u)$ where

$$\begin{gathered}\max_{\alpha\in\Gamma}\left\|F(s)\begin{bmatrix}G_{y_0d}(s)\\0\end{bmatrix}-G_{y_0d}(s)\right\|_\infty \equiv \max_{\alpha\in\Gamma}\left\|T_{rd}^0\right\| \le \gamma_d\\ \max_{\alpha\in\Gamma}\left\|W(s)-(F(s)\begin{bmatrix}G_{y_0f}(s)\\0\end{bmatrix}-G_{y_0f}(s))\right\|_\infty \equiv \max_{\alpha\in\Gamma}\left\|W(s)-T_{rf}^0\right\| \le \gamma_f\\ \max_{\alpha\in\Gamma}\left\|F(s)\begin{bmatrix}G_{y_0u}(s)\\I\end{bmatrix}-G_{y_0u}(s)\right\|_\infty \equiv \max_{\alpha\in\Gamma}\left\|T_{ru}^0\right\| \le \gamma_u\end{gathered} \tag{51}$$

and

$$G_{yf} = \begin{bmatrix}A & B_f\\C & D_f\end{bmatrix},\ G_{yd} = \begin{bmatrix}A & B_d\\C & D_d\end{bmatrix} \text{ and } G_{yu} = \begin{bmatrix}A & B_u\\C & D_u\end{bmatrix},\ G_{y_0f} \equiv WG_{yf},\ G_{y_0d} \equiv WG_{yd},$$

$$G_{y_0u} \equiv WG_{yu},\ T_{rf}^0 \equiv F\begin{bmatrix}G_{y_0f}\\0\end{bmatrix}-G_{y_0f}\ T_{rd}^0 \equiv F\begin{bmatrix}G_{y_0d}\\0\end{bmatrix}-G_{y_0d} \text{ and } T_{ru}^0 \equiv F\begin{bmatrix}G_{y_0u}\\0\end{bmatrix}-G_{y_0u}$$ ■

The positive scalars $(\alpha_d, \alpha_f, \alpha_u)$ are used to weight the relative importance of tracking and filtering performance.

4.2 Main result

In this section, LMIs will be used to solve the problem formulated in Section 4.1 for an RFDF. To solve the optimization problem in (51), we need the state-space realization of each transfer matrix. The reference model $W(s)$ can be realized as $W(s) = \begin{bmatrix}A_w & B_w\\C_w & D_w\end{bmatrix}$. After some manipulation, the transfer matrices term in (51) can be realized as

$$F(s)\begin{bmatrix}G_{y_0d}(s)\\0\end{bmatrix}-G_{y_0d}(s) = \left[\begin{array}{cccc:c}A & 0 & 0 & 0 & B_d\\ B_wC & A_w & 0 & 0 & B_wD_d\\ 0 & 0 & A_w & 0 & 0\\ 0 & B_F\begin{bmatrix}C_w\\0\end{bmatrix} & 0 & A_F & 0\\ \cdots & \cdots & \cdots & \cdots & \cdots\\ 0 & C_w - H_F\begin{bmatrix}C_w\\0\end{bmatrix} & C_w & -L_F & -H_F\begin{bmatrix}0\\I\end{bmatrix}\end{array}\right]$$

$$W(s)-(F(s)\begin{bmatrix} G_{y_0 f}(s) \\ 0 \end{bmatrix}-G_{y_0 f}(s))=$$

$$\left[\begin{array}{ccccc} A & 0 & 0 & 0 & \vdots & B_f \\ B_w C & A_w & 0 & 0 & \vdots & B_w D_f \\ 0 & 0 & A_w & 0 & \vdots & B_w \\ 0 & B_F\begin{bmatrix} C_w \\ 0 \end{bmatrix} & 0 & A_F & \vdots & 0 \\ \cdots & \cdots & \cdots & \cdots & \vdots & \cdots \\ 0 & C_w - H_F\begin{bmatrix} C_w \\ 0 \end{bmatrix} & C_w & -L_F & \vdots & -H_F\begin{bmatrix} 0 \\ I \end{bmatrix} \end{array}\right]$$

$$F(s)\begin{bmatrix} G_{y_0 u}(s) \\ I \end{bmatrix}-G_{y_0 u}(s)=$$

$$\left[\begin{array}{cccccc} A & 0 & 0 & 0 & \vdots & B_u \\ B_w C & A_w & 0 & 0 & \vdots & B_w D_u \\ 0 & 0 & A_w & 0 & \vdots & 0 \\ 0 & B_F\begin{bmatrix} C_w \\ 0 \end{bmatrix} & 0 & A_F & \vdots & B_F\begin{bmatrix} 0 \\ I \end{bmatrix} \\ \cdots & \cdots & \cdots & \cdots & \vdots & \cdots \\ 0 & C_w - H_F\begin{bmatrix} C_w \\ 0 \end{bmatrix} & C_w & -L_F & \vdots & -H_F\begin{bmatrix} 0 \\ I \end{bmatrix} \end{array}\right] \tag{52}$$

Here, we note that the order of filter is $n_F = 2n_w + n$, where n_F is the filter order, n_w is the tracking filter order, and n is the plant order.

To simplify (52), we let

$$\tilde{A} \equiv \begin{bmatrix} A & 0 & 0 \\ B_w C & A_w & 0 \\ 0 & 0 & A_w \end{bmatrix} \; \tilde{C} \equiv \begin{bmatrix} 0 & \begin{bmatrix} C_w \\ 0 \end{bmatrix} & 0 \end{bmatrix} \; \tilde{B}_f \equiv \begin{bmatrix} B_f \\ B_w D_f \\ B_w \end{bmatrix} \; \tilde{B}_u \equiv \begin{bmatrix} B_u \\ B_w D_u \\ 0 \end{bmatrix} \tag{53}$$

$$\tilde{L}_F \equiv \begin{bmatrix} 0 & C_w - H_F\begin{bmatrix} C_w \\ 0 \end{bmatrix} & C_w \end{bmatrix}$$

Using (53), (52) can be rewritten as

$$F(s)\begin{bmatrix} G_{y_0 d}(s) \\ 0 \end{bmatrix} - G_{y_0 d}(s) = \left[\begin{array}{cc:c} \tilde{A} & 0 & \tilde{B}_d \\ B_F\tilde{C} & A_F & 0 \\ \cdots & \cdots & \cdots \\ \tilde{L}_F & -L_F & -H_F\begin{bmatrix} 0 \\ I \end{bmatrix} \end{array}\right]$$

$$W(s) - (F(s)\begin{bmatrix} G_{y_0 f}(s) \\ 0 \end{bmatrix} - G_{y_0 f}(s)) = \left[\begin{array}{cc:c} \tilde{A} & 0 & \tilde{B}_f \\ B_F\tilde{C} & A_F & 0 \\ \cdots & \cdots & \cdots \\ \tilde{L}_F & -L_F & -H_F\begin{bmatrix} 0 \\ I \end{bmatrix} \end{array}\right]$$

$$F(s)\begin{bmatrix} G_{y_0 u}(s) \\ I \end{bmatrix} - G_{y_0 u}(s) = \left[\begin{array}{cc:c} \tilde{A} & 0 & \tilde{B}_u \\ B_F\tilde{C} & A_F & B_F\begin{bmatrix} 0 \\ I \end{bmatrix} \\ \cdots & \cdots & \cdots \\ \tilde{L}_F & -L_F & -H_F\begin{bmatrix} 0 \\ I \end{bmatrix} \end{array}\right] \tag{54}$$

Applying the BRL to (51) results in

$$\begin{bmatrix} \begin{bmatrix} \tilde{A} & 0 \\ B_F\tilde{C} & A_F \end{bmatrix}^T P + P\begin{bmatrix} \tilde{A} & 0 \\ B_F\tilde{C} & A_F \end{bmatrix} & P\begin{bmatrix} \tilde{B}_d \\ 0 \end{bmatrix} & \begin{bmatrix} \tilde{L}_F^T \\ -L_F{}^T \end{bmatrix} \\ * & -\gamma_d I & -H_F\begin{bmatrix} 0 \\ I \end{bmatrix} \\ * & * & -\gamma_d I \end{bmatrix} < 0$$

$$\begin{bmatrix} \begin{bmatrix} \tilde{A} & 0 \\ B_F\tilde{C} & A_F \end{bmatrix}^T P + P\begin{bmatrix} \tilde{A} & 0 \\ B_F\tilde{C} & A_F \end{bmatrix} & P\begin{bmatrix} \tilde{B}_f \\ 0 \end{bmatrix} & \begin{bmatrix} \tilde{L}_F^T \\ -L_F{}^T \end{bmatrix} \\ * & -\gamma_f I & -H_F\begin{bmatrix} 0 \\ I \end{bmatrix} \\ * & * & -\gamma_f I \end{bmatrix} < 0$$

$$\begin{bmatrix} \begin{bmatrix} \tilde{A} & 0 \\ B_F\tilde{C} & A_F \end{bmatrix}^T P + P\begin{bmatrix} \tilde{A} & 0 \\ B_F\tilde{C} & A_F \end{bmatrix} & P\begin{bmatrix} \tilde{B}_u \\ B_F\begin{bmatrix} 0 \\ I \end{bmatrix} \end{bmatrix} & \begin{bmatrix} \tilde{L}_F^T \\ -L_F{}^T \end{bmatrix} \\ * & -\gamma_u I & -H_F\begin{bmatrix} 0 \\ I \end{bmatrix} \\ * & * & -\gamma_u I \end{bmatrix} < 0 \tag{55}$$

where P is $2n_F \times 2n_F$ Lyapunov function matrix and $P > 0$. Using the same procedure as in (Kim & Watkins, 2006), the matrix inequalities in (55) can be rewritten as (56)

$$\begin{bmatrix} -(V_{11}+V_{11}^T) & -(S_2^T+S_1^T) & V_{11}^T\tilde{A}+\hat{B}_F\tilde{C}+\hat{P}_1 & \hat{A}_F+\hat{P}_3^T & V_{11}^T\tilde{B}_d & 0 & V_{11}^T & S_1^T \\ * & -(S_1+S_1^T) & S_2\tilde{A}+\hat{B}_F\tilde{C}+\hat{P}_3 & \hat{A}_F+\hat{P}_2 & S_2\tilde{B}_d & 0 & S_2 & S_1^T \\ * & * & -\mu\hat{P}_1 & -\mu\hat{P}_3^T & 0 & \tilde{L}_F^T & 0 & 0 \\ * & * & * & -\mu\hat{P}_2 & 0 & -\hat{L}_F^T & 0 & 0 \\ * & * & * & * & -\gamma_d I & -\begin{bmatrix} 0 & I \end{bmatrix}H_F^T & 0 & 0 \\ * & * & * & * & * & -\gamma_d I & 0 & 0 \\ * & * & * & * & * & * & -\frac{1}{\mu}\hat{P}_1 & -\frac{1}{\mu}\hat{P}_3^T \\ * & * & * & * & * & * & * & -\frac{1}{\mu}\hat{P}_2 \end{bmatrix} < 0$$

$$\begin{bmatrix} -(V_{11}+V_{11}^T) & -(S_2^T+S_1^T) & V_{11}^T\tilde{A}+\hat{B}_F\tilde{C}+\hat{P}_1 & \hat{A}_F+\hat{P}_3^T & V_{11}^T\tilde{B}_f & 0 & V_{11}^T & S_1^T \\ * & -(S_1+S_1^T) & S_2\tilde{A}+\hat{B}_F\tilde{C}+\hat{P}_3 & \hat{A}_F+\hat{P}_2 & S_2\tilde{B}_f & 0 & S_2 & S_1^T \\ * & * & -\mu\hat{P}_1 & -\mu\hat{P}_3^T & 0 & \tilde{L}_F^T & 0 & 0 \\ * & * & * & -\mu\hat{P}_2 & 0 & -\hat{L}_F^T & 0 & 0 \\ * & * & * & * & -\gamma_f I & -\begin{bmatrix} 0 & I \end{bmatrix}H_F^T & 0 & 0 \\ * & * & * & * & * & -\gamma_f I & 0 & 0 \\ * & * & * & * & * & * & -\frac{1}{\mu}\hat{P}_1 & -\frac{1}{\mu}\hat{P}_3^T \\ * & * & * & * & * & * & * & -\frac{1}{\mu}\hat{P}_2 \end{bmatrix} < 0$$

$$\begin{bmatrix} -(V_{11}+V_{11}^T) & -(S_2^T+S_1^T) & V_{11}^T\tilde{A}+\hat{B}_F\tilde{C}+\hat{P}_1 & \hat{A}_F+\hat{P}_3^T & V_{11}^T\tilde{B}_u+\hat{B}_F\begin{bmatrix}0\\I\end{bmatrix} & 0 & V_{11}^T & S_1^T \\ * & -(S_1+S_1^T) & S_2\tilde{A}+\hat{B}_F\tilde{C}+\hat{P}_3 & \hat{A}_F+\hat{P}_2 & S_2\tilde{B}_u+\hat{B}_F\begin{bmatrix}0\\I\end{bmatrix} & 0 & S_2 & S_1^T \\ * & * & -\mu\hat{P}_1 & -\mu\hat{P}_3^T & 0 & \tilde{L}_F^T & 0 & 0 \\ * & * & * & -\mu\hat{P}_2 & 0 & -\hat{L}_F^T & 0 & 0 \\ * & * & * & * & -\gamma_u I & -\begin{bmatrix}0 & I\end{bmatrix}H_F^T & 0 & 0 \\ * & * & * & * & * & -\gamma_u I & 0 & 0 \\ * & * & * & * & * & * & -\frac{1}{\mu}\hat{P}_1 & -\frac{1}{\mu}\hat{P}_3^T \\ * & * & * & * & * & * & * & -\frac{1}{\mu}\hat{P}_2 \end{bmatrix} < 0 \tag{56}$$

where

$S_1 = V_{21}^T V_{22}^{-T} V_{21}$, $S_2 = V_{21}^T V_{22}^{-T} V_{12}^T$, $\hat{B}_F = V_{21}^T B_F$, $\hat{P}_1 = P_1$, $\hat{P}_2 = V_{21}^T V_{22}^{-T} P_2 V_{22}^{-1} V_{21}$, $\hat{P}_3^T = P_3^T V_{22}^{-1} V_{21}$,
$\hat{A}_F = V_{21}^T A_F V_{22}^{-1} V_{21}$, $\hat{L}_F = L_F V_{22}^{-1} V_{21}$ and $\hat{P} = \begin{bmatrix} \hat{P}_1 & \hat{P}_3^T \\ \hat{P}_3 & \hat{P}_2 \end{bmatrix}$.

To reduce the filter order, we'll need partitions of the V and P variables as in (Kim & Watkins, 2007),

$$V = \begin{bmatrix} V_{11} & V_{12} \\ V_{21} & V_{22} \end{bmatrix}, \; P = \begin{bmatrix} P_1 & P_3^T \\ P_3 & P_2 \end{bmatrix} \tag{57}$$

Using the same method as in (Kim & Watkins, 2007), we can solve for the reduced-order filter whose filter order is $k(k \le n_F)$. The partitioned sub-matrices in (57) will be dimensionalized as $V_{11}(n_F \times n_F)$, $V_{12}(n_F \times k)$, $V_{21}(k \times n_F)$, and $V_{22}(k \times k)$. We need to enforce some special structure on V_{21} as

$$V_{21} = \begin{bmatrix} \tilde{V}_{21} & 0_{k\times(n_F-k)} \end{bmatrix} \tag{58}$$

where $\tilde{V}_{21}$ is a $k \times k$ matrix. Finally, we get three LMIs for the reduced-order RFDF with order k as (59)

$$\begin{bmatrix} -(V_{11}+V_{11}^T) & -(S_2^T+\tilde{S}_1^T) & V_{11}^T\tilde{A}+\tilde{B}_F\tilde{C}+\hat{P}_1 & \tilde{A}_F+\hat{P}_3^T & V_{11}^T\tilde{B}_d & 0 & V_{11}^T & \tilde{S}_1^T \\ * & -(S_1+S_1^T) & S_2\tilde{A}+\hat{B}_F\tilde{C}+\hat{P}_3 & \hat{A}_F+\hat{P}_2 & S_2\tilde{B}_d & 0 & S_2 & S_1^T \\ * & * & -\mu\hat{P}_1 & -\mu\hat{P}_3^T & 0 & \tilde{L}_F^T & 0 & 0 \\ * & * & * & -\mu\hat{P}_2 & 0 & -\hat{L}_F^T & 0 & 0 \\ * & * & * & * & -\gamma_d I & -\begin{bmatrix}0 & I\end{bmatrix}H_F^T & 0 & 0 \\ * & * & * & * & * & -\gamma_d I & 0 & 0 \\ * & * & * & * & * & * & -\frac{1}{\mu}\hat{P}_1 & -\frac{1}{\mu}\hat{P}_3^T \\ * & * & * & * & * & * & * & -\frac{1}{\mu}\hat{P}_2 \end{bmatrix} < 0$$

$$\begin{bmatrix} -(V_{11}+V_{11}^T) & -(S_2^T+\tilde{S}_1^T) & V_{11}^T\tilde{A}+\tilde{B}_F\tilde{C}+\hat{P}_1 & \tilde{A}_F+\hat{P}_3^T & V_{11}^T\tilde{B}_f & 0 & V_{11}^T & \tilde{S}_1^T \\ * & -(S_1+S_1^T) & S_2\tilde{A}+\hat{B}_F\tilde{C}+\hat{P}_3 & \hat{A}_F+\hat{P}_2 & S_2\tilde{B}_f & 0 & S_2 & S_1^T \\ * & * & -\mu\hat{P}_1 & -\mu\hat{P}_3^T & 0 & \tilde{L}_F^T & 0 & 0 \\ * & * & * & -\mu\hat{P}_2 & 0 & -\hat{L}_F^T & 0 & 0 \\ * & * & * & * & -\gamma_f I & -\begin{bmatrix}0 & I\end{bmatrix}H_F^T & 0 & 0 \\ * & * & * & * & * & -\gamma_f I & 0 & 0 \\ * & * & * & * & * & * & -\frac{1}{\mu}\hat{P}_1 & -\frac{1}{\mu}\hat{P}_3^T \\ * & * & * & * & * & * & * & -\frac{1}{\mu}\hat{P}_2 \end{bmatrix} < 0$$

$$\begin{bmatrix} -(V_{11}+V_{11}^T) & -(S_2^T+\tilde{S}_1^T) & V_{11}^T\tilde{A}+\tilde{B}_F\tilde{C}+\hat{P}_1 & \tilde{A}_F+\hat{P}_3^T & V_{11}^T\tilde{B}_u+\tilde{B}_F\begin{bmatrix}0\\ I\end{bmatrix} & 0 & V_{11}^T & \tilde{S}_1^T \\ * & -(S_1+S_1^T) & S_2\tilde{A}+\hat{B}_F\tilde{C}+\hat{P}_3 & \hat{A}_F+\hat{P}_2 & S_2\tilde{B}_u+\hat{B}_F\begin{bmatrix}0\\ I\end{bmatrix} & 0 & S_2 & S_1^T \\ * & * & -\mu\hat{P}_1 & -\mu\hat{P}_3^T & 0 & \tilde{L}_F^T & 0 & 0 \\ * & * & * & -\mu\hat{P}_2 & 0 & -\hat{L}_F^T & 0 & 0 \\ * & * & * & * & -\gamma_u I & -\begin{bmatrix}0 & I\end{bmatrix}H_F^T & 0 & 0 \\ * & * & * & * & * & -\gamma_u I & 0 & 0 \\ * & * & * & * & * & * & -\frac{1}{\mu}\hat{P}_1 & -\frac{1}{\mu}\hat{P}_3^T \\ * & * & * & * & * & * & * & -\frac{1}{\mu}\hat{P}_2 \end{bmatrix} < 0 \qquad (59)$$

where

$S_1 = \tilde{V}_{21}^T V_{22}^{-T} \tilde{V}_{21}$, $\tilde{S}_1 = \begin{bmatrix} S_1 & 0_{k\times(n_F-k)} \end{bmatrix}$, $S_2 = \tilde{V}_{21}^T V_{22}^{-T} V_{12}^T$, $\hat{B}_F = \tilde{V}_{21}^T B_F$, $\tilde{B}_F = \begin{bmatrix} \hat{B}_F \\ 0_{(n_F-k)\times m} \end{bmatrix}$, $\hat{P}_1 = P_1$,

$\hat{P}_2 = \tilde{V}_{21}^T V_{22}^{-T} P_2 V_{22}^{-1} \tilde{V}_{21}$, $\hat{P}_3^T = P_3^T V_{22}^{-1} \tilde{V}_{21}$, $\hat{A}_F = \tilde{V}_{21}^T A_F V_{22}^{-1} \tilde{V}_{21}$, $\tilde{A}_F = \begin{bmatrix} \hat{A}_F \\ 0_{(n_F-k)\times k} \end{bmatrix}$, $\hat{L}_F = L_F V_{22}^{-1} \tilde{V}_{21}$, and

$$\hat{P} = \begin{bmatrix} \hat{P}_1 & \hat{P}_3^T \\ \hat{P}_3 & \hat{P}_2 \end{bmatrix}.$$

Remark 4.2:
The filter matrices (A_F, B_F, L_F) can be derived by the following procedure:

(i) Compute $V_{22}, \tilde{V}_{21}$ by solving the factorization problem $S_1 = \tilde{V}_{21}^T V_{22}^{-T} \tilde{V}_{21}$ with the Schur decomposition,

(ii) Compute A_F, B_F, L_F : $A_F = \tilde{V}_{21}^{-T} \hat{A}_F \tilde{V}_{21}^{-1} V_{22}$ $B_F = \tilde{V}_{21}^{-T} \hat{B}_F$, $L_F = \hat{L}_F \tilde{V}_{21}^{-1} V_{22}$ ■

Now, we need to remember that we had a polytopic uncertain system. As explained earlier, the parameter dependent Lyapunov matrix $P(\alpha)$, $P(\alpha) = \sum_{i=1}^{s} \alpha_i P_i(\alpha), \alpha \in \Gamma$ is symmetric positive definite for all admissible values of α , if and only if, all the P_i's are positive definite. Therefore, we need to check (59) only at the vertices, $i = 1, 2, \cdots, s$. This gives us the final form (60)

$$\begin{bmatrix} -(V_{11}+V_{11}^T) & -(S_2^T+\tilde{S}_1^T) & V_{11}^T\tilde{A}_i+\tilde{B}_F\tilde{C}_i+\hat{P}_{1,i} & \tilde{A}_F+\hat{P}_{3,i}^T & V_{11}^T\tilde{B}_{d,i} & 0 & V_{11}^T & \tilde{S}_1^T \\ * & -(S_1+S_1^T) & S_2\tilde{A}_i+\hat{B}_F\tilde{C}_i+\hat{P}_{3,i} & \hat{A}_F+\hat{P}_{2,i} & S_2\tilde{B}_{d,i} & 0 & S_2 & S_1^T \\ * & * & -\mu\hat{P}_{1,i} & -\mu\hat{P}_{3,i}^T & 0 & \tilde{L}_F^T & 0 & 0 \\ * & * & * & -\mu\hat{P}_{2,i} & 0 & -\hat{L}_F^T & 0 & 0 \\ * & * & * & * & -\gamma_d I & -\begin{bmatrix} 0 & I \end{bmatrix} H_F^T & 0 & 0 \\ * & * & * & * & * & -\gamma_d I & 0 & 0 \\ * & * & * & * & * & * & -\frac{1}{\mu}\hat{P}_{1,i} & -\frac{1}{\mu}\hat{P}_{3,i}^T \\ * & * & * & * & * & * & * & -\frac{1}{\mu}\hat{P}_{2,i} \end{bmatrix} < 0$$

$$\begin{bmatrix} -(V_{11}+V_{11}^T) & -(S_2^T+\tilde{S}_1^T) & V_{11}^T\tilde{A}_i+\tilde{B}_F\tilde{C}_i+\hat{P}_{1,i} & \tilde{A}_F+\hat{P}_{3,i}^T & V_{11}^T\tilde{B}_{f,i} & 0 & V_{11}^T & \tilde{S}_1^T \\ * & -(S_1+S_1^T) & S_2\tilde{A}_i+\hat{B}_F\tilde{C}_i+\hat{P}_{3,i} & \hat{A}_F+\hat{P}_{2,i} & S_2\tilde{B}_{f,i} & 0 & S_2 & S_1^T \\ * & * & -\mu\hat{P}_{1,i} & -\mu\hat{P}_{3,i}^T & 0 & \tilde{L}_F^T & 0 & 0 \\ * & * & * & -\mu\hat{P}_{2,i} & 0 & -\hat{L}_F^T & 0 & 0 \\ * & * & * & * & -\gamma_f I & -\begin{bmatrix}0 & I\end{bmatrix}H_F^T & 0 & 0 \\ * & * & * & * & * & -\gamma_f I & 0 & 0 \\ * & * & * & * & * & * & -\frac{1}{\mu}\hat{P}_{1,i} & -\frac{1}{\mu}\hat{P}_{3,i}^T \\ * & * & * & * & * & * & * & -\frac{1}{\mu}\hat{P}_{2,i} \end{bmatrix} < 0$$

$$\begin{bmatrix} -(V_{11}+V_{11}^T) & -(S_2^T+\tilde{S}_1^T) & V_{11}^T\tilde{A}_i+\tilde{B}_F\tilde{C}_i+\hat{P}_{1,i} & \tilde{A}_F+\hat{P}_{3,i}^T & V_{11}^T\tilde{B}_{u,i}+\tilde{B}_F\begin{bmatrix}0\\I\end{bmatrix} & 0 & V_{11}^T & \tilde{S}_1^T \\ * & -(S_1+S_1^T) & S_2\tilde{A}_i+\hat{B}_F\tilde{C}_i+\hat{P}_{3,i} & \hat{A}_F+\hat{P}_{2,i} & S_2\tilde{B}_{u,i}+\hat{B}_F\begin{bmatrix}0\\I\end{bmatrix} & 0 & S_2 & S_1^T \\ * & * & -\mu\hat{P}_{1,i} & -\mu\hat{P}_{3,i}^T & 0 & \tilde{L}_F^T & 0 & 0 \\ * & * & * & -\mu\hat{P}_{2,i} & 0 & -\hat{L}_F^T & 0 & 0 \\ * & * & * & * & -\gamma_u I & -\begin{bmatrix}0 & I\end{bmatrix}H_F^T & 0 & 0 \\ * & * & * & * & * & -\gamma_u I & 0 & 0 \\ * & * & * & * & * & * & -\frac{1}{\mu}\hat{P}_{1,i} & -\frac{1}{\mu}\hat{P}_{3,i}^T \\ * & * & * & * & * & * & * & -\frac{1}{\mu}\hat{P}_{2,i} \end{bmatrix} < 0$$

$(i = 1, 2, \cdots, s)$

(60)

Theorem 4.3:
The reduced-order RFDF can be obtained by solving

$$\min_{V_{11}, S_1, S_2, \hat{A}_F, \hat{B}_F, \hat{L}_F, H_F, \gamma_d, \gamma_f, \gamma_u, \hat{P}_i} \{\alpha_d\gamma_d + \alpha_f\gamma_f + \alpha_u\gamma_u : (60)\} \tag{61}$$

where $\hat{P} > 0$ and α_d, α_f, and α_u are given as positive scalars. ∎

In summary, the reduced order RFDF filter in the multi-objective H_∞ formulation for polytopic uncertain system can be solved if and only if (60) holds for all vertices, $i = 1, 2, \cdots, s$. The minimum value can be found from (61). Also, the filter realization (A_F, B_F, L_F, H_F) defining the k_{th} order filter is obtained from (i) and (ii) in ***Remark 4.2*** and (61).

Another important task for fault detection is the evaluation of the generated residual. An adaptive threshold will be used in this work. The disadvantage of a fixed threshold is that if we fix the threshold too low, it can increase the rate of false alarms. Thus, the optimal choice of the magnitude of the threshold depends upon the nature of the system uncertainties and varies with the system input. That is called an adaptive threshold. The first step of residual evaluation is to choose an evaluation function and to determine the corresponding threshold. Among a number of residual evaluation functions, the so-called time windowed root mean square (RMS) is often used. The time windowed RMS is represented as

$$\|r(t)\|_{RMS,T} = \sqrt{\frac{1}{T}\int_{t-T}^{t} r^T(\tau)r(\tau)d\tau} \tag{62}$$

where T is the length of the finite time window. Since an evaluation of the residual signal over the whole time range is impractical, the time windowed RMS evaluation method is used in practice to detect faults as early as possible.
After selecting the evaluation function, we are able to determine the threshold. A major requirement on the fault detection is to reduce or prevent false alarms. Thus, in the absence of any faults, $\|r(t)\|_{RMS,T}$ should be less than the threshold value J_{th}, i.e.,

$$J_{th} = \sup_{\alpha\in\Gamma, f=0} \|r(t)\|_{RMS,T} \tag{63}$$

Under fault-free conditions, (63) can be described as

$$r(s) = T_{ru}^0(s)u(s) + T_{rd}^0(s)d(s) \tag{64}$$

From Parseval's Theorem (Zhou et al., 1996) and the RMS norm relationship (Boyd & Barratt, 1991), the threshold J_{th} can be found as follows

$$\begin{aligned} J_{th} &\equiv \|r\|_{RMS,T,f=0} \\ &= \|T_{ru}^0 u + T_{rd}^0 d\|_{RMS,T} \\ &\le \|T_{ru}^0\|_\infty \|u\|_{RMS,T} + \|T_{rd}^0\|_\infty \|d\|_{RMS,T} \\ &= \gamma_u \|u\|_{RMS,T} + \gamma_d \|d\|_{RMS,T} \end{aligned} \tag{65}$$

where γ_u and γ_d come from the optimization problem in (61) and $\|d\|_{RMS,T}$ is calculated or bounded by the worst disturbance acting on the plant. Thus in (65), $\|d\|_{RMS,T}$ is evaluated off-line, while u(t) is assumed to be known and $\|u\|_{RMS,T}$ is calculated on-line.

5. Example

5.1 H_∞ Filter design

In this example, we handle two cases, the full order and the reduced order filter design, and demonstrate the advantages of this study. We use the example from (Tuan et al., 2000) with the plant data

$$\dot{x} = \begin{bmatrix} 0 & -1+0.3\alpha \\ 1 & -0.5 \end{bmatrix} x + \begin{bmatrix} -2 & 0 \\ 1 & 0 \end{bmatrix} w$$
$$y = \begin{bmatrix} -100+10\beta & 100 \end{bmatrix} x + \begin{bmatrix} 0 & 1 \end{bmatrix} w \tag{66}$$
$$z = \begin{bmatrix} 1 & 0 \end{bmatrix} x$$

where the uncertainty parameters, α and β, have four types of uncertainty sets given by

$$|\alpha| \le 1, |\beta| \le 1 \tag{67}$$
$$|\alpha| \le 1, \beta = \alpha \tag{68}$$
$$|\alpha| \le 3, |\beta| \le 1 \tag{69}$$
$$|\alpha| \le 3, \beta = \alpha \tag{70}$$

Table 1 compares results between the filters in (Tuan, 2000, Geromel, 1999, De Souza, 1999) and (45) for the four types of uncertainties in (67)-(70). For the conservative case, a single Lyapunov matrix P is used and for the nonconservative case, four different Lyapunov matrices P are used for each vertex. The calculations were done using (Gahinet). An arbitrary positive scalar for μ should be chosen and the feasibility of the formulation should be checked. If it is feasible, the minimum value of (45) can be found.

Type	Filter order	(Geromel, 1999): $\|T_{ed}\|_2$	(De Souza, 1999): $\|T_{ed}\|_2$	(Tuan, 2000): $\|T_{ed}\|_2$ Non Conservative	(Tuan, 2000): $\|T_{ed}\|_2$ Conservative	(45): $\|T_{ed}\|_\infty$ Non Conservative	(45): $\|T_{ed}\|_\infty$ Conservative
$\lvert\alpha\rvert \le 1, \lvert\beta\rvert \le 1$	full	5.728	4.867	2.382	5.7495	2.965	3.346
	reduced		4.946	3.001	5.8467	3.179	4.304
$\lvert\alpha\rvert \le 1, \beta = \alpha$	full	4.819	4.373	2.382	4.8841	2.948	3.165
	reduced		4.556	3.079	5.0704	3.163	4.023
$\lvert\alpha\rvert \le 3, \lvert\beta\rvert \le 1$	full	$+\infty$	$+\infty$	93.365	$+\infty$	29.106	$+\infty$
	reduced			106.493	$+\infty$	29.679	$+\infty$
$\lvert\alpha\rvert \le 3, \beta = \alpha$	full	$+\infty$	$+\infty$	100.963	$+\infty$	28.808	$+\infty$
	reduced			106.517	$+\infty$	29.732	$+\infty$

Table 1. Performance comparison according to each uncertainty type

The feasibility must be checked because the product term of μ and Lyapunov variable P is nonlinear. We need to tune μ until the LMI solver returns a feasible solution. For any choice of $\mu > 0$, if solvable, it gives a unique solution because of the convexity. For case (67), we found that when $\mu = 7$ (45) results in $\gamma = 2.965$ for the full order case, and when $\mu = 1.451$ (45) results in $\gamma = 3.179$ for the reduced order case. From Table 1, we can see that the optimization in (45) using parameter-dependent Lyapunov functions does not fail in the uncertainty cases of (69) and (70), but the conservative approaches in (De Souza, 1999, Gahinet) failed for the same cases. In the uncertainty cases of (69) and (70) cases, we also found that the conservative application of (Tuan et al., 2000) and our approach in (45), i.e., the usage of a single parameter-independent Lyapunov function, also failed.
The filter data in the full and reduced order cases in (67) were found from (40)-(42). The Schur Decomposition method was used to solve the factorization problem in (40). The result is shown in Table 2 for the uncertainty case in (67).

Type	Filter order	A_F	B_F	L_F
$\lvert\alpha\rvert \le 1, \lvert\beta\rvert \le 1$	full	$\begin{bmatrix} -23.5163 & 0.6242 \\ 6.3142 & -3.0403 \end{bmatrix}$	$\begin{bmatrix} 0.0006 \\ -0.0031 \end{bmatrix}$	$\begin{bmatrix} 588.5549 & 6.1988 \end{bmatrix}$
	reduced	-0.8396	0.0022	4.0742

Table 2. Filter synthesis

As we already know, the H_∞ approach in (45) is more robust to uncertainties than the H_2 approach in (Tuan et al., 2000). To illustrate this, we found the filter data from (Tuan et al., 2000) and (45) and used (18) to calculate $\|T_{ed}\|_\infty$ for the uncertainty case in (67). Table 3 shows the result. Therefore, we find that for this example our approach is more robust to model uncertainty than former approaches, (Tuan et al., 2000, Geromel, 1999, Gahinet) and gives a non-conservative result.

Type	Filter order	$\|(Tuan\ et\ al., 2000)\|_\infty$	(45)
$\lvert\alpha\rvert \le 3, \lvert\beta\rvert \le 1$	Full	11.1939	9.8752
	Reduced	11.3486	10.7081

Table 3. Performance comparison between (Tuan et al., 2000) and (45)

5.2 Fault detection filter design

In this section, a numerical example is given. Consider the uncertain LTI plant that is borrowed from (Nobrega et al., 2000), but is modified to include uncertainties. The plant is

$$\dot{x} = \begin{bmatrix} 0 & -1+0.3\alpha \\ 1 & -0.5 \end{bmatrix} x + \begin{bmatrix} 1 \\ 1 \end{bmatrix} u + \begin{bmatrix} 0 \\ 1 \end{bmatrix} f + \begin{bmatrix} 0.1 \\ 0.2 \end{bmatrix} d$$
$$y = \begin{bmatrix} -100+10\beta & 100 \end{bmatrix} x + 0.1u + f + 0.05d \tag{71}$$

where the uncertainty parameters, α and β, have the uncertainty set given by $|\alpha| \le 1, |\beta| \le 1$. The selected reference model is $W(s) = \frac{2}{s+2}$. The fault signal f is simulated as a pulse of unit amplitude that occurs from 20 to 25 seconds and is zero elsewhere. The input u is taken as $1-e^{-0.01t}$. The unknown input d is assumed to be band-limited white noise with power 0.0005. The upper bound of $\|d\|_{RMS,T}$ is given as 0.15. This value is used to calculate J_{th} of (70).

With this information, we can get the RFDF data from (66). For the full order case, we checked the detection time with weights set at ($\alpha_d = \alpha_f = \alpha_u = 1$) and ($\alpha_d = \alpha_f = 1, \alpha_u = 100$). The detection time is defined as the time span where the J -residual (time-windowed residual RMS value) exceeds the threshold, J_{th}. The full order FDF is 4(= n_F). As the order k of the RFDF is reduced from 4 to 1, the effectiveness of the filter is compared by monitoring the detection time. The results are shown in Figures 3 and 4 and Table 4. Table 4 shows a comparison of the results. In this example, we can see that the fault detection time does not change much as the order of the RFDF is reduced.

RFDF Order (k)		Detection time
$k = n_F = 4$	$\alpha_d = \alpha_f = \alpha_u = 1$	20.705~34.96
(full)	$\alpha_d = \alpha_f = 1, \alpha_u = 100$	22~34.7
$k = 3$, ($\alpha_d = \alpha_f = \alpha_u = 1$)		22.99~31.36
$k = 2$, ($\alpha_d = \alpha_f = \alpha_u = 1$)		23.115~31.45
$k = 1$, ($\alpha_d = \alpha_f = \alpha_u = 1$)		23.58~31.157

Table 4. Comparison of the detection time as the order of RFDF is reduced

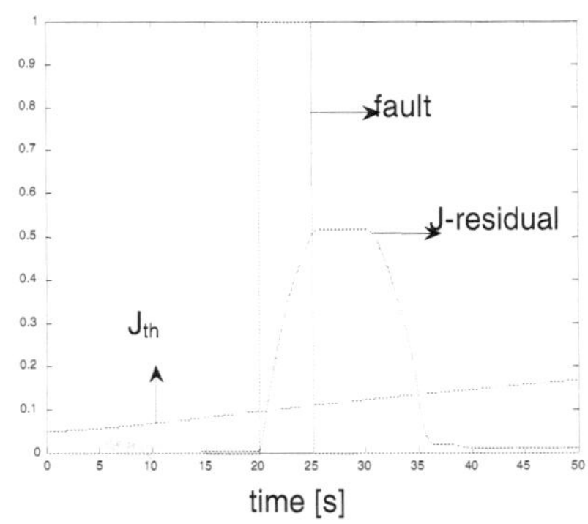

Fig. 3. Full order signals ($\alpha_d = \alpha_f = \alpha_u = 1$)

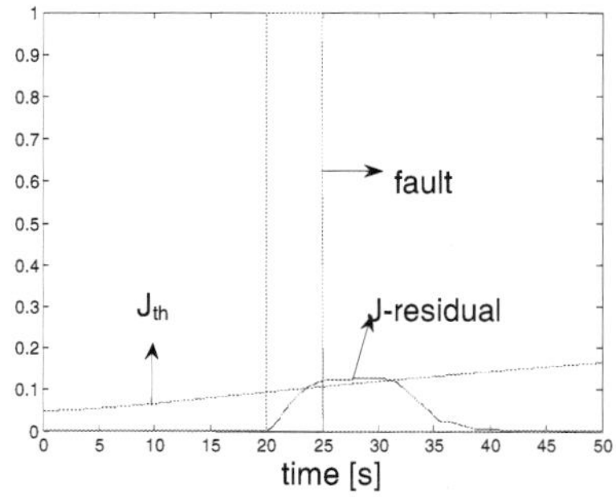

Fig. 4. Reduced order ($k = 1$) signals ($\alpha_d = \alpha_f = \alpha_u = 1$)

6. Conclusion

In this paper, we developed in detail a practical approach for solving the H_∞ reduced filter synthesis problem in an LMI framework and its application for fault detection filter design. The LMI characteristics of the filtering problem were described and formulated with the BRL. To overcome conservatism, parameter-dependent Lyapunov functions were used. A congruence transformation and a change of variable technique were used to linearize the problem. This formulation had formerly been applied to the H_2 problem. As is well-known, the H_∞ approach provides better robustness than the H_2 approach.

The order reduction problem for a robust fault detection filter was considered, and the result was illustrated with a simple example. A reference model was introduced and formulated as a multi-objective optimization for a polytopic uncertain system. The order of the RFDF is reduced using LMI techniques and the detection performance is compared with the full order filter. Using the LMI solver from Mathworks Robust Control Toolbox, the minimum value is found from (61). For the example considered, the detection time did not change much as the filter order was reduced. Thus, our reduced-order RFDF design shows promising results.

7. References

Bettayeb, M. & Kavranoglu, D. (1994). Reduced order H_∞ filtering, *Proceedings of American Control Conference*, pp. 1884-1888

Boyd, S. & Barratt, C. (1991). *Linear Controller Design, Limits of Performance*, Prentice Hall, pp. 98-99

Casavola, A.; Famularo, D. & Franze, G. (2003). Robust fault isolation of uncertain linear systems: An LMI approach, *Proceedings of the IFAC Fault Detection Supervision and Safety of Technical Processes*

Casavola, A.; Famularo, D. & Franze, G. (2005a). A robust deconvolution scheme for fault detection and isolation of uncertain linear systems: an LMI approach, *Automatica*, 41, pp. 1463-1472

Casavola, A.; Famularo, D.; & Franze, G. (2005b). Robust fault detection of uncertain linear systems via quasi-LMIs, *Proceedings of American Control Conference*, pp. 1654-1659

Chen, J. & Zhang, H. Y. (1991). Robust detection of faulty actuators via unknown input observers, *International Journal of Systems Science*, Vol. 22, No. 10, pp. 1829-1839

Chen, J. & Patton, R. J. (1997). Observer-based fault detection and isolation: robustness and applications, *Control Engineering Practice*, Vol. 5, No. 5, pp. 671-682

Chen, J. & Patton, R. J. (1999). *Robust Model-Based Fault Diagnosis for Dynamic Systems*, Kluwer Academic Publishers

De Souza, C. E. & Trofino, A. (1999). An LMI approach to the design of robust H_2 filters, *Recent Advances in Linear Matrix Inequality Methods in Control*, SIAM, Philadelphia, PA

Ding, S. X.; Frank, P. M.; Ding, E. L. & Jeinsch, T. (2000). Fault detection system design based on a new trade-off strategy, *Proceedings of 39th IEEE Conference on Decision and Control*, pp. 4144-4149

Ding, S. X. & Frank, P. M. (2003). Threshold calculation using LMI- technique and its integration in the design of fault detection systems, *Proceedings of 42nd IEEE Conference on Decision and Control*, pp. 469-474

Frisk, E. (2001). *Residual Generation for Fault Diagnosis*, Ph.D. dissertation, Dept. of Electrical Engineering, Linkoping University, Sweden

Fu, M.; De Souza, C. E. & Xie, L. (1994). H_∞ estimation for uncertain systems, *International Journal of Robust and Nonlinear Control*, Vol. 4, pp. 421-448

Gahinet, P. & Apkarian, P. (1994). A linear matrix inequality approach to H_∞ control, *International Journal of Robust and Nonlinear Control*, Vol. 4, pp. 421-448

Gahinet, P.; Apkarian, P. & Chilali, M. (1996). Parameter dependent Lyapunov functions for real parametric uncertainty, *IEEE Transactions on Automatic Control*, Vol. 41, pp. 436-442

Gahinet, P.; Nemirovski, A.; Laub, A. J. & Chilali, M. *The LMI Control Toolbox Tutorial, MATLAB v7.0*, Mathworks Inc.

Geromel, J. C. (1999). Optimal linear filtering under parameter uncertainty, *IEEE Transactions on Signal Processing*, Vol. 47, pp. 168-175

Grigoriadis, K. M. & Watson, J. T. (1997). Reduced order H_∞ and $L_2 - L_\infty$ filtering via linear matrix Inequalities, *IEEE Transactions on Aerospace and Electronic Systems*, Vol. 33, pp. 1326-1338

Kim, Y-M. & Watkins, J. M. (2006). A New Approach for Robust and Reduced Order H_∞ Filtering, *Proceedings of American Control Conference*, pp. 5171-5176, Minneapolis, MN

Kim, Y-M. & Watkins, J. M. (2007). A new approach for robust and reduced order fault detection filter design, *Proceedings of American Control Conference*, New York, NY

Nagpal, K. M.; Helmick, R. E. & Sims, C. S. (1987). Reduced order estimation, I. Filtering, *International Journal of Control*, Vol. 45, pp. 1867-1888

Niemann, H. & Stoustrup, J. (2000). Design of fault detectors using H_∞ optimization, *Proceedings of the 39th IEEE Conference on Decision and Control*, pp. 4327-4328

Nobrega, E. G. & Abdalla, M. O. (2000). Grigoriadis, K. M.; LMI-based filter design for fault detection and isolation, *Proceedings of 39th IEEE Conference on Decision and Control*, pp. 4329-4334

O'Brien Jr., R. T. & Kiriakidis, K. (2004). Reduced-order H_∞ filtering for discrete time linear time varying systems, *Proceedings of American Control Conference*, pp. 4120-4125
Patton, R. J. & Hou, M. (1999). On sensitivity of robust fault detection observers, *14th IFAC World Congress*, pp. 67-72
Tuan, H. D.; Apkarian, P. & Nguyen, T. Q. (2000). Robust and reduced order filtering: New characterizations and methods, *Proceedings of American Control Conference*, pp. 1327-1331
Zhong, M.; Ding, S. X.; Lam, J. & Wang, H. (2003). An LMI approach to design robust fault detection filter for uncertain LTI systems, *Automatica*, 39, pp. 543-550
Zhou, K.; Doyle, J. C. & Glover, K. (1996). *Robust and Optimal Control*, Prentice Hall

Mechanical fault detection in induction motor drives through stator current monitoring - Theory and application examples

Martin Blödt
Siemens AG
Germany

Pierre Granjon and Bertrand Raison
Grenoble-INP
France

Jérémi Regnier
INP Toulouse
France

1. Introduction

1.1 General Introduction

In a wide variety of industrial applications, an increasing demand exists to improve the reliability and availability of electrical systems. Popular examples include systems in aircraft, electric railway traction, power plant cooling or industrial production lines. A sudden failure of a system in these examples may lead to cost expensive downtime, damage to surrounding equipment or even danger to humans. Monitoring and failure detection improve the reliability and availability of an existing system. Since various failures degrade relatively slowly, there is a potential for fault detection followed by corrective maintenance at an early stage. This avoids the sudden, total system failure which can have serious consequences.
Electric machines are a key element in many electrical systems. Amongst all types of electric motors, induction motors are a frequent example due to their simplicity of construction, robustness and high efficiency. Common failures occurring in electrical drives can be roughly classified into:

Electrical faults: stator winding short circuit, broken rotor bar, broken end-ring, inverter failure

Mechanical faults: rotor eccentricity, bearing faults, shaft misalignment, load faults (unbalance, gearbox fault or general failure in the load part of the drive)

A reliability survey on large electric motors (>200 HP) revealed that most failures are due to bearing (≈ 44%) and winding faults (≈ 26%) (IEEE motor reliability working group (1985a))

(Engelmann & Middendorf (1995)). Similar results were obtained in an EPRI (Electric Power Research Institute) sponsored survey (Albrecht et al. (1987)). These studies concerned only the electric motor and not the whole drive including the load, but they show that mechanical fault detection is of great concern in electric drives.
A growing number of induction motors operates in variable speed drives. In this case, the motor is not directly connected to the power grid but supplied by an inverter. The inverter provides voltage of variable amplitude and frequency in order to vary the mechanical speed. Hence, this work addresses the problem of condition monitoring of mechanical faults in variable speed induction motor drives. A signal based approach is chosen i.e. the fault detection and diagnosis are only based on processing and analysis of measured signals and not on real-time models.

1.2 Motor Current Signature Analysis

A common approach for monitoring mechanical failures is vibration monitoring. Due to the nature of mechanical faults, their effect is most straightforward on the vibrations of the affected component. Since vibrations lead to acoustic noise, noise monitoring is also a possible approach. However, these methods are expensive since they require costly additional transducers. Their use only makes sense in case of large machines or highly critical applications. A cost effective alternative is stator current based monitoring since a current measurement is easy to implement. Moreover, current measurements are already available in many drives for control or protection purposes. However, the effects of mechanical failures on the motor stator current are complex to analyze. Therefore, stator current based monitoring is undoubtedly more difficult than vibration monitoring.
Another advantage of current based monitoring over vibration analysis is the limited number of necessary sensors. An electrical drive can be a complex and extended mechanical system. For complete monitoring, a large number of vibration transducers must be placed on the different system components that are likely to fail e.g. bearings, gearboxes, stator frame, load. However, a severe mechanical problem in any component influences necessarily the electric machine through load torque and shaft speed. This signifies that the motor can be considered as a type of intermediate transducer where various fault effects converge together. This strongly limits the number of necessary sensors. However, since numerous fault effects come together, fault diagnosis and discrimination become more difficult or sometimes even impossible.
A literature survey showed a lack of analytical models that account for the mechanical fault effect on the stator current. Most authors simply give expressions of additional frequencies but no precise stator current signal model. In various works, numerical machine models accounting for the fault are used. However, they do not provide analytical stator current expressions which are important for the choice of suitable signal analysis and detection strategies.
The most widely used method for stator current processing in this context is spectrum estimation. In general, the stator current power spectral density is estimated using Fourier transform based techniques such as the periodogram. These methods require stationary signals i.e. they are inappropriate when frequencies vary with respect to time such as during speed transients. Advanced methods for non-stationary signal analysis are required.
The organization of the present work is the following. Section 2 analyses the effects of load torque oscillations and dynamic eccentricity on the stator current. In section 3, suitable signal processing methods for stator current analysis are introduced. Experimental results under laboratory conditions are presented in section 4. Section 5 examines the detection of misalign-

ment faults in electric winches including analysis of experimental data from a real winch. Bearing faults are investigated apart in section 6 from a theoretical and practical point of view since they can introduce particular eccentricities and load torque oscillations.

2. Theoretical study of mechanical fault effects on stator current

The key assumption for the development of the theoretical models is that mechanical faults mainly produce two effects on induction machines: additional load torque oscillations at characteristic frequencies and/or airgap eccentricity.
Load torque oscillations can be caused by the following faults:

- load unbalance (not necessarily a fault but can also be inherent to the load type)
- shaft misalignment
- gearbox fault e.g. broken tooth
- bearing faults

Airgap eccentricity i.e. a non-uniform airgap can be for example the consequence of bearing wear or bearing failure, bad motor assembly with rotor unbalance or a rotor which is not perfectly centered. In general, eccentricity will be a sign for a mechanical problem within the electric motor whereas load torque oscillations point to a fault that is located outside of the motor.
The method used to study the influence of the periodic load torque variation and the rotor eccentricity on the stator current is the magnetomotive force (MMF) and permeance wave approach (Yang (1981)) (Timár (1989)) (Heller & Hamata (1977)). This approach is traditionally used for the calculation of the magnetic airgap field with respect to rotor and stator slotting or static and dynamic eccentricity (Cameron & Thomson (1986)) (Dorrell et al. (1997)).
First, the rotor and stator MMF are calculated which are directly related to the current flowing in the windings. The second important quantity is the airgap permeance Λ which is directly proportional to the inverse of the airgap length g. The magnetic field in the airgap can then be determined by multiplying the permeance by the sum of rotor and stator MMFs. The equivalent magnetic flux in one phase is obtained by integration of the magnetic field in each turn of the phase winding. The induced phase voltage, related to the current by the stator voltage equation, is then deduced from the magnetic flux.
As this work also considers variable speed drives, the supply frequency f_s and the characteristic fault frequency f_c may vary. Note that f_c can be for example the time-varying rotational frequency f_r. The theoretical stator current analysis during transients, however, is identical to the steady state if relatively slow frequency variations of f_s and f_c are considered.

2.1 Load torque oscillations

The influence of load torque oscillations on the stator current has been published for a general case by the authors in (Blödt, Chabert, Regnier & Faucher (2006)) (Blödt (2006)). The development will be shortly resumed in the following.

2.1.1 Effect on Rotor and Stator MMF

Under a mechanical fault, the load torque as a function of time is modeled by a constant component Γ_{const} and an additional component varying at the characteristic frequency f_c, depending on the fault type. It can be for example the rotational frequency f_r with load unbalance or a particular gearbox frequency in case of a gearbox fault. The first term of the variable component Fourier series is a cosine with frequency f_c. For the sake of clarity and since they are usually of smaller amplitude, higher order terms at kf_c are neglected in the following and only the fundamental term is considered. The load torque can therefore be described by:

$$\Gamma_{load}(t) = \Gamma_{const} + \Gamma_c \cos(\omega_c t) \tag{1}$$

where Γ_c is the amplitude of the load torque oscillation and $\omega_c = 2\pi f_c$.
The machine mechanical equation relates the torque oscillation to the motor speed ω_r and to the mechanical rotor position θ_r as follows:

$$\sum \Gamma(t) = \Gamma_{motor} - \Gamma_{load}(t) = J\frac{d\omega_r}{dt} = J\frac{d^2\theta_r}{dt^2} \tag{2}$$

where Γ_{motor} is the constant electromagnetic torque produced by the machine, J is the total inertia of the machine and the load.
After integrating twice, $\theta_r(t)$ is obtained as:

$$\theta_r(t) = \int_{t_0}^{t} \omega_r(\tau)\, d\tau = \frac{\Gamma_c}{J\omega_c^2}\cos(\omega_c t) + \omega_{r0} t \tag{3}$$

where ω_{r0} is the constant part of the motor speed. This equation shows that in contrast to the healthy machine where $\theta_r(t) = \omega_{r0} t$, oscillations at the characteristic frequencies are present on the mechanical rotor position.
The oscillations of the mechanical rotor position θ_r act on the rotor MMF. In a healthy state without faults, the fundamental rotor MMF in the rotor reference frame (R) is a wave with p pole pairs and frequency sf_s, given by:

$$F_r^{(R)}(\theta', t) = F_r \cos(p\theta' - s\omega_s t) \tag{4}$$

where θ' is the mechanical angle in the rotor reference frame (R) and s is the motor slip. Higher order space and time harmonics are neglected.

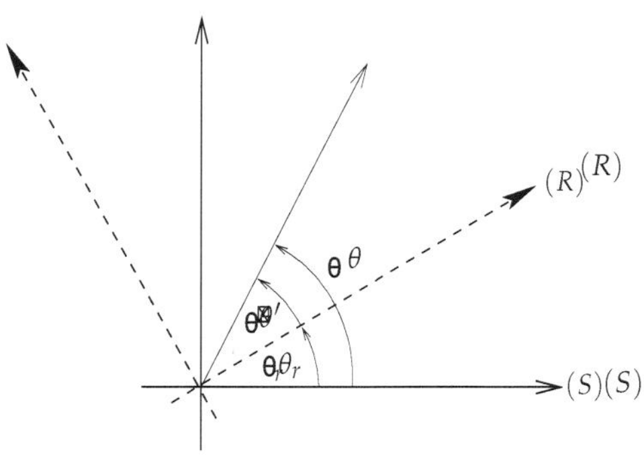

Fig. 1. Stator (S) and rotor (R) reference frame

Figure 1 illustrates the transformation between the rotor and stator reference frame, defined by $\theta = \theta' + \theta_r$. Using (3), this leads to:

$$\theta' = \theta - \omega_{r0} t - \frac{\Gamma_c}{J\omega_c^2}\cos(\omega_c t) \tag{5}$$

Thus, the rotor MMF given in (4) can be transformed to the stationary stator reference frame using (5) and the relation $\omega_{r0} = \omega_s(1-s)/p$:

$$F_r(\theta,t) = F_r \cos\left(p\theta - \omega_s t - \beta\cos\left(\omega_c t\right)\right) \tag{6}$$

with:

$$\beta = p\frac{\Gamma_c}{J\omega_c^2} \tag{7}$$

Equation (6) clearly shows that the load torque oscillations at frequency f_c lead to a phase modulation of the rotor MMF in the stator reference frame. This phase modulation is characterized by the introduction of the term $\beta\cos(\omega_c t)$ in the phase of the MMF wave. The parameter β is generally called the modulation index. For physically reasonable values J, Γ_c and ω_c, the approximation $\beta \ll 1$ holds in most cases.
The fault has no direct effect on the stator MMF and so it is considered to have the following form:

$$F_s(\theta,t) = F_s \cos\left(p\theta - \omega_s t - \varphi_s\right) \tag{8}$$

φ_s is the initial phase difference between rotor and stator MMF. As in the case of the rotor MMF, only the fundamental space and time harmonic is taken into account; higher order space and time harmonics are neglected.

2.1.2 Effect on Flux Density and Stator Current

The airgap flux density $B(\theta,t)$ is the product of total MMF and airgap permeance Λ. The airgap permeance is supposed to be constant because slotting effects and eccentricity are not taken into account for the sake of clarity and simplicity.

$$\begin{aligned} B(\theta,t) &= \left[F_s(\theta,t) + F_r(\theta,t)\right]\Lambda \\ &= B_s \cos\left(p\theta - \omega_s t - \varphi_s\right) \\ &\quad + B_r \cos\left(p\theta - \omega_s t - \beta\cos\left(\omega_c t\right)\right) \end{aligned} \tag{9}$$

The phase modulation of the flux density $B(\theta,t)$ exists for the flux $\Phi(t)$ itself, as $\Phi(t)$ is obtained by simple integration of $B(\theta,t)$ with respect to the winding structure. The winding structure has only an influence on the amplitudes of the flux harmonic components, not on their frequencies. Therefore, $\Phi(t)$ in an arbitrary phase can be expressed in a general form:

$$\Phi(t) = \Phi_s \cos\left(\omega_s t + \varphi_s\right) + \Phi_r \cos\left(\omega_s t + \beta\cos\left(\omega_c t\right)\right) \tag{10}$$

The relation between the flux and the stator current in a considered phase is given by the stator voltage equation:

$$V(t) = R_s I(t) + \frac{d\Phi(t)}{dt} \tag{11}$$

With $V(t)$ imposed by the voltage source, the resulting stator current will be in a linear relation to the time derivative of the phase flux $\Phi(t)$ and will have an equivalent frequency content. Differentiating (10) leads to:

$$\begin{aligned} \frac{d}{dt}\Phi(t) = &-\omega_s\Phi_s \sin\left(\omega_s t + \varphi_s\right) \\ &-\omega_s\Phi_r \sin\left(\omega_s t + \beta\cos\left(\omega_c t\right)\right) \\ &+\omega_c\beta\Phi_r \sin\left(\omega_s t + \beta\cos\left(\omega_c t\right)\right)\sin(\omega_c t) \end{aligned} \tag{12}$$

The amplitude of the last term is smaller than that of the other terms because $\beta \ll 1$. Thus, the last term in (12) will be neglected in the following.
As a consequence, the stator current in an arbitrary phase can be expressed in a general form:

$$\begin{aligned} I_{to}(t) &= i_{st}(t) + i_{rt}(t) \\ &= I_{st}\sin(\omega_s t + \varphi_s) + I_{rt}\sin(\omega_s t + \beta\cos(\omega_c t)) \end{aligned} \tag{13}$$

Therefore the stator current $I(t)$ can be considered as the sum of two components. The term $i_{st}(t)$ results from the stator MMF and it is not modulated. The term $i_{rt}(t)$, which is a direct consequence of the rotor MMF shows the phase modulation due to the considered load torque oscillations. The healthy case is obtained for $\beta = 0$.
In this study, the time harmonics of rotor MMF and the non-uniform airgap permeance have not been considered. However, the harmonics of supply frequency f_s and the rotor slot harmonics will theoretically show the same phase modulation as the fundamental component.

2.2 Airgap Eccentricity

Airgap eccentricity leads to an airgap length that is no longer constant with respect to the stator circumference angle θ and/or time. In general, three types of airgap eccentricity can be distinguished (see Fig. 2):

Static eccentricity: The rotor geometrical and rotational centers are identical, but different from the stator center. The point of minimal airgap length is stationary with respect to the stator.

Dynamic eccentricity: The rotor geometrical center differs from the rotational center. The rotational center is identical with the stator geometrical center. The point of minimal airgap length is moving with respect to the stator.

Mixed eccentricity: The two effects are combined. The rotor geometrical and rotational center as well as the stator geometrical center are different.

In the following theoretical development, static and dynamic eccentricity will be considered.

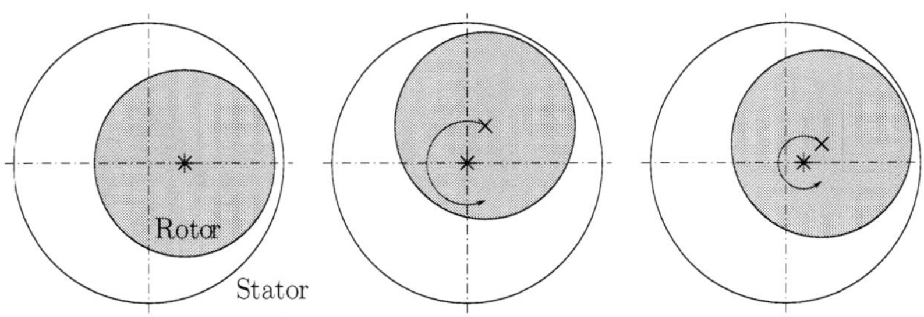

(a) Static eccentricity (b) Dynamic eccentricity (c) Mixed eccentricity

Fig. 2. Schematic representation of static, dynamic and mixed eccentricity. $\times$ denotes the rotor geometrical center, $*$ the rotor rotational center

The airgap length $g(\theta,t)$ can be approximated for a small airgap and low levels of static or dynamic eccentricity by the following expression (Dorrell et al. (1997)):

$$\begin{aligned} g_{se}(\theta,t) &\approx g_0(1 - \delta_s\cos(\theta)) \\ g_{de}(\theta,t) &\approx g_0(1 - \delta_d\cos(\theta - \omega_r t)) \end{aligned} \tag{14}$$

where δ_s, δ_d denote the relative degrees of static or dynamic eccentricity and g_0 the mean airgap length without eccentricity. Note that static eccentricity can be considered as a special case of dynamic eccentricity since $g_{se}(\theta,t)$ corresponds to $g_{de}(\theta,t)$ with $\omega_r = 0$, i.e. the point of minimum airgap length is stationary. Since dynamic eccentricity is more general, it will mainly be considered in the following.
The airgap permeance $\Lambda(\theta,t)$ is obtained as the inverse of $g(\theta,t)$ multiplied by the permeability of free space μ_0. Following a classical approach, the permeance is written as a Fourier series (Cameron & Thomson (1986)):

$$\Lambda_{de}(\theta,t) = \Lambda_0 + \sum_{i_{ecc}=1}^{\infty} \Lambda_{i_{ecc}} \cos(i_{ecc}\theta - i_{ecc}\omega_r t) \tag{15}$$

where $\Lambda_0 = \mu_0/g_0$ is the permeance without eccentricity. The higher order coefficients of the Fourier series can be written as (Cameron & Thomson (1986)):

$$\Lambda_{i_{ecc}} = \frac{2\mu_0(1-\sqrt{1-\delta^2})^{i_{ecc}}}{g_0\delta_d^{i_{ecc}}\sqrt{1-\delta^2}} \tag{16}$$

Dorrell has shown in (Dorrell (1996)) that the coefficients with $i_{ecc} \geq 2$ are rather small for $\delta_d < 40\%$. For the sake of simplicity, they are neglected in the following considerations.
The airgap flux density is the product of permeance with the magnetomotive force (MMF). The total fundamental MMF wave can be written as:

$$F_{tot}(\theta,t) = F_1 \cos(p\theta - \omega_s t - \varphi_t) \tag{17}$$

with φ_t the initial phase. Hence, the flux density in presence of dynamic eccentricity is:

$$B_{de}(\theta,t) \approx B_1\left[1 + 2\frac{\Lambda_1}{\Lambda_0}\cos(\theta - \omega_r t)\right]\cos(p\theta - \omega_s t - \varphi_t) \tag{18}$$

with $B_1 = \Lambda_0 F_1$
The fraction $2\Lambda_1/\Lambda_0$ equals approximately δ_d for small levels of eccentricity. The airgap flux density can therefore be written as:

$$B_{de}(\theta,t) = B_1\left[1 + \delta_d\cos(\theta - \omega_r t)\right]\cos(p\theta - \omega_s t - \varphi_t) \tag{19}$$

This equation shows the fundamental effect of dynamic eccentricity on the airgap magnetic flux density : the modified airgap permeance causes an amplitude modulation of the fundamental flux density wave with respect to time and space. The AM modulation index is approximately the degree of dynamic eccentricity δ.
In case of static eccentricity, the fundamental flux density expresses as:

$$B_{se}(\theta,t) = B_1\left[1 + \delta_s\cos(\theta)\right]\cos(p\theta - \omega_s t - \varphi_t) \tag{20}$$

which shows that static eccentricity leads only to flux density AM with respect to space.
Consequently, the amplitude modulation can also be found on the stator current $I(t)$ (see section 2.1.2) that is expressed as follows in case of dynamic eccentricity:

$$I_{de}(t) = I_1\left[1 + \alpha\cos(\omega_r t)\right]\cos(\omega_s t - \varphi_i) \tag{21}$$

In this expression, I_1 denotes the amplitude of the stator current fundamental component, α the AM modulation index which is proportional to the degree of dynamic eccentricity δ_d. Static eccentricity does not lead to frequencies different from ω_s since the corresponding additional flux density waves are also at the supply pulsation ω_s. It can be concluded that theoretically, pure static eccentricity cannot be detected by stator current analysis.

3. Signal processing tools for fault detection and diagnosis

The previous section has shown that load torque oscillations cause a phase modulation on one stator current component according to (13). On the other hand, dynamic airgap eccentricity leads to amplitude modulation of the stator current (see (21)). In this section, signal processing methods for detection of both modulation types in the stator current will be presented and discussed.

In order to simplify calculations, all signals will be considered in their complex form, the so-called analytical signal (Boashash (2003)) (Flandrin (1999)). The analytical signal $z(t)$ is related to the real signal $x(t)$ via the Hilbert Transform $H\{.\}$:

$$z(t) = x(t) + jH\{x(t)\} \tag{22}$$

The analytical signal contains the same information as the real signal but its Fourier transform is zero at negative frequencies.

3.1 Power Spectral Density

3.1.1 Definition

The classical method for signal analysis in the frequency domain is the estimation of the Power Spectral Density (PSD) based on the discrete Fourier transform of the signal $x[n]$. The PSD indicates the distribution of signal energy with respect to the frequency. The common estimation method for the PSD is the periodogram $P_{xx}(f)$ (Kay (1988)), defined as the square of the N-point Fourier transform divided by N:

$$P_{xx}(f) = \frac{1}{N}\left|\sum_{n=0}^{N-1} x(n)e^{-j2\pi fn}\right|^2 \tag{23}$$

3.1.2 Application

The PSD represents the basic signal analysis tool for stationary signals i.e. it can be used in case of a constant or quasi-constant supply frequency during the observation interval.

The absolute value of the Fourier transform $|I(f)|$ of the stator current PM signal (13) is obtained as follows (see Blödt, Chabert, Regnier & Faucher (2006) for details):

$$\begin{aligned} |I_{to}(f)| &= (I_{st} + I_{rt}J_0(\beta))\,\delta(f - f_s) \\ &\quad + I_{rt}\sum_{n=-\infty}^{+\infty} J_n(\beta)\,\delta(f - (f_s \pm nf_c)) \end{aligned} \tag{24}$$

where J_n denotes the n-th order Bessel function of the first kind and $\delta(f)$ is the Dirac delta function. For small modulation index β, the Bessel functions of order $n \geq 2$ are very small and may be neglected (the so-called narrowband approximation). It becomes clear through this expression that the fault leads to sideband components of the fundamental at $f_s \pm nf_c$. When the modulation index β is small, only the first order sidebands at $f_s \pm f_c$ will be visible and their amplitudes will be approximately $J_1(\beta)I_{rt} \approx 0.5\beta I_{rt}$.

The Fourier transform magnitude of the AM stator current signal according to (21) is:

$$|I_{de}(f)| = I_1\,\delta(f - f_s) + \frac{1}{2}\alpha I_1\delta(f - (f_s \pm f_c)) \tag{25}$$

The amplitude modulation leads to two sideband components at $f_s \pm f_c$ with equal amplitude $\alpha I_1/2$. Therefore, the spectral signature of the AM and PM signal is identical if the modulation

frequency is equal and the PM modulation index is small. This can be the case when e.g. load unbalance and dynamic rotor eccentricity are considered as faults.
It can be concluded that the PSD is a simple analysis tool for stationary drive conditions. It is not suitable for analysis when the drive speed is varying. Another drawback is that PM and AM cannot be clearly distinguished.

3.2 Instantaneous Frequency

3.2.1 Definition

For a complex monocomponent signal $z(t) = a(t)e^{j\varphi(t)}$, the instantaneous frequency $f_i(t)$ is defined by (Boashash (2003)):

$$f_i(t) = \frac{1}{2\pi}\frac{d}{dt}\varphi(t) \tag{26}$$

where $\varphi(t)$ is the instantaneous phase and $a(t)$ the instantaneous amplitude of the analytical signal $z(t)$.

3.2.2 Application

The instantaneous frequency (IF) of a monocomponent phase modulated signal can be calculated using the definition (26). For the phase modulated stator current component $i_{rt}(t)$ (see second term of equation (13)), it can be expressed as:

$$f_{i,i_{rt}}(t) = f_s - f_c\beta\sin(\omega_c t) \tag{27}$$

The fault has therefore a direct effect on the IF of the stator current component $i_{rt}(t)$. In the healthy case, its IF is constant; in the faulty case, a time varying component with frequency f_c appears.
If the complex multicomponent PM signal according to (13) is considered, the calculation of its IF leads to the following expression:

$$f_{i,I}(t) = f_s - f_c\beta\sin(\omega_c t)\frac{1}{1+a(t)} \tag{28}$$

with

$$a(t) = \frac{I_{st}^2 + I_{st}I_{rt}\cos\left(\beta\cos\left(\omega_c t\right) - \varphi_s\right)}{I_{rt}^2 + I_{st}I_{rt}\cos\left(\beta\cos\left(\omega_c t\right) - \varphi_s\right)} \tag{29}$$

Using reasonable approximations, it can be shown that $1/\left(1 + a(t)\right)$ is composed of a constant component with only small oscillations. Hence, the IF of (13) may be approximated by:

$$f_{i,I}(t) \approx f_s - Cf_c\beta\sin(\omega_c t) \tag{30}$$

where C is a constant, $C < 1$. Numerical evaluations confirm this approximation. It can therefore be concluded, that the multicomponent PM signal IF corresponding to the stator current also shows fault-related oscillations at f_c which may be used for detection.
The IF of an AM stator current signal according to (21) is simply a constant at frequency f_s. In contrast to the PM stator current signal, no time-variable component is present. The AM modulation index α is not reflected in the IF. Consequently, the stator current IF cannot be used for amplitude modulation detection i.e. airgap eccentricity related faults.

3.3 Wigner Distribution

The Wigner Distribution (WD) belongs to the class of time-frequency signal analysis tools. It provides a signal representation with respect to time and frequency which can be interpreted as a distribution of the signal energy.

3.3.1 Definition

The WD is defined as follows (Flandrin (1999)):

$$W_x(t,f) = \int_{-\infty}^{+\infty} x\left(t+\frac{\tau}{2}\right) x^*\left(t-\frac{\tau}{2}\right) e^{-j2\pi f\tau} d\tau \tag{31}$$

This formula can be seen as the Fourier transform of a kernel $K_x(\tau,t)$ with respect to the delay variable τ. The kernel is similar to an autocorrelation function.

An interesting property of the WD is its perfect concentration on the instantaneous frequency in the case of a linear frequency modulation. However, other types of modulations (e.g. in our case sinusoidal phase modulations) produce so-called inner interference terms in the distribution (Mecklenbräuker & Hlawatsch (1997)). Note that the interferences may however be used for detection purposes as it will be shown in the following.

Another important drawback of the distribution is its non-linearity due to the quadratic nature. When the sum of two signals is considered, so-called outer interference terms appear in the distribution at time instants or frequencies where there should not be any signal energy (Mecklenbräuker & Hlawatsch (1997)). The interference terms can be reduced by using e.g. the Pseudo Wigner Distribution which includes an additional smoothing window (see section 3.4).

3.3.2 Application

The stator current in the presence of load torque oscillations can be considered as the sum of a pure frequency and a phase modulated signal (see (13)). The detailed calculations of the stator current WD can be found in (Blödt, Chabert, Regnier & Faucher (2006)). The following approximate expression is obtained for small β:

$$\begin{aligned} W_{i_{pm}}(t,f) \approx & \left(I_{rt}^2 + I_{st}^2\right) \delta(f - f_s) \\ & - I_{rt}^2 \beta \sin(\omega_c t)\, \delta(f - f_s - \frac{f_c}{2}) \\ & + I_{rt}^2 \beta \sin(\omega_c t)\, \delta(f - f_s + \frac{f_c}{2}) \end{aligned} \tag{32}$$

The WD of the PM stator current is therefore a central frequency at f_s with sidebands at $f_s \pm f_c/2$. These components have time-varying amplitudes at frequency f_c. It is important to note that the lower sideband has the opposed sign to the upper sideband for a given point in time i.e. a phase shift of π exists theoretically between the two sidebands.

The WD of the AM signal according to (21) is calculated in details in (Blödt, Regnier & Faucher (2006)). The following approximate expression is obtained for small modulation indices α:

$$W_{i_{am}}(t,f) \approx I_1^2\, \delta(f - f_s) + \alpha \cos(\omega_r t)\, I_1^2\, \delta\left(f - f_s \pm \frac{f_r}{2}\right) \tag{33}$$

The AM signature on the WD is therefore sidebands at $f_s \pm f_r/2$. The sidebands oscillate at shaft rotational frequency f_r, their amplitude is αI_1^2. It should be noted that the signature is similar to the PM signal but with the important difference that the upper and lower sideband oscillations have the same amplitudes for a given point in time i.e. they are in phase.

3.4 Illustration with Synthesized Signals

In order to validate the preceding theoretical considerations, the periodogram and WD of AM and PM signals are calculated numerically with synthesized signals. The signals are discrete versions of the continuous time signals in (13) and (21) with the following parameters: $I_{st} = I_{rt} = \sqrt{2}/2$, $I_1 = \sqrt{2}$, $\alpha = \beta = 0.1$, $\varphi_s = -\pi/8$, $f_s = 0.25$ and $f_c = f_r = 0.125$ normalized frequency. These parameters are coherent with a realistic application, apart from the strong modulation indices which are used for demonstration purposes. White zero-mean Gaussian noise is added with a signal to noise ration of 50 dB. The signal length is $N = 512$ samples.
First, the periodogram of both signals is calculated (see Fig. 3). Both spectra show the fundamental component with sidebands at $f_s \pm f_r$. The higher order sidebands of the PM signal are buried in the noise floor so that both spectral signatures are identical.

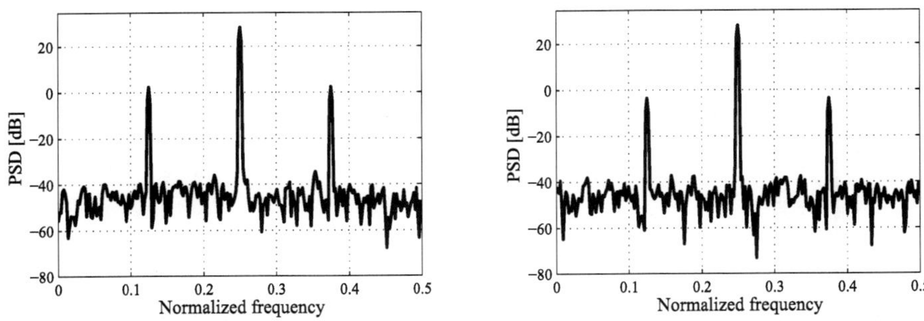

(a) PM signal (b) AM signal

Fig. 3. Power spectral density of synthesized PM and AM signals.

The WD is often replaced in practical applications with the Pseudo Wigner Distribution (PWD). The PWD is a smoothed and windowed version of the WD, defined as follows: (Flandrin (1999)):

$$PW_x(t,f) = \int_{-\infty}^{+\infty} p(\tau) x\left(t + \frac{\tau}{2}\right) x^*\left(t - \frac{\tau}{2}\right) e^{-j2\pi f \tau} d\tau \tag{34}$$

where $p(\tau)$ is the smoothing window. In the following, a Hanning window of length $N/4$ is used. The time-frequency distributions are calculated using the Matlab® Time-Frequency Toolbox (Auger et al. (1995/1996)). The PWD of the PM and AM stator current signals is displayed in Fig. 4. A constant frequency at $f_s = 0.25$ is visible in each case. Sidebands resulting from modulation appear at $f_s \pm f_r/2$ in both cases. The zoom on the interference structure shows that the sidebands are oscillating at f_r. According to the theory, the sidebands are phase-shifted by approximately π in the PM case whereas they are in phase with the AM signal.
For illustrating the stator current IF analysis, a simulated transient stator current signal is used. The supply frequency $f_s(t)$ varies from 0.05 to 0.25 normalized frequency. The modulation frequency $f_c(t)$ is half the supply frequency. The IF of the transient PM and AM stator

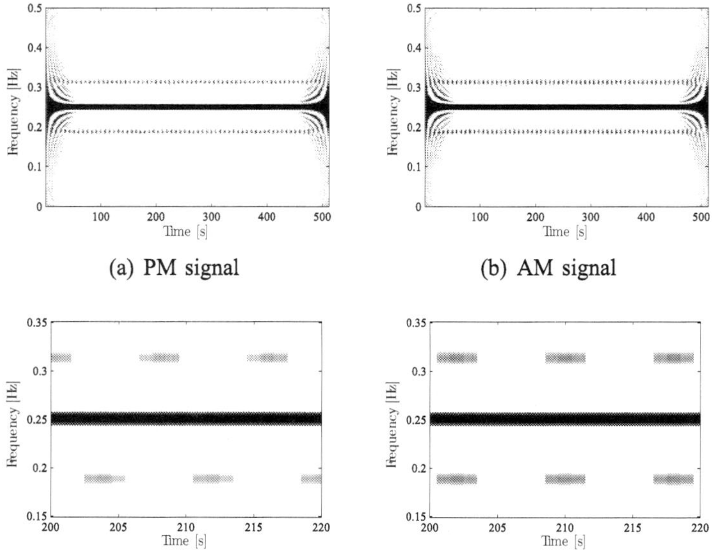

(a) PM signal (b) AM signal

(c) Zoom on PM signal (d) Zoom on AM signal

Fig. 4. Pseudo Wigner Distribution of synthesized PM and AM signals with zoom on interference structure.

current signal is shown in Fig. 5. The linear evolution of the supply frequency is clearly visible apart from border effects. With the PM signal, oscillations at varying fault frequency $f_c(t)$ can be recognized. In case of the AM signal, no oscillations are present. Further IF and PWD analysis with automatic extraction of fault indicators is described in (Blödt, Bonacci, Regnier, Chabert & Faucher (2008)).

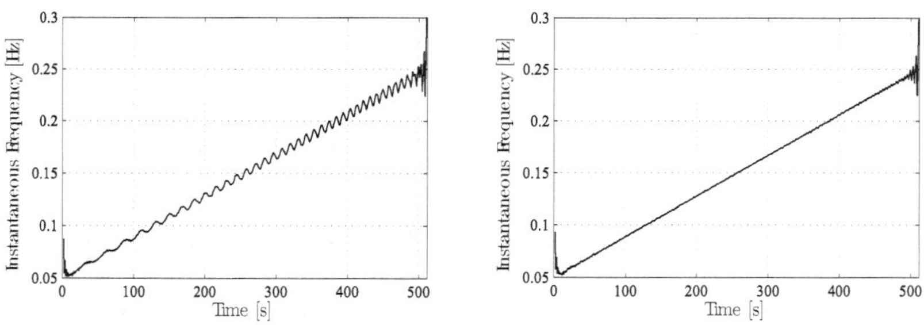

(a) PM signal (b) AM signal

Fig. 5. Instantaneous frequency of simulated transient PM and AM signals.

3.5 Summary

Several signal processing methods suitable for the detection of mechanical faults by stator current analysis have been presented. Classical spectral analysis based on the PSD can give a first indication of a possible fault by an increase of sidebands at $f_s \pm f_r$. This method can only be applied in case of stationary signals without important variations of the supply frequency. The IF can be used to detect phase modulations since they lead to a time-varying IF. A global time-frequency signal analysis is possible using the WD or PWD where a characteristic interference structure appears in presence of the phase or amplitude modulations. The three methods have been illustrated with simulated signals.

4. Detection of dynamic airgap eccentricity and load torque oscillations under laboratory conditions

4.1 Experimental Setup

Laboratory tests have been performed on an experimental setup (see Fig.6) with a three phase, 400 V, 50 Hz, 5.5 kW Leroy Somer induction motor (motor A). The motor has $p = 2$ pole pairs and its nominal torque Γ_n is about 36 Nm. The machine is supplied by a standard industrial inverter operating in open-loop condition with a constant voltage to frequency ratio. The switching frequency is 3 kHz.

The load is a DC motor with separate, constant excitation connected to a resistor through a DC/DC buck converter. A standard PI controller regulates the DC motor armature current. Thus, using an appropriate current reference signal, a constant torque with a small additional oscillating component can be introduced. The sinusoidal oscillation is provided through a voltage controlled oscillator (VCO) linked to a speed sensor.

Since the produced load torque oscillations are not a realistic fault, load unbalance is also examined. Thereto, a mass is fixed on a disc mounted on the shaft. The torque oscillation produced by such a load unbalance is sinusoidal at shaft rotational frequency. With the chosen mass and distance, the torque oscillation amplitude is $\Gamma_c = 0.04$ Nm. If the motor bearings are healthy, the additional centrifugal forces created by the mass will not lead to airgap eccentricity.

A second induction motor with identical parameters has been modified to introduce dynamic airgap eccentricity (motor B). Therefore, the bearings have been replaced with bearings having a larger inner diameter. Then, eccentrical fitting sleeves have been inserted between the shaft and the inner race. The obtained degree of dynamic eccentricity is approximately 40%.

Measured quantities in the experimental setup include the stator voltages and currents, torque and shaft speed. The signals are simultaneously acquired through a 24 bit data acquisition board at 25 kHz sampling frequency. Further signal processing is done off-line with Matlab®.

4.2 Stator Current Spectrum Analysis

For illustration purposes, the stator current spectral signatures of a machine with dynamic eccentricity (motor B) are compared to an operation with load torque oscillations at frequency $f_c = f_r$ (motor A). In Fig. 7 the current spectrum of a motor with 40% dynamic eccentricity is compared to an operation with load torque oscillations of amplitude Γ_c=0.14 Nm. This corresponds to only 0.4% of the nominal torque. The healthy motor spectrum is also displayed and the average load is 10% of nominal load during this test. The stator current spectra show identical fault signatures around the fundamental frequency i.e. an increasing amplitude of the peaks at $f_s \pm f_r \approx 25$ Hz and 75 Hz. This behavior is identical under different load conditions.

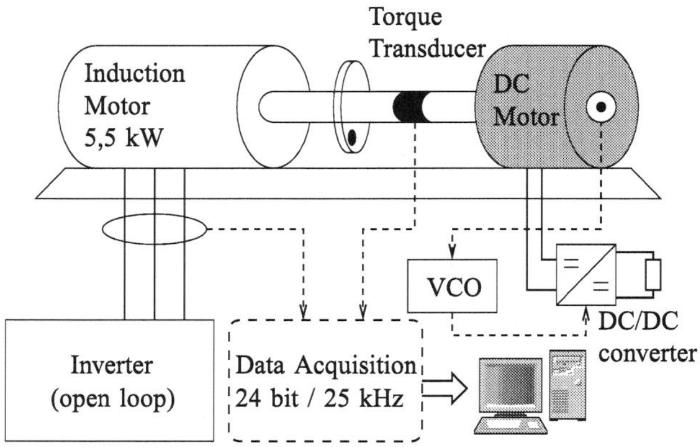

Fig. 6. Scheme of experimental setup

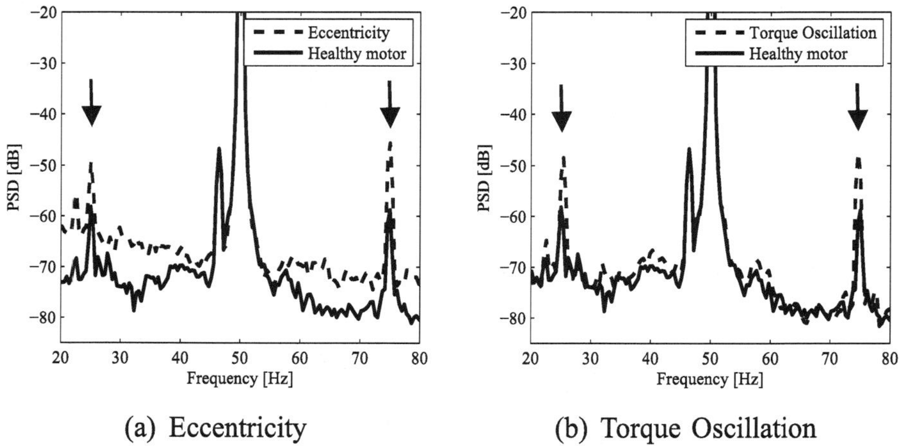

Fig. 7. Comparison of experimental motor stator current spectra: 40 % eccentricity (B) vs. healthy machine (A) and 0.14 Nm load torque oscillation (A) vs. healthy machine (A) at 10% average load.

The stator current with load unbalance is analyzed in Fig. 8. A small weight has been fixed on the disc on the shaft and the amplitude of the introduced torque oscillation is $\Gamma_c = 0.04$ Nm. The load unbalance as a realistic fault also leads to a rise in sideband amplitudes at $f_s \pm f_r$. These examples show that a monitoring strategy based on the spectral components $f_s \pm f_r$ can be used efficiently for detection purposes. In all three cases, these components show a considerable rise. However, this monitoring approach cannot distinguish between dynamic eccentricity and load torque oscillations.
In the following, transient stator current signals are also considered. They are obtained during motor startup between standstill and nominal supply frequency. The frequency sweep rate

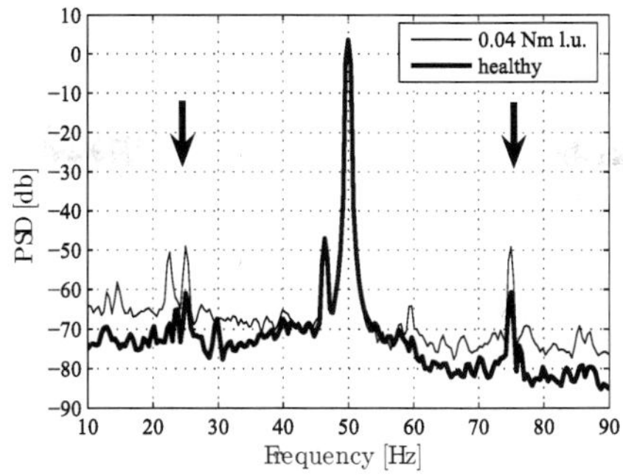

Fig. 8. PSD of stator current with load unbalance Γ_c = 0.04 Nm vs. healthy case

is 10 Hz per second i.e. the startup takes 5 seconds. For the following analysis, the transient between $f_s = 10$ Hz and 48 Hz is extracted. The PSD of a healthy and faulty transient signal are displayed in Fig. 9. This example illustrates that classic spectral estimation is not appropriate for transient signal analysis. The broad peak due to the time-varying supply frequency masks all other phenomena. The faulty and healthy case cannot be distinguished.

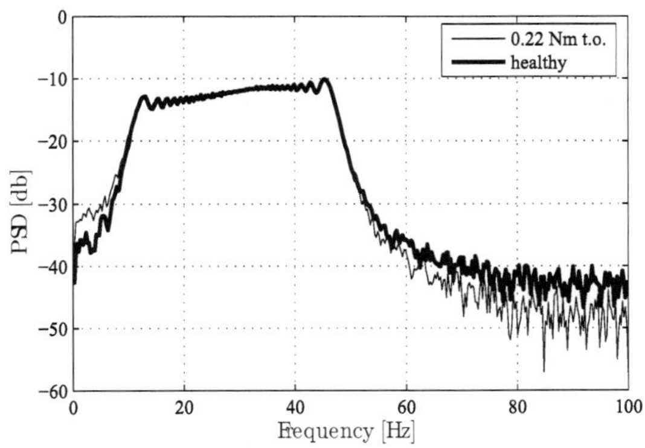

Fig. 9. PSD of stator current during speed transient with load torque oscillation $\Gamma_c = 0.22$ Nm vs. healthy case.

4.3 Stator Current Instantaneous Frequency Analysis

In this section, instantaneous frequency analysis will be applied to the stator current signals. The original signal has been lowpass filtered and downsampled to 200 Hz in order to remove high frequency content before time-frequency analysis. Only the information in a frequency range around the fundamental is conserved.

First, a transient stator current IF is shown in Fig. 10 for the healthy case and with a load torque oscillation $\Gamma_c = 0.5$ Nm. When load torque oscillations are present, the IF oscillations increase. The oscillation frequency is approximately half the supply frequency which corresponds to the shaft rotational frequency f_r.

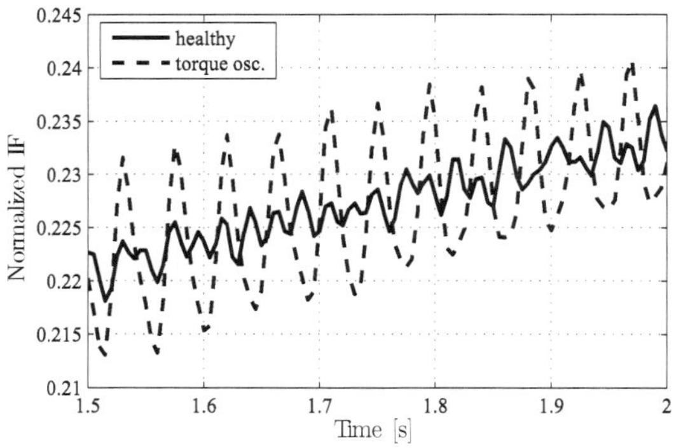

Fig. 10. Example of transient stator current IF with load torque oscillation ($\Gamma_c = 0.5$ Nm) vs. healthy case, 25% load.

For further analysis, the IF spectrogram can be employed. The spectrogram is a time-frequency signal analysis based on sliding short time Fourier transforms. More information can be found in (Boashash (2003)) (Flandrin (1999)). The two spectrograms depicted in Fig. 11 analyze the stator current IF during a motor startup with a small oscillation of $\Gamma_c = 0.22$ Nm and 10% average load. Besides the strong DC level at 0 Hz in the spectrogram, time varying components can already be noticed in the healthy case (a). They correspond to the supply frequency $f_s(t)$ and its second harmonic. Comparing the spectrogram of the healthy IF to the one with load torque oscillations (b), a fault-related component at $f_r(t)$ becomes clearly visible. More information about IF analysis can be found in (Blödt (2006)).

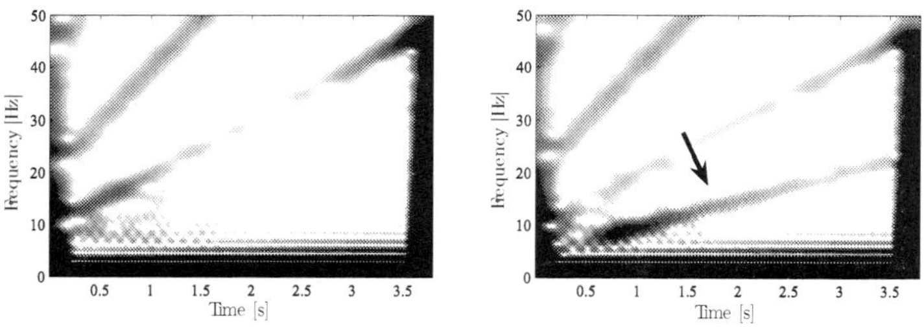

(a) healthy (b) $\Gamma_c = 0.22$ Nm

Fig. 11. Spectrogram of transient stator current IF with load torque oscillation $\Gamma_c = 0.22$ Nm vs. healthy case, 10% load.

4.4 Pseudo Wigner Distribution of Stator Current

The previously considered transient signals are also analyzed with the PWD. Figure 12 shows an example of the stator current PWD during a motor startup. Comparing the healthy case to 0.22 Nm load torque oscillations, the characteristic interference signature becomes visible around the time-varying fundamental frequency. Since the fault frequency is also time variable, the sideband location and their oscillation frequency depend on time (Blödt et al. (2005)).

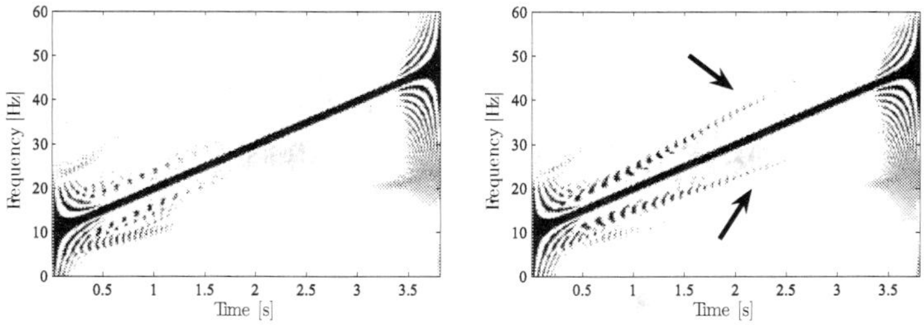

(a) healthy (b) 0.22 Nm torque oscillation

Fig. 12. PWD of transient stator current in healthy case and with load torque oscillation, 10% average load.

It is thereafter verified if dynamic eccentricity and load torque oscillations can be distinguished through the stator current PWD. The stator current PWDs with dynamic eccentricity and with 0.14 Nm load torque oscillation are shown in Fig. 13 for 10% average load. The characteristic fault signature is visible in both cases at $f_s \pm f_r/2 = 37.5$ Hz and 62.5 Hz. The phase shift between the upper and lower sideband seems closer to zero with eccentricity whereas with torque oscillations, it is closer to π. Nevertheless, it is difficult to determine the exact value from a visual analysis. However, the phase difference between the upper and lower sidebands can be automatically extracted from the PWD (see (Blödt, Regnier & Faucher (2006)). The result is about 125° with load torque oscillations and around 90° with dynamic eccentricity. These values differ from the theoretical ones (180° and 0° respectively) but this can be explained with load torque oscillations occurring as a consequence of dynamic eccentricity. A detailed discussion can be found in (Blödt, Regnier & Faucher (2006)). However, the phase shifts are sufficient to distinguish the two faults.

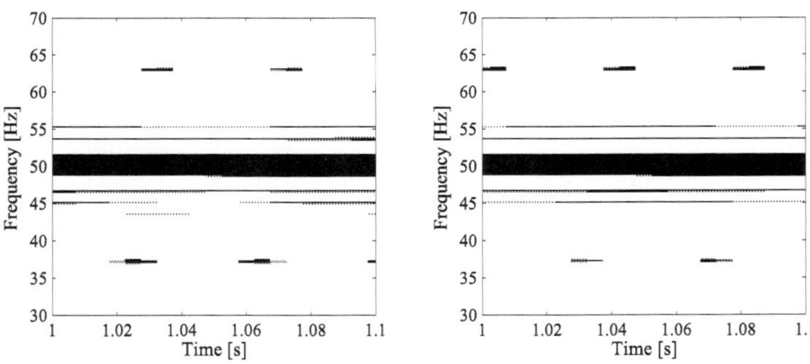

(a) Eccentricity (b) Torque oscillation

Fig. 13. Detail of stator current PWD with 40% dynamic eccentricity (B) and 0.14 Nm load torque oscillation (A) at small average load

5. Detection of shaft misalignment in electric winches

5.1 Problem statement

Electric winches are widely used in industrial handling systems such has cranes, overhead cranes or hoisting gears. As illustrated in Fig. 14, they are usually composed of an induction machine driving a drum through gears. Different faults can occur on such systems, leading

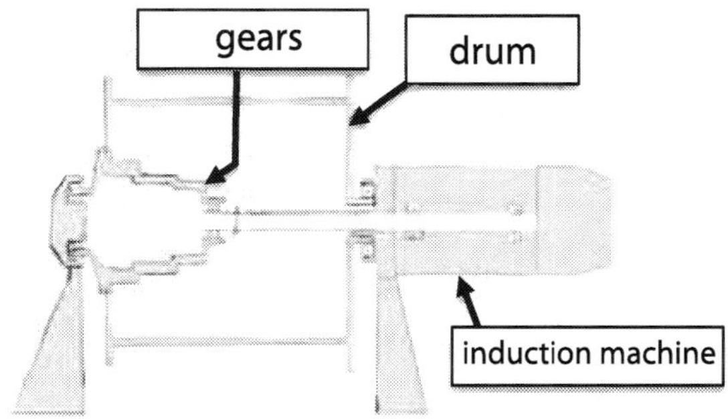

Fig. 14. Schematic representation of an electric winch.

to performance, reliability, and safety deterioration. A usual fault is the misalignment between the induction machine and the drum, generally due to strong radial forces applied to the drum by the handled load. Theoretical and experimental studies (see for example Xu & Marangoni (1994a;b)) show that such misalignments produce mechanical phenomena, which lead to torque oscillations and dynamic airgap eccentricity in the induction machine. It has been shown in section 2 that such phenomena generate amplitude and phase modulations in the supply currents of the machine. The goal of this part is to apply to these currents some of the signal processing tools presented in section 3 in order to detect a mechanical misalignment in the system. Therefore, this part is organized as follows : section 5.2 describes more precisely the fault and the necessary signal processing tools, and experimental results are finally presented in section 5.3.

5.2 Misalignment detection by stator current analysis

5.2.1 Shaft misalignment

Shaft misalignment is a frequent fault in rotating machinery, for which the shafts of the driving and the driven parts are not in the same centerline. The most general misalignment is a combination of angular misalignement (shaft centerlines do meet, but are not parallel) and offset misalignement (shaft centerlines are parallel, but do not meet). This type of fault generates reaction forces and torques in the coupling, and finally torque oscillations and dynamic airgap eccentricity in the driving machine. Moreover, these mechanical phenomena appear at even harmonics of the rotational frequencies of the driven parts (Xu & Marangoni (1994a)) (Xu & Marangoni (1994b)) (Sekhar & Prabhu (1995)) (Goodwin (1989)). For example, in the case of a misalignment of the winch shafts described in Fig. 14, torque oscillations and dynamic eccentricity are generated at even harmonics of the rotational frequencies of the induction machine, the gearbox and also the drum.

The theoretical model developed in sections 2.1 and 2.2 describes how mechanical phenomena are "seen" by an induction machine in its supply currents. More precisely, it has been shown that torque oscillations cause phase modulation of the stator current components (see Eq. (13)), while airgap eccentricity causes amplitude modulation (see Eq. (21)).
Therefore, in the case of a shaft misalignment in a system similar to Fig. 14, amplitude and phase modulations appear in the supply currents of the induction machine, and these modulations have frequencies equal to even harmonics of the rotational frequencies of the driving machine, the gearbox and the drum. Finally, the modulations generated by the drum are much more easy to detect since its rotational frequency is generally much lower than the supply and rotational frequencies of the machine f_s and f_r. In the following, only such low frequency modulations will be examined.

5.2.2 Shaft misalignment detection

The previous section has described that it should be possible to detect a shaft misalignment in an electric winch by analyzing its supply currents. Indeed, one only has to detect a significant increase in the amplitude and phase modulations of their fundamental component. A simple possibility is to analyze the variations of its instantaneous amplitude and instantaneous frequency, defined in section 3. These quantities can be easily real-time estimated as shown in (A. Reilly & Boashash (1994)), and Fig. 15 briefly describes the principle of this approach.

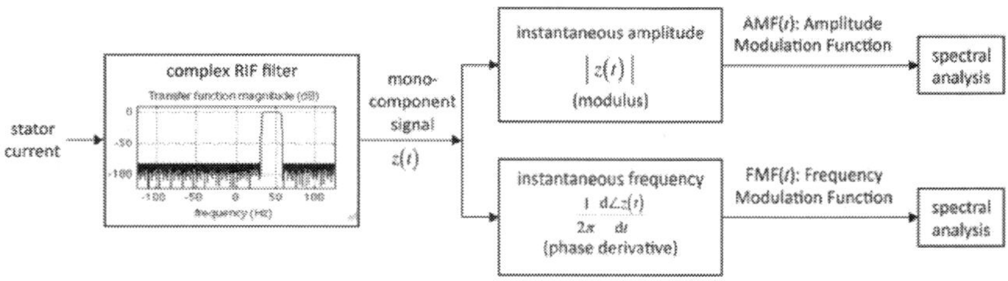

Fig. 15. Stator current analysis method for shaft misalignment detection.

First of all, one of the stator currents is filtered by a bandpass filter in order to obtain the fundamental component only, without any other component. The filter used in this application has a passband only situated in the positive frequency domain around $+f_s$ with $f_s = 45.05$ Hz (see the transfer function in Fig. 15) in order to directly obtain the analytical part of the analyzed signal, as explained in (A. Reilly & Boashash (1994)). The transfer function $H(f)$ of this filter is therefore not hermitian ($H(f) \neq H^*(-f)$), and its impulse response is complex-valued. This particularity is not problematic concerning the real-time implementation of this filter, since its finite impulse response only needs twice as many coefficients as a classical real-valued finite impulse response filter. Finally, the output of this filter is a complex-valued monocomponent signal $z(t)$ which represents the analytical part of the supply current fundamental component. In a second step, the modulus and the phase derivative of this complex signal lead to the instantaneous amplitude and frequency to estimate. Once their mean value substracted, these quantities are called amplitude modulation function (AMF) and frequency modulation function (FMF). They correspond to instantaneous amplitude and frequency variations of the current fundamental component and contain the researched modulations. Finally, power spectral densities of the AMF and FMF are estimated in order to detect and identify such modulations.

5.3 Experimental results

5.3.1 Test bench and operating conditions

A test bench has been designed by the CETIM (French CEntre Technique des Industries Mécaniques) in order to simulate different types of faults occuring in industrial handling systems (Sieg-Zieba & Tructin (2008)). It is composed of two 22 kW Potain electric winches, and one cable winding up from one winch to the other through a pulley as shown in Fig. 16.

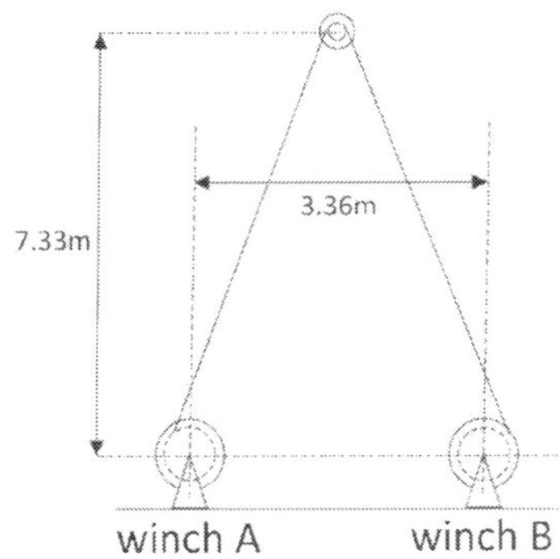

Fig. 16. Schematic description of the test bench.

The two winches are constituted as shown in Fig. 14, and the winch A is controlled through an inverter as a driving winch, while the winch B is only used to apply a predetermined mechanical load. The winch A is equipped with current probes in order to measure the stator currents of its induction machine. Moreover, an angular misalignment can be obtained on the same winch by inserting a shim with a slope between the motor flange and the drum bearing housing, thus creating an angle of 0.75° while the tolerance of the coupling is 0.5°.
During the experiments, the signals were recorded during 80 s at a sampling frequency of 25 kHz under stationary working conditions with and without misalignment. The constant mechanical load applied by the winch B was 2000 daN, and the rotational frequency reference value of the induction machine of the winch A was $f_r = 23$ Hz. These conditions resulted in a fundamental supply frequency of $f_s = 45.05$ Hz, and a drum rotational frequency of $f_{rd} = 0.29$ Hz.

5.3.2 Results

Experimental results presented in this section have been obtained by applying the method developed in section 5.2 to a supply current measured under the operating conditions described in the previous section. The proposed method leads to the estimation of the AMF and FMF of the stator current fundamental component, and the performance of this method is illustrated by the power spectral densities of these two functions. Low-frequency spectral contents of the AMF and FMF are respectively represented in Fig. 17 and 18. The power spectral densities obtained without any misalignment are in dashed line, while they are in solid line in case of misalignment.
As expected, amplitude and frequency modulations strongly increase in the low-frequency range when a shaft misalignment occurs, and more precisely at even harmonics of the drum rotational frequency (see arrows in Fig. 17 and 18 around $2 \times f_{rd} = 0.58$ Hz and $4 \times f_{rd} =$

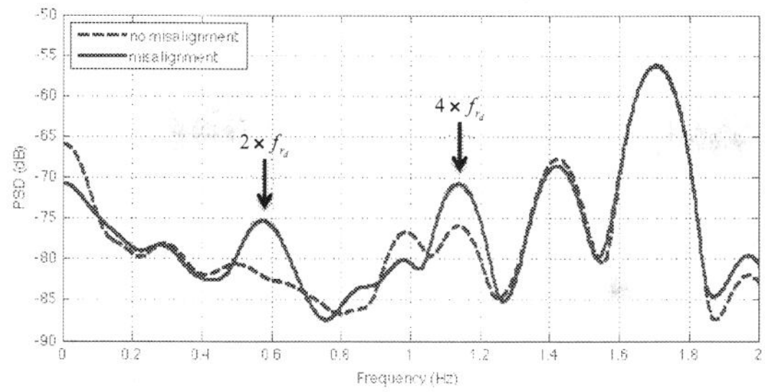

Fig. 17. Power spectral density of the AMF with (-) and without (- -) misalignment.

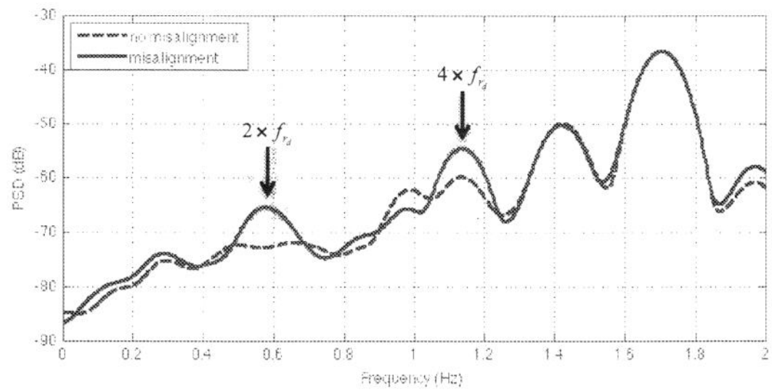

Fig. 18. Power spectral density of the FMF with (-) and without (- -) misalignment.

1.16 Hz). These results clearly show that a potential misalignment in an electric winch can be detected by analyzing its supply currents.

Furthermore, a simple and efficient misalignment detector can be derived from AMF and FMF power spectral densities. For example, the power increase of AMF and FMF in a frequency band corresponding to the expected fault can easily be estimated from these quantities by integration. In the present case, the power of these two modulation functions between 0.25 Hz and 1.25 Hz (a frequency band containing $2 \times f_{rd}$ and $4 \times f_{rd}$) increases of about 80% (for AMF) and 120% (for FMF) when a misalignment occurs. Finally, it can be noted that such a detector can be easily real-time implemented since it is based on real-time operations only (see Fig. 15).

5.4 Conclusion

This section has shown that a shaft misalignment in an electric winch can be detected by analyzing its supply currents. Experimental results confirm that a misalignment generates additional amplitude and frequency modulations in stator currents. These phenomena can be easily detected and characterized by analyzing the spectral content of amplitude and frequency modulation functions of the stator current fundamental component. For example, Fig. 17 and 18 show that these additional modulations occur at even harmonics of the drum rotational frequency, *i.e.* in a very low-frequency range. Finally, a simple and efficient real-time detector has been proposed, based on the integration of the power spectral densities of the AMF and FMF.

6. Detection of single point bearing defects

The following section considers the detection of single point bearing defects in induction motors. Bearing faults are the most frequent faults in electric motors (41%) according to an IEEE motor reliability study for large motors above 200 HP (IEEE Motor reliability working group (1985b)), followed by stator (37%) and rotor faults (10%). Therefore, their detection is of great concern. First, some general information about bearing geometry and characteristic frequencies will be given. Then, a theoretical study of bearing fault effects on the stator current is presented (see Blödt, Granjon, Raison & Rostaing (2008)). Finally, experimental results illustrate and validate the theoretical approach.

6.1 Bearing Fault Types

This paper considers rolling-element bearings with a geometry shown in Fig. 19. The bearing consists mainly of the outer and inner raceway, the balls and the cage which assures equidistance between the balls. The number of balls is denoted N_b, their diameter is D_b and the pitch or cage diameter is D_c. The point of contact between a ball and the raceway is characterized by the contact angle β.

Bearing faults can be categorized into distributed and localized defects (Tandon & Choudhury (1997)) (Stack et al. (2004b)). Distributed defects affect a whole region and are difficult to characterize by distinct frequencies. In contrast, single-point defects are localized and can be classified according to the affected element:

- outer raceway defect
- inner raceway defect
- ball defect

A single point defect could be imagined as a small hole, a pit or a missing piece of material on the corresponding element. Only these are considered in the following.

6.2 Characteristic Fault Frequencies

With each type of bearing fault, a characteristic fault frequency f_c can be associated. This frequency is equivalent to the periodicity by which an anomaly appears due to the existence of the fault. Imagining for example a hole on the outer raceway: as the rolling elements move over the defect, they are regularly in contact with the hole which produces an effect on the machine at a given frequency.

The characteristic frequencies are functions of the bearing geometry and the mechanical rotor frequency f_r. A detailed calculation of these frequencies can be found in (Li et al. (2000)). For the three considered fault types, f_c takes the following expressions:

Outer raceway: $$f_o = \frac{N_b}{2} f_r \left(1 - \frac{D_b}{D_c}\cos\beta\right) \quad (35)$$

Inner raceway: $$f_i = \frac{N_b}{2} f_r \left(1 + \frac{D_b}{D_c}\cos\beta\right) \quad (36)$$

Ball: $$f_b = \frac{D_c}{D_b} f_r \left(1 - \frac{D_b^2}{D_c^2}\cos^2\beta\right) \quad (37)$$

It has been statistically shown in (Schiltz (1990)) that the vibration frequencies can be approximated for most bearings with between six and twelve balls by :

$$f_o = 0.4\,N_b f_r \tag{38}$$
$$f_i = 0.6\,N_b f_r \tag{39}$$

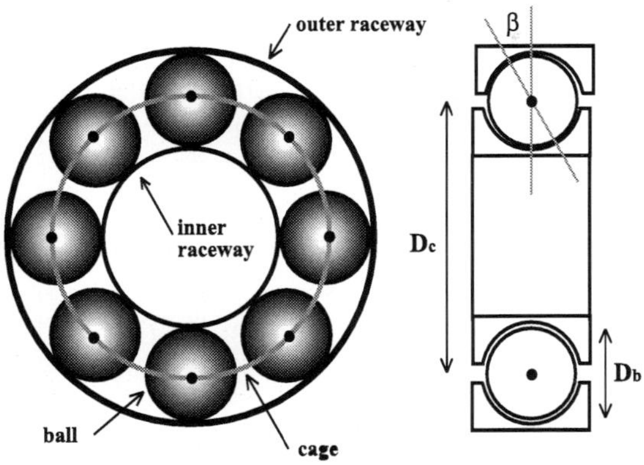

Fig. 19. Geometry of a rolling-element bearing.

6.3 Short Literature Survey on Bearing Fault Detection by Stator Current Analysis

Vibration measurement is traditionally used to detect bearing defects. Analytical models describing the vibration response of a bearing with single point defects can be found in (Tandon & Choudhury (1997)) (MacFadden & Smith (1984)) (Wang & Kootsookos (1998)). The most often quoted model studying the influence of bearing damage on the induction machine's stator current was proposed by R. R. Schoen et al. in (Schoen et al. (1995)). The authors consider the generation of rotating eccentricities at bearing fault characteristic frequencies f_c which leads to periodical changes in the machine inductances. This should produce additional frequencies f_{bf} in the stator current given by:

$$f_{\mathrm{bf}} = |f_s \pm k f_c| \tag{40}$$

where f_s is the electrical stator supply frequency and $k = 1,2,3,\ldots$.

A general review about bearing fault detection by stator current monitoring can be found in (Zhou et al. (2007)). Stack examines in (Stack et al. (2004b)) single point defects and generalized roughness. In (Stack et al. (2004a)), the stator current is analyzed using parametric spectrum analysis such as autoregressive modelling. Neural network techniques and the wavelet transform are used in (Eren et al. (2004)) for bearing fault detection and wavelet decomposition is applied in the case of variable speed drives in (Teotrakool et al. (2006)). In (Zhou et al. (2008)) stator current noise cancellation techniques are described for bearing fault detection.

In the following, a detailed theoretical study will be conducted to analyze the physical effects of bearing faults on the induction machine and the stator current. This will yield additional stator current frequencies with respect to the existing model and will give insight on the modulation type.

6.4 Theoretical Study of Single Point Bearing Defects

Two physical effects are considered in the theoretical study when the single point defect comes into contact with another bearing element:

1. the introduction of a radial movement of the rotor center,
2. the apparition of load torque variations.

The method used to study the influence of the rotor displacement on the stator current is again based on the MMF (magnetomotive force) and permeance wave approach (see section 2). The following model relies on several simplifying assumptions. First, load zone effects in the bearing are not considered. The fault impact on the airgap length is considered by a series of Dirac generalized functions. In reality, the fault generates other pulse shapes, but this alters only the harmonic amplitudes. Since this modeling approach focusses on the frequency combinations and modulation types and not on exact amplitudes, this assumption is reasonable. The calculation of the airgap magnetic field does not take into account higher order space and time harmonics for the sake of simplicity. However, the calculated modulation effects affect higher harmonics in the same way as the fundamental. As before, higher order armature reactions are also neglected.

6.4.1 Airgap Length Variations

The first step in the theoretical analysis is the determination of the airgap length g as a function of time t and angular position θ in the stator reference frame. The radial rotor movement causes the airgap length to vary as a function of the defect, which is always considered as a hole or a point of missing material in the corresponding bearing element.

6.4.1.1 Outer Race Defect

Without loss of generality, the outer race defect can be assumed to be located at the angular position $\theta = 0$. When there is no contact between a ball and the defect, the rotor is perfectly centered. In this case, the airgap length g is supposed to take the constant value g_0, neglecting rotor and stator slotting effects. On the contrary, every $t = k/f_o$ (with k integer), the contact between a ball and the defect leads to a small movement of the rotor center in the stator reference frame (see Fig. 20). In this case, the airgap length can be approximated by $g_0(1 - e_o \cos\theta)$, where e_o is the relative degree of eccentricity. In order to model the fault impact on the airgap length as a function of time, a series of Dirac generalized functions can then be used as it is common in models for vibration analysis (MacFadden & Smith (1984)).
These considerations lead to the following expression for the airgap length:

$$g_o(\theta,t) = g_0 \left[1 - e_o \cos(\theta) \sum_{k=-\infty}^{+\infty} \delta\left(t - \frac{k}{f_o}\right) \right] \tag{41}$$

where e_o is the relative degree of eccentricity introduced by the outer race defect. This equation can be interpreted as a temporary static eccentricity of the rotor, appearing only at $t = k/f_o$. The function $g_o(\theta,t)$ is represented in Fig. 21 for $\theta = 0$ as an example.

6.4.1.2 Inner Race Defect

In this case, the situation is slightly different from the outer race defect. The fault occurs at the instants $t = k/f_i$. As the defect is located on the inner race, the angular position of the minimal airgap length moves with respect to the stator reference frame as the rotor turns at

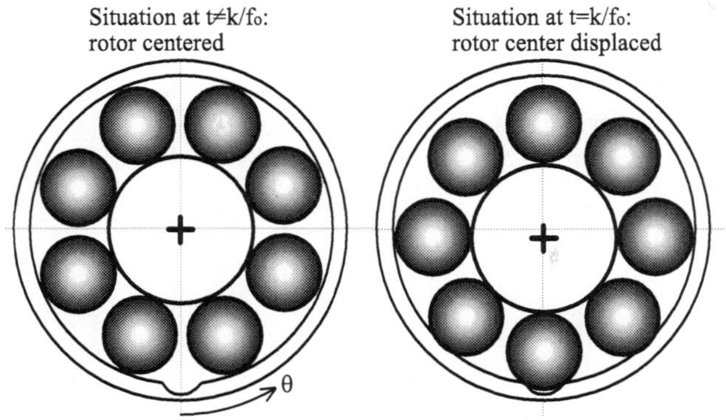

Fig. 20. Radial rotor movement due to an outer bearing raceway defect.

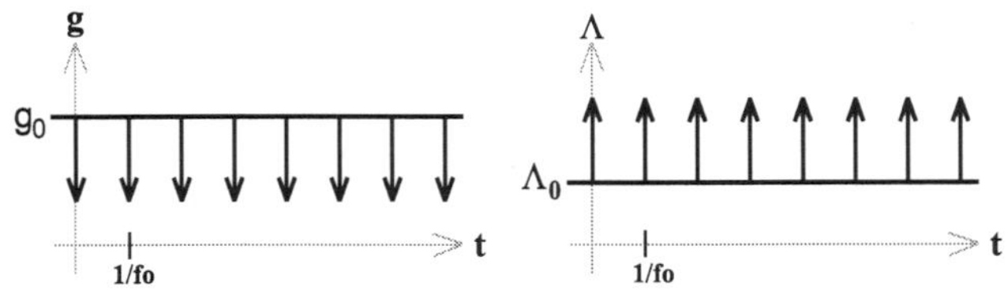

Fig. 21. Airgap length g and permeance Λ in the presence of an outer bearing raceway defect for $\theta = 0$.

the angular frequency ω_r (see Fig. 22). Between two contacts with the defect, the defect itself has moved by an angle described by:

$$\Delta\theta_i = \omega_r \Delta t = \frac{\omega_r}{f_i} \tag{42}$$

Hence, equation (41) becomes:

$$g_i(\theta,t) = g_0 \left[1 - e_i \sum_{k=-\infty}^{+\infty} \cos(\theta + k\Delta\theta_i)\, \delta\left(t - \frac{k}{f_i}\right)\right] \tag{43}$$

where e_i is the relative degree of eccentricity introduced by the inner race defect.
This equation can be simplified for further calculations by extracting the cosine-term of the sum so that the series of Dirac generalized functions may be later developed into a Fourier series. One fundamental property of the Dirac generalized function is given by the following equation (Max & Lacoume (2000)):

$$h(k) \cdot \delta\left(t - \frac{k}{f_i}\right) = h(tf_i) \cdot \delta\left(t - \frac{k}{f_i}\right) \tag{44}$$

This formula becomes obvious when one considers that $\delta(t - k/f_i)$ always equals 0, except for $t = k/f_i$. After combining (44), (43) and (42), the airgap length becomes:

$$g_i(\theta,t) = g_0 \left[1 - e_i \cos(\theta + \omega_r t) \sum_{k=-\infty}^{+\infty} \delta\left(t - \frac{k}{f_i}\right)\right] \tag{45}$$

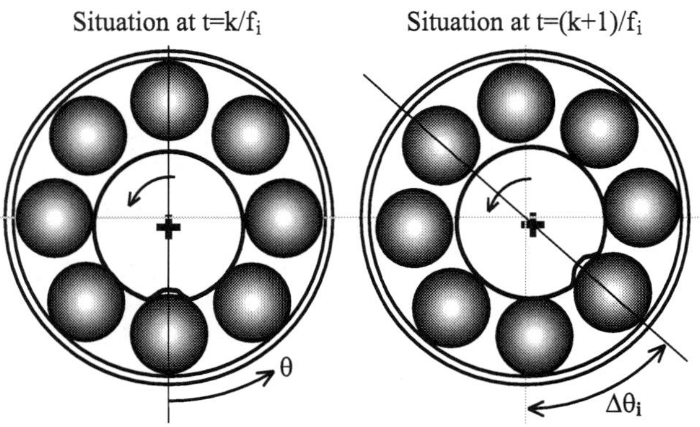

Fig. 22. Radial rotor movement due to an inner bearing raceway defect.

6.4.1.3 Ball Defect

In presence of ball defect, the defect location moves in a similar way as the inner raceway fault. The fault causes an anomaly on the airgap length at the instants $t = k/f_b$. The angular position of minimal airgap length changes in function of the cage rotational frequency. Actually, the balls are all fixed in the cage which rotates at the fundamental cage frequency ω_{cage}, given by (Li et al. (2000)):

$$\omega_{cage} = \frac{1}{2}\omega_r \left(1 - \frac{D_b}{D_c}\cos\beta\right) \tag{46}$$

The angle $\Delta\theta_b$ by which the fault location has moved between two fault impacts becomes:

$$\Delta\theta_b = \omega_{cage}\Delta t = \frac{\omega_{cage}}{f_b} \tag{47}$$

By analogy with (45), the expression of airgap length in presence of a ball defect becomes:

$$g_b(\theta,t) = g_0 \left[1 - e_b \cos\left(\theta + \omega_{cage}t\right) \sum_{k=-\infty}^{+\infty} \delta\left(t - \frac{k}{f_b}\right)\right] \tag{48}$$

where e_b is the relative degree of eccentricity introduced by the ball defect.

6.4.1.4 Generalization

In order to simplify the following considerations, equations (41), (45) and (48) can be combined in a generalized expression for the airgap length g in presence of a bearing fault:

$$g(\theta,t) = g_0 \left[1 - e\cos\left(\theta + \psi(t)\right) \sum_{k=-\infty}^{+\infty} \delta\left(t - \frac{k}{f_c}\right)\right] \tag{49}$$

where f_c is the characteristic bearing fault frequency given by (35), (36) or (37), and $\psi(t)$ is defined as follows:

$$\psi(t) = \begin{cases} 0 & \text{for an outer race defect} \\ \omega_r t & \text{for an inner race defect} \\ \omega_{cage} t & \text{for a ball defect} \end{cases} \tag{50}$$

6.4.2 Airgap Permeance

The airgap permeance Λ is proportional to the inverse of the airgap length g and is defined as follows:

$$\Lambda = \frac{\mu}{g} \tag{51}$$

where $\mu = \mu_r \mu_0$ is the magnetic permeability of the airgap. In the case of a bearing fault, the permeance becomes with (49):

$$\Lambda(\theta,t) = \Lambda_0 \frac{1}{\left[1 - e\cos\left(\theta + \psi(t)\right) \sum_{k=-\infty}^{+\infty} \delta\left(t - \frac{k}{f_c}\right)\right]} \tag{52}$$

where $\Lambda_0 = \mu / g_0$. The relationship between airgap length $g(\theta,t)$ and airgap permeance $\Lambda(\theta,t)$ is illustrated on Fig. 21 at the position $\theta = 0$ for an outer raceway defect.
Firstly, in order to simplify this expression, the fraction $1/(1-x)$ is approximated for small airgap variations by the first order term of its series development:

$$\begin{aligned} \frac{1}{1-x} &= 1 + x + x^2 + x^3 + \ldots \qquad \text{for } |x| < 1 \\ &\approx 1 + x \end{aligned} \tag{53}$$

The condition $|x| < 1$ is always satisfied because the degree of eccentricity verifies $|e| < 1$ in order to avoid contact between rotor and stator.
Secondly, the series of Dirac generalized functions is expressed as a complex Fourier series development (Max & Lacoume (2000)):

$$\begin{aligned} \sum_{k=-\infty}^{+\infty} \delta\left(t - \frac{k}{f_c}\right) &= \sum_{k=-\infty}^{+\infty} c_k e^{-j2\pi k f_c t} \\ &= c_0 + 2\sum_{k=1}^{+\infty} c_k \cos\left(2\pi k f_c t\right) \end{aligned} \tag{54}$$

with the Fourier series coefficients $c_k = f_c \;\; \forall k$.
Equations (52), (53) and (54) can be combined into a simplified expression for the airgap permeance wave:

$$\begin{aligned} \Lambda(\theta,t) \approx \Lambda_0 \Bigg\{ & 1 + e c_0 \cos\left(\theta + \psi(t)\right) \\ & + e \sum_{k=1}^{+\infty} c_k \cos\left(\theta + \psi(t) + k\omega_c t\right) \\ & + e \sum_{k=1}^{+\infty} c_k \cos\left(\theta + \psi(t) - k\omega_c t\right) \Bigg\} \end{aligned} \tag{55}$$

6.4.3 Airgap Flux Density

The total fundamental MMF wave F_{tot} is assumed:

$$F_{tot}(\theta,t) = F\cos(p\theta - \omega_s t + \varphi) \tag{56}$$

Multiplication of (55) and (56) leads to the expression of the flux density distribution $B_{tot}(\theta,t)$:

$$\begin{aligned} B_{tot}(\theta,t) &= F_{tot}(\theta,t) \cdot \Lambda(\theta,t) \\ &= F\Lambda_0 \cos(p\theta - \omega_s t + \varphi) \\ + \sum_{k=0}^{\infty} B_k &\left[\cos\left((p \pm 1)\theta \pm \psi(t) \pm k\omega_c t - \omega_s t + \varphi\right) \right] \end{aligned} \tag{57}$$

where B_k are the amplitudes of the fault-related flux density waves. The notation $\pm$ is used to write all possible frequency combinations in a compact form.
Equation (57) clearly shows the influence of the rotor displacement caused by the bearing fault on the flux density. In addition to the fundamental sine wave (term B_0), a multitude of fault-related sine waves appear in the airgap. These supplementary waves have $p \pm 1$ pole pairs and a frequency content f_{ecc} given by:

$$f_{ecc} = \frac{1}{2\pi}\left(\pm\frac{d\psi(t)}{dt} \pm k\omega_c - \omega_s \right) \tag{58}$$

6.4.4 Stator Current

The additional flux density components according to (57) are equivalent to an additional magnetic flux $\Phi(\theta,t)$. By considering the realization of the winding and the geometry of the machine, the additional flux $\Phi(t)$ in each stator phase can be obtained. If the stator voltages are imposed, the time varying flux causes additional components in the motor stator current according to the stator voltage equation for the phase m:

$$V_m(t) = R_s I_m(t) + \frac{d\Phi_m(t)}{dt} \tag{59}$$

The frequency content of the flux in each phase is supposed to be equal to the frequency content of the airgap field according to (58). Under the hypothesis of imposed stator voltages, the stator current in each phase is given by the derivative of the corresponding flux. This leads to the following expression for the stator current $I_m(t)$ with ω_r supposed constant:

$$I_m(t) = \sum_{k=0}^{\infty} I_k \cos\left[\pm \psi(t) \pm k\omega_c t - \omega_s t + \varphi_m \right] \tag{60}$$

It becomes thus obvious, that the radial rotor movement due to the bearing fault results in additional frequencies in the stator current. With the three fault types, these frequencies are obtained from (50) and (60):

Outer race defect: $$f_{\mathrm{ecc\ or}} = f_s \pm k f_o \tag{61}$$
Inner race defect: $$f_{\mathrm{ecc\ ir}} = f_s \pm f_r \pm k f_i \tag{62}$$
Ball defect: $$f_{\mathrm{ecc\ ball}} = f_s \pm f_{cage} \pm k f_b \tag{63}$$

where $k = 1,2,3,\ldots$. In terms of signal processing, it can be noticed that the effect of the fault related rotor movement on the stator current is an amplitude modulation of the fundamental sine wave, due to the effect of the modified permeance on the fundamental MMF wave.

	Eccentricity	Torque oscillations
Outer raceway	$f_s \pm kf_o$	$f_s \pm kf_o$
Inner raceway	$f_s \pm f_r \pm kf_i$	$f_s \pm kf_i$
Ball defect	$f_s \pm f_{cage} \pm kf_b$	$f_s \pm kf_b$

Table 1. Summary of bearing fault related frequencies in the stator current spectrum

6.4.5 Load torque oscillations

In this section, the second considered effect of a bearing fault on the machine is studied. Imagining for example a hole in the outer race: each time a ball passes in a hole, a mechanical resistance will appear when the ball tries to leave the hole. The consequence is a small increase of the load torque at each contact between the defect and another bearing element. The bearing fault related torque variations appear at the previously mentioned characteristic vibration frequencies f_c (see section 6.2) as they are both of same origin: a contact between the defect and another element.

The effect of load torque oscillations on the stator current has already been studied in section 2.1. The torque oscillations resulting from single point bearing defects will result in the same stator current phase modulations as described in equation (13). Note that the fault characteristic frequency f_c will be take values depending on the fault type defined in section 6.2.

6.5 Summary

The results from the preceding theoretical study enlarge the existing model of the effects of bearing faults on stator current. The frequencies that can be found when the stator current PSD is analyzed, are resumed in Table 1.

6.6 Experimental Results

6.6.1 Description of Experimental Setup

The experimental tests were carried out on a test rig with a standard 1.1 kW, 2-pole pair, Y-coupled induction motor. A DC-machine was used to simulate different load levels. In order to reduce harmonic content in the supply voltage, the induction motor is directly fed by a synchronous generator (100 kVA) working as a generator. Measured quantities are the three line currents, the stator voltages, motor speed, torque and two vibration signals issued from piezoelectric accelerometers mounted on the stator core. Data are sampled at 16 kHz and processed using Matlab®.

Two classes of faulty bearings (NSK 6205) are available. First, new bearings have been damaged artificially to produce defects on the outer and inner raceway. The defects consist of holes that have been drilled axially through the raceways (see Fig. 23). Secondly, bearings with realistic damage were tested, issued from industrial maintenance. The faulty bearings are mounted at the load-end of the induction machine.

The characteristic vibration frequencies take the following values at no-load operation: outer raceway frequency $f_o = 89.6$ Hz, inner raceway frequency $f_i = 135.4$ Hz, ball frequency $f_b = 58.8$ Hz. The contact angle has been assumed to be $\beta = 0$.

6.6.2 Outer Raceway Defect

The defect on the outer raceway has already been experimentally studied in (Schoen et al. (1995)), so that it will be discussed very shortly. During the tests, the characteristic vibration

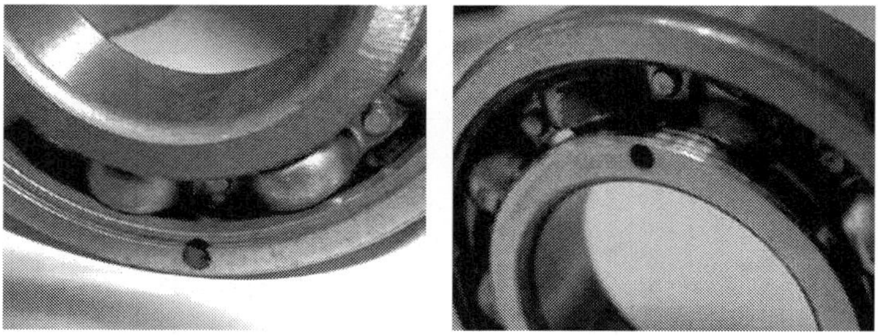

(a) outer raceway defect (b) inner raceway defect

Fig. 23. Photo of bearings with single point defects

frequency and its multiples were clearly visible on the vibration spectrum of the machine. There also appeared torque oscillations at the characteristic vibration frequencies.
The current spectrum shows a characteristic component at 125 Hz which corresponds to the frequency combination $|f_s - 2f_o|$ (see Fig. 24). It is interesting to note, that the same frequency combination appeared in (Bonaldi et al. (2002)) where a bearing with an outer race defect was tested experimentally.

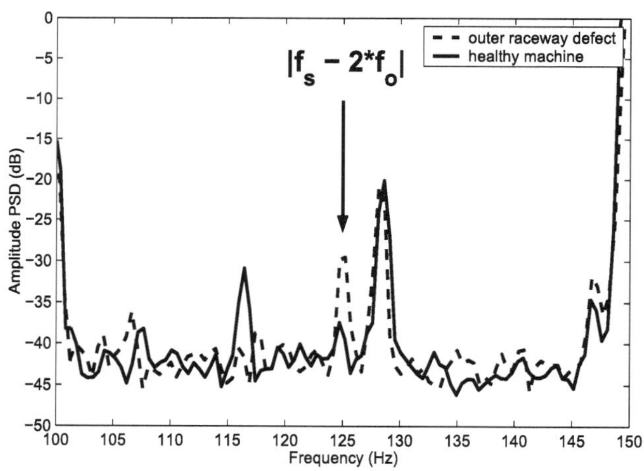

Fig. 24. Stator current spectrum of loaded machine with outer raceway defect.

6.6.3 Inner Raceway Defect

In a first step, the vibration signal is analyzed. A logarithmic plot of the vibration spectrum with a damaged bearing in comparison with the healthy machine condition is shown in Fig. 25. The characteristic frequency of the inner raceway defect f_i and its multiples (e.g. $2f_i$) are the components with the largest magnitude. Multiple tests with different load levels permitted to observe slight variations of the characteristic vibration frequency according to equation (36). Additional components due to other mechanical effects e.g. the cage rotational frequency

($\approx$ 10 Hz) and a general rise of the vibration level can also be noticed on the vibration spectrum.

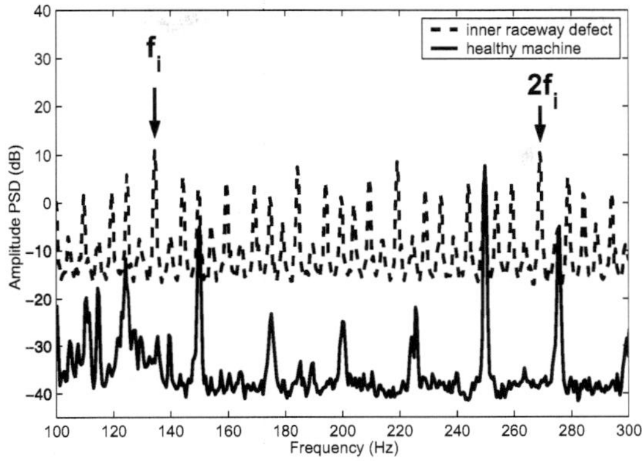

Fig. 25. Vibration spectrum of unloaded machine with inner raceway defect.

A spectral analysis of the measured load torque is shown in Fig. 26. The characteristic fault frequency f_i clearly appears on the torque spectrum with an amplitude of +15 dB in comparison to the healthy case. This validates the proposed theoretical approach which assumes torque variations at the characteristic frequency as a consequence of the bearing fault. Higher harmonics of f_i can also be observed. In addition to the mentioned components, other frequencies appear in the torque spectrum at e.g. 110 and 115 Hz, but they have no direct link to a predicted characteristic frequency.

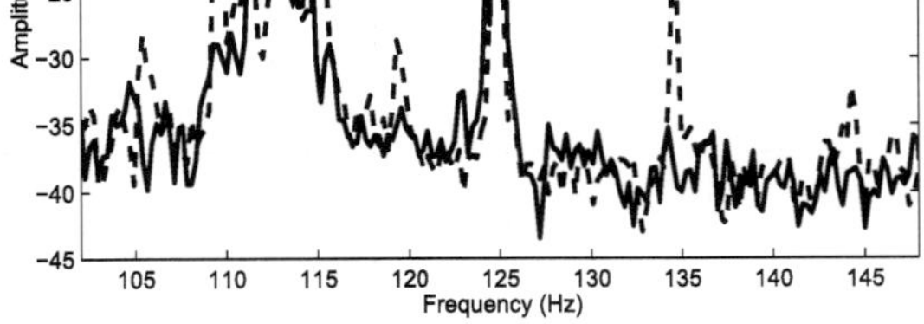

Fig. 26. Torque spectrum of unloaded machine with inner raceway defect.

The stator current spectrum (see Fig. 27) shows, on the one hand, a rise of eccentricity related components. The frequency components at $5f_s + f_r$ and $7f_s - f_r$ are already present in the spectrum of the healthy machine due to an inherent level of dynamic eccentricity. The fault

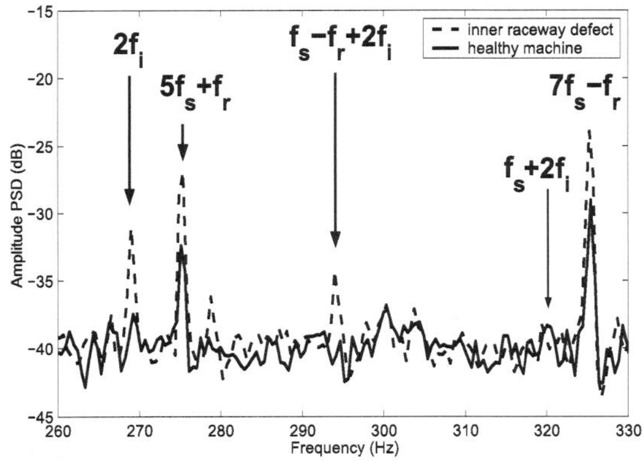

Fig. 27. Stator current spectrum of unloaded machine with inner raceway defect.

related eccentricity increases these components according to (62) (with k=0). The component at $f_s - f_r + 2f_i$ does not appear in the healthy spectrum but in case of the fault as it is the consequence of the particular form of eccentricity introduced by the inner raceway fault. Another fault related component at $2f_i$ can be noticed. The obtained results for this fault validate therefore the precedent theoretical development.

6.6.4 Realistic bearing fault

After artificially produced single point bearing defects, tests were conducted with industrially used bearings that have been changed due to a problem with an unknown fault type. The tested bearing shows only small effects on the vibration spectrum such as a small peak at 33 Hz and a slight general increase of the vibration level for frequencies higher than 150 Hz. Characteristic vibration frequencies could not have been clearly identified.

However, the measured machine torque shows considerable changes in comparison to the healthy case (see Fig. 28). At nominal load level, torque oscillations of great amplitude can be identified at 33 Hz and its multiples.

These torque variations have a considerable effect on the stator current. In Fig. 29, the stator current spectrum with the faulty bearing can be compared to the healthy machine. Sideband components to the fundamental appear at $50 \pm k \cdot 33$ Hz. This is the characteristic signature on the spectrum of a phase modulation of the fundamental component (see (24)).

In order to analyze the phase modulation, the stator current IF is computed. In Fig. 30, the power spectral density (PSD) of the stator current IF is shown for the healthy and faulty case. The significant difference is a component at 33 Hz in the faulty case. The spectral peak at this frequency indicates IF oscillations at 33 Hz. These IF oscillations are a sign of phase modulations present in the faulty case. Considering the measured torque oscillations with the faulty bearing (see Fig. 28), it can be concluded that the observed phase modulation is the consequence of the recognized torque oscillations as it has been developed in section 2.1.

6.6.5 Summary of Experimental Results

The previous experimental results have validated several theoretical aspects. The produced single point defects showed the expected effects on the vibration spectrum i.e. the apparition

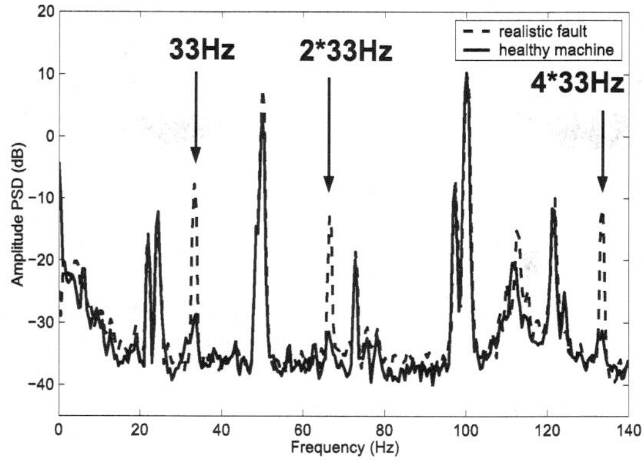

Fig. 28. Torque spectrum of loaded machine with realistic bearing fault.

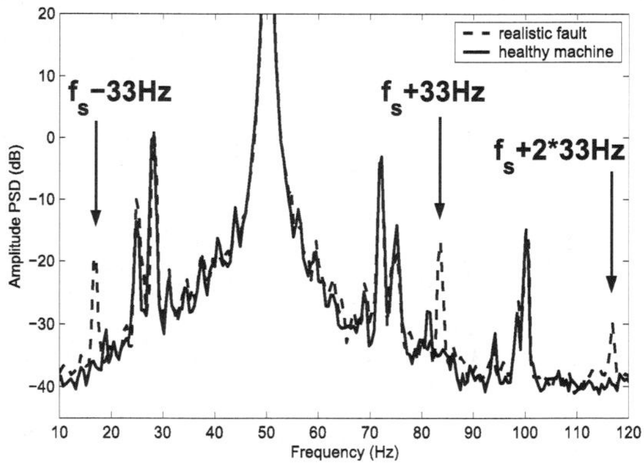

Fig. 29. Stator current spectrum of loaded machine with realistic bearing fault.

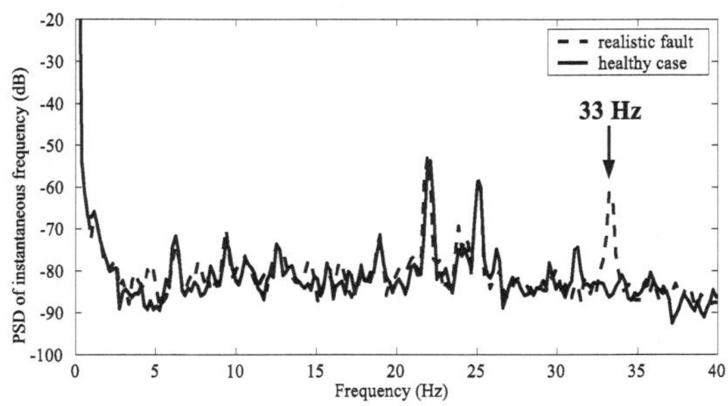

Fig. 30. Spectrum of stator current instantaneous frequency, loaded machine with realistic bearing fault.

of the characteristic vibration frequencies. In case of inner and outer race faults, the measured load torque showed oscillations at these frequencies confirming that the assumption of bearing fault-related torque oscillations is valid. Furthermore, significant effects could be observed in the stator current spectrum i.e. the apparition of some of the theoretically predicted frequencies.
Moreover, low frequency load torque oscillations have been observed with a realistic bearing fault. The resulting stator current shows phase modulations which validates the second part of the theoretical study.
However, it must be noted that the amplitudes of the additional stator current frequencies may depend heavily on the considered bearing and the load condition. Therefore, a systematic bearing monitoring using only the stator current is difficult to realize. Nevertheless, in some cases, the stator current showed more significant effects than the vibration data which suggests that a combined approach using vibration and current analysis could be reasonable.

7. Conclusion

This work has investigated mechanical fault detection in induction motors by stator current monitoring. Through a theoretical approach, stator current models were derived that contain different modulations, AM and PM, caused by eccentricity and torque oscillations. Three signal processing methods suitable for stator current analysis and fault detection were described: classical spectrum analysis, suitable only for stationary signals, instantaneous frequency analysis and the Wigner distributions as time-frequency analysis methods that can also be applied with varying supply frequency as can be found with variable-speed drives. It was also demonstrated that the Wigner distribution is a tool for distinguishing eccentricity related faults from faults causing load torque oscillation.
The theoretically derived fault signatures are then validated in experimental setups: first, dynamic eccentricity and load torque oscillations are investigated under laboratory conditions. Then, the detection of misalignment faults in electric winches was presented as an example in a realistic drive system.
As a particular class of faults, bearing faults were also considered. A theoretical development has shown that these faults lead to particular forms of eccentricity and load torque oscillations. After deriving the fault frequencies and modulation types, faulty bearing were examined under laboratory conditions. The obtained results are examples that bearing faults can lead to detectable changes in the stator current.
It can be concluded, that stator current monitoring is a suitable tool for mechanical fault detection in electrical drives. Stator current analysis is certainly more complex than traditional vibration signal analysis due to the fact that the vibration signals are physically closer to the fault. However, it offers advantages over vibration analysis in terms of cost and possibilities of application.

8. References

A. Reilly, G. F. & Boashash, B. (1994). *IEEE Trans. on Signal Processing* **42**(11): 3241–3245.
Albrecht, P. F., Appiairius, J. C., Cornell, E. P., Houghtaling, D. W., McCoy, R. M., Owen, E. L. & Sharma, D. K. (1987). Assessment of the reliability of motors in utility applications - part 1, **EC-2**(3): 396–406.

Auger, F., Flandrin, P., Gonçalvès, P. & Lemoine, O. (1995/1996). Time-frequency toolbox, CNRS / Rice University, France.
URL: *http://tftb.nongnu.org*

Blödt, M. (2006). *Condition Monitoring of Mechanical Faults in Variable Speed Induction Motor Drives - Application of Stator Current Time-Frequency Analysis and Parameter Estimation (available online)*, PhD thesis, Institut National Polytechnique de Toulouse, Toulouse, France.

Blödt, M., Bonacci, D., Regnier, J., Chabert, M. & Faucher, J. (2008). On-line monitoring of mechanical faults in variable-speed induction motor drives using the wigner distribution, *IEEE Transactions on Industrial Electronics* **55**(2).

Blödt, M., Chabert, M., Regnier, J. & Faucher, J. (2006). Mechanical load fault detection in induction motors by stator current time-frequency analysis, *IEEE Transactions on Industry Applications* **42**(6): 1454 – 1463.

Blödt, M., Chabert, M., Regnier, J., Faucher, J. & Dagues, B. (2005). Detection of mechanical load faults in induction motors at variable speed using stator current time-frequency analysis, *Proc. IEEE International Symposium on Diagnostics for Electric Machines, Power Electronics and Drives (SDEMPED '05)*, Vienna, Austria.

Blödt, M., Granjon, P., Raison, B. & Rostaing, G. (2008). Models for bearing damage detection in induction motors using stator current monitoring, *IEEE Transactions on Industrial Electronics* **55**(4): 1813 – 1822.

Blödt, M., Regnier, J. & Faucher, J. (2006). Distinguishing load torque oscillations and eccentricity faults in induction motors using stator current wigner distribution, *Proc. IEEE Industry Applications Society Annual Meeting 2006*, Tampa, Florida.

Boashash, B. (2003). *Time Frequency Signal Analysis and Processing - A Comprehensive Reference*, first edn, Elsevier, Oxford, UK.

Bonaldi, E. L., da Silva, L. E. B., Lambert-Torres, G., Oliveira, L. E. L. & Assuncao, F. O. (2002). Using rough sets techniques as a fault diagnosis classifier for induction motors, *Proc. IEEE IECON'02*, Vol. 4, Sevilla, Spain, pp. 3383–3388.

Cameron, J. R. & Thomson, W. T. (1986). Vibration and current monitoring for detecting airgap eccentricities in large induction motors, *IEE Proceedings* **133**(3): 155–163.

Dorrell, D. G. (1996). Calculation of unbalanced magnetic pull in small cage induction motors with skewed rotors and dynamic rotor eccentricity, *IEEE Transactions on Energy Conversion* **11**(3): 483–488.

Dorrell, D. G., Thomson, W. T. & Roach, S. (1997). Analysis of airgap flux, current, and vibration signals as a function of the combination of static and dynamic airgap eccentricity in 3-phase induction motors, *IEEE Transactions on Industry Applications* **33**(1): 24–34.

Engelmann, R. H. & Middendorf, W. H. (1995). *Handbook of Electric Motors*, Marcel Dekker, New York.

Eren, L., Karahoca, A. & Devaney, M. J. (2004). Neural network based motor bearing fault detection, *Proc. Instrumentation and Measurement Technology Conference (IMTC '04)*, Vol. 3, Como, Italy, pp. 1657–1660.

Flandrin, P. (1999). *Time-Frequency/Time-Scale Analysis*, Academic Press, San Diego.

Goodwin, M. J. (1989). *Dynamics of rotor-bearing systems*, Unwin Hyman, London.

Heller, B. & Hamata, V. (1977). *Harmonic Field Effects in Induction Machines*, Elsevier, Amsterdam, Netherlands.

IEEE motor reliability working group (1985a). Report on large motor reliability survey of industrial and commercial installations, **IA-21**(4): 853–872.

IEEE Motor reliability working group (1985b). Report on large motor reliability survey of industrial and commercial installations, **IA-21**(4): 853–872.

Kay, S. M. (1988). *Modern Spectral Estimation: Theory and Application*, Prentice Hall, Englewood Cliffs, New Jersey.

Li, B., Chow, M., Tipsuwan, Y. & Hung, J. (2000). Neural-network based motor rolling bearing fault diagnosis, **47**(5): 1060–1069.

MacFadden, P. D. & Smith, J. D. (1984). Model for the vibration produced by a single point defect in a rolling element bearing, *Journal of Sound and Vibration* **96**(1): 69–82.

Max, J. & Lacoume, J.-L. (2000). *Méthodes et techniques de traitement du signal*, 5th edn, Dunod, Paris.

Mecklenbräuker, W. & Hlawatsch, F. (eds) (1997). *The Wigner Distribution — Theory and Applications in Signal Processing*, Elsevier, Amsterdam (The Netherlands).

Schiltz, R. L. (1990). Forcing frequency identification of rolling element bearings, *Sound and Vibration* pp. 16–19.

Schoen, R. R., Habetler, T. G., Kamran, F. & Bartheld, R. (1995). Motor bearing damage detection using stator current monitoring, *IEEE Transactions on Industry Applications* **31**(6): 1274–1279.

Sekhar, A. S. & Prabhu, B. S. (1995). Effects of coupling misalignment on vibrations of rotating machinery, *Journal of Sound and Vibrations* **185**(4): 655–671.

Sieg-Zieba, S. & Tructin, E. (2008). Handling systems condition monitoring using vibrations and motor current, *Proc. Condition Monitoring*, Edinburgh, Scotland.

Stack, J., Habetler, T. G. & Harley, R. G. (2004a). Bearing fault detection via autoregressive stator current modeling, *IEEE Transactions on Industry Applications* **40**(3): 740–747.

Stack, J., Habetler, T. G. & Harley, R. G. (2004b). Fault classification and fault signature production for rolling element bearings in electric machines, *IEEE Transactions on Industry Applications* **40**(3): 735–739.

Tandon, N. & Choudhury, A. (1997). An analytical model for the prediction of the vibration response of rolling element bearings due to a localized defect, *Journal of Sound and Vibration* **205**(3): 275–292.

Teotrakool, K., Devaney, M. J. & Eren, L. (2006). Adjustable speed drive bearing fault detection via wavelet packet decomposition, *Proc. Instrumentation and Measurement Technology Conference (IMTC '06)*, Sorrento, Italy, pp. 22–25.

Timár, P. L. (1989). *Noise and Vibration of Electrical Machines*, Elsevier, North-Holland.

Wang, Y. F. & Kootsookos, P. J. (1998). Modeling of low shaft speed bearing faults for condition monitoring, *Mechanical Systems and Signal Processing* **12**(3): 415–426.

Xu, M. & Marangoni, R. D. (1994a). Vibration analysis of a motor-flexible coupling-rotor system subject to misalignment and unbalance, part i: theoretical model and analysis, *Journal of Sound and Vibrations* **176**(5): 663–679.

Xu, M. & Marangoni, R. D. (1994b). Vibration analysis of a motor-flexible coupling-rotor system subject to misalignment and unbalance, part ii: experimental validation, *Journal of Sound and Vibrations* **176**(5): 681–691.

Yang, S. J. (1981). *Low-noise electrical motors*, Clarendon Press, Oxford.

Zhou, W., Habetler, T. G. & Harley, R. G. (2007). Stator current-based bearing fault detection techniques: A general review, *IEEE International Symposium on Diagnostics for Electric Machines, Power Electronics and Drives (SDEMPED)*, Cracow, Poland, pp. 7–10.

Zhou, W., Habetler, T. G. & Harley, R. G. (2008). Bearing fault detection via stator current noise cancellation and statistical control, *IEEE Transactions on Industrial Electronics* **55**(12): 4260–4269.

22

Fault Detection Method based on Extended Locally Linear Embedding

Wei Zhang and Xu Yang
Shenyang Institute of Automation (SIA), Chinese Academy of Science (CAS)
China

1. Introduction

To enhance the autonomous capability, a system must be equipped with adequate numbers of the sensors responsible for the detection of system's input/output states, operating status, environment conditions, etc. However, more sensors may introduce other problems - the certainty, reliability, and robustness. Therefore, the reliable detection of sensor faults becomes extremely critical in order to assure system certainty, reliability, stability, and robustness. In general, the raw data obtained from the sensors of the complex systems have high dimensionalities and severe nonlinearity. High-dimensional raw data typically bear considerable redundancies and correlations hiding important relationships. It is, therefore, highly desirable to obtain more descriptive, informative, and useful data representations for the subsequent operations by eliminating the redundancies from the raw data. This process is referred to as dimensionality reduction. Dimensionality reduction can be achieved by creating a set of low-dimensional data representation with new features based on the transformations and/or combinations of the original data features. This operation is often referred to as "feature extraction".

By far, the most frequently-used feature extraction method is Principle Component Analysis (PCA), which was proposed by Pearson in 1901[1] when he was studying regression analysis, and was mathematically developed by Hotelling in 1933[2]PCA has been wildly used in pattern recognition and machine learning. However, early dimensionality reduction methods were developed to primarily target the linear data systems [3][4]. When data system is high-dimensional and severely non-linear, these methods become ineffective.

In 1984 Hastie proposed concept of the principle curves[5], which extended the PCA analysis to nonlinear fields. The principle curves can more accurately represent the nonlinear structures of data in linear dimensions. In 2000, Roweis proposed a manifold learning algorithm called Local Linear Embedding (LLE), which is an unsupervised non-linear technique that analyzes the high-dimensional data sets and reduces their dimensionalities with preserved local topology[6]. Today, LLE has been widely used in cluster analysis, image processing, biological informatics, etc.

2. Locally Linear Embedding Algorithm

The basic concept of LLE is to find a weight vector between a sample and its neighbors, and to keep this relationship in a feature space[6]. It assumes that even if the manifold embedded in a high-dimensional space is nonlinear, it still can be considered locally linear if each data point and its neighbors lie on or close to a locally linear patch of the manifold, i.e., the manifold can be covered with a set of locally linear patches which, when analyzed together, can yield information about the global geometry of the manifold. The weight vector expressing the intrinsic geometrical properties of the local patch can be obtained in three steps: (a) to find the neighbors of every sample in the high-dimensional space, (b) to obtain the reconstruction weight and a sparse matrix of the weight vectors, and (c) to compute the low dimensional embedding -- the bottom nonzero eigenvectors of the sparse matrix are the low dimensional embeddings of high dimensional samples.

Firstly, given a dataset $X = \{x_1, \cdots x_n\}, x \in R^d$, assuming the data lies on a nonlinear manifold which locally can be approximated linearly, the cost function can be written as:

$$\min \varepsilon(W) = \sum_{i=1}^{n} \left| (x_i - \sum_{j=1}^{k} w_j^i x_{ij}) \right|^2 \tag{1}$$

Where x_{ij} is the neighbor of x_i, and w_j^i is the weight which subject to the constraint of $\sum_{j=1}^{k} w_j^i = 1$. It can be seen that the cost function means how well each x_i can be linearly reconstructed in terms of its neighbors $(x_{i1}, \cdots, x_{ik})$. Because weights of w_j^i are sum up to 1, so eqn.1 can also be written as

$$\min \varepsilon(W) = \sum_{i=1}^{N} \left| \sum_{j=1}^{k} w_j^i (x_i - x_{ij}) \right|^2 \tag{2}$$

For a specific x_i

$$\varepsilon^i(W) = \left| \sum_{j=1}^{k} w_j^i (x_i - x_{ij}) \right|^2 = \sum_{j=1}^{k} \sum_{m=1}^{k} w_j^i w_m^i Q_{jm}^i \tag{3}$$

where $Q_{jm}^i \in R^{k \times k}$, is the covariance matrix,

$$Q_{jm}^i = (x_i - x_{ij})^T (x_i - x_{im}) \tag{4}$$

Let $R^i = (Q^i)^{-1}$. Solving the least squares problem with constraint $\sum_{j=1}^{k} w_j^i = 1$ gives:

$$w_{ij} = \frac{\sum_{m=1}^{k} (Q^i)^{-1}_{jm}}{\sum_{p=1}^{k}\sum_{q=1}^{k} (Q^i)^{-1}_{pq}} \tag{5}$$

In practice, a regularization parameter r will have to be used for Q^i before inversion:

$$Q^i = Q^i + rI \tag{6}$$

Secondly, the weights w are fixed and new m-dimensional vectors y_i are sought which minimize the criterion:

$$\min \varepsilon(Y) = \sum_{i=1}^{n} \left| y_i - \sum_{j=1}^{k} w_{ij} y_{ij} \right|^2 \tag{7}$$

Re-writing eqn.6 gives

$$\min \varepsilon(Y) = \sum_{i=1}^{n}\sum_{j=1}^{n} M_{i,j} y_i^T y_j \tag{8}$$

where M is a $n \times n$ positive semidefinite matrix found as $M = (I - W)^T (I - W)$, and Y contains the $y_i^{'}s$ as its columns.

To be able to solve this problem, the covariance matrix of the $y^{'}s$ can be constrained to be identity. Finding Y then becomes a well-known problem: minimize $tr(YMY^T)$ with constraint $\frac{1}{n}YY^T = I$. Using Lagrange multipliers and setting the derivative to zero give $(M - \Lambda)Y^T = 0$, where Λ is the diagonal Lagrange multiplier matrix. Clearly, this is an eigenvalue problem. All eigenvectors of M are solutions, but the eigenvectors corresponding to the smallest eigenvalues minimize $\varepsilon(Y)$.

The eigenvector with the smallest eigenvalue corresponds to the mean of Y and can be discarded to enforce $\sum_{i=1}^{n} y_i = 0$. The next m eigenvectors then give the Y which minimizes eqn.8.

3. Extended Locally Linear Embedding Algorithm

3.1. Neighbor Selection Based on Tangent Space Distance

From the definition of LLE we can see that the point and its neighbors must lie on or close to a locally linear patch of the manifold. Usually the correlation of data is computed after neighbors have been decided in Euclidean distance, but sometimes the nearest neighbors in Euclidean distance do not lie on the approximate linear curved face, as shown in figure 1,

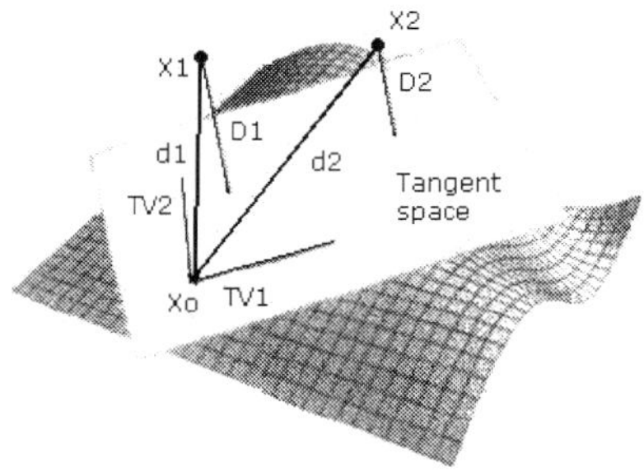

Fig 1. Euclidean distance and tangent space distance from neighbors to point

Where X_0 is the point and lies on a curved face, X_1 and X_2 are its neighbors. TV_1 and TV_2 are tangent vectors of X_0, and construct the tangent space. d1 , d2 are Euclidean distance from neighbors to X_0, and D1,D2 are distance from neighbors to tangent space of X_0. According to the definition of LLE we know that shorter D means better of the locally linear quality of point and its neighbors. From figure 1 we can see that although X_1 is the nearest neighbor of point X_0 in Euclidean distance sense, but in fact X_2 is more suitable to be the nearest neighbor in locally linear sense.

So we introduce tangent space into the algorithm. Firstly, determine tangent space of every point, and then compute distance from neighbors to these tangent spaces, so we can get nearest neighbors which satisfy the hypothesis of a locally linear patch better.

To compute tangent space, we must know differential of the explicit formulation of the function. But in reality problem, there is no explicit formulation of original dataset, but only some discrete samples, so how to compute the tangent space is a question. Fortunately, a step in tangent space alignment method provides a resolvent[7]. For every data in higher dimensional space, it can get the approximate tangent space through local PCA, or a set of orthogonal basis.

Suppose F is a d dimensional manifold in m dimensional space mapped by an unknown function $f(\tau), \tau \in R^d$, and data $X = (x_1, x_{2,} \cdots x_n), x_i \in R^m$ is the image of the unknown function $x_i = f(\tau_i), i = 1,2,\cdots,n$. To get the tangent space of x_i, it is equivalent to get the differential of function $f(\tau)$ at τ_i. Suppose f is smooth enough, and we can get the first Taylor expansion at some τ,

$$f(\overline{\tau}) = f(\tau) + J_f(\tau) \cdot (\overline{\tau} - \tau) + O(\|\overline{\tau} - \tau\|_2^2) \tag{9}$$

where $J_f(\tau) \in R^{m \times d}$ is the Jacobi matrix of f at τ.

We record $f(\tau) = \begin{pmatrix} f_1(\tau) \\ \vdots \\ f_m(\tau) \end{pmatrix}$, and then $J_f(\tau) = \begin{pmatrix} \partial f_1 / \partial \tau_1 & \cdots & \partial f_1 / \partial \tau_d \\ \vdots & \vdots & \vdots \\ \partial fm / \partial \tau_1 & \cdots & \partial fm / \partial \tau_d \end{pmatrix}$, so the tangent space Γ_τ of f at τ is expanded by d column vectors of $J_f(\tau)$, $\Gamma_\tau = span(J_f(\tau))$. Because we do not know function f, so we can not compute $J_f(\tau)$ directly, but we know it is a d dimensional space, so if we know a set of orthogonal basis of $J_f(\tau)$,we can get $J_f(\tau)\cdot(\overline{\tau}-\tau)=Q_\tau\theta_\tau^*$, where θ_τ^* is the coordinate of $\overline{\tau}-\tau$ at tangent space corresponding to Q_τ. Because $f(\overline{\tau})-f(\tau)\approx J_f(\tau)\cdot(\overline{\tau}-\tau)$, so we can get $f(\overline{\tau})-f(\tau)\approx Q_\tau\theta_\tau^*$, that is $f(\overline{\tau})\approx Q_\tau\theta_\tau^*+f(\tau)$.

We need to compute all the tangent space, so for a certain point x_i, its neighbor $x_{ij}\approx x_i+Q_i\theta_i$. Construct followed optimum function,

$$\min_{x_i^*,Q_i,\theta}\sum_{j=1}^{k}\left\|x_{ij}-(x_i^*+Q_i\theta_i)\right\|_2^2=\min_{x_i^*,Q_i,\theta}\left\|X_i-(x_i^*e^T+Q_i\theta_i)\right\|_F^2 \tag{10}$$

Where $X_i=(x_{i1},x_{i2},\cdots x_{ik})$, $Q_i\in R^{m\times d}$, $\theta_i=(\theta_1,\theta_2\cdots\theta_k)\in R^{d\times k}$, so Q_i can be seen as d approximate orthogonal basis of tangent space of x_i. The question can be resolved by singular value decomposition (SVD) to $X_i(I-\frac{1}{k}ee^T)$, where Q_i is the d eigenvectors corresponding to d maximum eigenvalues of $X_i(I-\frac{1}{k}ee^T)$. So we can get every tangent space of each point.

Because the tangent space we get is a set of orthogonal basis but not an analytic expression, so we can not compute the distance directly. Firstly we consider three dimensional condition.

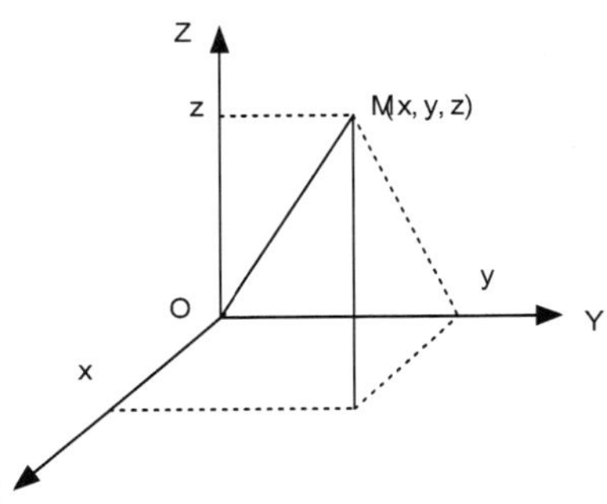

Fig. 2. Distance from point to plain

From figure 2 we can see that, the distance from point M to plain XOY is the coordinate of M in Z axis, and we extend it to higher dimensional space. Suppose data set is m dimension, and eigenvectors of $X_i(I-\frac{1}{k}ee^T)$ is $E=(e_1,e_2,\cdots,e_m)$, so a set of orthogonal basis of tangent space of point is $Q=(e_1,e_2,\cdots,e_d)$, then the distance of neighbors to the tangent space can be written as $\sqrt{x_{d+1}^2+x_{d+2}^2+\cdots x_m^2}$. So we can get tangent space distance dataset to each point, and then nearest neighbors in tangent space can be arrived.

To evaluate the performance of dimension reduction, some researchers propose residual variance according to Input/Output mapping quality, and that is the description effectiveness of the original data in higher dimensional space. It is defined as $1-\rho^2_{D_x D_y}$, where ρ is the standard linear correlation coefficient, taken over all entries of D_x and D_y , where D_x and D_y are the matrices of Euclidean distances (between pairs of points) in the high-dimensional and corresponding low-dimensional spaces, respectively. According to the definition, we can see that the lower the residual variance is, the better high-dimensional data is represented in the embedded space. S-curve dataset is uniform sampled from noiseless three dimensional S-curve curved face. We will reduce the dimension of these two datasets, and compare the effectiveness by residual variance in Euclidean distance and tangent space distance respectively.

S-curve dataset is shown in figure 3. We choose 30 neighbors, and the nearest 30 neighbors of the first point distribution are shown in figure 4. Where the yellow points are neighbors in tangent space distance and green points are neighbors in Euclidean distance, and the red line is the tangent space of the first point. According to the character of S-curve, we can see from the figure that yellow points are distributed in tangent space of the point, and green points are distributed in curve face in relative sense.

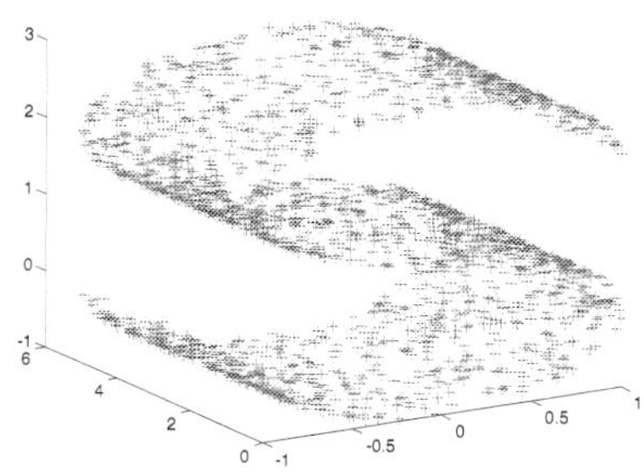

Fig. 3. S-curve dataset

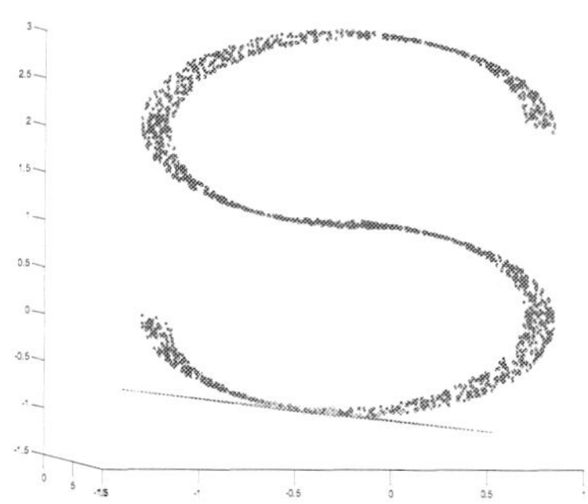

Fig. 4. Neighbors in Euclidean distance and tangent space distance

To test the performance of the algorithm, we compute residual variances under 1 to 50 neighbors, and the result is shown in figure 5,

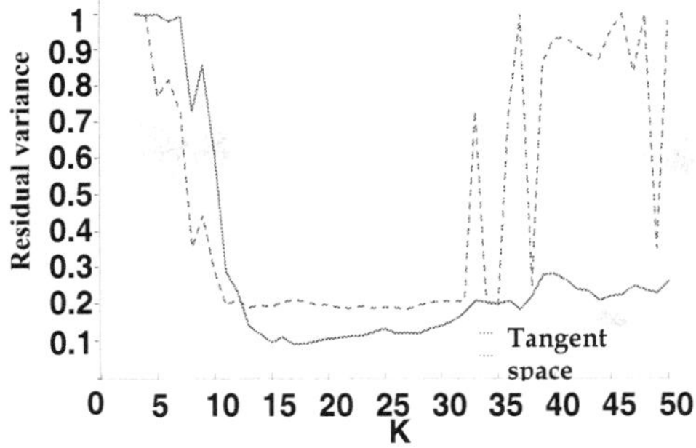

Fig. 5. S-curve dataset residual variance

It can be seen from the figure that when there are few neighbors, the locally structure can not be expressed well, so the residual variance is much big. With more neighbors, performances of two methods both increase, but with more and more neighbors, the method in Euclidean distances can not guarantee the local linear characteristic, so the result is not good as in tangent space distance method obviously.

3.2. Intrinsic Dimensionality Estimation

According LLE algorithm the projection data is the bottom d nonzero eigenvectors of sparse matrix and d is the dimension of the feature space. In all the studies conducted by other authors, d is mostly selected based on the experience, through plenty of tests, which restricts the autonomy and adaptivity of the algorithm to certain extent.

Considerable efforts have been invested by other authors into the improvement of the autonomy and adaptivity of the algorithm. For example, Kegl computes the correlation dimension using Grassberger and Procaccia (G-P) algorithm, which is efficient but still suffers from several drawbacks In this paper, we will show certain improvements made to the G-P algorithm using linear fit, which not only realizes the adaptation of the algorithm to a large extent, but also improves the precision of the dimension estimation.

According to the concept of topological dimension, the ID of X is the topological dimension of the support of the distribution of X [8]. However, when data set is finite, it is difficult to apply the original definition to the ID estimate. The most popular definition, among many proposed, is called correlation dimension, which states that given a data set X with dimension D and a distance r, the number of samples coupled with the distance less than r is in direct ratio to r^D. At present, the most frequently used algorithm for estimating the correlation dimension is G-P algorithm, which is summarized as follows:

Given a finite data set $\Phi = \{x_1, \cdots, x_n\}$ in metric space X, let

$$C_n(r) = \frac{2}{n(n-1)} \sum_{i=1}^{n} \sum_{j=i+1}^{n} H(r - |x_i - x_j|) \tag{11}$$

where $H(s)$ is an indicator function with $H(s) = 1$ when $s > 0$ and $H(s) = 0$ when $s < 0$. For a countable set $\Phi = \{x_1, \cdots, x_n\}$, the correlation integral is defined as

$C(r) = \lim_{r \to 0} C_n(r)$. If the limit exists, the correlation dimension D_{cor} of Φ is equal to

$$D_{cor} = \lim_{r \to 0} \frac{\log C(r)}{\log(r)} \tag{12}$$

It has been proven that for a finite sample, the zero limit cannot be achieved. Thus, the original estimate procedure proposed by Grassberger and Procaccia involves plotting $\log C(r)$ versus $\log r$ and then measuring the slope of the linear region of the curve[9][10]. The correlation dimension of the finite data set Φ can then be expressed as

$$D_{cor} = \frac{\log C(r_2) - \log C(r_1)}{\log(r_2) - \log(r_1)} \tag{13}$$

The algorithm for computing ID can be summarized in the following steps: (a) to select a series of r with different scales, (b) to compute the corresponding correlation integrals, (c) to identify the linear part of the curve and then to measure the slope of the identified linear region, and (d) to compute the correlation dimension of Φ. Among these steps, the selection of the linear part is the most crucial.

Given a countable data set $\Phi = \{x_1, \cdots, x_n\}$ and different scales of distance $\{r_1, \cdots, r_m\}$, we can obtain a series of corresponding correlation integrals $\{\log C(r_1), \cdots, \log C(r_m)\}$ and plot $\log C(r)$ versus $\log r$ curves. The objective is to identify the linear region of this curve. To do so, a third order linear fit is applied to the data series ($\log C(r_i)$, $\log r_i$) $(i = 1 \cdots m)$ to yield:

$$y = a_1 x^3 + a_2 x^2 + a_3 x + a_4 (x \in \log r_i) \tag{14}$$

From Taylor Expansion, in order to make eqn.14 linear, the high order terms must approach to zero, that is, $f(x) = a_1 x^3 + a_2 x^2 \approx 0$. Hence, the linear approximation of eqn.13 can be achieved. In other words, the linear part of the curve can be identified and then the correlation dimension can be obtained by computing the slope of linear part of the curve.

Tennessee Eastman (TE) process is a standard test process proposed by J. J. Downs and E. F. Vogel[11]. Here we use Tennessee Eastman Process(TEP) data to test the algorithm, which involves selecting a series of r with different scales and computing the corresponding correlation integrals. TEP has been a well-known benchmark process for comparing various process monitoring methods. The details on the process description can be found in Ref[12]. The curve is plotted in figure 6-1. Applying the third order linear fit to the curve yields

$$Y = a_1(\log r_i)^3 + a_2(\log r_i)^2 + a_3(\log r_i) + a_4 \tag{15}$$

Let $F = a_1(\log r_i)^3 + a_2(\log r_i)^2$, the absolute value of F is plotted in figure 6-2.

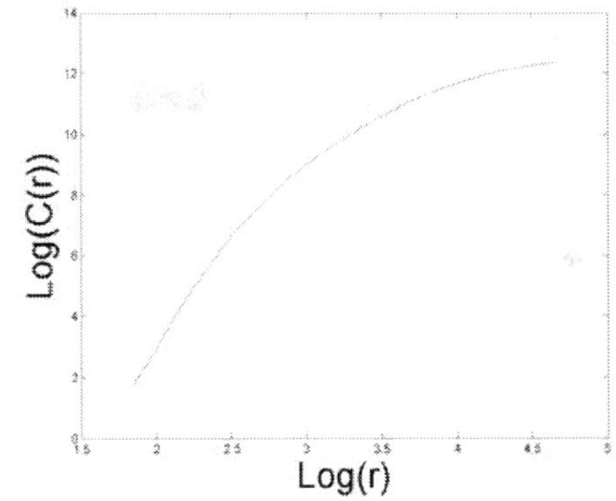

Fig. 6-1. $L\text{og}\,C(r) - \log r$ curve

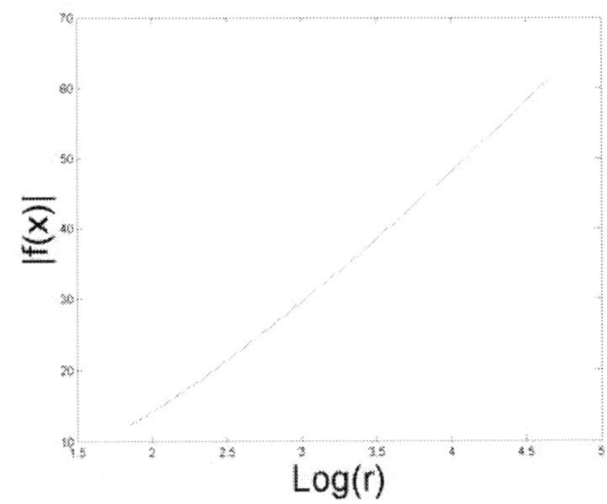

Fig. 6-2. Nonlinear part of fitting curve

Given a threshold h, the bonds of r can be obtained. The intrinsic dimension of the data set can be obtained by computing the slope of the linear part of this curve.

4. Feature Extraction Comparison

To test the performance of LLE to feature extraction, we considered the classification problem. Iris data set is a standard database for classification, which consists of three classes, with each class containing m=4 measurements and n=50 observations (see table 1). Here PCA and LLE feature extraction techniques are compared via theoretical and graphical analysis.

The projections of the experimental data onto the first two PCA loading vectors are shown in Figure 7. From which we can see that, separation degree between class 1 and class 2 is small, but aggregation degree of each class is low. And the projections of Iris data by LLE are shown in Figure 8. It can be seen that the separation degree between class 1 and class 2 is much bigger than PCA and the aggregation degree of class 3 is much bigger.

Class 1 : Iris Virginica	Mean	Std. Deviation	Range
Sepal length	6.59	0.64	4.9-7.9
Sepal width	2.98	0.32	2.2-3.8
Petal length	5.55	0.55	4.5-6.9
Petal width	2.03	0.27	1.4-2.5
Class 2 : Iris Versicolor	**Mean**	**Std. Deviation**	**Range**
Sepal length	5.94	0.52	4.9-7.0
Sepal width	2.77	0.31	2.0-3.4
Petal length	4.29	0.47	3.0-5.1
Petal width	1.33	0.20	1.0-1.8
Class 3 : Iris Setosa	**Mean**	**Std. Deviation**	**Range**
Sepal length	5.01	0.35	4.3-5.8
Sepal width	3.43	0.38	2.3-4.4
Petal length	1.46	0.17	1.0-1.9
Petal width	0.30	0.40	0.1-3.0

Table 1. Statistics of Iris data

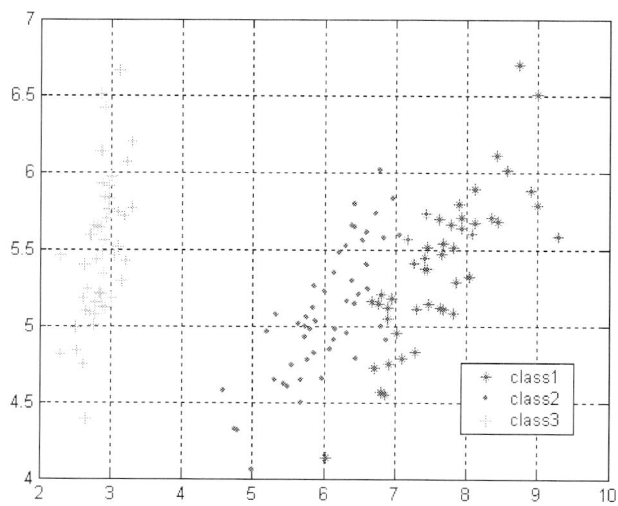

Fig. 7. Projection of Iris data onto the first two PCA loading vectors

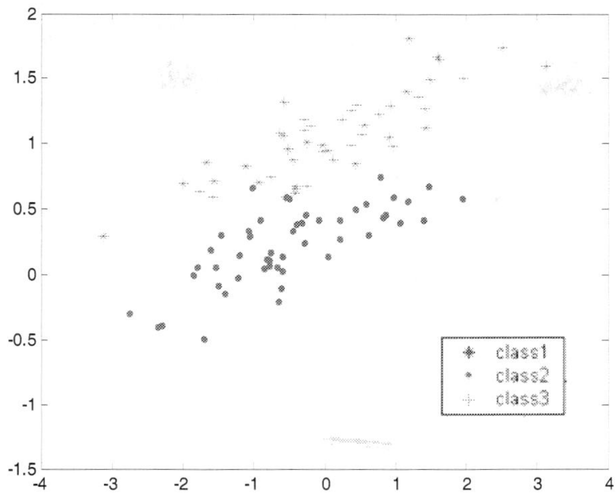

Fig. 8. Two dimension projection of Iris data by LLE

Figure 9 and Figure 10 give the projection of Iris data onto the first three PCA loading vectors and the three dimensional projection by LLE. According to separability criterion, the LLE projection performance of inner-class distance and intra-class distance is much better than PCA. It indicates that LLE algorithm has good description performance and consistency with the data feature.

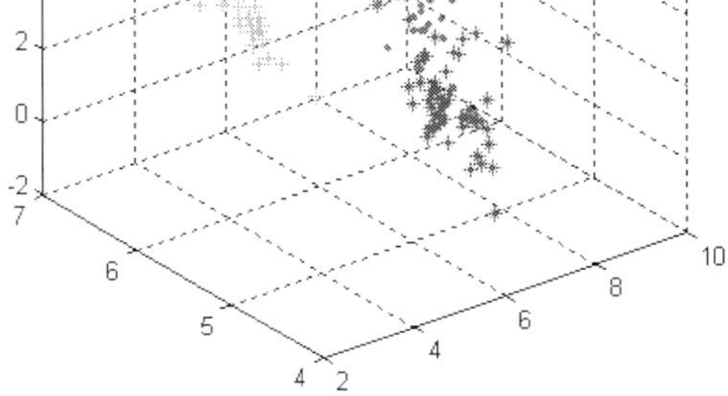

Fig. 9. Projection of Iris data onto the first two PCA loading vectors

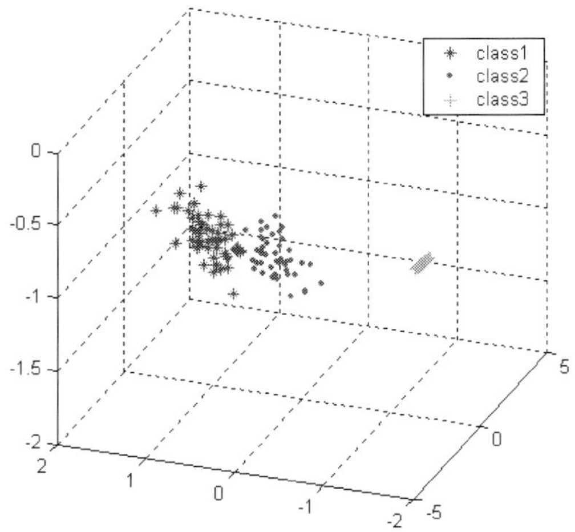

Fig. 10. Three dimensional projections of Iris data by LLE

Table 2 gives the quantitative analysis of inner-class distance and intra-class distance of the two methods, where S_w is inner-class distance; S_b is intra-class distance and $J_d = S_b / S_w$. The table demonstrates that compared with PCA, LLE can capture the data feature more effectively.

	S_b	S_w	J_d
PCA	11.8152	75.4084	0.157
LLE	2.8005	9.9729	0.281

Table 2. Performance comparison of PCA and LLE

5. Fault detection

LLE cannot compare the projection data with the original data like PCA. But after the projection, LLE can keep the topological structure of the original data as well as the similarity of normal data and illed data. Therefore, we can perform fault detection by computing the inter-class distance between the testing data and the training data[13][14].

Inter-class distance is an important index in pattern recognition. For a multi-class estimation, suppose $x_k^{(i)}$ and $x_l^{(j)}$ are vectors of class w_i and class w_j, respectively, and $\delta(x_k^{(i)}, x_l^{(j)})$ is the distance between the two vectors. The average distance between all classes of the vectors is

$$J_d(x) = \frac{1}{2}\sum_{i=1}^{c} P_i \sum_{j=1}^{c} P_j \frac{1}{n_i n_j} \sum_{k=1}^{n_i} \sum_{l=1}^{n_j} \delta(x_k^{(i)}, x_l^{(j)}) \tag{16}$$

where c is the number of classes, n_i is the number of samples of class w_i, and n_j is the number of samples of class w_j. P_i and P_j are prior probabilities of the corresponding classes, respectively. Here, we assume $P_i = P_j = 1/c$.

After have been projected to the feature space, the distance between the testing data and training data can be computed. Thus, we can get the similarity of the testing data and training data.

Suppose $x_k = \{x_1, x_2 \cdots x_n\}$ is the training data and x_j is the testing data. eqn.16 can be simplified as

$$J'_d(x_j) = \frac{1}{2}\sum_{p=1}^{2} P_p \sum_{q=1}^{2} P_q \frac{1}{n}\sum_{k=1}^{n} \delta(x_k^{(i)}, x_j) \quad (17)$$

The fault will be alerted if $J_d < h$, where h is a threshold.

Case	Disturbance	Type
1	*A*/*C* feed ratio, *B* composition constant	Step
2	*B* composition, *A*/*C* ratio constant	Step
3	*D* feed temperature	Step
4	Reactor cooling water inlet temperature	Step
5	Condenser cooling water inlet temperature	Step
6	*A* feed loss	Step
7	*C* header pressure loss – reduced availability	Step
8	*A*, *B*, *C* feed composition	Random variation
9	*D* feed temperature	Random variation
10	*C* feed temperature	Random variation
11	Reactor cooling water inlet temperature	Random variation
12	Condenser cooling water inlet temperature	Random variation
13	Reaction kinetics	Slow drift
14	Reactor cooling water valve	Sticking
15	Condenser cooling water valve	Sticking
16–20	Unknown	Unknown
21	The valve for stream 4 was fixed at the steady state position	Constant position

Table 3. Process faults for the Tennessee Eastman process

To validate the fault detection performance of LLE method, Tennessee Eastman process data is used. A set of programmed faults (Fault 1–21) is listed in Table 3. In addition, one training data set (Fault 0) consisting of 500 samples is generated with no fault. Twenty-one testing data sets corresponding to the twenty-one classes of faults in Table 3 are generated, and each set consists of 960 samples. The fault in each testing data set is introduced from sample 160 and the number of variables is m = 52.

Figures 11-13 show fault detection results of Tennessee Eastman Process in case of fault 4, fault 10 and fault 16, respectively. The confidence limits are also shown in these figures.

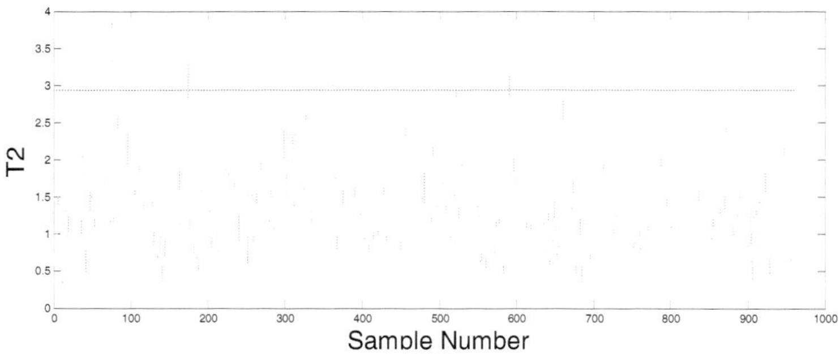

(a) PCA

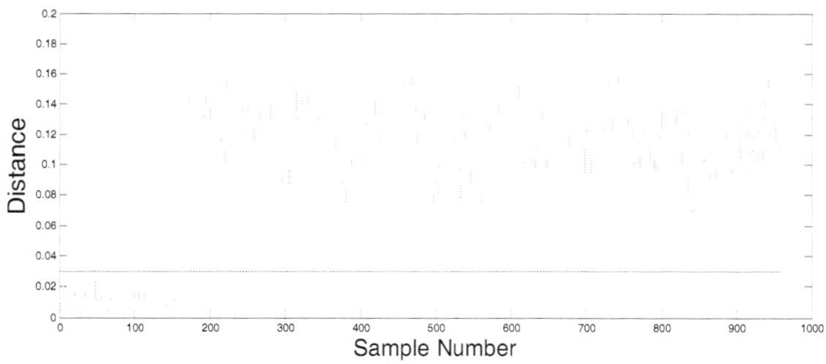

(b) LLE

Fig. 11. Fault detection results of Tennessee Eastman process in case of fault 4: (a) PCA and (b) LLE

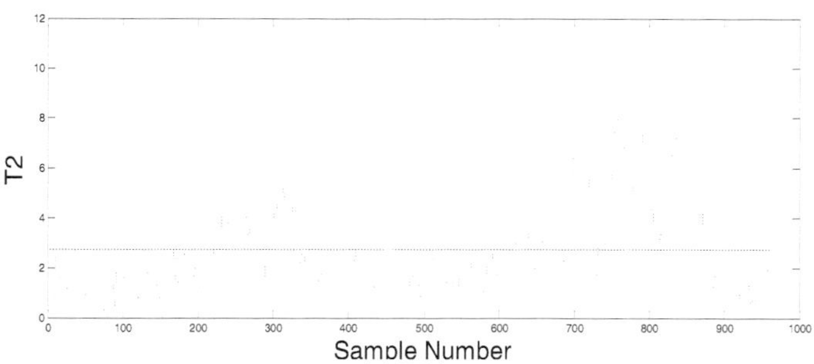

(a) PCA

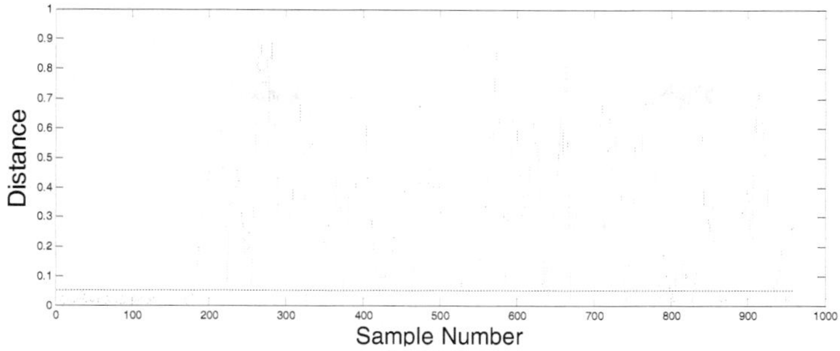

(b) LLE

Fig. 12. Fault detection results of Tennessee Eastman process in case of fault 10: (a) PCA and (b) LLE

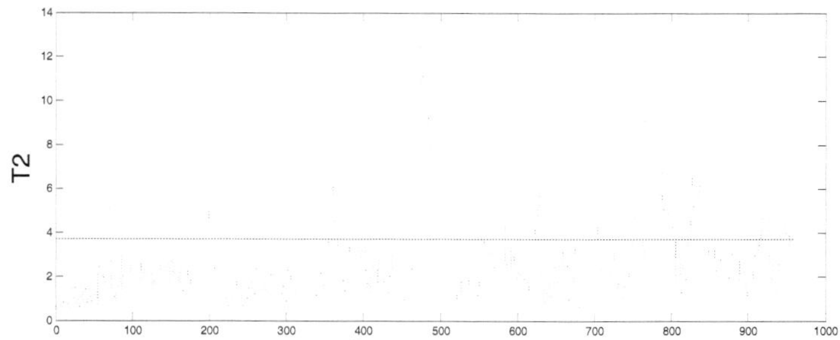

(a) PCA

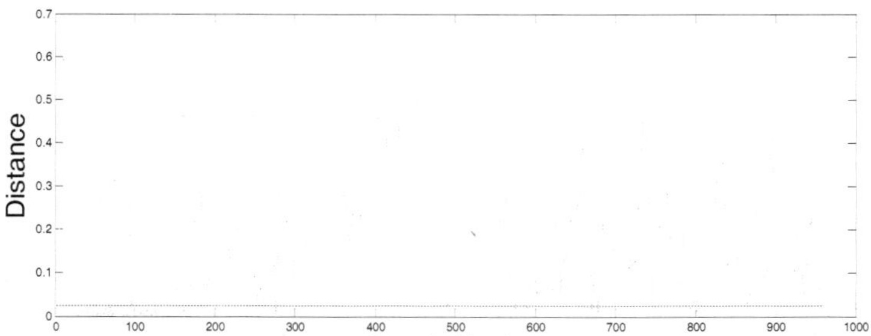

(b) LLE

Fig. 13. Fault detection results of Tennessee Eastman process in case of fault 16: (a) PCA and (b) LLE

The false positive(FP) rate and false negative(FN) rate of figures 11-13 are listed in table 4. From table 4, it can be concluded that the FP rate of LLE is a little higher than PCA, which is mostly because the PCA is insensitivity to the change of data. But the FN rate of LLE is much lower than PCA, which is 4.498% in average. On the other hand, the FN rate of PCA achieves 77.53% in average, which means that the approach based on LLE can detect each type of the faults, but PCA fails to alert these faults.

		False Positive	False Negative
Fault 4	PCA	1 . 25 %	99 . 25%
	LLE	8 . 75%	0
Fault 10	PCA	1 . 25%	60 . 25%
	LLE	7 . 75%	7 . 75%
Fault 16	PCA	0	81 . 75%
	LLE	5 . 63%	7 . 62%

Table 4. FN and FP of fault 4、 10 and 16

6. References

[1] PEARSON K. On lines and planes of closest fit to systems of points in space[J]. Philos Mag, 1901, 6(2):559 - 572.

[2] SAMUEL A L. Some studies in machine learning using the game of checkers[J]. IBM Journal Research and Development, 1967,11(4): 601-617.

[3] DUNIA R, QIN S Joe, EDGAR T F, MCAVOY T J. Use of principal component analysis for sensor fault identification[J]. Computers chem. Eng, 1996, 20(7):13-718

[4] JACKSON J E. Principal component and factor analysis: Part I - principal component[J]. Journal of Quality Technology, 1980,12:201-213

[5] HASTIE T. Principle curves and surfaces[R]. Technical Report, Laboratory for Computational Statistics, Stanford University,1984

[6] ROWEIS S T, SAUL L K. Nonlinear dimensionality reduction by locally linear embedding[J]. Science, 2000, 290(22): 2323-2326

[7] Zhang Zhen-yue, Zha Hong-yuan. Principal manifolds and nonlinear dimensionality reduction via tangent space alignment[J]. SIAM J.Sci. Comput. 2004, 26(1): 313- 338.

[8] KEGL B. Intrinsic dimension estimation using packing numbers[C].//Neural Information Processing Systems 15 (NIPS'2002), 2002.

[9] BELUSSIA A, FALOUTSOS C. Spatial join selectivity estimation using fractal concept[J]. ACM Transaction on Information Systems, 1998, 16(2): 161-201.

[10] HASTAD J. Clique is hard to approximate within $n^{1-\varepsilon}$ [C].// Proceedings of the 37th Annual Symposium on Foundations of Computer Science FOCS'96,1996,627-636.

[11] DOWNS J J, VOGEL E F. Plant-wide industrial process control problem[J]. Computers&Chemical Engineering. 1993, 17(3): 245-255.

[12] Tracy, N. D., Young, J. C., & Mason, R. L. Multivariate control charts for individual observations. Journal of Quality Technology, 1992, 24:88-95.

[13] MARQUES J P. Pattern Recognition: Concepts, Methods and Applications[M]. Berlin: Springer, 2001.

[14] DEVIJVER P A, KITTLER J. Pattern recognition theory and applications[M]. Berlin: Springer-Verlag, 1987.